Studies in Logic

Volume 115

Four Functors

The 'ancestor' of Mundici's equivalence between unital commutative *l*-groups and MV algebras

Volume 105
Logic as a Tool. A Guide to Formal Logical Reasoning
Valentin Goranko

Volume 106
New Directions in Term Logic
George Englebretsen, ed.

Volume 107
Non-commutative Algebras. Pseudo-BCK Algebreas versus m-pseudo-BCK Algebras
Afrodita Iorgulescu

Volume 108
Semitopology: decentralised collaborative action via topology, algebra, and logic
Murdoch J. Gabbay

Volume 109
The Cognitive Dimension of Social Argumentation. Proceedings of the 4th European Conference on Argumentation, Volume I. Fabio Paglieri, Alessandro Ansani and Marco Marini, eds.

Volume 110
The Cognitive Dimension of Social Argumentation. Proceedings of the 4th European Conference on Argumentation, Volume II. Fabio Paglieri, Alessandro Ansani and Marco Marini, eds.

Volume 111
The Cognitive Dimension of Social Argumentation. Proceedings of the 4th European Conference on Argumentation, Volume III. Fabio Paglieri, Alessandro Ansani and Marco Marini, eds.

Volume 112
Implicative-groups vs. Groups and Generalizations. Second Edition
Afrodita Iorgulescu

Volume 113
Meaning as a Set-theoretic Object. A Gentle Introduction to the Ideas Behind Formal Semantics
Jaroslav Peregrin

Volume 114
Logic and Language: New Frontiers in Analysis and Interpretation
José David García-Cruz, ed.

Volume 115
Four Functors. The 'ancestor' of Mundici's equivalence between unital commutative *l*-groups and MV algebras
Afrodita Iorgulescu

Studies in Logic Series Editor
Dov Gabbay

dov.gabbay@kcl.ac.uk

Four Functors

The 'ancestor' of Mundici's equivalence between unital commutative *l*-groups and MV algebras

Afrodita Iorgulescu

ISBN 978-1-84890-496-5

College Publications
Scientific Director: Dov Gabbay
Managing Director: Jane Spurr

http://www.collegepublications.co.uk

Cover prepared by Laraine Welch

To the memory
of my professors at the University of Bucharest
Nicolae Dinculeanu, Solomon Marcus, Gr.C. Moisil, Sergiu Rudeanu

Preface

Daniele Mundici has proved in 1986 [31], [10] that the category of *lattice-ordered commutative groups with strong unit* (or unital commutative l-groups, for short) is equivalent to the category of *MV algebras*, by proving the existance of the quasi-inverse functors $\mathbf{\Gamma}$ and Ξ. He has made the proof in one step, but the proof can be made in two steps - we sketch a proof in two steps in Introduction - by introducing the intermediary category of *positive-unital commutative l-groups*: thus, the category of unital commutative l-groups is equivalent to the category of positive-unital commutative l-groups (step 1) and the category of positive-unital commutative l-groups is equivalent to the category of MV algebras (step 2).

Marco Abbadini has presented in 2021 [1] his equivalence 'a la Mundici for commutative l-monoids, in two steps.

I have introduced and studied in 2020 [20], [21] a new framework, containing, among many other (left-) algebras, the commutative groups and the m_1-ME algebras and the MV algebras, m-MEL algebras, m-BCK algebras and the Boolean algebras; a commutative group $(G, +, -, 0)$ is a m_0-ME algebra verifying the property (m_0-Re) and an (right-) MV algebra $(A, \oplus, ^-, 0)$ is an involutive (right-) m-MEL algebra verifying the property ($\vee_m$-comm).

Hence, I have had sufficient background material to generalize Mundici's equivalence, to find its 'ancestor'. The role of *unital commutative l-groups* is played by the *strong m_0-ME structures*, which are involutive m_0-ME algebras verifying some properties, the role of *positive-unital commutative l-groups* is played by the *strong m_0-ME positive cones*, which are involutive m_0-ME algebras too verifying some properties, and the role of *MV algebras* is played by the *XY algebras*, which are involutive m-MEL algebras verifying the properties (XX) and (YY). I present in this book this 'ancestor', in two steps: the category of strong m_0-ME structures is not equivalent to the category of strong m_0-ME positive cones (step 1), but the category of strong m_0-ME positive cones is equivalent to the category of XY algebras (step 2). I show how the original Mundici's equivalence is a particular case of my result.

This book is organized in ten chapters and four parts, as follows:

Part 1 (The categories involved) contains chapters 2, 3. In Part 1, in the new framework created in [20], [21], I introduce and study the categories $\mathcal{C}(\mathbf{XY}^R)$, $\mathcal{C}(\mathbf{s\text{-}m_0\text{-}ME})$ and $\mathcal{C}(\mathbf{s\text{-}m_0\text{-}ME^+})$ of XY algebras, of strong m_0-ME structures and of strong m_0-ME positive cones, respectively, the 'ancestors' of MV algebras, of unital commutative l-groups and of positive-unital commutative l-groups, respectively. I also introduce two new equivalent definitions of MV algebras. I prove that XY algebras verifying the property (m-Re^R) coincide with MV algebras, that the unital commutative l-groups verify most of the properties of strong m_0-ME structures and the positive-unital commutative l-groups verify most of the properties of strong m_0-ME positive cones.

Part 2 (The first step: the functors $\mathbf{c}^+_{\mathbf{m}}$ and $\mathbf{T_m}$) contains chapters 4 - 6. In Part 2, I prove that the categories $\mathcal{C}(\mathbf{s\text{-}m_0\text{-}ME})$ and $\mathcal{C}(\mathbf{s\text{-}m_0\text{-}ME^+})$ are not equivalent (Corollary 6.3.3). In order to reach this goal, I define the functor $\mathbf{c}^+_{\mathbf{m}}$ from $\mathcal{C}(\mathbf{s\text{-}m_0\text{-}ME})$ to $\mathcal{C}(\mathbf{s\text{-}m_0\text{-}ME^+})$ and the functor $\mathbf{T_m}$ from $\mathcal{C}(\mathbf{s\text{-}m_0\text{-}ME^+})$ to $\mathcal{C}(\mathbf{s\text{-}m_0\text{-}ME})$ (following [1]) and I prove that these functors are not quasi-inverse.

Part 3 (The second step: the functors $\mathbf{G}^+_{\mathbf{m}}$ and $\mathbf{U_m}$) contains chapters 7 - 9. In Part 3, I prove that the categories $\mathcal{C}(\mathbf{s\text{-}m_0\text{-}ME^+})$ and $\mathcal{C}(\mathbf{XY}^R)$ are equivalent (Corollary 9.3.3). In order to reach this goal, I define the functor $\mathbf{G}^+_{\mathbf{m}}$ from $\mathcal{C}(\mathbf{XY}^R)$ to $\mathcal{C}(\mathbf{s\text{-}m_0\text{-}ME^+})$ and the functor $\mathbf{U_m}$ from $\mathcal{C}(\mathbf{s\text{-}m_0\text{-}ME^+})$ to $\mathcal{C}(\mathbf{XY}^R)$ (following [1]) and I prove that these functors are quasi-inverse.

Part 4 (Connections between the four functors) contains Chapter 10 and an **Appendix**. In Part 4, I present the main result of the book (Theorem 10.1.1) and many other results. The book ends with an **Appendix**, as Chapter 11, containing: in **Appendix A**, three Figures from [21], [22] connecting the classes of algebras involved in this book; in **Appendix B**, a long proof from Chapter 2; in **Appendix C**, some long proofs from Chapter 3.

The **Bibliography** has a small number of titles, only those used directly in this book, otherwise it would be huge.

Some proofs and all the examples presented in the book are obtained by PROVER9-MACE4. Note that the proofs provided by PROVER9 need to be 'humanized' and, in the book, they are, in general, integrally 'humanized'. But, in order to save space, the very long proofs are 'hybrid': the proofs of the necessary results for proving the goal are 'humanized', while the proof of the goal is not 'humanized' (it is just copied from the PROVER9 proof) - see the 'hybrid' proofs from **Appendix B** and **Appendix C**.

The book is written in a **unifying way**, which consist in fixing unique names for the defining properties.

The book was originally intended to be a paper, but, because of the long definitions and proofs etc., became a book, a 'paper-book' - we could say. It contains an original work, motivated by the curiosity to find out the 'ancestor' of the famous Mundici's equivalence; everything is new, never published elsewhere. The intended audience is made by researchers, academics and students in computer science and in mathematics, mainly in the topics of algebras of logic and group theory.

As the readers will see, the strong m_0-ME structures, the strong m_0-ME positive cones and the XY algebras - the 'ancestors' of unital commutative l-groups, positive-unital commutative l-groups and MV algebras, respectively - are not looking very nice, some are even 'ugli'. But, they are - as all ancestors are - very important, because they help us to better understand their descendants. 'Digging' for these 'ancestors' was a very difficult work, work most probably impossible without the help of PROVER9-MACE4 tool given us by William W. McCune.

*

This research was started on October 3, 2024 and was finished theoretically on April 15, 2025; since then, I was working on the connections from Chapter 3 and **Appendix C**, on Introduction and on Preface.

I due many, many thanks to George Georgescu, my colleague as students of the University of Bucharest, who, hearing that I wish to generalize Mundici's equivalence, has sent to me Marco Abbadini's paper [1], on October 22, 2024, and has encouraged me all this time. He has suggested me then to make, in Introduction, summaries of Mundici's equivalence, of Abbadini's equivalence and of my results.

The wonderful Abbadini's paper was of major help in writting this book.

I was helped with some proofs and all the examples presented in the book by the wonderful instrument which is PROVER9-MACE4 (version Dec-2007) on Windows, developed by William W. McCune (1953-2011) [28].

I due many, many thanks to Michael Kinyon, Department of Mathematics, University of Denver, Denver, CO, USA, for his friendly help with some proofs by his more powerfull PROVER9-MACE4 on Linux (when my PROVER9 failed to provide a proof, I used to ask Michael to try with his PROVER9); he helped me mainly at the begining of this research to establish that the appropriate generalization of MV algebras for my goal are the XY algebras. I never met Michael (a friend of William McCune), who was born when I started my studies in mathematics - computer science - at the University of Bucharest. He earned all his degrees from the University of Utah, finishing his PhD in 1991. His first academic position was at Indiana University South Bend. In 2006, he moved to the University of Denver. Although his PhD work was in differential equations, within a few years of graduating, he changed to being an algebraist. In particular, he became very interested in nonassociative algebra, particularly quasigroups. In 2000, thanks to a collaboration with Ken Kunen, he became interested in using automated deduction, especially Prover9, in algebra. Since then, that has been the focus of his work. What he says about PROVER9?

"Prover9 was developed by William McCune. After his untimely death in 2011, Prover9 was left mostly untouched except for updates that some individuals did for their own private use. In recent years, Prover9 has been transformed or extended in various ways, such as being rewritten in C++ or integrated into suites of other tools. This has mostly been the work of students and colleagues of Joao Araujo at NOVA University of Lisbon; see https://sites.google.com/fct.unl.pt/laitep/home"

Finally, I due many thanks to Dov Gabbay and Jane Spurr from *College Publications* for their confidence and for the absolute freedom given to me.

Afrodita Iorgulescu
afrodita.iorgulescu@ase.ro

Bucharest,
September 15, 2025

Contents

Chapter 1

Introduction and summaries

This chapter contains an introduction to the subject of the book and three summaries: of Abbadini's equivalence, of our results from this book and a sketch of Mundici's equivalence in two steps.

1.1 Introduction

MV algebras were introduced in 1958, by C.C. Chang [6], in order to prove the completeness of infinite-valued Lukasiewicz logic, as a model of $\aleph_0$-valued Łukasiewicz logic whose axioms are the following [6]:
A.1 $P \to (Q \to P)$
A.2 $(P \to Q) \to ((Q \to R) \to (P \to R))$
A.3 $((P \to Q) \to Q) \to ((Q \to P) \to P)$
A.4 $(\neg P \to \neg Q) \to (Q \to P)$.

Definition 1.1.1 [6]

An MV *algebra* is an algebra $\mathcal{A} = (A, \oplus, \odot, ^-, 0, 1)$ of type $(2, 2, 1, 0, 0)$, where the following axioms are verified: for every $x, y, z \in A$,

(M1) $x \oplus y = y \oplus x$,
(M2) $x \oplus (y \oplus z) = (x \oplus y) \oplus z$,
(M3) $x \oplus x^- = 1$,
(M4) $x \oplus 1 = 1$,
(M5) $x \oplus 0 = x$,
(M6) $(x \oplus y)^- = x^- \odot y^-$,
(M7) $(x^-)^- = x$,
(M8) $0^- = 1$;

(M1') $x \odot y = y \odot x$,
(M2') $x \odot (y \odot z) = (x \odot y) \odot z$,
(M3') $x \odot x^- = 0$,
(M4') $x \odot 0 = 0$,
(M5') $x \odot 1 = x$,
(M6') $(x \odot y)^- = x^- \oplus y^-$,

in order to write the remaining axioms in a compact form, Chang introduced the following definition:

$$x \vee y \overset{def}{=} (x \odot y^-) \oplus y,$$
$$x \wedge y \overset{def}{=} (x^- \vee y^-)^- = (x \oplus y^-) \odot y;$$

(M9) $x \vee y = y \vee x$, (M9') $x \wedge y = y \wedge x$,
(M10) $x \vee (y \vee z) = (x \vee y) \vee z$, (M10') $x \wedge (y \wedge z) = (x \wedge y) \wedge z$,
(M11) $x \oplus (y \wedge z) = (x \oplus y) \wedge (x \oplus z)$, (M11') $x \odot (y \vee z) = (x \odot y) \vee (x \odot z)$.

For historical information on MV algebras see [8].

Another equivalent definition of MV algebras, due to P. Mangani, is the following.

Definition 1.1.2 [29]
An *MV algebra* is an algebra $\mathcal{A} = (A, \oplus, \odot, ^-, 0, 1)$ of type (2,2,1,0,0) satisfying the following axioms: for all $x, y, z \in A$,
(M1-R) $(x \oplus y) \oplus z = x \oplus (y \oplus z)$,
(M2-R) $x \oplus 0 = x$,
(M3-R) $x \oplus y = y \oplus x$,
(M4-R) $x \oplus 1 = 1$,
(M5-R) $(x^-)^- = x$,
(M6-R) $0^- = 1$,
(M7-R) $x \oplus x^- = 1$,
(M8-R) $(x^- \oplus y)^- \oplus y = (x \oplus y^-)^- \oplus x$,
(M9-R) $x \odot y = (x^- \oplus y^-)^-$.

A third, shorter, equivalent definition of MV algebras, is the following.

Definition 1.1.3 ([10], Definition 1.1.1)
An *MV algebra* is an algebra $(A, \oplus, ^-, 0)$ of type (2,1,0) satisfying the following axioms: for all $x, y, z \in A$,
(MV1-R) $x \oplus (y \oplus z) = (x \oplus y) \oplus z$,
(MV2-R) $x \oplus y = y \oplus x$,
(MV3-R) $x \oplus 0 = x$,
(MV4-R) $(x^-)^- = x$,
(MV5-R) $x \oplus 0^- = 0^-$,
(MV6-R) $(x^- \oplus y)^- \oplus y = (y^- \oplus x)^- \oplus x$.

The reader has at least [10] and [33] to find more about MV algebras.

Equivalences of MV algebras are known in the literature under several names: Lacava's *Ł-algebras* [26], Rodríguez's *Wajsberg algebras* [32], [13], Bosbach's *bricks* [4], Komori's *CN-algebras* [25], Buff's *S-algebras* [5], *bounded commutative BCK algebras* [30] - where the BCK algebras were introduced in 1966 by K. Iseki [23] (see also [24], [17], [21]).

An important result in MV algebras is *Chang's subdirect representation theorem* ([10], Theorem 1.3.3), stating that an equation is valid in any MV algebra iff it is valid in any totally ordered MV algebra; the proof of this theorem uses *Zorn's lemma* (If S is any nonempty partially ordered set in which every chain (i.e., totally ordered subset) has an upper bound, then S has at least a maximal element), which is equivalent to the *axiom of choice* (Given any collection of non-empty sets, it is possible to construct a new set by choosing one element from each set, even if the collection is infinite).

In 1986 [31], Daniele Mundici has proved that the category of *lattice-ordered commutative groups with strong unit* (or unital commutative l-groups, for short) is equivalent to the category of *MV algebras*, by proving the existance of the quasi-inverse functors $\mathbf{\Gamma}$ and $\mathbf{\Xi}$ ([10], Corollary 7.1.8). He uses the axiom of choice (by using Chang's subdirect representation theorem) in his proof and makes the proof in one step, namely:

- In ([10], Chapter 2), it is proved that, for any MV algebra A, there exists a unital commutative l-group that envelops A. First, a lattice ordered monoid M_A of good sequences of A is constructed from A; then, a unital commutative l-group G_A is constructed from M_A (in a way that is analogous to the definition of $\mathbf{Z}$ from $\mathbf{N}$[1]).

- Conversely, in ([10], Chapter 2) also, it is proved that, for any unital commutative l-group G, there exists an MV algebra, $\Gamma(G, u)$, that envelops G (see [10], Definition 2.1.1): Let $(G, \vee, \wedge, +, -, 0)$ $(x \leq y \Longleftrightarrow x \vee y = y \Longleftrightarrow x \wedge y = x)$ be an l-group. For any element $u \in G$, $u > 0$ (not necessarily u being a strong unit of G), let $[0, u] \overset{def.}{=} \{x \in G \mid 0 \leq x \leq u\}$ and for each $x, y \in [0, u]$,

$$x \oplus y \overset{def.}{=} u \wedge (x + y),$$

$$\neg x \overset{def.}{=} u - x. \tag{1.1}$$

The structure $([0, u], \oplus, \neg, 0)$ is denoted $\Gamma(G, u)$. Then, it is proved ([10], Proposition 2.1.2) that $\Gamma(G, u)$ is an MV algebra. By ([10], Proposition 2.1.5), $\mathbf{\Gamma}$ is a functor. The functor $\mathbf{\Gamma}$ defines a natural equivalence between the category of commutative l-groups with strong unit and the category of (right-) MV algebras ([10], Corollary 7.1.8).

Remarks 1.1.4

The idea of associating a totally ordered commutative group to any MV algebra is due to Chang, who in [6] and [7] gave the first purely algebraic proof of the completeness of the Lukasiewicz axioms for the infinite-valued calculus, using quantifier elimination for totally ordered divisible commutative groups, cf. [10].

Good sequences and the $\mathbf{\Gamma}$ *functor were first introduced in [31].*

In 2021 ([1], Theorem 8.21), Marco Abbadini has established the following generalization of Mundici's equivalence (hence of functors $\mathbf{\Gamma}$ and $\mathbf{\Xi}$): The category of *unital lattice-ordered commutative monoids*, or unital commutative l-monoids for short (roughly speaking, unital commutative l-groups without the unary operation $x \longmapsto -x$) is equivalent to the category of *MV-monoidal algebras* (roughly speaking, MV algebras without the unary operation $x \longmapsto x^-$), by proving the existance of the quasi-inverse functors $\mathbf{\Gamma}_\star$ and $\mathbf{\Xi}_\star$. He does not use the axiom of choice and makes the proof in two steps, i.e. he introduces the intermediary category of *positive-unital commutative l-monoids* and four intermediary functors: the quasi-inverse functors $\mathbf{c}_\star^+$ and $\mathbf{T}_\star$, in step 1, and the quasi-inverse functors $\mathbf{G}_\star^+$

[1]Our LaTeX has no commands for the usual notations of these sets.

and $\mathbf{U}_\star$, in step 2, such that $\mathbf{\Gamma}_\star = \mathbf{U}_\star \circ \mathbf{c}^+_\star$ and $\mathbf{\Xi}_\star = \mathbf{T}_\star \circ \mathbf{G}^+_\star$. He obtains the original Mundici's equivalence as a corollary of his equivalence (in [1], Appendix A).

Recall ([27], Chapter IV, Section 4 - conform [1]) that:
Two functors $\mathbf{F} : \mathcal{C}(\mathbf{X}) \longrightarrow \mathcal{C}(\mathbf{Y})$ and $\mathbf{G} : \mathcal{C}(\mathbf{Y}) \longrightarrow \mathcal{C}(\mathbf{X})$ are called *quasi-inverse*, if the two functors $\mathbf{G} \circ \mathbf{F} : \mathcal{C}(\mathbf{X}) \longrightarrow \mathcal{C}(\mathbf{X})$ and $\mathbf{F} \circ \mathbf{G} : \mathcal{C}(\mathbf{Y}) \longrightarrow \mathcal{C}(\mathbf{Y})$ are naturally isomorphic to the identity functors on $\mathcal{C}(\mathbf{X})$ and $\mathcal{C}(\mathbf{Y})$, respectively.
Two categories $\mathcal{C}(\mathbf{X})$ and $\mathcal{C}(\mathbf{Y})$ are *equivalent*, if there exists two quasi-inverse functors $\mathbf{F} : \mathcal{C}(\mathbf{X}) \longrightarrow \mathcal{C}(\mathbf{Y})$ and $\mathbf{G} : \mathcal{C}(\mathbf{Y}) \longrightarrow \mathcal{C}(\mathbf{X})$.

Starting from the **old commutative framework centered on BCK algebras** begun in papers [16], containing BCK algebras, BE algebras and many other commutative algebras of logic, we have introduced, in the monograph [19], the *M, ME, ML* and *MEL algebras* (in the non-commutative case) and many other (left-) algebras, thus completing the **old framework centered on BCK algebras.** A *(left-) MEL algebra* is an algebra $(A, \to, 1)$ of type $(2, 0)$ verifying the properties (M) $(1 \to x = x)$, (Ex) $(x \to (y \to z) = y \to (x \to z))$ and (La') $(x \leq 1)$, where $x \leq y \overset{def.}{\Longleftrightarrow} x \to y = 1$. We also established a general 'bridge theorem' ([19], Theorem 5.3.1) saying that, in the involutive case (i.e. in an algebra $(A, \to, 0, 1)$ with $x^- \overset{def.}{=} x \to 0$ verifying (DN) $((x^-)^- = x)$), the properties (M) and (Ex) are equivalent with the properties (Pcomm) $(x \odot y = y \odot x)$, (PU) $(1 \odot x = x)$ and (Pass) $(x \odot (y \odot z) = (x \odot y) \odot z)$, respectively, of an algebra $(A, \odot, ^-, 1)$ of type $(2, 1, 0)$ verifying (DN).

Then, searching around *(left-) MV algebras* - algebras $(A, \odot, ^-, 1)$ verifying the properties (PU), (Pcomm), (Pass), $x \odot 0 = 0$ $(0 \overset{def.}{=} 1^-)$, (DN) and ($\wedge_m$-comm) $((x^- \odot y)^- \odot y = (y^- \odot x)^- \odot x)$, where the binary relation $x \leq_m y \overset{def.}{\Longleftrightarrow} x \odot y^- = 0$ can be defined - on April 21, 2019, we have realized that the property $x \odot 0 = 0$ means $x \leq_m 1$, hence it is the analogous property, called (m-La') ('m' coming from 'magma'), of the above property (La'); thus, we were able to define a new algebra $(A, \odot, ^-, 1)$, the *m-MEL algebra*, verifying (PU), (Pcomm), (Pass), (m-La') and, hence, we have realized that **the (left-) MV algebra is just the involutive (left-) m-MEL algebra verifying ($\wedge_m$-comm)** ([21], Remark 6.5.2 (i)). So, April 21, 2019, was the starting day of the paper [20] and then of the monograph [21]. Finally, we have introduced many other new commutative (left-) algebras, including the *m-BCK algebra* (which is always involutive), thus obtaining a **new framework, centered on m-BCK algebras**, which contains the commutative groups, the MV algebras, the Boolean algebras and many other commutative (left-) algebras (see **Appendix A**, in the dual (right-) case). A commutative group $(G, \cdot, ^{-1}, 1)$ is an m_1-ME algebra verifying the property (m_1-Re) ([21], Remarks 7.1.1 (4)) and the (left-) MV algebra $(A, \odot, ^-, 1)$ is just the involutive (left-) m-MEL algebra verifying the property ($\wedge_m$-comm).

The old framework centered on (left-) BCK algebras and the new framework centered on (left-) m-BCK algebras are connected in the involutive case ([21], Theorem 17.1.1); thus, the involutive MEL algebras are definitionally equivalent (d.e.)

to the involutive m-MEL algebras ([21], Theorem 17.1.2), the involutive BE algebras are d.e. to the involutive m-BE algebras ([21], Corollary 17.1.3), the involutive BCK algebras are d.e. to the (involutive) m-BCK algebras ([21], Corollary 17.1.4), the Wajsberg algebras are d.e. to the MV algebras ([21], Corollary 17.1.5), the implicative-Boolean algebras are d.e. to the Boolean algebras ([21], Corollary 17.1.7).

Similarly, starting from the old non-commutative framework centered on (left-) pseudo-BCK algebras, we have introduced a new non-commutative framework centered on (left-) m-pseudo-BCK algebras, in monograph [22].

The *algebraists* work usually with the commutative groups defined additively and with the positive (right) cone of a partially-ordered commutative group $(G, \leq, +, -, 0)$, where there are essentially a sum $\oplus = +$ and an element 0. They work with algebras that have associated an order relation, which usually does not appear explicitely in the definitions. The presence of the order relation implies the presence of the duality principle. Thus, each algebra has a dual one, the order relation has a dual one. We have given names to the dual algebras [17], [19], [20]: "left" algebra and "right" algebra, names related to the left-continuity of a t-norm $\odot$ and to the right-continuity of a t-conorm $\oplus$, respectively. Hence, the algebraists usually work with *right-algebras*.

By contrary, the *logicians* work with the logic of *truth*, where the *truth* is represented by 1, and there is essentially one implication, $\to$ (two, in the non-commutative case); we could name this logic "left-logic". One can imagine also a "right-logic", as a logic of *false*, where the *false* is represented by 0, and there is essentially one implication, $\to^R$ (two, in the non-commutative case). Hence, the logicians usually work with *left-algebras of logic* (or *algebras of left-logic*).

Summarizing, for algebraists, the appropriate algebras are the *unital commutative magmas* and, among these, the appropriate algebras are the *right-algebras*. For logicians, by contrary, the appropriate algebras are the *unital commutative implicative-magmas* and, among these, the appropriate algebras are the *left-algebras*. This explains why, for examples, the MV algebras were initially introduced [6] as right-algebras, while the Wajsberg algebras were initially introduced [13] as left-algebras of logic.

In monographs [19], [21], [22], **we have worked unitarily with left-algebras**, since we are coming from (algebras of) logic side.

In this book, we shall work unitarily with right-algebras, since MV algebras were initially defined as right-MV algebras and Mundici worked with right-MV algebras.

*

The natural ideea then came, in October 2024, to generalize, in the new commutative framework centered on m-BCK algebras, Mundici's equivalence between commutative l-groups with strong unit and MV algebras.

In this book, we generalize Mundici's equivalence (hence the functors $\mathbf{\Gamma}$ and $\mathbf{\Xi}$). The role of *unital commutative l-groups* is played by *strong m_0-ME structures* and the role of *MV algebras* is played by *XY algebras*. Both *strong* m_0*-ME structures* and *XY algebras* are new notions belonging to the new framework centered on m-BCK algebras (see **Appendix A**). We do not follow the proof of Mundici's equivalence in [10], but the proof of Abbadini's equivalence in [1]: we do not use the axiom of choice in the proof (in fact, it is the only choice in our case, since there are no proper XY chains - the XY chains are in fact MV chains) and we make the proof in two steps (in fact, it is the only choice in our case too, since the corresponding functors $\mathbf{\Gamma_m}$ and $\mathbf{\Xi_m}$ are not quasi-inverse, but the intermediary two functors $\mathbf{G^+_m}$ and $\mathbf{U_m}$ - of step 2 - are quasi-inverse, the intermediary category being that of *strong* m_0*-ME positive cones*). So, we establish the following generalization of Mundici's equivalence (Theorem 10.1.1): The category of strong m_0-ME structures is not equivalent to the category of strong m_0-ME positive cones, but the category of strong m_0-ME positive cones is equivalent to the category of XY algebras. We show how the original Mundici's equivalence is a particular case of our results (see Corollary 2.2.68 and also Connections from Chapter 3 and **Appendix C**).

The book is written in a **unifying way**, which consist in fixing unique names for the defining properties.

We shall make below a summary of the proof in two steps of Abbadini's equivalence from [1].

We shall make below a summary of our results presented in this book.

Since we had to follow Abbadini' proof in two steps, while generalizing Mundici's proof in one step, we are able finally, in order to make better understandable our generalization, to sketch below a summary of the Mundici's proof in two steps, by introducing the intermediary category of *positive-unital commutative l-groups* and four intermediary functors: the quasi-inverse functors $\mathbf{c^+}$ and $\mathbf{T}$, in step 1, and the quasi-inverse functors $\mathbf{G^+}$ and $\mathbf{U}$, in step 2, such that $\mathbf{\Gamma} = \mathbf{U} \circ \mathbf{c^+}$ and $\mathbf{\Xi} = \mathbf{T} \circ \mathbf{G^+}$.

Note that these three summaries are written also in a **unifying way**, including the same names ($\mathbf{c^+}$, $\mathbf{T}$, $\mathbf{G^+}$, $\mathbf{U}$) for the intermediary four functors.

1.2 Summary of Abbadini's equivalence

Abbadini has proved that the category of *unital commutative l-monoids* is equivalent to the category of *MV-monoidal algebras*. In his paper [1], he proceeds as follows:

- the role of *unital commutative l-groups* is played by *unital commutative l-monoids*,
- the intermediary role of *positive-unital commutative l-groups* (Definition 1.4.1) is played by *positive-unital commutative l-monoids*,
- the role of *MV algebras* is played by *MV-monoidal algebras*,

all defined as follows.

An *MV-monoidal algebra* is an algebra $(A, \vee, \wedge, \oplus, \odot, 0, 1)$ of type $(2, 2, 2, 2, 0, 0)$ satisfying the following equational axioms (see [1], Definition 2.5.):
(A1) $(A, \vee, \wedge)$ is a distributive lattice,
(A2) $(A, \oplus, 0)$ and $A, \odot, 1)$ are commutative monoids,
(A3) both the operations $\oplus$ and $\odot$ distribute over both $\vee$ and $\wedge$,

(A4) $(x \oplus y) \odot ((x \odot y) \oplus z) = (x \odot (y \oplus z)) \oplus (y \odot z)$,
(A5) $(x \odot y) \oplus ((x \oplus y) \odot z) = (x \oplus (y \odot z)) \odot (y \oplus z)$,
(A6) $(x \odot y) \oplus z = ((x \oplus y) \odot ((x \odot y) \oplus z)) \vee z$,
(A7) $(x \oplus y) \odot z = ((x \odot y) \oplus ((x \oplus y) \odot z)) \wedge z$.

On $[0, 1]$, consider the elements 0 and 1 and the operations $x \vee y \overset{def.}{=} \max\{x, y\}$, $x \wedge y \overset{def.}{=} \min\{x, y\}$, $x \oplus y \overset{def.}{=} \min\{x + y, 1\}$, and $x \odot y \overset{def.}{=} \max\{x + y - 1, 0\}$. This gives a first example of MV-monoidal algebra, cf. [1].

Abbadini remarks that MV-monoidal algebras form a variety of algebras whose primitive operations are finitely many and of finite arity, and which is axiomatised by a finite number of equations.

Bounded distributive lattices form a subvariety of the variety of MV-monoidal algebras, obtained by adding the axioms $x \oplus y = x \vee y$ and $x \odot y = x \wedge y$ ([1], Remark 2.6.).

We denote by $\mathcal{C}(\mathbf{MVM})$ the category of MV-monoidal algebras with homomorphisms.

Definition 1.2.1 (See [1], Definition 2.1.)
A *lattice-ordered commutative monoid*, or a *commutative l-monoid* for short, is an algebra $(M, \vee, \wedge, +, 0)$ of type $(2, 2, 2, 0)$ with the following properties:
(M1) $(M, \vee, \wedge)$ is a distributive lattice, with $x \leq y \iff x \vee y = y \iff x \wedge y = x$,
(M2) $(M, +, 0)$ is a commutative monoid,
(M3) $+$ distributes over $\vee$ and $\wedge$.

A *unital commutative lattice-ordered monoid*, or a *unital commutative l-monoid*, for short, is an algebra $(M, \vee, \wedge, +, -1, 0, 1)$ of type $(2, 2, 2, 0, 0, 0)$ with the following properties (see [1], Definition 2.2.):
(U0) $(M, \vee, \wedge, +, 0)$ is a commutative l-monoid,
(U1) $-1 + 1 = 0$,
(U2) $0 \leq 1$,
(U3) for every $x \in M$, there exists $n \in \mathbf{N}$ such that

$$\underbrace{(-1) + \ldots + (-1)}_{n \text{ times}} \leq x \leq \underbrace{1 + \ldots + 1}_{n \text{ times}}.$$

Abbadini writes $z - 1$ for $z + (-1)$.
Given $n \in \mathbf{N}$, he writes n for $\underbrace{1 + \ldots + 1}_{n \text{ times}}$ and $-n$ for $\underbrace{(-1) + \ldots + (-1)}_{n \text{ times}}$.

The set $\Re$, endowed with the binary operations $+$ (addition), $\vee$ (maximum), $\wedge$ (minimum), and the constants 0, 1 and -1 is a prototypical example of a unital commutative lattice-ordered monoid, cf. [1].

For every topological space X equipped with a preorder, the set of bounded continuous order-preserving functions from X to $\Re$ is a unital commutative l-monoid ([1], Example 2.4.).

We denote with $\mathcal{C}(\mathbf{u}l\mathbf{M})$ the category of unital commutative l-monoids with homomorphisms.

A *positive-unital commutative l-monoid* is an algebra $(M, \vee, \wedge, +, - \ominus 1, 0, 1)$ of type $(2,2,2,1,0,0)$ such that, for every $x \in M$, the following properties hold (see [1], Definition 4.1.):
(P0) $(M, \vee, \wedge, +, 0)$ is a commutative l-monoid,
(P1) $x \geq 0$,
(P2) $(x+1) \ominus 1 = x$,
(P3) $(x \ominus 1) + 1 = x \vee 1$,
(P4) there exists $n \in \mathbf{N}$ such that $x \leq \underbrace{1 + \ldots + 1}_{n \text{ times}}$.

We denote with $\mathcal{C}(\mathbf{u}l\mathbf{M}^+)$ the category of positive-unital commutative l-monoids with homomorphisms.

"Roughly speaking, if we think of a unital commutative l-monoid as the interval $(-\infty, \infty)$, then an MV-monoidal algebra is the interval $[0,1]$, whereas a positive-unital commutative l-monoid is the interval $[0, \infty)$".

1.2.1 The step 1 at Abbadini: the functors $\mathbf{c}_\star^+$ and $\mathbf{T}_\star$

The algebraic structures involved in the first step are the *unital commutative l-monoids* and the *positive-unital commutative l-monoids.*

- **The functor $\mathbf{c}_\star^+$**

Given a unital commutative l-monoid $(M, \vee, \wedge, +, -1, 0, 1)$, Abbadini sets $M^+ := \{x \in M \mid x \geq 0\}$. He endows M^+ with the operations $\vee, \wedge, +, 0, 1$ defined by restriction and with $- \ominus 1$ defined by $x \ominus 1 \stackrel{def.}{=} (x-1) \vee 0$.

For every unital commutative l-monoid M, the algebra $(M^+, \vee, \wedge, +, - \ominus 1, 0, 1)$ is a positive-unital commutative l-monoid ([1], Proposition 4.4.).

Given a morphism $f : M \longrightarrow N$ of unital commutative l-monoids, f restricts to a function f^+ from M^+ to N^+. Moreover, f preserves $\vee, \wedge, +$ and 1 and so f^+ is a morphism of positive-unital commutative l-monoids. This establishes a functor $\mathbf{c}_\star^+$ ($(-)^+$, in the notation of Abbadini), $\mathbf{c}_\star^+ : \mathcal{C}(\mathbf{u}l\mathbf{M}) \longrightarrow \mathcal{C}(\mathbf{u}l\mathbf{M}^+)$ that maps M to M^+, and maps a morphism $f : M \longrightarrow N$ to its restriction $f^+ : M^+ \longrightarrow N^+$.

- **The functor $\mathbf{T}_\star$**

Let M^+ be a positive-unital commutative l-monoid. Abbadini wants to construct "a unital commutative l-monoid $\mathbf{T}_\star(M^+)$ such that, if N is a unital commutative l-monoid and $N^+ \cong M^+$, then $\mathbf{T}_\star(M^+) \cong N$. Every element of a unital commutative l-monoid N can be expressed as $x - n$ for some $x \in \mathbf{N}^+$ and $n \in \mathbf{N}$. Roughly speaking, we will obtain $\mathbf{T}_\star(N^+) \cong N$ by translating the elements of N^+ by negative integers. (In fact, $\mathbf{T}$ stands for 'translations'.)"

Given a positive-unital commutative l-monoid $(M^+, \vee, \wedge, +, - \ominus 1, 0, 1)$, Abbadini considers the binary relation $\sim$ defined on $M^+ \times \mathbf{N}$ as follows:

$$(x,n) \sim (y,p) \overset{def.}{\iff} x+p = y+n. \tag{1.2}$$

The equivalence class of an element (x,n) of $M^+ \times \mathbf{N}$ is denoted by $[(x,n)]$, or simply by $[x,n]$. The set of all equivalence classes is denoted by $\mathbf{T}_\star(M^+)$:

$$\mathbf{T}_\star(M^+) \overset{def.}{=} \frac{M^+ \times \mathbf{N}}{\sim}$$

and is endowed with the operations of a unital commutative l-monoid:

$$-1 \overset{def.}{=} [0,1],$$

$$0 \overset{def.}{=} [0,0],$$

$$1 \overset{def.}{=} [1,0],$$

$$[x,n] \; + \; [y,m] \overset{def.}{=} [x+y, n+m],$$

$$[x,n] \; \vee \; [y,m] \overset{def.}{=} [(x+m) \vee (y+n), n+m],$$

$$[x,n] \; \wedge \; [y,m] \overset{def.}{=} [(x+m) \wedge (y+n), n+m].$$

The algebra $(\mathbf{T}_\star(M^+) = \frac{M^+ \times \mathbf{N}}{\sim}, \vee, \wedge, +, -1 = [0,1], 0 = [0,0], 1 = [1,0])$ is a unital commutative l-monoid ([1], Proposition 4.7).

- **The connections between the functors $\mathbf{c}_\star^+$ and $\mathbf{T}_\star$**

The functors $\mathbf{c}_\star^+ : \mathcal{C}(\mathbf{u}l\mathbf{M}) \longrightarrow \mathcal{C}(\mathbf{u}l\mathbf{M}^+)$ and $\mathbf{T}_\star : \mathcal{C}(\mathbf{u}l\mathbf{M}^+) \longrightarrow \mathcal{C}(\mathbf{u}l\mathbf{M})$ are quasi-inverses. Thus, the categories of unital commutative l-monoids and positive-unital commutative l-monoids are equivalent ([1], Theorem 4.16.).

1.2.2 The step 2 at Abbadini: the functors $\mathbf{G}_\star^+$ and $\mathbf{U}_\star$

The algebraic structures involved in the second step are the above *positive-unital commutative l-monoids* and the *MV-monoidal algebras.*

- **The functor $\mathbf{G}_\star^+$**

Let $\mathcal{A} = (A, \vee, \wedge, \oplus, \odot, 0, 1)$ be an MV-monoidal algebra.

A *good pair* of A is a pair (x_0, x_1) of elements of A such that $x_0 \oplus x_1 = x_0$ and $x_0 \odot x_1 = x_1$. A *good sequence* of A is a sequence $(x_0, x_1, x_2, \ldots)$ of elements of A which is eventually 0 and such that, for each $n \in \mathbf{N}$, (x_n, x_{n+1}) is a good pair (See [1], Definition 5.1.).

In Abbadini's definition of good pair, he has included both the condition $x_0 \oplus x_1 = x_0$ and the condition $x_0 \odot x_1 = x_1$ because, in general, they are not equivalent ([1], Remark 5.2.).

He denotes with $G(A)$ the set of good sequences of A. ("In fact, G stands for 'good'.") He endows $G(A)$ with the structure of a positive-unital commutative l-monoid, as follows.

He lets $\mathbf{0}$ denote the good sequence $(0, 0, 0, \ldots)$ and $\mathbf{1}$ denote the good sequence $(1, 0, 0, 0, \ldots)$. For good sequences $\mathbf{a} = (a_0, a_1, a_2, \ldots)$ and $\mathbf{b} = (b_0, b_1, b_2, \ldots)$ of A, he sets

$$\mathbf{a} \vee \mathbf{b} \stackrel{def.}{=} (a_0 \vee b_0, a_1 \vee b_1, a_2 \vee b_2, \ldots)$$

and

$$\mathbf{a} \wedge \mathbf{b} \stackrel{def.}{=} (a_0 \wedge b_0, a_1 \wedge b_1, a_2 \wedge b_2, \ldots).$$

We have a partial order $\leq$ on $G(A)$, induced by the lattice operations. Since the lattice operations are defined componentwise, we have the following. For all good sequences $\mathbf{a} = (a_0, a_1, a_2, \ldots)$ and $\mathbf{b} = (b_0, b_1, b_2, \ldots)$ of A, we have $\mathbf{a} \leq \mathbf{b}$ if, and only if, for all $n \in \mathbf{N}$, $a_n \leq b_n$ ([1], Remark 7.4.).

Given two good sequences $\mathbf{a} = (a_0, a_1, a_2, \ldots)$ and $\mathbf{b} = (b_0, b_1, b_2, \ldots)$ of A, Abbadini sets

$$\mathbf{a} + \mathbf{b} \stackrel{def.}{=} (c_0, c_1, c_2, \ldots),$$

where:

$$c_n \stackrel{def.}{=} (a_0 \oplus b_n) \odot (a_1 \oplus b_{n-1}) \odot \ldots \odot (a_{n-1} \oplus b_1) \odot (a_n \oplus b_0)$$

or, equivalently,

$$c_n \stackrel{def.}{=} b_n \oplus (a_0 \odot b_{n-1}) \oplus (a_1 \odot b_{n-2}) \oplus \ldots \oplus (a_{n-2} \odot b_1) \oplus (a_{n-1} \odot b_0) \oplus a_n.$$

For a good sequence $\mathbf{a} = (a_0, a_1, a_2, \ldots)$, he sets $\mathbf{a} \ominus 1 \stackrel{def.}{=} (a_1, a_2, a_3, \ldots)$. The sequence $\mathbf{a} \ominus 1$ is a good sequence.

For every MV-monoidal algebra $\mathcal{A}$, the algebra

$$\mathbf{G}_\star^+(\mathcal{A}) = (G(A), \vee, \wedge, +, - \ominus 1, \mathbf{0}, \mathbf{1})$$

is a positive-unital commutative l-monoid ([1], Proposition 7.24.).

"It is easy to see that $\mathbf{G}_\star^+ : \mathcal{C}(\mathbf{MVM}) \longrightarrow \mathcal{C}(\mathbf{u}l\mathbf{M}^+)$ is a functor."

- **The unit interval functor $\mathbf{U}_\star$**

Let $(M^+, \vee, \wedge, - \ominus 1, 0, 1)$ be a positive-unital commutative l-monoid.

Abbadini sets $[0, 1] \stackrel{def.}{=} \{x \in M^+ \mid 0 \leq x \leq 1\}$ and endows $[0, 1]$ with the operations of MV-monoidal algebra, as follows.

The operations $\vee, \wedge, 0, 1$ are defined by restriction. For $x, y \in [0, 1]$, he sets $x \oplus y \stackrel{def.}{=} (x + y) \wedge 1$ and $x \odot y \stackrel{def.}{=} (x + y) \ominus 1$.

Then, $\mathbf{U}_\star(M^+) \stackrel{def.}{=} ([0, 1], \vee, \wedge, \oplus, \odot, 0, 1)$ is an MV-monoidal algebra, by the previous results from [1].

Given a morphism $f^+ : M^+ \longrightarrow N^+$ of positive-unital commutative l-monoids, he sets $\mathbf{U}_\star(f^+) : \mathbf{U}_\star(M^+) \longrightarrow \mathbf{U}_\star(N^+)$ as the restriction of f^+.

This assignment establishes a functor $\mathbf{U}_\star : \mathcal{C}(\mathbf{u}l\mathbf{M}^+) \longrightarrow \mathcal{C}(\mathbf{MVM})$. ($\mathbf{U}$ stands for 'unit interval'.)

• **The connections between the functors $\mathbf{G}^+_\star$ and $\mathbf{U}_\star$**

The functors $\mathbf{U}_\star : \mathcal{C}(\mathbf{u}l\mathbf{M}^+) \longrightarrow \mathcal{C}(\mathbf{MVM})$ and $\mathbf{G}^+_\star : \mathcal{C}(\mathbf{MVM}) \longrightarrow \mathcal{C}(\mathbf{u}l\mathbf{M}^+)$ are quasi-inverses. Thus, the categories of positive-unital commutative l-monoids and MV-monoidal algebras are equivalent ([1], Theorem 8.20.).

1.2.3 Connections between the four functors at Abbadini

The main result of Abbadini is ([1], Theorem 8.21):
The functor $\Gamma_\star = \mathbf{U}_\star \circ \mathbf{c}^+_\star : \mathcal{C}(\mathbf{u}l\mathbf{M}) \longrightarrow \mathcal{C}(\mathbf{MVM})$ is an equivalence of categories.
A quasi-inverse of $\mathbf{\Gamma}_\star$ is $\mathbf{\Xi}_\star = \mathbf{T}_\star \circ \mathbf{G}^+_\star$.

Resuming, the two steps and four functors of Abbadini's equivalence are:

$$(M,\vee,\wedge,+,-1,0,1) \quad \overset{\mathbf{c}^+_\star}{\longrightarrow} \quad (M^+,\vee,\wedge,+,-\ominus 1,0,1) \quad \overset{\mathbf{U}_\star}{\longrightarrow} \quad ([0,1],\vee,\wedge,\oplus,\odot,0,1)$$

$$(G(A),\vee,\wedge,+,-\ominus 1,(0),(1)) \quad \overset{\mathbf{G}^+_\star}{\longleftarrow} \quad (A,\vee,\wedge,\oplus,\odot,0,1)$$

$$(\tfrac{M^+\times\mathbf{N}}{\star},\vee,\wedge,+,-1,0,1) \quad \overset{\mathbf{T}_\star}{\longleftarrow} \quad (M^+,\vee,\wedge,+,-\ominus 1,0,1)$$

and we have:

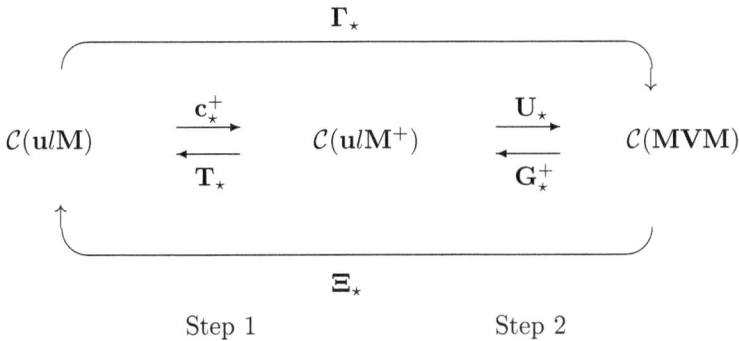

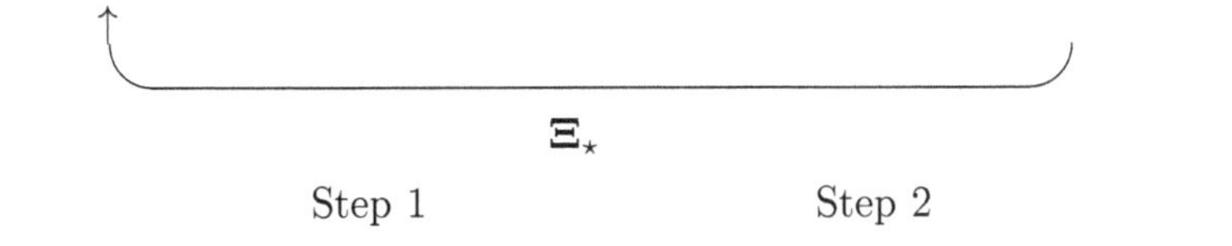

where:
$\mathcal{C}(\mathbf{u}l\mathbf{M})$ is the category of unital commutative l-monoids,
$\mathcal{C}(\mathbf{u}l\mathbf{M}^+)$ is the category of positive-unital commutative l-monoids,
$\mathcal{C}(\mathbf{MVM})$ is the category of MV-monoidal algebras.

Abbadini obtains the original Mundici's equivalence as a corollary of his equivalence (in [1], **Appendix A**).

1.3 Summary of our results from this book

In this book, we also generalize Mundici's equivalence. We proceed in two steps, analogously as Abbadini did:
- the role of *unital commutative l-groups* is played by *strong* m_0*-ME structures*,
- the role of *positive-unital commutative l-groups* (see Definition 1.4.1) is played by *strong* m_0*-ME positive cones*,
- the role of *MV algebras* is played by *XY algebras*.

1.3.1 The step 1 in this book: the functors $\mathbf{c}^+_{\mathbf{m}}$ and $\mathbf{T}_{\mathbf{m}}$

The algebraic structures involved in the first step are the *strong* m_0*-ME structures* (Definition 3.2.6) and the *strong* m_0*-ME positive cones* (Definition 3.3.1), which both are involutive m_0-ME algebras verifying some properties.

We denote with $\mathcal{C}(\mathbf{s\text{-}m_0\text{-}ME})$ the category of strong m_0-ME structures with homomorphisms.

We denote with $\mathcal{C}(\mathbf{s\text{-}m_0\text{-}ME^+})$ the category of strong m_0-ME positive cones with homomorphisms.

- **The functor $\mathbf{c}^+_{\mathbf{m}}$**

Let $\mathcal{M}_u = (M, \leq_m, +, \smile, 0_M, u_M)$ be a strong m_0-ME structure and let define the positive cone as follows:

$$M^+ \overset{def.}{=} \{x \in M \mid x \geq_m 0_M,\ such\ that\ (Nu^{+1R}),\ (S^{+1R})\ hold\}.$$

If we denote with the same symbols $+$ and $\smile$ the restrictions on M^+ of the operations $+$ and $\smile$ on M and if we denote with $\leq_m$ the restriction on M^+ of the binary relation $\leq_m$ on M, then we obtain the structure

$$\mathcal{M}_u^+ = (M^+, \leq_m, +, \smile, 0_M, u_M)$$

which, obviously, is a *strong* m_0*-ME positive cone*, where $0_{M^+} = 0_M$ and $u_{M^+} = u_M$.

We put:

$$\mathbf{c}^+_{\mathbf{m}}(\mathcal{M}_u) = \mathcal{M}_u^+.$$

Given a morphism $f : M \longrightarrow N$ of strong m_0-ME structures
$\mathcal{M}_u = (M, \leq_m, +, \smile, 0_M, u = u_M)$ and $\mathcal{N}_v = (N, \leq_m, +, \smile, 0_N, v = u_N)$,
f restricts to a function f^+ from M^+ to N^+ and, obviously, f^+ is a morphism of strong m_0-ME positive cones
$\mathcal{M}_u^+ = (M^+, \leq_m, +, \smile, 0_{M^+}, u = u_{M^+})$ and $\mathcal{N}_v^+ = (N^+, \leq_m, +, \smile, 0_{N^+}, v = u_{N^+})$.

This establishes that $\mathbf{c}^+_{\mathbf{m}}$ is a functor $\mathbf{c}^+_{\mathbf{m}} : \mathcal{C}(\mathbf{s\text{-}}m_0\mathbf{\text{-}ME}) \longrightarrow \mathcal{C}(\mathbf{s\text{-}}m_0\mathbf{\text{-}ME^+})$ that maps $\mathcal{M}_u$ to $\mathcal{M}_u^+$ and maps a morphism $f : M \longrightarrow N$ to its restriction

$$f^+ = \mathbf{c}^+_{\mathbf{m}}(f) : \mathbf{c}^+_{\mathbf{m}}(M) = M^+ \longrightarrow \mathbf{c}^+_{\mathbf{m}}(N) = N^+.$$

- **The functor $\mathbf{T_m}$**

The name '$\mathbf{T_m}$' comes here from 'turning/tipping' (versus 'translations', from [1]), because the entire construction made by this functor is a turning/tipping point (or an inflexion point): it differs essentially from the construction from [10], because **the strong m_0-ME positive cone does not verify the cancellation property (C)**; it is analogous to that from [1].

Let $\mathcal{M}_u^+ = (M^+, \leq_m, +, \smile, 0_{M^+}, u = u_{M^+})$ be a strong m_0-ME positive cone, **which is not cancellative, as happens in [10]**.

Denote $M^+ \times \mathbf{N} = \{(x, n) \mid x \in M^+, n \in \mathbf{N}\}$. We consider the binary relation $\sim$ defined on $M^+ \times \mathbf{N}$ as follows (see (1.10), (1.2) and (5.1)):

$$(x, n) \sim (y, p) \overset{def.}{\Longleftrightarrow} x + pu_{M^+} = y + nu_{M^+}.$$

We shall denote the equivalence class of $(x, n) \in M^+ \times \mathbf{N}$ by $\overbrace{(x, n)}$. The set of all equivalence classes (the quotient set) is denoted by $\mathbf{T_m}(M^+)$:

$$\mathbf{T_m}(M^+) := \frac{M^+ \times \mathbf{N}}{\sim}$$

and is endowed with the operations of a strong m_0-ME structure:

$$\mathbf{0} \overset{def.}{=} \overbrace{(0_{M^+}, 0)} \quad (5.2),$$

$$\mathbf{u} \overset{def.}{=} \overbrace{(u_{M^+}, 0)} \quad (5.3),$$

$$\smile \overbrace{(x + mu_{M^+}, n)} \overset{def.}{=} \overbrace{(\smile x + mu_{M^+}, n)} \quad (5.5),$$

$$\overbrace{(X, n)} + \overbrace{(Y, p)} \overset{def.}{=} \overbrace{(X + Y, n + p)} \quad (5.4),$$

$$\overbrace{(X, n)} \preceq_m \overbrace{(Y, p)} \overset{def.}{\Longleftrightarrow} X + pu_{M^+} \leq_m Y + nu_{M^+} \quad (5.6).$$

We define the infimum related to $\preceq_m$, $\inf_m$, by:

$$\inf_m(\overbrace{(X, n)}, \overbrace{(Y, p)}) \overset{def.}{=} \overbrace{(\inf_m(X + pu_{M^+}, Y + nu_{M^+}), n + p)} \quad (5.7).$$

Then, $(\mathbf{T_m}(M^+) = \frac{M^+ \times \mathbf{N}}{\sim}, \preceq_m, +, \smile, \mathbf{0}, \mathbf{u})$ is a strong m_0-ME structure (Theorem 5.1.27). $\mathbf{T_m}$ is a functor (by Proposition 5.2.1).

- **The connections between the functors $\mathbf{c_m^+}$ and $\mathbf{T_m}$**

The functors $\mathbf{c_m^+}$ and $\mathbf{T_m}$ are not quasi-inverses, thus the categories $\mathcal{C}(\mathbf{s\text{-}m_0\text{-}ME})$ and $\mathcal{C}(\mathbf{s\text{-}m_0\text{-}ME^+})$ are not equivalent (the functor $\mathbf{c_m^+}$ is not an equivalence) (Corollary 6.3.3).

1.3.2 The step 2 in this book: the functors $\mathbf{G_m^+}$ and $\mathbf{U_m}$

The algebraic structures involved in the second step are the above *strong* m_0*-ME positive cones* and the *right-XY algebras* (Definition 2.2.13).

We denote with $\mathcal{C}(\mathbf{XY}^R)$ the category of right-XY algebras with homomorphisms.

- **The functor $\mathbf{G_m^+}$**

Let $\mathcal{A} = (A, \oplus, ^-, 0 = 0_A)$ be a right-XY algebra, with $1 = 1_A \stackrel{def.}{=} 0^-$.

(1) A *good pair* of A is a pair (x, y) of elements of A such that $x \oplus y = x$.
(2) A *good sequence* of A is a sequence $\mathbf{a} = (a_0, a_1, \ldots)$ of elements of A such that:
(i) for each $i = 0, 1, \ldots$, the pair (a_i, a_{i+1}) is good and
(ii) there is $n \in \mathbf{N}$, such that $a_k = 0$ for all $k > n$, i.e. $\mathbf{a} = (a_0, a_1, \ldots, a_n, 0, 0, \ldots)$.
We shall simply write: $\mathbf{a} = (a_0, a_1, \ldots, a_n)$ (Definitions 7.1.1).

Note that, like in MV algebras, (x, y) is a good pair iff $x \odot y = y$.

We shall denote by $G^+(A)$ the set of good sequences of $\mathcal{A}$ ('G' comes from 'good'). We endow $G^+(A)$ with the structure of a strong m_0-ME positive cone, as follows.

We let (x) denote the good sequence $(x, 0, 0, \ldots)$, hence (0) denotes the good sequence $(0, 0, 0, \ldots)$ and (1) denotes the good sequence $(1, 0, 0, 0, \ldots)$. We put $0_{G^+(A)} = (0)$ and $u_{G^+(A)} = (1)$.

Given two good sequences $\mathbf{a} = (a_0, a_1, a_2, \ldots)$ and $\mathbf{b} = (b_0, b_1, b_2, \ldots)$ of $\mathcal{A}$, like Abbadini in [1], we define their sum

$$\mathbf{a} + \mathbf{b} \stackrel{def.}{=} (c_0, c_1, c_2, \ldots),$$

where:

$$c_n \stackrel{def.}{=} (a_0 \oplus b_n) \odot (a_1 \oplus b_{n-1}) \odot \ldots \odot (a_{n-1} \oplus b_1) \odot (a_n \oplus b_0)$$

or, equivalently,

$$c_n \stackrel{def.}{=} b_n \oplus (a_0 \odot b_{n-1}) \oplus (a_1 \odot b_{n-2}) \oplus \ldots \oplus (a_{n-2} \odot b_1) \oplus (a_{n-1} \odot b_0) \oplus a_n.$$

Given the good sequence $\mathbf{a} = (a_0, a_1, \ldots, a_{n-1}, a_n)$ of $\mathcal{A}$, we define (see (1.11)):
$\smile \mathbf{a} = \smile (a_0, a_1, \ldots, a_{n-1}, a_n) \stackrel{def.}{=} (a_n^-, a_{n-1}^-, \ldots, a_1^-, a_0^-)$ (7.10).

Let $\mathbf{a} = (a_0, a_1, \ldots, a_{n-1}, a_n)$ and $\mathbf{b} = (b_0, b_1, \ldots, b_{n-1}, b_n)$ be two good sequences of $\mathcal{A}$. We define the external (i.e. not defined by the operations on $G^+(A)$) binary relation, $\leq_m$, on $G^+(A)$ by:
$\mathbf{a} \leq_m \mathbf{b} \stackrel{def.}{\Longleftrightarrow} a_i \leq_m b_i,\ for\ i = 0, 1, \ldots, n, \quad in\ \mathcal{A}.$

For the good sequences $\mathbf{a} = (a_0, a_1, \ldots, a_{n-1}, a_n)$ and $\mathbf{b} = (b_0, b_1, \ldots, b_{n-1}, b_n)$ and the binary relation $\leq_m$ on $G^+(A)$, we set
$\inf_m(\mathbf{a}, \mathbf{b}) \stackrel{def.}{=} (\inf_m(a_0, b_0), \inf_m(a_1, b_1), \ldots, \inf_m(a_n, b_n))$ (7.13),
$\sup_m(\mathbf{a}, \mathbf{b}) \stackrel{def.}{=} (\sup_m(a_0, b_0), \sup_m(a_1, b_1), \ldots, \sup_m(a_n, b_n))$ (7.14),
and they exist, if $\inf_m(a_0, b_0), \inf_m(a_1, b_1), \ldots, \inf_m(a_n, b_n)$ and

$\sup_m(a_0, b_0), \sup_m(a_1, b_1), \ldots, \sup_m(a_n, b_n)$, respectively, do exist. When they exist, the problem is if they are good sequences.

For every right-XY algebra $\mathcal{A}$, the structure $(G^+(A), \leq_m, +, \smile, 0_{G^+(A)}, u_{G^+(A)})$ is a strong $\mathrm{m_0}$-ME positive cone (Theorem 7.1.77).

We put $\mathbf{G^+_m}(\mathcal{A}) = (G^+(A), \leq_m, +, \smile, 0_{G^+(A)}, u_{G^+(A)})$. $\mathbf{G^+_m}$ is a functor (by Proposition 7.2.1).

- **The unit interval functor $\mathbf{U_m}$**

Let $\mathcal{M}^+_u = (M^+, \leq_m, +, \smile, 0_{M^+}, u = u_{M^+})$ be a strong $\mathrm{m_0}$-ME positive cone.

Let $[0_{M^+}, u_{M^+}] \stackrel{def.}{=} \{x \in M^+ \mid 0_{M^+} \leq_m x \leq_m u_{M^+}\} \subset M^+$ be the unit interval in $\mathcal{M}^+_u$. We endow $[0_{M^+}, u_{M^+}]$ with a structure of right-XY algebra, as follows.

For all $x, y \in [0_{M^+}, u_{M^+}]$, we define (Definitions 8.1.1):

$x \oplus y \stackrel{def.}{=} \inf_m(u_{M^+}, x + y)$, $\neg x \stackrel{def.}{=} \smile x$, $\quad x \neq 0_{M^+}$, $\neg 0_{M^+} \stackrel{def.}{=} u_{M^+}$.

Then, $([0_{M^+}, u_{M^+}], \oplus, \neg, 0_{M^+})$ is a right-XY algebra (Theorem 8.1.2).

We put: $\mathbf{U_m}(\mathcal{M}^+_u) = ([0_{M^+}, u_{M^+}], \oplus, \neg, 0_{M^+})$. $\mathbf{U_m}$ is a functor (by Proposition 8.2.1) ($\mathbf{U_m}$ stands for 'unit interval').

- **The connections between the functors $\mathbf{G^+_m}$ and $\mathbf{U_m}$**

The functors $\mathbf{U_m} : \mathcal{C}(\mathbf{s\text{-}m_0\text{-}ME^+}) \longrightarrow \mathcal{C}(\mathbf{XY}^R)$ and $\mathbf{G^+_m} : \mathcal{C}(\mathbf{XY}^R) \longrightarrow \mathcal{C}(\mathbf{s\text{-}m_0\text{-}ME^+})$ are quasi-inverses. Thus, the categories of strong $\mathrm{m_0}$-ME positive cones and of right-XY algebras are equivalent (Corollary 9.3.3).

1.3.3 Connections between the four functors in this book

We put $\mathbf{\Gamma_m} = \mathbf{U_m} \circ \mathbf{c^+_m}$ and $\mathbf{\Xi_m} = \mathbf{T_m} \circ \mathbf{G^+_m}$.

The main result of this book is the following (Theorem 10.1.1):
The functor $\mathbf{\Gamma_m} = \mathbf{U_m} \circ \mathbf{c^+_m} : \mathcal{C}(\mathbf{s\text{-}m_0\text{-}ME}) \longrightarrow \mathcal{C}(\mathbf{XY}^R)$ is not an equivalence of categories, but the functor $\mathbf{U_m} : \mathcal{C}(\mathbf{s\text{-}m_0\text{-}ME^+}) \longrightarrow \mathcal{C}(\mathbf{XY}^R)$ is an equivalence of categories.

Resuming, the two steps and the four functors are:

$$
\begin{array}{lcl}
(M, \leq_m, +, \smile, 0_M, u_M) & \stackrel{\mathbf{c^+_m}}{\longrightarrow} & (M^+, \leq_m, +, \smile, 0_{M^+} = 0_M, u_{M^+} = u_M) \\
(\frac{M^+ \times \mathbf{N}}{\sim}, \preceq_m, +, \smile, \mathbf{0}, \mathbf{u}) & \stackrel{\mathbf{T_m}}{\longleftarrow} & (M^+, \leq_m, +, \smile, 0_{M^+}, u_{M^+})
\end{array}
$$

$$
\begin{array}{lcl}
(M^+, \leq_m, +, \smile, 0_{M^+}, u_{M^+}) & \stackrel{\mathbf{U_m}}{\longrightarrow} & ([0_{M^+}, u_{M^+}], \oplus, \neg, 0_{M^+}) \\
(G^+(A), \leq_m, +, \smile, (0), (1)) & \stackrel{\mathbf{G^+_m}}{\longleftarrow} & (A, \oplus, ^-, 0)
\end{array}
$$

and we have:

$$\Gamma_{\mathbf{m}}$$

$$\mathcal{C}(\textbf{s-m}_0\textbf{-ME}) \quad \xrightarrow[\textbf{T}_\textbf{m}]{\textbf{c}_\textbf{m}^+} \quad \mathcal{C}(\textbf{s-m}_0\textbf{-ME}^+) \quad \xrightarrow[\textbf{G}_\textbf{m}^+]{\textbf{U}_\textbf{m}} \quad \mathcal{C}(\textbf{XY}^R)$$

$$\Xi_{\mathbf{m}}$$

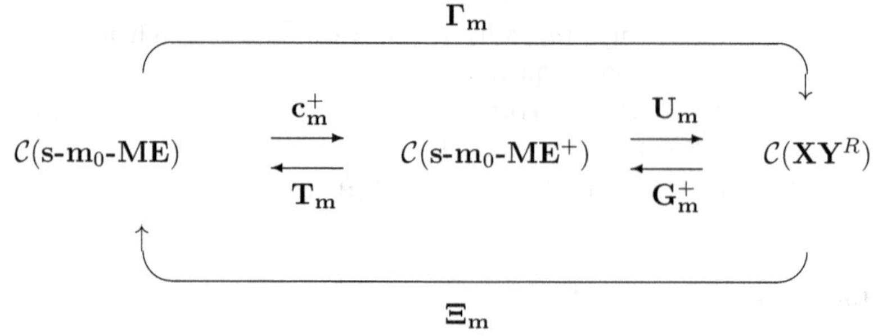

Step 1   Step 2

where 'm' comes from 'magma' and:

$\mathcal{C}(\textbf{s-m}_0\textbf{-ME})$ is the category of strong m_0-ME structures,

$\mathcal{C}(\textbf{s-m}_0\textbf{-ME}^+)$ is the category of strong m_0-ME positive cones,

$\mathcal{C}(\textbf{XY}^R)$ is the category of right-XY algebras.

We show how the original Mundici's equivalence is a particular case of our results (see Corollary 2.2.68 and also Connections from Chapter 3 and **Appendix C**).

1.4   A sketch of Mundici's equivalence in two steps

Mundici has proved in one step that the category of *unital commutative l-groups* is equivalent to the category of *MV algebras*.

In order to better see that our construction in two steps from this book generalizes the Mundici's equivalence, it is necessary to transform Mundici's one step proof into a two steps proof, by introducing the intermediary notion of *positive-unital commutative l-group* (the positive cone algebra of a unital commutative l-group) and by involving four corresponding functors, called $\mathbf{c}^+$, $\mathbf{T}$, $\mathbf{G}^+$, $\mathbf{U}$, such that $\mathbf{\Gamma} = \mathbf{U} \circ \mathbf{c}^+$ and $\mathbf{\Xi} = \mathbf{T} \circ \mathbf{G}^+$.

The intermediary notion of *positive-unital commutative l-group* is obtained from Definition 3.3.1 of *strong m_0-ME positive cone* from this book, by replacing: $\leq_m$ by $\leq$, $\inf_m(u, x + y)$ by $u \wedge (x + y)$, the two properties $(\textbf{XX}_u^R)$, $(\textbf{YY}_u^R)$ by $(\textbf{mv}_\vee)$ and so on:

Definition 1.4.1 A *positive-unital commutative l-group* is an algebra

$$\mathcal{G}_u^+ = (G^+, \vee, \wedge, +, \smile, 0_{G^+}, u = u_{G^+}),$$

of type $(2, 2, 2, 1, 0, 0)$, with $x \leq y \Longleftrightarrow x \vee y = y \Longleftrightarrow x \wedge y = x$, such that:

(i) $(G^+, \vee, \wedge)$ is a lattice and $(G^+, +, \smile, 0_{G^+})$ is a m_0-ME algebra,

(ii) $0_{G^+} < u$ is an element of G^+ verifying:

$(\textbf{s0}^R)$ $0_{G^+} \leq x$ for each $x \in G^+$ and $\smile$ verifies (DN) on $[0_{G^+}, u]$;

$(\textbf{s1}^R)$ for all $x \in G^+$, there exists $n_x \in \mathbf{N}$ such that $x, \smile x \leq n_x u$;

$(\textbf{s2}^{+R})$ (i) for all $x \in G^+$, there exists $n_x \in \mathbf{N}$

and there exist the unique elements of G^+:

$x_1, x_2, \ldots, x_{n_x} \leq u$ (hence $0_{G^+} \leq \smile x_1, \smile x_2, \ldots, \smile x_{n_x} \leq u$)

s.t. $x = x_1 + \ldots + x_{n_x}$ $(\leq n_x u)$, $\smile x = \smile x_1 \smile \ldots \smile x_{n_x}$ $(\leq n_x u)$,
(ii) if $x = x_1 + \ldots + x_n$ and $y = y_1 + \ldots + y_n$,
then $x = y \Longleftrightarrow x_i = y_i$, $i = 1, \ldots, n$,
(iii) if $x = x_1 + \ldots + x_n$ and $y = y_1 + \ldots + y_n$,
then $x \leq y \Longleftrightarrow x_i \leq y_i$, $i = 1, \ldots, n$;

($s3^R$) for $0_{M^+} \leq x, y \leq u$ (i.e. $x, y \in [0_{G^+}, u]$),
(i) there exist $u \wedge (x + y)$, $\smile (u \wedge (\smile x \smile y))$ and belong to $[0_{G^+}, u]$,
(ii) $u \wedge (x + u) = u$, $\smile (u \wedge (\smile x \smile 0_{G^+})) = 0_{G^+}$, $\smile (u \wedge (\smile x \smile u)) = x$,
(iii) $x \leq y$ $(x, y \neq 0_{G^+}) \Longleftrightarrow u \leq y \smile x$;

($s4^R$) for $0_{G^+} \leq x, y, z \leq u$,
(i) there exist and are equal $u \wedge (x + u \wedge (y + z)) = u \wedge (x + (y + z))$,
(ii) $u \wedge (\smile (u \wedge (\smile x \smile y))) = \smile (u \wedge (\smile x \smile y))$;

($s5^R$) for $0_{G^+} \leq x_0, x_1, \ldots, x_n, y_0 \leq u = u_{G^+}$,
(V^{00}) $x_0 + y_0 = u \wedge (x_0 + y_0) \smile (u \wedge (\smile x_0 \smile y_0))$,
(V^{10}) if $u \wedge (x_0 + x_1) = x_0$, then $x_0 + x_1 + y_0 = u \wedge (x_0 + y_0)$
$\smile (u \wedge (\smile x_0 \smile u \wedge (x_1 + y_0))) \smile (u \wedge (\smile x_1 \smile y_0))$,
(V^{n0}) if $u \wedge (x_i + x_{i+1}) = x_i$, for $i = 0, \ldots, n-1$, then
$\sum_{i=0}^{n} x_i + y_0 = \sum_{i=0}^{n+1} c_i$, where
$c_i = \smile (u \wedge (\smile x_{i-1} \smile u \wedge (x_i + y_0)))$, with $x_{-1} = u$, $x_{n+1} = 0_{G^+}$;

($mv_\vee$) for any $0_{G^+} \leq x, y \leq u = u_{G^+}$, we have:
$u \wedge (y + \smile (u \wedge (y + \smile x))) = u \wedge (x + \smile (u \wedge (x + \smile y)))$;

(C) $z + x = z + y \Longrightarrow x = y$, for all $x, y, z \in G^+$;

(Cp) $x \leq y \Longleftrightarrow x + z \leq y + z$, for all $x, y, z \in G^+$;

($m\text{-}Tr^R$) $x \leq y$ and $y \leq z \Longrightarrow x \leq z$, for each $x, y, z \in G^+$;

(Nu^{+1R}) $\smile u = 0_{G^+}$;

(NNu^{1R}) $\smile (\smile u) = u$;

(S^{+1R}) $\smile (x + u) = \smile x$, for each $x \in G^+$,

where $\smile x \smile y$ means $\smile x + \smile y$.

1.4.1 The step 1 at Mundici: the functors c^+ and T

The algebraic structures involved in the first step are the *commutative l-groups with strong unit* (see Definitions 3.2.1), or *unital commutative l-group* for short, and the *positive cone of an unital commutative l-group*, or *positive-unital commutative l-group* for short, defined above.

We denote with $\mathcal{C}(\mathbf{u}l\mathbf{G})$ the category of unital commutative l-groups with homomorphisms and with $\mathcal{C}(\mathbf{u}l\mathbf{G}^+)$ the category of positive-unital commutative l-groups with homomorphisms.

Recall that any commutative group $(G, +, -, 0_G)$ verifies the cancellation property (C): for all $x, y, z \in G$,
(C) $z + x = z + y \Longrightarrow x = y$.

Hence, **the unital commutative l-group and the positive-unital commutative l-group verify the cancellation property (C).**

- **The functor $\mathbf{c}^+$**

Given a commutative l-group with strong unit $\mathcal{G}_u = (G, \vee, \wedge, +, -, 0_G, u = u_G)$, with $x \leq y \Longleftrightarrow x \vee y = y \Longleftrightarrow x \wedge y = x$, let $G^+ \stackrel{def.}{=} \{x \in G \mid x \geq 0_G\}$ be its positive cone. We endow G^+ with the operations $\vee, \wedge, +, 0_G, u_G$ defined by restriction.

We endow also G^+ with a negation $\smile$ defined by (see (1.1)): for all $x \in G^+$,

$$(1.3) \quad \smile x \stackrel{def.}{=} \begin{cases} u - x, & \text{if} \quad u - x \geq 0_G (\Longleftrightarrow x \leq u), \\ x - u, & \text{if} \quad u - x < 0_G (\Longleftrightarrow u < x), \quad x \neq 0_G, \quad \smile 0_G \stackrel{def.}{=} 0_G, \\ u \wedge x, & \text{otherwise,} \end{cases}$$

$$\smile (x + u) \stackrel{def.}{=} \smile x.$$

Note that $x \in [0_G, u] \Longleftrightarrow \smile x \in [0_G, u]$; $\smile (\smile x) = x$ and $x + \smile x = u$, for all $x \in [0_G, u]$; $\smile u = 0_G$.

Proposition 1.4.2 *Let $\mathcal{G}_u = (G, \vee, \wedge, +, -, 0 = 0_G, u = u_G)$ be a commutative l-group with strong unit u, with $x \leq y \Longleftrightarrow x \vee y = y \Longleftrightarrow x \wedge y = x$. Then, for all $0 \leq x, y \leq u$, we have, by (1.3):*
($mv_\vee$) $u \wedge (y + (u + -(u \wedge (y + (u + -x))))) = u \wedge (x + (u + -(u \wedge (x + (u + -y)))))$.

Proof. (Based on a proof by PROVER9 of Length 25)

The following properties are immediate:

$$(1.4) \qquad x + (y + z) = y + (x + z),$$

$$(1.5) \qquad -x + (y + x) = y,$$

$$(1.6) \qquad x + (y + (-x + z)) = y + z.$$

Now, in (g3) $((x + y) \wedge (x + z) = x + (y \wedge z))$, take $Y := 0$, to obtain, by (SU):

$$(1.7) \qquad x \wedge (x + z) = x + (0 \wedge z).$$

Then, in (g3) again, take $Z := -x$, to obtain:
$(x + y) \wedge (x + -x) = x + (y \wedge -x)$,
which, by (m_0-Re) and (Wcomm), becomes:

$$(1.8) \qquad 0 \wedge (x + y) = x + (y \wedge -x).$$

Now, in (1.7), take $X := -x$, $Z := y + x$, to obtain:
$-x \wedge (-x + (y + x)) = x + (0 \wedge (y + x))$,
which, by (1.5), becomes:

$$(1.9) \qquad -x \wedge y = x + (0 \wedge (y + x)).$$

Now, we are ready to prove that $u \wedge (x + (u + -(u \wedge (x + (u + -y))))) = u \wedge (y + (u + -(u \wedge (y + (u + -x)))))$.

$A \overset{notation}{=} u \wedge (x + (u + -(u \wedge (x + (u + -y)))))$
$= u \wedge (x + (u + -(u \wedge (u + (x + -y)))))$, by (1.4),
$= u \wedge (u + (x + -(u \wedge (u + (x + -y)))))$, by (1.4),
$= u \wedge (u + (x + -(u + (0 \wedge (x + -y)))))$, by (1.7),
$= u + (0 \wedge (x + -(u + (0 \wedge (x + -y)))))$, by (1.7),
$= u + (0 \wedge (x + (-(0 \wedge (x + -y)) + -u)))$, by (NegS),
$= u + (0 \wedge (x + (-u + -(0 \wedge (x + -y)))))$, by (Scomm),
and
$B \overset{notation}{=} u \wedge (y + (u + -(u \wedge (y + (u + -x)))))$
$= u \wedge (y + (u + -(u \wedge (u + (y + -x)))))$, by (1.4),
$= u \wedge (u + (y + -(u \wedge (u + (y + -x)))))$, by (1.4),
$= u \wedge (u + (y + -(u + (0 \wedge (y + -x)))))$, by (1.7),
$= u + (0 \wedge (y + -(u + (0 \wedge (y + -x)))))$, by (1.7),
$= u + (0 \wedge (y + -(u + (y + (-x \wedge -y)))))$, by (1.8),
$= u + (0 \wedge (y + -(u + (y + (-x + (0 \wedge (-y + x)))))))$, by (1.9),
$= u + (0 \wedge (y + -(u + (y + (-x + (0 \wedge (x + -y)))))))$, by (Scomm),
$= u + (0 \wedge (y + (-(y + (-x + (0 \wedge (x + -y)))) + -u)))$, by (NegS),
$= u + (0 \wedge (y + ((-(-x + (0 \wedge (x + -y))) + -y) + -u)))$, by (NegS),
$= u + (0 \wedge (y + (((-(0 \wedge (x + -y)) + --x) + -y) + -u)))$, by (NegS),
$= u + (0 \wedge (y + (((-(0 \wedge (x + -y)) + x) + -y) + -u)))$, by (DN),
$= u + (0 \wedge (y + (((x + -(0 \wedge (x + -y))) + -y) + -u)))$, by (Scomm),
$= u + (0 \wedge (y + ((-y + (x + -(0 \wedge (x + -y)))) + -u)))$, by (Scomm),
$= u + (0 \wedge (y + ((x + (-y + -(0 \wedge (x + -y)))) + -u)))$, by (1.4),
$= u + (0 \wedge (y + (-u + (x + (-y + -(0 \wedge (x + -y)))))))$, by (Scomm),
$= u + (0 \wedge (y + (x + (-u + (-y + -(0 \wedge (x + -y)))))))$, by (1.4),
$= u + (0 \wedge (x + (y + (-u + (-y + -(0 \wedge (x + -y)))))))$, by (1.4),
$= u + (0 \wedge (x + (-u + -(0 \wedge (x + -y)))))$, by (1.6).
Hence, $A = B$. □

Then, the algebra $\mathcal{G}_u^+ = (G^+, \vee, \wedge, +, \smile, 0_G, u = u_G)$ is a positive-unital commutative l-group.

Given a morphism $f : G \longrightarrow H$ of unital commutative l-groups, f restricts to a function f^+ from G^+ to H^+. Moreover, f preserves $\vee, \wedge, +$ and $0_G, u_G$ and so f^+ is a morphism of positive-unital commutative l-groups. This establishes a functor $\mathbf{c}^+ : \mathcal{C}(\mathbf{ulG}) \longrightarrow \mathcal{C}(\mathbf{ulG}^+)$ that maps G to G^+, and maps a morphism $f : G \longrightarrow H$ to its restriction $f^+ : G^+ \longrightarrow H^+$.

- **The functor T**

Let $\mathcal{G}_u^+ = (G^+, \vee, \wedge, +, \smile, 0_G, u = u_G)$, with $x \leq y \Longleftrightarrow x \vee y = y \Longleftrightarrow x \wedge y = x$, be a positive-unital commutative l-group. Then, **the cancellation property (C) holds.**

Note that, in [10], Mundici starts with M_A, the lattice ordered monoid

of good sequences of an MV algebra A which verifies the cancellation property (C).

Let us consider the binary relation $\sim$ defined on $G^+ \times G^+$ as follows [10]:

$$(a, b) \sim (a', b') \overset{def.}{\Longleftrightarrow} a + b' = a' + b. \tag{1.10}$$

The cancellation property (C) of G^+ (of M_A in [10]) ([10], Proposition 2.3.1 (i)) makes the binary relation $\sim$ be transitive, hence an equivalence relation.

The equivalence class of an element $(a, b) \in G^+ \times G^+$ is denoted by $[(a, b)]$, or simply by $[a, b]$. The set of all equivalent classes, denoted here by $\mathbf{T}(G^+)$:

$$\mathbf{T}(G^+) \overset{def.}{=} \frac{G^+ \times G^+}{\sim}$$

is endowed with the operations of a unital commutative l-group:

$$\mathbf{0} \overset{def.}{=} [0_G, 0_G],$$

$$\mathbf{u} \overset{def.}{=} [u_G, 0_G],$$

$$-[a, b] \overset{def.}{=} [b, a],$$

$$[a, b] + [c, d] \overset{def.}{=} [a + c, b + d],$$

$$[a, b] \vee [c, d] \overset{def.}{=} [(a + d) \vee (c + b), b + d],$$

$$[a, b] \wedge [c, d] \overset{def.}{=} [(a + d) \wedge (c + b), b + d].$$

We say that the equivalence class $[c, d]$ *dominates* the equivalence class $[a, b]$, in symbols ([10], Definition 2.4.1):

$$[a, b] \preceq [c, d]$$

if, and only if, $[c, d] + -[a, b] = [e, 0]$ for some $e \in G^+$.
Equivalently, $[a, b] \preceq [c, d]$ if, and only if, $a + d \leq c + b$,
where $\leq$ is the lattice order of G^+ given by ([10], Definition 2.3.3).

The algebra $(\mathbf{T}(G^+) = \frac{G^+ \times G^+}{\sim}, \vee, \wedge, +, -, \mathbf{0}, \mathbf{u})$ is a unital commutative l-group ([10], Propositions 2.4.2 and 2.4.4).

Note that, in [10], Mundici ends with the particular unital commutative l-group G_A.

We have thus a functor $\mathbf{T} : \mathcal{C}(\mathbf{ulG^+}) \longrightarrow \mathcal{C}(\mathbf{ulG})$.

- **The connections between the functors $\mathbf{c}^+$ and $\mathbf{T}$**

The functors $\mathbf{c}^+ : \mathcal{C}(\mathbf{ulG}) \longrightarrow \mathcal{C}(\mathbf{ulG^+})$ and $\mathbf{T} : \mathcal{C}(\mathbf{ulG^+}) \longrightarrow \mathcal{C}(\mathbf{ulG})$ are quasi-inverses. Thus, the categories of unital commutative l-groups and positive-unital commutative l-groups are equivalent.

1.4.2 The step 2 at Mundici: the functors $\mathbf{G}^+$ and $\mathbf{U}$

The algebraic structures involved in the second step are the *positive-unital commutative l-groups* and the *MV algebras.*

An MV algebra (see Definition 2.2.1) is an involutive m-MEL algebra $(A, \oplus, ^-, 0)$ verifying the property:
($\vee_m$-comm) $y \oplus (y \oplus x^-)^- = x \oplus (x \oplus y^-)^-$.

By ([21], Remark 6.5.9) (in the dual case), in MV algebras $\mathcal{A} = (A, \oplus, ^-, 0)$, the two binary relations $\leq_m \stackrel{def.}{\Longleftrightarrow} y \oplus x^- = 1$ and $\leq_m^M \stackrel{def.}{\Longleftrightarrow} y \vee_m^M x = y \Longleftrightarrow x \oplus (x \oplus y^-)^- = y$ are lattice orders with respect to $\vee_m^M$, $\wedge_m^M$ and they coincide, i.e. $\leq_m \Longleftrightarrow \leq_m^M$, the resulting unique lattice order being denoted by $\leq$ (with respect to $\vee = \vee_m^M$, $\wedge = \wedge_m^M$). The lattice $(A, \vee, \wedge)$ is distributive. If the lattice order $\leq$ is totally ordered (or linearly ordered, here), then we say that we have an *MV chain.*

In [10], the authors use in some proofs the *Chang's subdirect representation theorem* ([10], Theorem 1.3.3): "Every nontrivial MV algebra is a subdirect product of MV chains."

We denote by $\mathcal{C}(\mathbf{MV}^R)$ the category of (right-) MV algebras with homomorphisms.

- **The functor $\mathbf{G}^+$**

Let $\mathcal{A} = (A, \oplus, ^-, 0)$ be an MV algebra.

A sequence $\mathbf{a} = (a_1, a_2, \ldots)$ of elements of A is said to be *good* if, and only if, for each $i = 1, 2, \ldots$ ([10], Section 2.2),

$$a_i \oplus a_{i+1} = a_i$$

and there is an integer n such that $a_r = 0$ for all $r > n$. Instead of $\mathbf{a} = (a_1, a_2, \ldots, a_n, 0, 0, \ldots)$ we shall often write, more concisely,

$$\mathbf{a} = (a_1, a_2, \ldots, a_n).$$

For each $a \in A$, the good sequence $(a, 0, \ldots, 0, \ldots)$ will be denoted by (a). Hence, (0) denotes the good sequence $(0, 0, 0, \ldots)$ and (1) denotes the good sequence $(1, 0, 0, \ldots)$.

The set of all good sequences of A was denoted with M_A in [10]. Then, M_A was endowed, in fact, with the structure of a positive-unital commutative l-group, as follows.

$0_{M_A} \stackrel{def.}{=} (0)$, $u_{M_A} \stackrel{def.}{=} (1)$.

For good sequences $\mathbf{a} = (a_1, a_2, \ldots)$ and $\mathbf{b} = (b_1, b_2, \ldots)$, let

$$\mathbf{a} \vee \mathbf{b} \stackrel{def.}{=} (a_1 \vee b_1, a_2 \vee b_2, \ldots),$$

$$\mathbf{a} \wedge \mathbf{b} \stackrel{def.}{=} (a_1 \wedge b_1, a_2 \wedge b_2, \ldots)$$

and prove they are good sequences too (see [10], Lemma 2.2.3).

For good sequences $\mathbf{a} = (a_1, a_2, \ldots)$ and $\mathbf{b} = (b_1, b_2, \ldots)$,

$$\mathbf{a} \leq \mathbf{b} \overset{def.}{\Longleftrightarrow} a_i \leq b_i, \quad i = 1, 2, \ldots.$$

Given two good sequences $\mathbf{a} = (a_1, a_2, \ldots, a_n)$ and $\mathbf{b} = (b_1, b_2, \ldots, b_m)$ of A, their *sum* $\mathbf{c} = \mathbf{a} + \mathbf{b}$ is defined by $\mathbf{c} = (c_1, c_2, \ldots, c_{n+m})$, where for all $i = 1, 2, \ldots$

$$c_i \overset{def.}{=} a_i \oplus (a_{i-1} \odot b_1) \oplus \ldots \oplus (a_1 \odot b_{i-1}) \oplus b_i$$

([10], Definition 2.2.4).

The sum of two good sequences is a good sequence (by [10], Theorem 1.3.3 and Lemma 2.2.2).

Then, the algebra $(M_A, +, (0))$ is a commutative monoid with the following additional properties ([10], Proposition 2.3.1):
(i) (**cancellation**) For any good sequences **a**, **b**, **c**, if $\mathbf{a} + \mathbf{b} = \mathbf{a} + \mathbf{c}$, then $\mathbf{b} = \mathbf{c}$;
(ii) (**zero-law**) If $\mathbf{a} + \mathbf{b} = (0)$, then $\mathbf{a} = \mathbf{b} = (0)$.

Let $\mathbf{a} = (a_1, a_2, \ldots, a_n)$ and $\mathbf{b} = (b_1, b_2, \ldots, b_n)$ be good sequences of A. Then, the following are equivalent ([10], Proposition 2.3.2):
(i) There is a good sequence **c** such that $\mathbf{b} + \mathbf{c} = \mathbf{a}$;
(ii) $b_i \leq a_i$, for all $i = 1, \ldots, n$.

Given any two good sequences **a** and **b** of A, we define ([10], Definition 2.3.3): $\mathbf{b} \leq \mathbf{a}$ iff **b** and **a** satisfy the equivalent conditions from ([10], Proposition 2.3.2).

Let $\mathbf{a} = (a_1, a_2, \ldots, a_n)$ be a good sequence of A. Define the *negation* $\smile$ by:

$$\smile \mathbf{a} \overset{def.}{=} (a_n^-, \ldots, a_2^-, a_1^-),\ \mathbf{a} \neq 0_{MA}, \qquad \smile 0_{MA} \overset{def.}{=} 0_{MA}. \tag{1.11}$$

Then, $\smile \mathbf{a}$ is a good sequence.

We have $\smile (\smile \mathbf{a}) = \mathbf{a}$, $\smile u = 0_{MA}$, $\smile (\mathbf{a} + u) = \smile \mathbf{a} + \smile u = \smile \mathbf{a}$.

Remarks 1.4.3 $0_{MA} = (0) \leq \mathbf{a} \leq u_{MA} = (1) \Longleftrightarrow \mathbf{a} = (a)$ *and we have:*
$\smile (a) = (a^-)$, $\smile u_{MA} = \smile (1) = (1^-) = (0) = 0_{MA}$,
$\smile (\smile (a)) = \smile (a^-) = ((a^-)^-) \overset{(DN)}{=} (a)$,
$(a) + \smile (a) = (a) + (a^-) = (a \oplus a^-, a \odot a^-) = (1, 0) = (1) = u_{MA}$.

Let **a** and **b** be good sequences of A. Then, we have ([10], Proposition 2.3.4):
(i) If $\mathbf{b} \leq \mathbf{a}$, then there is a unique good sequence **c** such that $\mathbf{b} + \mathbf{c} = \mathbf{a}$. This **c**, denoted $\mathbf{a} \smile \mathbf{b}$ is given by:
$\mathbf{c} = (a_1, \ldots, a_n) + (b_n^-, \ldots, b_1^-)$ omitting the first n terms.
(ii) In particular, for each $a \in A$, we have:
$(a^-) = (1) \smile (a)$.
(iii) The order is translation invariant, in the sense that $\mathbf{b} \leq \mathbf{a}$ implies $\mathbf{b} + \mathbf{d} \leq \mathbf{a} + \mathbf{d}$ for every good sequence **d**.

Let $\mathbf{a} = (a_1, a_2, \ldots, a_n, \ldots)$ and $\mathbf{b} = (b_1, b_2, \ldots, b_n, \ldots)$ be good sequences of A. Then, we have ([10], Proposition 2.3.5):
(i) The good sequence

$$\mathbf{a} \vee \mathbf{b} \stackrel{def.}{=} (a_1 \vee b_1, \ldots, a_n \vee b_n, \ldots)$$

is in fact the supremum of $\mathbf{a}$ and $\mathbf{b}$ with respect to the order defined by Definition 2.3.3.
(ii) Analogously, the sequence

$$\mathbf{a} \wedge \mathbf{b} \stackrel{def.}{=} (a_1 \wedge b_1, \ldots, a_n \wedge b_n, \ldots)$$

is good and is in fact the infimum of $\mathbf{a}$ and $\mathbf{b}$.
(iii) For all $a, b, c \in A$, we have: $((a) + (b)) \wedge (1) = (a \oplus b)$.

The commutative monoid M_A, enriched with the lattice order of ([10], Proposition 2.3.5), becomes a commutative l-monoid with strong unit $(M_A, \vee, \wedge, +, 0_{M_A} = (0), u_{M_A} = (1))$.

Moreover, we have the following additional result.

Corollary 1.4.4 *For any* $0_{M_A} \leq (x), (y) \leq u = u_{M_A}$ *in* M_A*, we have:*
($mv_\vee$) $u \wedge ((y) + \smile (u \wedge ((y) + \smile (x)))) = u \wedge ((x) + \smile (u \wedge ((x) + \smile (y))))$.

Proof. $A \stackrel{notation}{=} u \wedge ((y) + \smile (u \wedge ((y) + \smile (x))))$
$= u \wedge ((y) + \smile (u \wedge ((y) + (x^-))))$, by definition of $\smile$,
$= u \wedge ((y) + \smile (y \oplus x^-))$, by ($s3^R$) (i),
$= u \wedge ((y) + ((y \oplus x^-)^-))$, by definition of $\smile$,
$= (y \oplus (y \oplus x^-)^-)$, by ($s3^R$) (i),
and, similarly,
$B \stackrel{notation}{=} u \wedge ((x) + \smile (u \wedge ((x) + \smile (y)))) = (x \oplus (x \oplus y^-)^-)$.
But, $y \oplus (y \oplus x^-)^- = x \oplus (x \oplus y^-)^-$ is just ($\vee_m$-comm), that is true in $\mathcal{A}$; it follows that $A = B$. □

Consequently, $(M_A, \vee, \wedge, +, \smile, 0_{M_A} = (0), u_{M_A} = (1))$ is a positive-unital commutative l-group.

Note that, in [10], this particular positive-unital commutative l-group M_A is the starting point to obtain the particular unital commutative l-group G_A - by the functor named here T.

We put $\mathbf{G}^+(A) = M_A$. This establishes a functor $\mathbf{G}^+ : \mathcal{C}(\mathbf{MV}^R) \longrightarrow \mathcal{C}(\mathbf{u}l\mathbf{G}^+)$.

- **The unit interval functor U**

Let $(G^+, \vee, \wedge, +, \smile, 0_{G^+}, u = u_{G^+})$ be a positive-unital commutative l-group.

For any element $u \in G^+$ $(u > 0_{G^+})$, put $[0_{G^+}, u] := \{x \in G^+ \mid 0_{G^+} \leq x \leq u\}$ and endow it with the operations of an MV algebra, as follows.

For each $x, y \in [0_{G^+}, u]$, put:

$$x \oplus y \stackrel{def.}{=} u \wedge (x + y),$$

and (see (1.1) and (1.3)):

$$\neg x \overset{def.}{=} \smile x, \; x \neq 0_{G^+}, \quad \neg 0_{G^+} \overset{def.}{=} u.$$

Then, the algebra $\mathbf{U}(G^+) \overset{def.}{=} ([0_{G^+}, u], \oplus, \neg, 0_{G^+})$ is an MV algebra. To prove this result, we replace (in the proof from [10] of this result) the proof of ($\vee_m$-comm) by the following proof:

($\vee_m$-comm): $y \oplus \neg(y \oplus \neg x) = x \oplus \neg(x \oplus \neg y)$. Indeed,
$A \overset{notation}{=} y \oplus \neg(y \oplus \neg x) = u \wedge (y + \smile (y \oplus \smile x)) = u \wedge (y + \smile (u \wedge (y + \smile x)))$
and, similarly,
$B \overset{notation}{=} x \oplus \neg(x \oplus \neg y) = u \wedge (x + \smile (u \wedge (x + \smile y)))$.
But, $u \wedge (y + \smile (u \wedge (y + \smile x))) = u \wedge (x + \smile (u \wedge (x + \smile y)))$, by ($mv_\vee$) in $\mathcal{G}_u^+$; it follows that $A = B$.

Given a morphism $f^+ : G^+ \longrightarrow H^+$ of positive-unital commutative l-groups, we set $\mathbf{U}(f^+) : \mathbf{U}(G^+) \longrightarrow \mathbf{U}(H^+)$ as the restriction of f^+. This assignment establishes a functor $\mathbf{U} : \mathcal{C}(\mathbf{u}l\mathbf{G}^+) \longrightarrow \mathcal{C}(\mathbf{MV})$ ($\mathbf{U}$ stands for 'unit interval').

By ([10], Proposition 2.1.5), $\mathbf{\Gamma}$ ($= \mathbf{U} \circ \mathbf{c}^+$) : $\mathcal{C}(\mathbf{u}l\mathbf{G}) \longrightarrow \mathcal{C}(\mathbf{MV}^R)$ is a functor.

- **The connections between the functors $\mathbf{G}^+$ and $\mathbf{U}$**

The functors $\mathbf{U} : \mathcal{C}(\mathbf{u}l\mathbf{G}^+) \longrightarrow \mathcal{C}(\mathbf{MV}^R)$ and $\mathbf{G}^+ : \mathcal{C}(\mathbf{MV}^R) \longrightarrow \mathcal{C}(\mathbf{u}l\mathbf{G}^+)$ are quasi-inverses. Thus, the categories of positive-unital commutative l-groups and (right-) MV algebras are equivalent.

1.4.3 Connections between the four functors at Mundici

The main result of Mundici is ([10], Corollary 7.1.8):

The functor $\mathbf{\Gamma}$ ($= \mathbf{U} \circ \mathbf{c}^+$) : $\mathcal{C}(\mathbf{u}l\mathbf{G}) \longrightarrow \mathcal{C}(\mathbf{MV}^R)$ defines a natural equivalence between the category of commutative l-groups with strong unit and the category of (right-) MV algebras.

Resuming, the two steps and four functors of the proof of Mundici's equivalence are:

$$\begin{array}{lcl} (G, \leq, +, -, 0_G, u_G) & \overset{\mathbf{c}^+}{\longrightarrow} & (G^+, \leq, +, \smile, 0_{G^+} = 0_G, u_{G^+} = u_G) \\ (\frac{G^+ \times G^+}{\sim}, \preceq, +, -, \mathbf{0}, \mathbf{u}) & \overset{\mathbf{T}}{\longleftarrow} & (G^+, \leq, +, \smile, 0_{G^+}, u_{G^+}) \end{array}$$

$$\begin{array}{lcl} (G^+, \leq, +, \smile, 0_{G^+}, u_{G^+}) & \overset{\mathbf{U}}{\longrightarrow} & ([0_{G^+}, u_{G^+}], \oplus, \neg, 0_{G^+}) \\ (M_A, \leq, +, \smile, (0), (1)) & \overset{\mathbf{G}^+}{\longleftarrow} & (A, \oplus, ^-, 0) \end{array}$$

and we have:

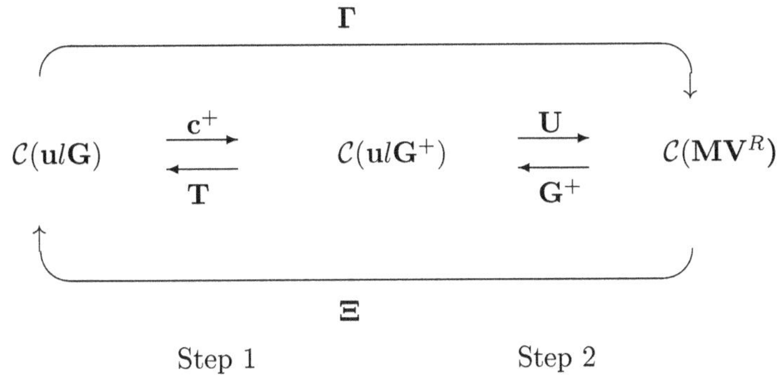

where:
$\mathcal{C}(\mathbf{u}l\mathbf{G})$ is the category of commutative l-groups with strong unit, or unital commutative l-groups,
$\mathcal{C}(\mathbf{u}l\mathbf{G}^{+})$ is the category of positive-unital commutative l-groups,
$\mathcal{C}(\mathbf{MV}^{R})$ is the category of (right-) MV algebras.

Part I

The categories involved

Chapter 2

XY algebras

Recall that, starting from the old framework centered on BCK algebras, we have introduced the m-BCK algebras ('m' coming from 'magma') and other more general algebras, thus obtaining an analogous framework centered on m-BCK algebras, such that, for example, the involutive BCK algebras are definitionally equivalent to (involutive) m-BCK algebras. The m_0-ME algebras, the commutative groups and the m-MEL algebras, the MV algebras, the Boolean algebras belong to this new framework, presented in the paper [20], in 2020, and in the monograph [21], in 2022 (see **Appendix A**).

In this chapter, we introduce and study, in this new framework, the XY algebras as the 'ancestor' (generalization) of MV algebras; they can be non-cancellative or cancellative. We present also two new equivalent definitions of MV algebras.

2.1 Involutive m-MEL algebras

In this section, we recall from the framework centered on m-BCK algebras the involutive m-MEL algebras and the MV algebras. The ancestor (generalization) of MV algebras called *XY algebras*, introduced in the second section, will belong to this framework too. Some new results are obtained.

2.1.1 Definition of involutive m-MEL algebras

Recall the following definitions from [20], [21].

Definitions 2.1.1 (See [21], subsection 6.1.1)

(1) A *right-m-MEL algebra* is an algebra $\mathcal{A} = (A, \oplus, ^-, 0 = 0_A)$ of type $(2, 1, 0)$ verifying the following axioms: for all $x, y, z \in A$,

(SU) $x \oplus 0 = x = 0 \oplus x$,
(Scomm) $x \oplus y = y \oplus x$,
(Sass) $(x \oplus y) \oplus z = x \oplus (y \oplus z)$,
(Neg0-1) $1 = 1_A \stackrel{def.}{=} 0^-$,
(Neg1-0) $1^- = 0$,
(m-La^R) $x \oplus 1 = 1$.

(1') Dually, a *left-m-MEL algebra* is an algebra $\mathcal{A} = (A, \odot, ^-, 1)$ of type $(2,1,0)$ verifying the following axioms: for all $x, y, z \in A$,
(PU) $x \odot 1 = x = 1 \odot x$,
(Pcomm) $x \odot y = y \odot x$,
(Pass) $(x \odot y) \odot z = x \odot (y \odot z)$,
(Neg1-0) $0 \stackrel{def.}{=} 1^-$,
(Neg0-1) $0^- = 1$,
(m-La) $x \odot 0 = 0$.

Denote by $\mathbf{m\text{-}MEL}^R$ the class of all right-m-MEL algebras and by **m-MEL** the class of all left-m-MEL algebras (see the first and the second Figure from **Appendix A**).

Recall [20], [21] that 'm-' comes from 'magma'.

Definitions 2.1.2 ([21], Definition 6.4.1)

(1) An *involutive right-m-MEL algebra* is a right-m-MEL algebra $\mathcal{A}$ verifying the additional axiom: for each $x \in A$,
(DN) $(x^-)^- = x$ (or, better, $x^= = x$).

(1') An *involutive left-m-MEL algebra* is a left-m-MEL algebra $\mathcal{A}$ verifying the additional axiom (DN).

Denote by $\mathbf{m\text{-}MEL}^R_{(DN)}$ the class of all involutive right-m-MEL algebras and by $\mathbf{m\text{-}MEL}_{(DN)}$ the class of all involutive left-m-MEL algebras (see the third Figure from **Appendix A**).

Let $\mathcal{A} = (A, \oplus, ^-, 0)$ be an involutive right-m-MEL algebra. Let us introduce a new operation, $\odot$, defined by ([21], (6.5)): for all $x, y \in A$,

$$x \odot y \stackrel{def.}{=} (x^- \oplus y^-)^-. \tag{2.1}$$

Then, we have the following result.

Proposition 2.1.3 *(See ([21], Corollary 6.4.3))*

Let $\mathcal{A} = (A, \oplus, ^-, 0)$ be an involutive right-m-MEL algebra. Then, the algebra $(A, \odot, ^-, 1)$ is an involutive left-m-MEL algebra.

It follows that $\mathcal{A} = (A, \oplus, ^-, 0)$ verifies (PU), (Pcomm), (Pass), (m-La) too and

$$x \oplus y \stackrel{def.}{=} (x^- \odot y^-)^-.$$

2.1.2 Two binary relations, $\leq_m$ and $\leq_m^M$, on $\mathcal{A}$

First, let $\mathcal{A} = (A, \oplus, ^-, 0)$ be a right-m-MEL algebra. We define the following basic internal binary relation ([21], Chapter 6, section 6.1): for all $x, y \in A$,

$$x \leq_m y \stackrel{def.}{\Longleftrightarrow} y \oplus x^- = 1 \ (\Longleftrightarrow x \odot y^- = 0). \tag{2.2}$$

We shall denote:
- the *reflexivity* of $\leq_m$, by (m-Re^R),
- the *antisymmetry* of $\leq_m$, by (m-An^R) and
- the *transitivity* of $\leq_m$, by (m-Tr^R),
hence we have the following properties:

(m-Re^R) $x \oplus x^- = 1$,
(m-Re'^{R}) $x \leq_m x$;

(m-An^R) $y \oplus x^- = 1$ and $x \oplus y^- = 1$ imply $x = y$,
(m-An'^{R}) $x \leq_m y$ and $y \leq_m x$ imply $x = y$;

(m-Tr^R) $y \oplus x^- = 1$ and $z \oplus y^- = 1$ imply $z \oplus x^- = 1$,
(m-Tr'^{R}) $x \leq_m y$ and $y \leq_m z$ imply $x \leq_m z$.

Note also that the property (m-La^R) ($x \oplus 1 = 1$) means $0 \leq_m x$, denoted by (m-La'^{R}).

Now, let $\mathcal{A} = (A, \oplus, ^-, 0)$ be an involutive right-m-MEL algebra. We define the following new operations and a new binary relation ([21], Chapter 6, subsection 6.4.3): for all $x, y \in A$,
$x \vee_m^M y \stackrel{def.}{=} y \oplus (y \oplus x^-)^-$, $\quad x \wedge_m^M y \stackrel{def.}{=} (x^- \vee_m^M y^-)^- = y \odot (y \odot x^-)^-$ and

$$x \leq_m^M y \stackrel{def.}{\Longleftrightarrow} y \vee_m^M x = y \Longleftrightarrow x \oplus (x \oplus y^-)^- = y. \tag{2.3}$$

Note that, if $x \leq_m^M y$, then there exists $c = (x \oplus y^-)^- \in A$ such that $x \oplus c = y$. The converse does not hold (as in MV algebras case), as the following example shows.

Example 2.1.4 Involutive right-m-MEL algebra verifying $a \oplus c = b$ and not verifying $a \leq_m^M b$, i.e. $a \oplus (a \oplus b^-)^- \neq b$.

We put $g(x, y) \stackrel{def.}{=} x \oplus (x \oplus y^-)^-$.
Consider the involutive right-m-MEL algebra $(A = \{0, 2, 3, 1\}, \oplus, ^-, 0)$ with the following tables:

x	0	2	3	1
x^-	1	3	2	0

and

$\oplus$	0	2	3	1
0	0	2	3	1
2	2	1	2	1
3	3	2	0	1
1	1	1	1	1

,

$\odot$	0	2	3	1
0	0	0	0	0
2	0	1	3	2
3	0	3	0	3
1	0	2	3	1

;

$\vee_m^M$	0	2	3	1
0	0	2	3	1
2	2	2	1	1
3	3	2	0	1
1	1	2	2	1

,

g	0	2	3	1
0	0	2	3	1
2	2	2	2	2
3	3	1	0	2
1	1	1	1	1

.

For $a = 2$, $b = 1$, $c = 2$, $a \oplus c = b$, but $a \not\leq_m^M b$ $(2 = g(a, b) \neq b = 1)$.

Note that, in general, in any right-m-MEL algebra, $\leq_m$ and $\leq_m^M$ are not reflexive, antisymmetric or transitive.

Dually, first let $\mathcal{A} = (A, \odot, ^-, 1)$ be a left-m-MEL algebra. We define the following basic internal binary relation ([21], Chapter 6, section 6.1): for all $x, y \in A$,

$$x \leq_m y \overset{def.}{\Longleftrightarrow} x \odot y^- = 0.$$

Note that the property (m-La) ($x \odot 0 = 0$) means that $x \leq_m 1$, denoted by (m-La').

Now, let $\mathcal{A} = (A, \odot, ^-, 1)$ be an involutive left-m-MEL algebra. We define the following new operations and new additional binary relation ([21], Chapter 6, subsection 6.4.3): for all $x, y \in A$,
$x \wedge_m^M y \overset{def.}{=} y \odot (y \odot x^-)^-$, $\quad x \vee_m^M y \overset{def.}{=} (x^- \wedge_m^M y^-)^- = y \oplus (y \oplus x^-)^-$ and

$$x \leq_m^M y \overset{def.}{\Longleftrightarrow} x \wedge_m^M y = x \Longleftrightarrow y \odot (y \odot x^-)^- = x.$$

Proposition 2.1.5 *(The dual one is omitted)*
Let $\mathcal{A} = (A, \oplus, ^-, 0)$ be an involutive right-m-MEL algebra. For all $x \in A$,

$$0 \leq_m x \leq_m 1 \quad and \quad 0 \leq_m^M x.$$

Proof. $0 \leq_m x \overset{def.}{\Longleftrightarrow} x \oplus 0^- = 1 \overset{(Neg0-1)}{\Longleftrightarrow} x \oplus 1 = 1$, that is true, by (m-La^R).
$x \leq_m 1 \overset{def.}{\Longleftrightarrow} 1 \oplus x^- = 1$, that is true, by (m-La^R).
$0 \leq_m^M x \overset{def.}{\Longleftrightarrow} x \vee_m^M 0 = x \Longleftrightarrow 0 \oplus (0 \oplus x^-)^- = x$, that is true, by (DN), (SU).
Note that $x \leq_m^M 1 \overset{def.}{\Longleftrightarrow} 1 \vee_m^M x = 1 \Longleftrightarrow x \oplus (x \oplus 1^-)^- = 1 \overset{(Neg1-0),(SU)}{\Longleftrightarrow} x \oplus x^- = 1$, that is not true, because (m-Re^R) does not hold. □

Lemma 2.1.6

$$(i)\ 0 \leq_m 0, \qquad (ii)\ 1 \leq_m 1, \qquad (iii)\ 1 \leq_m x \Longleftrightarrow x = 1.$$

Proof. (i): $0 \leq_m 0$ follows from (m-La^R), for $x = 0$.

(ii): $1 \leq_m 1$ means $1 \oplus 1^- = 1 \overset{(Neg1-0)}{\Longleftrightarrow} 1 \oplus 0 = 1 \overset{(SU)}{\Longleftrightarrow} 1 = 1$.

(iii): $1 \leq_m x$ means $x \oplus 1^- = 1 \overset{(Neg1-0)}{\Longleftrightarrow} x \oplus 0 = 1 \overset{(SU)}{\Longleftrightarrow} x = 1$. □

*

- **Totally ordered (involutive) m-MEL algebras**

We shall introduce now the following two definitions (the dual ones are omitted).

Definitions 2.1.7

(1) A right-m-MEL algebra $\mathcal{A}$ is *totally ordered by* $\leq_m$ if, for all $x, y \in A$, we have:

$$(tot - ord - m) \qquad x \leq_m y \quad or \quad y \leq_m x,$$

i.e. if we have: $y \oplus x^- = 1$ or $x \oplus y^- = 1$.

(2) An involutive right-m-MEL algebra $\mathcal{A}$ is *totally ordered by* $\leq_m^M$ if, for all $x, y \in A$, we have:

$$(tot - ord - m - M) \qquad x \leq_m^M y \quad or \quad y \leq_m^M x,$$

i.e. if we have: $x \oplus (x \oplus y^-)^- = y$ or $y \oplus (y \oplus x^-)^- = x$.

Note that, if a right-m-MEL algebra $\mathcal{A}$ is *totally ordered by* $\leq_m$, then $x \oplus x^- = 1$, for all $x \in A$, i.e. the property of reflexivity, (m-Re^R), holds.

Note also that, if an involutive right-m-MEL algebra $\mathcal{A}$ is *totally ordered by* $\leq_m^M$, then $x \oplus (x \oplus x^-)^- = x$, for all $x \in A$.

Lemma 2.1.8 *(SU) + (Scomm) + (Sass) + (m-La^R) + (Neg1-0) + (PU)+ (DN) + (tot-ord-m-M) $\Longrightarrow$ ($\vee_m$-comm).*

Proof. (Based on ideas of a proof by Prover9, Length of proof was 59, in 0.52 seconds.)

Put

$$(2.4) \qquad f(x, y) \overset{notation}{=} y \vee_m^M x = x \oplus (x^- \odot y) = x \oplus (x \oplus y^-)^-.$$

The hypothesis is:

$$(2.5) \qquad f(x, y) = y \quad or \quad f(y, x) = x.$$

We must prove:

($\vee_m$-comm) $x \vee_m^M y = y \vee_m^M x$,

i.e. $f(y, x) = f(x, y)$. Hence, by (2.5), we must prove that:

$$(2.6) \qquad \begin{array}{l} if \quad f(x, y) = y, \quad then \quad f(y, x) = y, \\ if \quad f(y, x) = x, \quad then \quad f(x, y) = x. \end{array}$$

First, from (2.4), for $X := 1$, we obtain:
(a) $f(1,y) = 1 \oplus (1^- \odot y) = 1$, by (m-La^R),
and, for $Y := 1$, we obtain:
(b) $f(x,1) = x \oplus (x^- \odot 1) = x \oplus x^-$, by (PU).
Now, we prove:

$$(2.7) \qquad f(x,1) = 1.$$

Indeed, by (2.5), for $X := 1$, we have:

$$f(1,y) = y \quad or \quad f(y,1) = 1;$$

- if $f(1,y) = y$, i.e. $y = 1$, by (a), then $f(y,1) \stackrel{(b)}{=} y \oplus y^- = 1 \oplus 1^- \stackrel{(m-La^R)}{=} 1 = y$;
- if $f(y,1) = 1$, i.e. $y \oplus y^- = 1$, by (b), then $f(1,y) \stackrel{(a)}{=} 1$.
Thus, $f(y,1) = 1$, i.e. (2.7) holds.
By (2.7), we obtain immediately, by (b):

$$(2.8) \qquad x \oplus x^- = 1,$$

i.e. (m-Re^R) holds.
Now, we prove:

$$(2.9) \qquad x^- \oplus f(x,y) = 1.$$

Indeed, $x^- \oplus f(x,y) \stackrel{(2.4)}{=} x^- \oplus (x \oplus (x^- \odot y)) \stackrel{(Sass)}{=} (x^- \oplus x) \oplus (x^- \odot y) \stackrel{(2.8)}{=}$ $1 \oplus (x^- \odot y) \stackrel{(m-La^R)}{=} 1$.
Now, we prove:

$$(2.10) \qquad x \oplus (y^- \oplus (x^- \odot y)) = 1.$$

Indeed, $x \oplus (y^- \oplus (x^- \odot y)) = x \oplus (y^- \oplus (x \oplus y^-)^-) \stackrel{(Sass)}{=} (x \oplus y^-) \oplus (x \oplus y^-)^- \stackrel{(2.8)}{=} 1$.
Now, we prove:

$$(2.11) \qquad x^- \odot (y \odot (x \oplus y^-)) = 0.$$

Indeed, from $x \odot y = (x^- \oplus y^-)^-$, by (2.1), we obtain, for $X := x^-$, $Y := y^-$ and by (DN):
(c) $x^- \odot y^- = (x \oplus y)^-$;
now, in (c), take $Y := y^- \oplus (x^- \odot y)$, to obtain:
$x^- \odot (y^- \oplus (x^- \odot y))^- = (x \oplus (y^- \oplus (x^- \odot y)))^-$,
which, by (2.10), becomes:
$x^- \odot (y^- \oplus (x^- \odot y))^- = 1^-$,
which, by (Neg1-0) and (2.1), becomes:
$x^- \odot (y \odot (x^- \odot y)^-) = 0$, hence
$x^- \odot (y \odot (x \oplus y^-)) = 0$, i.e. (2.11) holds.
Now, we prove:

$$(2.12) \qquad f(x, y \odot (x \oplus y^-)) = x.$$

Indeed, in (2.4) ($f(x,y) = x \oplus (x^- \odot y)$), take $Y := y \odot (x \oplus y^-)$, to obtain:
(d) $f(x, y \odot (x \oplus y^-)) = x \oplus (x^- \odot (y \odot (x \oplus y^-)))$;
but, in (d), the part $x^- \odot (y \odot (x \oplus y^-))$ equals 0, by (2.11); hence, (d) becomes:
$f(x, y \odot (x \oplus y^-)) = x \oplus 0$,
which, by (SU), becomes:
$f(x, y \odot (x \oplus y^-)) = x$, i.e. (2.12) holds.

Now, we prove:

$$f(f(x,y),x) = f(x,y). \tag{2.13}$$

Indeed, in (2.12), take $X := f(x,y)$, $Y := x$, to obtain:
(e) $f(f(x,y), x \odot (f(x,y) \oplus x^-)) = f(x,y)$;
but, in (e), the part $f(x,y) \oplus x^-$ equals 1, by (2.9); hence, (e) becomes:
$f(f(x,y), x \odot 1) = f(x,y)$,
which becomes, by (PU):
$f(f(x,y),x) = f(x,y)$, i.e. (2.13) holds.

Finally, we prove (2.6). Indeed,
- if $f(x,y) = y$, then $f(y,x) = f(f(x,y),x) \overset{(2.13)}{=} f(x,y) = y$;
- if $f(y,x) = x$, then $f(x,y) = f(f(y,x),y) \overset{(2.13)}{=} f(y,x) = x$. □

Thus, we obtain the following important result.

Proposition 2.1.9 *If an involutive right-m-MEL algebra $\mathcal{A}$ is totally ordered by $\leq_m^M$, then it is an MV algebra.*

Proof. By Lemma 2.1.8. □

2.1.3 Particular m-MEL algebras

Recall the following definitions from [20], [21].

Definitions 2.1.10 (The dual ones are omitted)

(1) A *right-m-BE algebra* is a right-m-MEL algebra $\mathcal{A}$ verifying additionally the axiom: for each $x \in A$,
(m-Re^R) $x \oplus x^- = 1$, i.e.
(m-Re'^{R}) $x \leq_m x$.

(2) A *right-m-pre-BCK algebra* is a right-m-BE algebra $\mathcal{A}$ verifying additionally the axiom: for all $x, y, z \in A$,
(m-B^R) $((x \oplus y^-)^- \oplus (x \oplus z) \oplus (y \oplus z)^- = 1$.

(3) A *right-m-aBE algebra* is a right-m-BE algebra $\mathcal{A}$ verifying the additional axiom: for all $x, y \in A$,
(m-An^R) $y \oplus x^- = 1$ and $x \oplus y^- = 1$ imply $x = y$, i.e.
(m-An'^{R}) $x \leq_m y$ and $y \leq_m x$ imply $x = y$.

(4) A *right-m-BCK algebra* is a right-m-aBE algebra $\mathcal{A}$ verifying additionally the above axiom (m-B^R).

Denote by **m-BE**^{R} the class of all right-m-BE algebras, by **m-pre-BCK**^{R} the class of all right-m-pre-BCK algebras, by **m-aBE**^{R} the class of all right-m-aBE algebras, by **m-BCK**^{R} the class of all right-m-BCK algebras, while denote by **m-BE**, **m-pre-BCK**, **m-aBE**, **m-BCK** the classes of all corresponding left-algebras.

Denote by $\mathbf{X}_{(DN)}$ the class of all involutive members (i.e. verifying (DN)) of the class **X** of algebras.

By ([20], Theorem 6.12), ([21], Theorem 6.3.5) (in the dual case), if $\mathcal{A} = (A, \oplus, ^-, 0)$ is an algebra of type $(2, 1, 0)$, $1 \stackrel{def.}{=} 0^-$, i.e. (Neg0-1) is verified, and if (Neg1-0), (SU), (Scomm), (Sass), (m-Re^R), (DN) hold, then

$$(m - B^R) \Longleftrightarrow (m - BB^R) \Longleftrightarrow (m - Tr^R),$$

where:
(m-BB^R) $((z \oplus x)^- \oplus (y \oplus x)) \oplus (y \oplus z^-)^- = 1.$

By ([20], Theorem 6.13), ([21], Theorem 6.3.6 (2)) (in the dual case), any right-m-BCK algebra $\mathcal{A} = (A, \oplus, ^-, 0)$ is involutive, so

$$\mathbf{m-BCK^R} = \mathbf{m-BCK^R_{(DN)}}.$$

By ([21], Proposition 6.4.23) (in the dual case), in any involutive right-m-BE algebra $\mathcal{A} = (A, \oplus, ^-, 0)$, if $x \leq_m^M y$, then $x \leq_m y$, for all $x, y \in A$.

By ([21], Corollary 6.4.25) (in the dual case), in any involutive right-m-BE algebra $\mathcal{A}$, the binary relation $\leq_m^M$ is reflexive and antisymmetric and $0 \leq_m^M x \leq_m^M 1$, for each $x \in A$.

By ([21], Theorem 12.2.3) (in the dual case), in any (involutive) right-m-BCK algebra $\mathcal{A} = (A, \oplus, ^-, 0)$, the binary relation $\leq_m^M$ is transitive.

By ([21], Remark 12.3.4) (in the dual case), in any (involutive) right-m-BCK algebra $\mathcal{A}$:
- the initial binary relation $\leq_m$ is an order (since (m-Re^R), (m-An^R), (m-Tr^R) hold);
- the binary relation $\leq_m^M$ is an order, by ([21], Corollary 6.4.25) and ([21], Theorem 12.2.3), but not a lattice order, in general, with respect to $\vee_m^M$; it is a distributive lattice order iff ($\vee_m$-comm) holds, i.e. in the case of MV algebras, when $\leq_m \Longleftrightarrow \leq_m^M$.

We recall now from [19], [21] the following corresponding definition (see [16] for the commutative case).

Definitions 2.1.11 Let $\mathcal{A} = (A, \oplus, ^-, 0)$ be a right-algebra and $\leq_m$ be a binary relation.

1) We shall say that $\mathcal{A}$ is *reflexive*, if $\leq_m$ is reflexive (i.e. it satisfies the property (m-Re^R)).

2) We shall say that $\mathcal{A}$ is *antisymmetric*, if $\leq_m$ is antisymmetric (i.e. it satisfies the property (m-An^R)).

3) We shall say that $\mathcal{A}$ is *transitive*, if $\leq_m$ is transitive (i.e. it satisfies property (m-Tr^R)).

4) We shall say that $\mathcal{A}$ is *pre-ordered*, if $\leq_m$ is a pre-order relation (i.e. it is reflexive and transitive).

5) We shall say that $\mathcal{A}$ is *ordered*, if $\leq_m$ is a partial order relation (i.e. it is reflexive, antisymmetrique and transitive).

6) We shall say that $\mathcal{A}$ is a *lattice*, if $\leq_m$ is an Ore lattice order (i.e. it is a partial order such that there exist $\inf_m(x, y)$ and $\sup_m(x, y)$ for each $x, y \in A$).

- **Hasse diagrams and Hasse-type diagrams**

Remarks 2.1.12 ([18], Remark 3.11) (see also ([19], Remark 2.1.21))

(1) If the right-algebra is ordered, then the usual *Hasse diagram* is used, where each element is represented by a bullet • and if $x \leq_m y$ and there is no z such that $x <_m z <_m y$, then x is reprezented below y and a line will connect them.

(2) If the right-algebra is not ordered (i.e. it is neither reflexive nor antisymmetric nor transitive, or it is only reflexive, or reflexive and transitive, or reflexive and antisymmetric), then a *Hasse-type diagram* is used [18], where each element is represented by a circ ∘ and if $x \leq_m y$ and $y \leq_m x$ and $x \neq y$ (i.e. x and y have the *same 'height'*, or are *'parallel'*), then a horizontal line will connect them.

Remark 2.1.13 [16] In the diagram of a hierarchy of classes of algebras, we shall represent:

- *reflexive* algebras by ◯
- *antisymmetric* algebras by ∘
- *transitive* algebras by •
- *reflexive* and *antisymmetric* algebras by ◎
- *reflexive* and *transitive* algebras by ⊙
- *ordered* algebras by ●
- any other algebra by □

The connections between the above right-algebras and many others are recalled in **Appendix A**.

Recall [20], [21] that MV algebras are particular cases of (involutive) m-BCK algebras.

2.2 XY algebras versus MV algebras

In this second section, we introduce and study, in the framework recalled in the first section, the *XY algebras*, the 'ancestor' (generalization) of MV algebras which will play a central role in this research. We present two new, equivalent definitions of MV algebras.

2.2.1 MV algebras. Two new equivalent definitions

Let $\mathcal{A} = (A, \oplus, ^-, 0)$ be an involutive right-m-MEL algebra, with $1 \stackrel{def.}{=} 0^-$.

Consider the property ($\vee_m$-comm) = (Vcomm$_m^M$) (commutativity of $\vee_m^M$ - see Definition 3.2.3) defined by (see for example [21]): for all $x, y \in A$,
($\vee_m$-comm) $\quad x \vee_m^M y = y \vee_m^M x,$

where $x \vee_m^M y \stackrel{def.}{=} y \oplus (y \oplus x^-)^-$.

Now recall the our days definition of an MV algebra.

Definitions 2.2.1 ([21], Definitions 6.5.1)

(i) A *right-MV algebra* is an algebra $\mathcal{A}^R = (A^R, \oplus, ^- = ^{-^R}, 0)$ verifying: for all $x, y, z \in A^R$,
(SU) $\quad 0 \oplus x = x (= x \oplus 0),$
(Scomm) $\quad x \oplus y = y \oplus x,$
(Sass) $\quad x \oplus (y \oplus z) = (x \oplus y) \oplus z,$
(m-La^R) $\quad x \oplus 1 = 1$, where $1 \stackrel{def.}{=} 0^-$ (i.e. (Neg0-1) holds),
(DN) $\quad (x^-)^- = x,$
($\vee_m$-comm) $\quad (x^- \oplus y)^- \oplus y = (y^- \oplus x)^- \oplus x.$

(i') Dually, a *left-MV algebra* is an algebra $\mathcal{A}^L = (A^L, \odot, ^- = ^{-^L}, 1)$ verifying: for all $x, y, z \in A^L$,
(PU) $\quad 1 \odot x = x (= x \odot 1),$
(Pcomm) $\quad x \odot y = y \odot x,$
(Pass) $\quad x \odot (y \odot z) = (x \odot y) \odot z,$
(m-La) $\quad x \odot 0 = 0$, where $0 \stackrel{def.}{=} 1^-$ (i.e. (Neg1-0) holds),
(DN) $\quad (x^-)^- = x,$
($\wedge_m$-comm) $\quad (x^- \odot y)^- \odot y = (y^- \odot x)^- \odot x.$

Denote by $\mathbf{MV}^R$ the class of all right-MV algebras and by **MV** the class of all left-MV algebras. Note that they are varieties.

Note that right-MV algebras verify the property (Neg1-0) too: $1^- \stackrel{def.}{=} (0^-)^- \stackrel{(DN)}{=} 0$; left-MV algebras verify (Neg0-1) too.

We recall the following important remark, which was the motivation of paper [20] and, hence, of the monograph [21]:

Remark 2.2.2
(i) Right-MV algebra is just the involutive right-m-MEL algebra verifying the additional axiom ($\vee_m$-comm), i.e.

$$\mathbf{MV}^R = \mathbf{m-MEL}^R_{(DN)} + (\vee_m - comm).$$

(i') Left-MV algebra is just the involutive left-m-MEL algebra verifying the additional axiom ($\wedge_m$-comm), i.e.

$$\mathbf{MV} = \mathbf{m-MEL}_{(DN)} + (\wedge_m - comm).$$

Proposition 2.2.3 *Let $\mathcal{A} = (A, \oplus, ^-, 0)$ be a right-MV algebra. Then, the cancellation property (C) holds, where: for all $x, y, z \in A$,*
(C) $z \oplus x = z \oplus y$ and $z \odot x = z \odot y$ imply $x = y$.

Proof. See ([10], Lemma 1.6.1 (ii)). A direct proof by PROVER9 has the Length 35 (the Length of expanded renumbered proof is 78). □

Right-MV algebras verify the properties (m-Re^R) and (m-An^R), by ([21], Proposition 6.2.2) (in the dual case); the property (m-BB^R), by ([21], Proposition 6.3.3) (in the dual case); the properties (m-Tr^R), (m-B^R), by above recalled ([21], Theorem 6.3.5) (in the dual case).

Right-MV algebras are (coincide with) (involutive) m-BCK^R algebras verifying the property ($\vee_m$-comm), by ([21], Theorem 6.5.10) (in the dual case).

By ([21], Proposition 6.4.19) (in the dual case), in any involutive right-m-BE algebra $\mathcal{A} = (A, \oplus, ^-, 0)$, if ($\vee_m$-comm) holds, then $x \leq_m y \Longleftrightarrow x \leq_m^M y$, for all $x, y \in A$.

By ([21], Remark 6.5.9) (in the dual case), in right-MV algebras $\mathcal{A} = (A, \oplus, ^-, 0)$, the two binary relations $\leq_m$ and $\leq_m^M$ are lattice orders with respect to $\vee_m^M$, $\wedge_m^M$ and they coincide, i.e. $\leq_m \Longleftrightarrow \leq_m^M$, the resulting unique lattice order being denoted by $\leq$ (with respect to $\vee = \vee_m^M$, $\wedge = \wedge_m^M$). The lattice $(A, \vee, \wedge)$ is distributive. If the lattice order $\leq$ is totally ordered (or linearly ordered, here), then we say that we have an *MV chain*.

- **A first equivalent definition of MV algebras**

Consider the following property:
($M_\vee$) $(x \oplus y) \odot ((x \odot y) \oplus y^-) = x$.
Note that ($M_\vee$) is the property (1.15) (for $Y := y^-$, by (DN)) verified by any MV algebra from ([10], Proposition 1.6.2).

Proposition 2.2.4 *Let $\mathcal{A} = (A, \oplus, ^-, 0)$ be an involutive right-m-MEL algebra. Then,*

$$(\vee_m - comm) \Longrightarrow (M_\vee),$$

where:
($\vee_m$-comm) $x \vee_m^M y = y \vee_m^M x$,
($M_\vee$) $(x \oplus y) \odot ((x \odot y) \oplus y^-) = x$.

Proof. (Based on a proof by PROVER9, Length of proof was 25; the length of the expanded proof was 54.)

Let $\mathcal{A}$ be an involutive right-m-MEL algebra verifying ($\vee_m$-comm), i.e. an MV algebra; then (SU), (Scomm), (Sass), (Neg0-1), (m-La^R), (DN), ($\vee_m$-comm) and (Neg1-0), (m-Re^R) hold.

Then, since $x^- \oplus (x \oplus y) \overset{(Sass)}{=} (x^- \oplus x) \oplus y \overset{(Scomm)}{=} (x \oplus x^-) \oplus y \overset{(m-Re^R)}{=} 1 \oplus y \overset{(m-La^R)}{=} 1$, it follows that we have:

$$x^- \oplus (x \oplus y) = 1. \tag{2.14}$$

Note now that ($\vee_m$-comm) means that $y \oplus (y \oplus x^-)^- = x \oplus (x \oplus y^-)^-$, i.e. we have:
$x \oplus (x \oplus y^-)^- = y \oplus (y \oplus x^-)^-$,
which, by (Scomm), becomes:

$$(2.15) \qquad x \oplus (y^- \oplus x)^- = y \oplus (y \oplus x^-)^-.$$

Then, in (2.15), take $Y := y^-$, to obtain, by (DN):
$x \oplus (y \oplus x)^- = y^- \oplus (y^- \oplus x^-)^-$,
i.e. we have:

$$(2.16) \qquad x^- \oplus (x^- \oplus y^-)^- = y \oplus (x \oplus y)^-.$$

Then, in (2.15) again, take $X := x \oplus y$, $Y := x$, to obtain:
$(x \oplus y) \oplus (x^- \oplus (x \oplus y))^- = x \oplus (x \oplus (x \oplus y)^-)^-$,
which, by (2.14), becomes:
$(x \oplus y) \oplus 1^- = x \oplus (x \oplus (x \oplus y)^-)^-$,
which, by (Neg1-0), (SU), becomes:
$x \oplus y = x \oplus (x \oplus (x \oplus y)^-)^-$, i.e.

$$(2.17) \qquad x \oplus (x \oplus (x \oplus y)^-)^- = x \oplus y.$$

Now, in (2.16), take $Y := (x \oplus (x \oplus y)^-)^-$, to obtain, by (DN):
$x^- \oplus (x^- \oplus (x \oplus (x \oplus y)^-))^- = (x \oplus (x \oplus y)^-)^- \oplus (x \oplus (x \oplus (x \oplus y)^-)^-)^-$,
which, by (2.17), becomes:
$x^- \oplus (x^- \oplus (x \oplus (x \oplus y)^-))^- = (x \oplus (x \oplus y)^-)^- \oplus (x \oplus y)^-$,
which, by (2.14), becomes:
$x^- \oplus 1^- = (x \oplus (x \oplus y)^-)^- \oplus (x \oplus y)^-$,
which, by (Neg1-0), (SU), becomes:
$x^- = (x \oplus (x \oplus y)^-)^- \oplus (x \oplus y)^-$,
which, by (Scomm), becomes:
$x^- = (x \oplus y)^- \oplus (x \oplus (x \oplus y)^-)^-$,
which, by (DN), becomes:
$((x \oplus y)^- \oplus (x \oplus (x \oplus y)^-)^-)^- = x$,
which, by definition of $\odot$, becomes:
$(x \oplus y) \odot (x \oplus (x \oplus y)^-) = x$,
which, by (DN), becomes:
$(x \oplus y) \odot (x \oplus (x \oplus y^{=})^-) = x$,
which, by (2.15), becomes:
$(x \oplus y) \odot (y^- \oplus (x^- \oplus y^-)^-) = x$,
which, by (Scomm), becomes:
$(x \oplus y) \odot ((x^- \oplus y^-)^- \oplus y^-) = x$,
which, by definition of $\odot$, becomes:
$(x \oplus y) \odot ((x \odot y) \oplus y^-) = x$, i.e. ($M_\vee$) holds. □

Proposition 2.2.5 *Let $\mathcal{A} = (A, \oplus, ^-, 0)$ be an involutive right-m-MEL algebra. Then,*

$$(M_\vee) \implies (\vee_m - comm),$$

where:
($\vee_m$-*comm*) $x \vee_m^M y = y \vee_m^M x$,
($M_\vee$) $(x \oplus y) \odot ((x \odot y) \oplus y^-) = x$.

Proof. (Based on a proof by PROVER9, Length of proof was 24; the length of the expanded proof was 52.)

Let $\mathcal{A} = (A, \oplus, ^-, 0)$ be an involutive right-m-MEL algebra, i.e. (SU), (Scomm), (Sass), (Neg0-1), (m-La^R), (DN) hold. Then, $1^- \overset{(Neg0-1)}{=} (0^-)^- \overset{(DN)}{=} 0$, i.e. (Neg1-0) holds too.

Also, $x \oplus (y \oplus z) \overset{(Sass)}{=} (x \oplus y) \oplus z \overset{(Scomm)}{=} (y \oplus x) \oplus z \overset{(Sass)}{=} y \oplus (x \oplus z)$, hence we have:

$$x \oplus (y \oplus z) = y \oplus (x \oplus z). \tag{2.18}$$

Suppose ($M_\vee$) holds. Since $x \odot y \overset{def.}{=} (x^- \oplus y^-)^-$, it follows that the property ($M_\vee$) means:
$(x \oplus y) \odot [(x^- \oplus y^-)^- \oplus y^-] = x$,
hence:
$((x \oplus y)^- \oplus [(x^- \oplus y^-)^- \oplus y^-]^-)^- = x$,
which, by (Scomm), becomes:

$$((x \oplus y)^- \oplus [y^- \oplus (x^- \oplus y^-)^-]^-)^- = x. \tag{2.19}$$

Now, in (2.19), take $X := 0$, $Y := x$, to obtain:
$((0 \oplus x)^- \oplus [x^- \oplus (0^- \oplus x^-)^-]^-)^- = 0$,
which, by (Scomm), (SU), (Neg0-1), becomes:
$(x^- \oplus [x^- \oplus (1 \oplus x^-)^-]^-)^- = 0$,
which, by (m-La^R), becomes:
$(x^- \oplus [x^- \oplus 1^-]^-)^- = 0$,
which, by (Neg1-0), (SU), becomes:
$(x^- \oplus [x^-]^-)^- = 0$,
which, by (DN), becomes:
$(x^- \oplus x)^- = 0$,
which, by (DN), Neg0-1), becomes:
$x^- \oplus x = 1$,
which, by (Scomm), becomes:
$x \oplus x^- = 1$, i.e. (m-Re^R) holds.

Then, $x \oplus (x^- \oplus y) \overset{(Sass)}{=} (x \oplus x^-) \oplus y \overset{(m-Re^R)}{=} 1 \oplus y \overset{(m-La^R)}{=} 1$, hence we have:

$$x \oplus (x^- \oplus y) = 1 \tag{2.20}$$

and $x \oplus (y \oplus (x^- \oplus z)) \overset{(2.18)}{=} y \oplus (x \oplus (x^- \oplus z)) \overset{(2.20)}{=} y \oplus 1 \overset{(m-La^R)}{=} 1$, hence we have:

$$x \oplus (y \oplus (x^- \oplus z)) = 1. \tag{2.21}$$

Note now that ($\vee_m$-comm) means $y \oplus (y \oplus x^-)^- = x \oplus (x \oplus y^-)^-$, hence we have to prove:

$$y \oplus (y \oplus x^-)^- = x \oplus (x \oplus y^-)^-. \tag{2.22}$$

Indeed, first, from (2.19), by (Scomm), we obtain:
$((x \oplus y)^- \oplus [y^- \oplus (y^- \oplus x^-)^-]^-)^- = x$,
which, for $Y := y^-$, becomes, by (DN):

$$((x \oplus y^-)^- \oplus [y \oplus (y \oplus x^-)^-]^-)^- = x. \tag{2.23}$$

Then, in (2.23), take $X := x \oplus (x \oplus y^-)^-$ and $Y := (y \oplus x^-)^-$, to obtain:
(a) $([(x \oplus (x \oplus y^-)^-) \oplus ((y \oplus x^-)^-)^-]^- \oplus$
$[(y \oplus x^-)^- \oplus ((y \oplus x^-)^- \oplus (x \oplus (x \oplus y^-)^-)^-)^-]^-)^- = x \oplus (x \oplus y^-)^-$;
but, in (a), the part $((y \oplus x^-)^- \oplus (x \oplus (x \oplus y^-)^-)^-)^-$ equals y, by (2.23); hence, (a) becomes:
$([(x \oplus (x \oplus y^-)^-) \oplus ((y \oplus x^-)^-)^-]^- \oplus [(y \oplus x^-)^- \oplus y]^-)^- = x \oplus (x \oplus y^-)^-$,
which, by (DN), becomes:
$([(x \oplus (x \oplus y^-)^-) \oplus (y \oplus x^-)]^- \oplus [(y \oplus x^-)^- \oplus y]^-)^- = x \oplus (x \oplus y^-)^-$,
which, by (Scomm), becomes:
$([(y \oplus x^-) \oplus (x \oplus (x \oplus y^-)^-)]^- \oplus [(y \oplus x^-)^- \oplus y]^-)^- = x \oplus (x \oplus y^-)^-$,
which, by (2.18), becomes:
$([x \oplus ((y \oplus x^-) \oplus (x \oplus y^-)^-)]^- \oplus [(y \oplus x^-)^- \oplus y]^-)^- = x \oplus (x \oplus y^-)^-$,
which, by (Sass), becomes:
$([x \oplus (y \oplus (x^- \oplus (x \oplus y^-)^-))]^- \oplus [(y \oplus x^-)^- \oplus y]^-)^- = x \oplus (x \oplus y^-)^-$,
which, by (2.21), becomes:
$([1]^- \oplus [(y \oplus x^-)^- \oplus y]^-)^- = x \oplus (x \oplus y^-)^-$,
which, by (Neg1-0), (SU), becomes:
$([(y \oplus x^-)^- \oplus y]^-)^- = x \oplus (x \oplus y^-)^-$,
which, by (DN), becomes:
$(y \oplus x^-)^- \oplus y = x \oplus (x \oplus y^-)^-$,
which, by (Scomm), becomes:
$y \oplus (y \oplus x^-)^- = x \oplus (x \oplus y^-)^-$, i.e. (2.22) holds. □

Corollary 2.2.6 *Let $\mathcal{A} = (A, \oplus, ^-, 0)$ be an involutive right-m-MEL algebra. Then,*

$$(\vee_m - comm) \iff (M_\vee),$$

where:
$(\vee_m$*-comm)* $\quad x \vee_m^M y = y \vee_m^M x$,
$(M_\vee)$ $\quad (x \oplus y) \odot ((x \odot y) \oplus y^-) = x$.

Proof. By Propositions 2.2.4, 2.2.5. □

Remark 2.2.7 *The above Corollary 2.2.4 says that an equivalent definition of MV algebras is the following:*
A right-MV algebra is an involutive right-m-MEL algebra $\mathcal{A} = (A, \oplus, ^-, 0)$ verifying: for all $x, y \in A$,
$(M_\vee)$ $(x \oplus y) \odot ((x \odot y) \oplus y^-) = x$, i.e.

$$\mathbf{MV}^R = \mathbf{m} - \mathbf{MEL}^R_{(DN)} + (M_\vee).$$

Dually,
a left-MV algebra is an involutive left-m-MEL algebra $\mathcal{A} = (A, \odot, ^-, 1)$ verifying:

for all $x, y \in A$,
$(M_\wedge)$ $(x \odot y) \oplus ((x \oplus y) \odot y^-) = x$, *i.e.*

$$\mathbf{MV} \ = \ \mathbf{m-MEL}_{(DN)} \ + \ (M_\wedge).$$

- **A second equivalent definition of MV algebras**

Consider the following two properties (see Definition 3.2.3):
($\vee_m$-comm) = (Vcomm_m^M) $\quad x \vee_m^M y = y \vee_m^M x \quad$ (commutativity of $\vee_m^M$),
(Vabs_m^M) $\quad x \vee_m^M (x \wedge_m^M y) = x \quad$ (absorbtion of $\vee_m^M$ over $\wedge_m^M$),
where:
$x \vee_m^M y \stackrel{def.}{=} y \oplus (y \oplus x^-)^-$, $\quad x \wedge_m^M y \stackrel{def.}{=} (x^- \vee_m^M y^-)^-$.

Proposition 2.2.8 *Let* $\mathcal{A} = (A, \oplus, ^-, 0)$ *be an involutive right-m-MEL algebra, with* $1 \stackrel{def.}{=} 0^-$. *Then,*

$$(\vee_m - comm) \implies (Vabs_m^M).$$

Proof. (Based on a proof by PROVER9 of Length 22.)
Let $\mathcal{A} = (A, \oplus, ^-, 0)$ be an involutive right-m-MEL algebra and suppose that ($\vee_m$-comm) holds, i.e $\mathcal{A}$ is a right-MV algebra. Then, (m-Re^R) holds, i.e. $x \oplus x^- = 1$, for all $x \in A$. Then,
$x \oplus (y \oplus (x \oplus y)^-) \stackrel{(Sass)}{=} (x \oplus y) \oplus (x \oplus y)^- \stackrel{(m-Re^R)}{=} 1$, hence, we have:

$$x \oplus (y \oplus (x \oplus y)^-) = 1. \tag{2.24}$$

We must prove that (Vabs_m^M) holds, i.e. that $x \vee_m^M (x \wedge_m^M y) = x$. Indeed,
$x \vee_m^M (x \wedge_m^M y) = x \vee_m^M (x^- \vee_m^M y^-)^- = x \vee_m^M (y^- \oplus (y^- \oplus x^=)^-)^- \stackrel{(DN)}{=} x \vee_m^M (y^- \oplus (y^- \oplus x)^-)^- \stackrel{(\vee_m-comm)}{=} (y^- \oplus (y^- \oplus x)^-)^- \vee_m^M x = x \oplus (x \oplus (y^- \oplus (y^- \oplus x)^-)^=)^- \stackrel{(DN)}{=} x \oplus (x \oplus (y^- \oplus (y^- \oplus x)^-))^- \stackrel{(Scomm)}{=} x \oplus (x \oplus (y^- \oplus (x \oplus y^-)^-))^- \stackrel{(2.24)}{=} x \oplus 1^- \stackrel{(Neg1-0)}{=} x \oplus 0 \stackrel{(SU)}{=} x.$ □

Proposition 2.2.9 *Let* $\mathcal{A} = (A, \oplus, ^-, 0)$ *be an involutive right-m-MEL algebra, with* $1 \stackrel{def.}{=} 0^-$. *Then,*

$$(Vabs_m^M) \implies (\vee_m - comm).$$

Proof. (By a proof by PROVER9 of Length 46, in 1867.89 seconds; expended renumbered proof of Length 161.)
Let $\mathcal{A} = (A, \oplus, ^-, 0)$ be an involutive right-m-MEL algebra verifying (Vabs_m^M). Then, we obtain immediately:

$$x \oplus (y \oplus z) = y \oplus (x \oplus z). \tag{2.25}$$

We have to prove that ($\vee_m$-comm) holds, i.e. that:

$$y \oplus (y \oplus x^-)^- = x \oplus (x \oplus y^-)^-. \tag{2.26}$$

Indeed, first note that (Vabs_m^M) means, by definitions:

$x \vee_m^M (x \wedge_m^M y) = x$

$\Longleftrightarrow x \vee_m^M (x^- \vee_m^M y^-)^- = x$

$\Longleftrightarrow x \vee_m^M (y^- \oplus (y^- \oplus x^=)^-)^- = x$

$\overset{(DN)}{\Longleftrightarrow} x \vee_m^M (y^- \oplus (y^- \oplus x)^-)^- = x$

$\Longleftrightarrow (y^- \oplus (y^- \oplus x)^-)^- \oplus [(y^- \oplus (y^- \oplus x)^-)^- \oplus x^-]^- = x$

$\overset{(Scomm)}{\Longleftrightarrow} (y^- \oplus (y^- \oplus x)^-)^- \oplus [x^- \oplus (y^- \oplus (y^- \oplus x)^-)^-]^- = x$, i.e.

$$(x^- \oplus (x^- \oplus y)^-)^- \oplus [y^- \oplus (x^- \oplus (x^- \oplus y)^-)^-]^- = y. \tag{2.27}$$

Then, in (2.27), take $Y := 1$, to obtain:
$(x^- \oplus (x^- \oplus 1)^-)^- \oplus [1^- \oplus (x^- \oplus (x^- \oplus 1)^-)^-]^- = 1$,
which, by (m-La^R), (Neg1-0), (SU), (DN), becomes:
(m-Re^R) $x \oplus x^- = 1$.

Consequently, by (m-Re^R), we obtain immediately:

$$x \oplus (x^- \oplus y) = 1, \tag{2.28}$$

$$x \oplus (y \oplus x^-) = 1, \tag{2.29}$$

$$x^- \oplus (y \oplus x) = 1, \tag{2.30}$$

$$x \oplus (y \oplus (x \oplus y)^-) = 1. \tag{2.31}$$

Now, from (2.27), by (Scomm), we obtain:

$$(x^- \oplus (y \oplus x^-)^-)^- \oplus [y^- \oplus (x^- \oplus (x^- \oplus y)^-)^-]^- = y \tag{2.32}$$

and

$$(x^- \oplus (x^- \oplus y)^-)^- \oplus [y^- \oplus (x^- \oplus (y \oplus x^-)^-)^-]^- = y. \tag{2.33}$$

Now, add z to both sides of (2.27), to obtain, by (Sass):

$$(x^- \oplus (x^- \oplus y)^-)^- \oplus ([y^- \oplus (x^- \oplus (x^- \oplus y)^-)^-]^- \oplus z) = y \oplus z. \tag{2.34}$$

Finally, in (2.27) again, take $X := x^-$, to obtain, by (DN):

$$(x \oplus (x \oplus y)^-)^- \oplus [y^- \oplus (x \oplus (x \oplus y)^-)^-]^- = y. \tag{2.35}$$

Now, in (2.33), take $X := x^-$, to obtain, by (DN):

$$(x \oplus (x \oplus y)^-)^- \oplus [y^- \oplus (x \oplus (y \oplus x)^-)^-]^- = y. \tag{2.36}$$

Now, in (2.34), take $Y := y \oplus (x^- \oplus y)^-$, to obtain:
(a) $(x^- \oplus (x^- \oplus [y \oplus (x^- \oplus y)^-])^-)^- \oplus$
$([[y \oplus (x^- \oplus y)^-]^- \oplus (x^- \oplus (x^- \oplus [y \oplus (x^- \oplus y)^-])^-)^-]^- \oplus z) = [y \oplus (x^- \oplus y)^-] \oplus z;$

but, in (a), the two parts $x^- \oplus [y \oplus (x^- \oplus y)^-]$ equal 1, by (2.31); then, (a) becomes, by (Neg1-0), (SU), (DN):
$x \oplus ([[y \oplus (x^- \oplus y)^-]^- \oplus x]^- \oplus z) = [y \oplus (x^- \oplus y)^-] \oplus z$,
which, by (Scomm), (Sass), becomes:

$$(2.37) \qquad x \oplus ([x \oplus [y \oplus (x^- \oplus y)^-]^-]^- \oplus z) = y \oplus [(x^- \oplus y)^- \oplus z].$$

Then, in (2.37), take $Y := (x^- \oplus y)^-$, $Z := [y^- \oplus (x^- \oplus (y \oplus x^-)^-)^-]^-$, to obtain:
(b) $x \oplus ([x \oplus [(x^- \oplus y)^- \oplus (x^- \oplus (x^- \oplus y)^-)^-]^-]^- \oplus [y^- \oplus (x^- \oplus (y \oplus x^-)^-)^-]^-)$
$= (x^- \oplus y)^- \oplus [(x^- \oplus (x^- \oplus y)^-)^- \oplus [y^- \oplus (x^- \oplus (y \oplus x^-)^-)^-]^-]$;
but, in (b), the part $(x^- \oplus (x^- \oplus y)^-)^- \oplus [y^- \oplus (x^- \oplus (y \oplus x^-)^-)^-]^-$ equals y, by (2.36) for $X := x^-$; then, (b) becomes:
(2.38)
$x \oplus ([x \oplus [(x^- \oplus y)^- \oplus (x^- \oplus (x^- \oplus y)^-)^-]^-]^- \oplus [y^- \oplus (x^- \oplus (y \oplus x^-)^-)^-]^-) = (x^- \oplus y)^- \oplus y$.

Now, in (2.32), take $X := x^-$, to obtain, by (DN):
(c) $(x \oplus (y \oplus x)^-)^- \oplus [y^- \oplus (x \oplus (x \oplus y)^-)^-]^- = y$;
then, in (c), take $X := x \oplus y$, $Y := y^-$, to obtain, by (DN):
(d) $((x \oplus y) \oplus (y^- \oplus (x \oplus y))^-)^- \oplus [y \oplus ((x \oplus y) \oplus ((x \oplus y) \oplus y^-)^-)^-]^- = y^-$;
but, in (d), the part $y^- \oplus (x \oplus y)$ equals 1 by (2.30) and the part $(x \oplus y) \oplus y^-$ equals 1, by (Scomm) and (2.30); then, (d) becomes, by (Neg1-0), (SU):
$(x \oplus y)^- \oplus [y \oplus (x \oplus y)^-]^- = y^-$,
which, by (Scomm) twice, becomes:
$(y \oplus x)^- \oplus [y \oplus (y \oplus x)^-]^- = y^-$, i.e.

$$(2.39) \qquad (x \oplus y)^- \oplus [x \oplus (x \oplus y)^-]^- = x^-.$$

Then, in (2.38), the part $(x^- \oplus y)^- \oplus (x^- \oplus (x^- \oplus y)^-)^-$ equals $x^=$, by (2.39) for $X := x^-$; it follows that (2.38) becomes:
$x \oplus ([x \oplus [x^=]^-]^- \oplus [y^- \oplus (x^- \oplus (y \oplus x^-)^-)^-]^-) = (x^- \oplus y)^- \oplus y$,
which, by (DN), (m-Re^R), (Neg1-0), (SU), becomes:
$x \oplus [y^- \oplus (x^- \oplus (y \oplus x^-)^-)^-]^- = (x^- \oplus y)^- \oplus y$,
which, by (Scomm), becomes:

$$(2.40) \qquad x \oplus [y^- \oplus (x^- \oplus (y \oplus x^-)^-)^-]^- = y \oplus (x^- \oplus y)^-.$$

Now, in (2.32), take $Y := y \oplus (x^- \oplus y)^-$, to obtain:
$(x^- \oplus ([y \oplus (x^- \oplus y)^-] \oplus x^-)^-)^- \oplus$
$[[y \oplus (x^- \oplus y)^-]^- \oplus (x^- \oplus (x^- \oplus [y \oplus (x^- \oplus y)^-])^-)^-]^- = y \oplus (x^- \oplus y)^-$,
which, by (Scomm), becomes:
(e) $(x^- \oplus (x^- \oplus [y \oplus (x^- \oplus y)^-])^-)^- \oplus$
$[[y \oplus (x^- \oplus y)^-]^- \oplus (x^- \oplus (x^- \oplus [y \oplus (x^- \oplus y)^-])^-)^-]^- = y \oplus (x^- \oplus y)^-$;
but, in (e), the two parts $x^- \oplus [y \oplus (x^- \oplus y)^-]$ equal 1, by (2.31); then, (e) becomes:
$(x^- \oplus 1^-)^- \oplus [[y \oplus (x^- \oplus y)^-]^- \oplus (x^- \oplus 1^-)^-]^- = y \oplus (x^- \oplus y)^-$,
which, by (Neg1-0), (SU), (DN), becomes:
$x \oplus [[y \oplus (x^- \oplus y)^-]^- \oplus x]^- = y \oplus (x^- \oplus y)^-$,
which, by (Scomm), becomes:

$$(2.41) \qquad x \oplus [x \oplus [y \oplus (x^- \oplus y)^-]^-]^- = y \oplus (x^- \oplus y)^-.$$

Then,
$x \oplus (z \oplus (x^- \oplus z)^-)$
$\overset{(2.41)}{=} x \oplus (z \oplus [z \oplus [y \oplus (z^- \oplus y)^-]^-]^-)$
$\overset{(2.25)}{=} z \oplus (x \oplus [z \oplus [y \oplus (z^- \oplus y)^-]^-]^-)$,
hence, we have:

$$(2.42) \qquad x \oplus (y \oplus [x \oplus [z \oplus (x^- \oplus z)^-]^-]^-) = y \oplus (z \oplus (x^- \oplus z)^-).$$

Now, in (2.34), take $Y := y \oplus x$, to obtain:
(f) $(x^- \oplus (x^- \oplus (y \oplus x))^-)^- \oplus ([(y \oplus x)^- \oplus (x^- \oplus (x^- \oplus (y \oplus x))^-)^-]^- \oplus z) = (y \oplus x) \oplus z$;
but, in (f), the two parts $x^- \oplus (y \oplus x)$ equal 1, by (2.30); then, (f) becomes:
$(x^- \oplus 1^-)^- \oplus ([(y \oplus x)^- \oplus (x^- \oplus 1^-)^-]^- \oplus z) = (y \oplus x) \oplus z$,
which, by (Neg1-0), (SU), (DN), becomes:
$x \oplus ([(y \oplus x)^- \oplus x]^- \oplus z) = (y \oplus x) \oplus z$,
which, by (Scomm), (Sass), becomes:
(f ') $x \oplus ([x \oplus (y \oplus x)^-]^- \oplus z) = y \oplus (x \oplus z)$.
Then, in (f '), take $Z := (x \oplus (z \oplus (x^- \oplus z)^-)^-)^-$, to obtain:
(f ") $x \oplus ([x \oplus (y \oplus x)^-]^- \oplus (x \oplus (z \oplus (x^- \oplus z)^-)^-)^-) = y \oplus (x \oplus (x \oplus (z \oplus (x^- \oplus z)^-)^-)^-)$;
but, in (f "), the part $x \oplus (x \oplus (z \oplus (x^- \oplus z)^-)^-)^-$ equals $z \oplus (x^- \oplus z)^-$, by (2.41);
hence, (f ") becomes:
$x \oplus ([x \oplus (y \oplus x)^-]^- \oplus (x \oplus (z \oplus (x^- \oplus z)^-)^-)^-) = y \oplus (z \oplus (x^- \oplus z)^-)$,
which, by (2.42) for $Y := [x \oplus (y \oplus x)^-]^-$, becomes:
(f "') $[x \oplus (y \oplus x)^-]^- \oplus (z \oplus (x^- \oplus z)^-) = y \oplus (z \oplus (x^- \oplus z)^-)$;
then, in (f "'), take $Z := y^-$, to obtain:
$[x \oplus (y \oplus x)^-]^- \oplus (y^- \oplus (x^- \oplus y^-)^-) = y \oplus (y^- \oplus (x^- \oplus y^-)^-)$,
which, by (2.28), becomes:
$[x \oplus (y \oplus x)^-]^- \oplus (y^- \oplus (x^- \oplus y^-)^-) = 1$,
which, by (Scomm), becomes:
$(y^- \oplus (x^- \oplus y^-)^-) \oplus [x \oplus (y \oplus x)^-]^- = 1$,
which, by (Sass), becomes:

$$(2.43) \qquad y^- \oplus ((x^- \oplus y^-)^- \oplus [x \oplus (y \oplus x)^-]^-) = 1.$$

Now, in (2.35), take $Y := y \oplus z$, to obtain:
$(x \oplus (x \oplus (y \oplus z))^-)^- \oplus [(y \oplus z)^- \oplus (x \oplus (x \oplus (y \oplus z))^-)^-]^- = y \oplus z$,
which, by (2.25), becomes:
(g) $(x \oplus (y \oplus (x \oplus z))^-)^- \oplus [(y \oplus z)^- \oplus (x \oplus (x \oplus (y \oplus z))^-)^-]^- = y \oplus z$;
then, in (g), take $X := (x^- \oplus y^-)^-$, $Y := y^-$, $Z := (x \oplus (y \oplus x)^-)^-$, to obtain:
$((x^- \oplus y^-)^- \oplus (y^- \oplus ((x^- \oplus y^-)^- \oplus (x \oplus (y \oplus x)^-)^-))^-)^- \oplus$
$[(y^- \oplus (x \oplus (y \oplus x)^-)^-)^- \oplus ((x^- \oplus y^-)^- \oplus ((x^- \oplus y^-)^- \oplus (y^- \oplus (x \oplus (y \oplus x)^-)^-))^-)^-]^-$
$= y^- \oplus (x \oplus (y \oplus x)^-)^-$,
which, by (2.25), becomes:
(g') $((x^- \oplus y^-)^- \oplus (y^- \oplus ((x^- \oplus y^-)^- \oplus (x \oplus (y \oplus x)^-)^-))^-)^- \oplus$
$[(y^- \oplus (x \oplus (y \oplus x)^-)^-)^- \oplus ((x^- \oplus y^-)^- \oplus (y^- \oplus ((x^- \oplus y^-)^- \oplus (x \oplus (y \oplus x)^-)^-))^-)^-]^-$
$= y^- \oplus (x \oplus (y \oplus x)^-)^-$;
but, in (g'), the two parts $y^- \oplus ((x^- \oplus y^-)^- \oplus (x \oplus (y \oplus x)^-)^-)$ equal 1, by (2.43);

hence, (g') becomes:
$((x^- \oplus y^-)^- \oplus 1^-)^- \oplus [(y^- \oplus (x \oplus (y \oplus x)^-)^-)^- \oplus ((x^- \oplus y^-)^- \oplus 1^-)^-]^-$
$= y^- \oplus (x \oplus (y \oplus x)^-)^-$,
which, by (Neg1-0), (SU), (DN), becomes:

$$(2.44)\quad (x^- \oplus y^-) \oplus [(y^- \oplus (x \oplus (y \oplus x)^-)^-)^- \oplus (x^- \oplus y^-)]^- = y^- \oplus (x \oplus (y \oplus x)^-)^-.$$

Now, in (2.35) again, take $Y := y \oplus (x \oplus y)^-$, to obtain:
(h) $(x \oplus (x \oplus [y \oplus (x \oplus y)^-])^-)^- \oplus$
$[[y \oplus (x \oplus y)^-]^- \oplus (x \oplus (x \oplus [y \oplus (x \oplus y)^-])^-)^-]^- = y \oplus (x \oplus y)^-$;
but, in (h), the two parts $x \oplus [y \oplus (x \oplus y)^-]$ equal 1, by (2.31); hence, (h) becomes:
$(x \oplus 1^-)^- \oplus [[y \oplus (x \oplus y)^-]^- \oplus (x \oplus 1^-)^-]^- = y \oplus (x \oplus y)^-$,
which, by (Neg1-0), (SU), (DN), becomes:
$x^- \oplus [[y \oplus (x \oplus y)^-]^- \oplus x^-]^- = y \oplus (x \oplus y)^-$,
which, by (Scomm), becomes:

$$(2.45)\qquad x^- \oplus [x^- \oplus [y \oplus (x \oplus y)^-]^-]^- = y \oplus (x \oplus y)^-.$$

Now, from (2.44), by (Scomm), we obtain:
$(x^- \oplus y^-) \oplus [(x^- \oplus y^-) \oplus (y^- \oplus (x \oplus (y \oplus x)^-)^-)^-]^- = y^- \oplus (x \oplus (y \oplus x)^-)^-$,
which, by (Sass), becomes:
(i) $(x^- \oplus y^-) \oplus [x^- \oplus (y^- \oplus (y^- \oplus (x \oplus (y \oplus x)^-)^-)^-)]^- = y^- \oplus (x \oplus (y \oplus x)^-)^-$;
but, in (i), the part $y^- \oplus (x \oplus (y \oplus x)^-)^-)^-$ equals $x \oplus (y \oplus x)^-$, by (2.45); then, (i) becomes:
(i ') $(x^- \oplus y^-) \oplus [x^- \oplus (y^- \oplus (x \oplus (y \oplus x)^-)]^- = y^- \oplus (x \oplus (y \oplus x)^-)^-$;
but, in (i '), the part $x^- \oplus (y^- \oplus (x \oplus (y \oplus x)^-)$, by (2.25) twice, becomes
$x^- \oplus (x \oplus (y^- \oplus (y \oplus x)^-)$ and then $x \oplus (x^- \oplus (y^- \oplus (y \oplus x)^-)$, which equals 1, by (2.28); hence, (i ') becomes:
$(x^- \oplus y^-) \oplus 1^- = y^- \oplus (x \oplus (y \oplus x)^-)^-$,
which, by (Neg1-0), (SU), becomes:
$x^- \oplus y^- = y^- \oplus (x \oplus (y \oplus x)^-)^-$, i.e.

$$(2.46)\qquad y^- \oplus (x \oplus (y \oplus x)^-)^- = x^- \oplus y^-.$$

Finally, from (2.40) $(x \oplus [y^- \oplus (x^- \oplus (y \oplus x^-)^-)^-]^- = y \oplus (x^- \oplus y)^-)$, by (2.46) for $X := x^-$, we obtain:
$x \oplus [x^= \oplus y^-]^- = y \oplus (x^- \oplus y)^-$,
which, by (DN), becomes:
$x \oplus (x \oplus y^-)^- = y \oplus (x^- \oplus y)^-$,
which, by (Scomm), becomes:
$y \oplus (y \oplus x^-)^- = x \oplus (x \oplus y^-)^-$, i.e. (2.26) holds. □

Corollary 2.2.10 *Let $\mathcal{A} = (A, \oplus, ^-, 0)$ be an involutive right-m-MEL algebra, with $1 \stackrel{def.}{=} 0^-$. Then,*

$$(\vee_m - comm) \iff (Vabs_m^M).$$

Proof. By Propositions 2.2.8 and 2.2.9. □

Remark 2.2.11 *The above Corollary 2.2.10 says that an equivalent definition of MV algebras is the following:*
A right-MV algebra is an involutive right-m-MEL algebra $\mathcal{A} = (A, \oplus, ^-, 0)$ *verifying: for all* $x, y \in A$,
($Vabs_m^M$*)* $x \vee_m^M (x \wedge_m^M y) = x$, *i.e.*

$$\mathbf{MV}^R \;=\; \mathbf{m} - \mathbf{MEL}^R_{(DN)} \;+\; (Vabs_m^M).$$

• Let $(A, \oplus, ^-, 0)$ be an involutive right-m-MEL algebra, with $1 \stackrel{def.}{=} 0^-$.
Consider the following two properties (see Definition 3.2.3):
($\vee_m$-comm) = (Vcomm$_m^M$) $x \vee_m^M y = y \vee_m^M x$
(commutativity of $\vee_m^M$),
(Vass$_m^M$) $(x \vee_m^M y) \vee_m^M z = x \vee_m^M (y \vee_m^M z)$
(associativity of $\vee_m^M$),
where:
$x \vee_m^M y \stackrel{def.}{=} y \oplus (y \oplus x^-)^-$.

Remarks 2.2.12 *We have:*
1. *($\vee_m$-comm)* $\Longrightarrow$ *(Vass$_m^M$)*
(the proof by PROVER9 *has Length 84 and is omitted).*
2. *(Vass$_m^M$) does not imply ($\vee_m$-comm),*
as the following example found by MACE4 *shows:*

x	0	2	1
x^-	1	2	0

and

$\oplus$	0	2	1
0	0	2	1
2	2	2	1
1	1	1	1

,

$\odot$	0	2	1
0	0	0	0
2	0	2	2
1	0	2	1

,

$\vee_m^M$	0	2	1
0	0	2	1
2	2	2	1
1	1	2	1

.

The property ($\vee_m$-comm) is not verified by $(x, y) = (1, 2)$ *(see the table of* $\vee_m^M$*).*

2.2.2 Definition of XY algebras. Cancellative XY algebras

We shall introduce now a new notion, which will be proved (Corollary 2.2.68) to be the 'ancestor' of MV algebras.

Definitions 2.2.13
(1) A *right-XY algebra* is an involutive right-m-MEL algebra
$\mathcal{A} = (A, \oplus, ^-, 0)$ verifying the additional axioms: for all $x, y, z \in A$,
(XX) $(x \oplus y) \odot ((x \odot y) \oplus z) = (y \oplus z) \odot ((y \odot z) \oplus x)$,
(YY) $x \oplus y = x \Longrightarrow x \odot (y \oplus z) = y \oplus (x \odot z)$.
(1') Dually, a *left-XY algebra* is an involutive left-m-MEL algebra $\mathcal{A} = (A, \odot, ^-, 1)$ verifying the additional axioms: for all $x, y, z \in A$,
(XXd) $(x \odot y) \oplus ((x \oplus y) \odot z) = (y \odot z) \oplus ((y \oplus z) \odot x)$,
(YYd) $x \odot y = x \Longrightarrow x \oplus (y \odot z) = y \odot (x \oplus z)$.

Note that:
- the property (XXd) appears (with a permutation) in ([10], Proposition 1.6.2) as property (1.16) verified by any MV algebra and the property (XX) appears, in [1] (see Lemma 6.3), as $\sigma_1(x,y,z) = \sigma_4(x,y,z)$;
- the property (YY) appears in [1] as Lemma 6.9.

Denote by $\mathbf{XY}^R$ the class of all right-XY algebras and by **XY** the class of all left-XY algebras. Note that these classes are not varieties.

We shall say that a XY algebra is *proper*, if it is not an MV algebra.

Note that a right-XY algebra does not verify the properties (m-Re^R), (m-An^R), (m-Tr^R) (hence, the binary relation $\leq_m$ is not reflexive, anti-symmetric or transitive), (m-B^R), (m-BB^R).

Note also that the properties (XX) and (YY) are independent, as the following examples prove (the non-commutative table of $\vee_m^M$ shows that $\mathcal{A}$ is not an MV algebra).

Example 2.2.14 Involutive right-m-MEL algebra verifying (XX) and not verifying (YY)
Consider the involutive right-m-MEL algebra $(A = \{0,2,3,1\}, \oplus, ^-, 0)$ with the following tables of $^-$, $\oplus$, $\odot$ and $\vee_m^M$:

x	0	2	3	1
x^-	1	2	3	0

and

$\oplus$	0	2	3	1
0	0	2	3	1
2	2	2	2	1
3	3	2	2	1
1	1	1	1	1

,

$\odot$	0	2	3	1
0	0	0	0	0
2	0	2	2	2
3	0	2	2	3
1	0	2	3	1

;

$\vee_m^M$	0	2	3	1
0	0	2	3	1
2	2	2	2	1
3	3	2	2	1
1	1	2	2	1

.

The property (YY) is not verified for $x = 2$, $y = 3$, $z = 0$.

Example 2.2.15 Involutive right-m-MEL algebra verifying (YY) and not verifying (XX)
Consider the involutive right-m-MEL algebra $(A = \{0,2,1\}, \oplus, ^-, 0)$ with the following tables of $^-$, $\oplus$, $\odot$ and $\vee_m^M$:

x	0	2	1
x^-	1	2	0

and

$\oplus$	0	2	1
0	0	2	1
2	2	0	1
1	1	1	1

,

$\odot$	0	2	1
0	0	0	0
2	0	1	2
1	0	2	1

;

$\vee_m^M$	0	2	1
0	0	2	1
2	2	1	1
1	1	0	1

.

The property (XX) is not verified for $x = 0$, $y = 2$, $z = 2$.

Note that it was a hard work to find that the *right-XY algebra* is that generalization of the right-MV algebra needed in the generalization of the Mundici's equivalence. After finding the properties (XX) and (YY) - able to assure the main properties of the set of good sequences, $G^+(A)$, according to [1] - a very difficult problem was to find which is the basic algebra: the (involutive) right-m-BCK algebra (the initial supposition), or the involutive right-m-BE algebra or, finally, the involutive right-m-MEL algebra. In order to solve this problem, we used PROVER9-MACE4 program for Windows and Michael Kinyon helped us with his most performant PROVER9-MACE4 program for Linux to check all kinds of variants.

- **Cancellative XY algebras**

An even better 'ancestor' (generalization) of MV algebras are those right-XY algebras verifying the cancellative property (C) (verified by any MV algebra).

Definition 2.2.16 (The dual one is omitted)
A right-XY algebra $\mathcal{A} = (A, \oplus, ^-, 0)$ is *cancellative*, if the following cancellative selfdual property (C) holds: for all $x, y, z \in A$,
(C) $z \oplus x = z \oplus y$ and $z \odot x = z \odot y$ imply $x = y$.

Denote by $\mathbf{CXY}^R$ the class of cancellative right-XY algebras. Hence, we have:

$$\mathbf{CXY}^R \subset \mathbf{XY}^R, \quad namely$$

$$\mathbf{CXY}^R = \mathbf{XY}^R + (C).$$

2.2.3 The property (Z) and the Z algebras

Let $\mathcal{A} = (A, \oplus, ^-, 0)$ be an involutive right-m-MEL algebra.

- **Consider the following property** (see Corollary 7.1.82 (ii), (iii)) (m=0):

(Z)=(Z^{00}) $(x \oplus y) \oplus (x \odot y) = x \oplus y$ and, dually,
(Zd) $(x \odot y) \odot (x \oplus y) = x \odot y$.

An alternative notation of (Z), which suggested the name (Z^{00}), is (see Corollary 7.1.82 (ii)):
$(x_0 \oplus y_0) \oplus (x_0 \odot y_0) = x_0 \oplus y_0$.

Note that the property (Z) appears as the property (1.14) verified by any MV algebra, in ([10], Proposition 1.6.2), and appears in [1] (its dual also), in the proof of Lemma 6.10.

Related to the property (Z), we shall introduce a new particular involutive m-MEL algebra, called *Z algebra*, as follows.

Definitions 2.2.17
(1) A *right-Z algebra* is an involutive right-m-MEL algebra $\mathcal{A} = (A, \oplus, ^-, 0)$ verifying the additional axiom (Z).
(1') Dually, a *left-Z algebra* is an involutive left-m-MEL algebra $\mathcal{A} = (A, \odot, ^-, 1)$ verifying the additional axiom (Zd).

Denote by $\mathbf{Z}^R$ the class of all right-Z algebras and by $\mathbf{Z}$ the class of all left-Z algebras. Note that these classes are varieties. We have:

$$\mathbf{Z}^R = \textbf{m-MEL}^R_{(DN)} + (\mathrm{Z}).$$

We shall say that an Z algebra is *proper*, if it is not an XY algebra or an MV algebra.

Proposition 2.2.18 *We have:*

$$(DN) \implies ((Z) \Leftrightarrow (Zd)).$$

Proof. Since $x \odot y = (x^- \oplus y^-)^-$,
the property (Zd) $(x \odot y) \odot (x \oplus y) = x \odot y$
$\Longleftrightarrow$ $(x^- \oplus y^-)^- \odot (x^- \odot y^-)^- = (x^- \oplus y^-)^-$
$\Longleftrightarrow$ $((x^- \oplus y^-)^= \oplus (x^- \odot y^-)^=)^- = (x^- \oplus y^-)^-$
$\Longleftrightarrow$ $(x^- \oplus y^-) \oplus (x^- \odot y^-) = x^- \oplus y^-$, by (DN).

Now, in (Z) $((x \oplus y) \oplus (x \odot y) = x \oplus y)$, take $X := x^-$, $Y := y^-$, to obtain: $(x^- \oplus y^-) \oplus (x^- \odot y^-) = x^- \oplus y^-$, i.e. (Zd).

Conversely, in (Zd) $((x^- \oplus y^-) \oplus (x^- \odot y^-) = x^- \oplus y^-)$, take $X := x^-$, $Y := y^-$, to obtain:
$(x^= \oplus y^=) \oplus (x^= \odot y^=) = x^= \oplus y^=$,
which, by (DN), becomes:
$(x \oplus y) \oplus (x \odot y) = x \oplus y$, i.e (Z). □

Proposition 2.2.19 *If $\mathcal{A} = (A, \oplus, ^-, 0)$ is a right-Z algebra, with $1 \stackrel{def.}{=} 0^-$, then $(A, \odot, ^-, 1)$ is a left-Z algebra.*

Proof. By Propositions 2.1.3, 2.2.18. □

The property (Z) (= (Z^{00})) was considered from the begining of this research, but it was refound in Corollary 7.1.82 (ii). Consequently, the ideea came of the following new properties ((Z^{10}), (Z^{20}) etc. from Corollaries 7.1.85 (ii), 7.1.90 (ii) etc.) introduced next.

• **Consider the following new property (the dual one is omitted)** (see Corollary 7.1.85 (ii)) (m=1):
(Z^{10}) $x \oplus y = x \Longrightarrow x \oplus y \oplus z = (x \oplus z) \oplus (x \odot (y \oplus z)) \oplus (y \odot z)$.

An alternative notation, suggesting the name (Z^{10}) is:
$x_0 \oplus x_1 = x_0 \Longrightarrow x_0 \oplus x_1 \oplus y_0 = (x_0 \oplus y_0) \oplus (x_0 \odot (x_1 \oplus y_0)) \oplus (x_1 \odot y_0)$.

Proposition 2.2.20 *We have:*

$$(Z) + (Scomm) + (Sass) \implies (Z^{10}).$$

Proof. By PROVER9, Length of proof is 16, in 0.11 seconds.

Suppose we have:

$$c1 \oplus c2 = c1. \tag{2.47}$$

We must prove that (since $c1 \oplus c2 \oplus c3 = c1 \oplus c3$, by (Sass) and (2.47)):
$c1 \oplus c3 = (c1 \oplus c3) \oplus (c1 \odot (c2 \oplus c3)) \oplus (c2 \odot c3)$,
which, by (Sass), becomes:

$$c1 \oplus c3 = c1 \oplus (c3 \oplus [(c1 \odot (c2 \oplus c3)) \oplus (c2 \odot c3)]). \tag{2.48}$$

Indeed, first, from (Z) ($(x \oplus y) \oplus (x \odot y) = x \oplus y$), by (Sass), we obtain:

$$x \oplus (y \oplus (x \odot y)) = x \oplus y; \tag{2.49}$$

then, in (2.49), take $Y := y \oplus z$, to obtain:
$x \oplus ((y \oplus z) \oplus (x \odot (y \oplus z))) = x \oplus (y \oplus z)$,
which, by (Sass), becomes:

$$x \oplus (y \oplus (z \oplus (x \odot (y \oplus z)))) = x \oplus (y \oplus z). \tag{2.50}$$

Now, in (Sass) ($(x \oplus y) \oplus z = x \oplus (y \oplus z)$), take $X := c1$, $Y := c2$, to obtain:
$(c1 \oplus c2) \oplus z = c1 \oplus (c2 \oplus z)$,
which, by (2.47), becomes:

$$c1 \oplus z = c1 \oplus (c2 \oplus z). \tag{2.51}$$

Now, in (2.51), take $Z := z \oplus (c2 \odot z)$, to obtain:
(a) $c1 \oplus (z \oplus (c2 \odot z)) = c1 \oplus (c2 \oplus (z \oplus (c2 \odot z)))$;
but, in (a), the part $c2 \oplus (z \oplus (c2 \odot z))$ equals $c2 \oplus z$, by (2.49); hence, (a) becomes:
$c1 \oplus (z \oplus (c2 \odot z)) = c1 \oplus (c2 \oplus z)$,
which, by (2.51), becomes:

$$c1 \oplus (z \oplus (c2 \odot z)) = c1 \oplus z. \tag{2.52}$$

Now, in (2.51), take $Z := z \oplus (c1 \odot (c2 \oplus z))$, to obtain:
(b) $c1 \oplus (z \oplus (c1 \odot (c2 \oplus z))) = c1 \oplus (c2 \oplus (z \oplus (c1 \odot (c2 \oplus z))))$;
but, in (b), the right side equals $c1 \oplus (c2 \oplus z)$, by (2.50); hence, (b) becomes:
(c) $c1 \oplus (z \oplus (c1 \odot (c2 \oplus z))) = c1 \oplus (c2 \oplus z)$;
but, in (c), the right part equals $c1 \oplus z$, by (2.51); hence, (c) becomes:

$$c1 \oplus (z \oplus (c1 \odot (c2 \oplus z))) = c1 \oplus z. \tag{2.53}$$

Now, from (2.53), by adding y to both sides, we obtain:
$[c1 \oplus (z \oplus (c1 \odot (c2 \oplus z)))] \oplus y = (c1 \oplus z) \oplus y$,
which, by (Sass) three times, becomes:

$$c1 \oplus [z \oplus ((c1 \odot (c2 \oplus z)) \oplus y)] = c1 \oplus (z \oplus y). \tag{2.54}$$

Finally, in (2.54), take $Z := c3$, $Y := c2 \odot c3$, to obtain:
(d) $c1 \oplus [c3 \oplus ((c1 \odot (c2 \oplus c3)) \oplus (c2 \odot c3))] = c1 \oplus (c3 \oplus (c2 \odot c3))$;
but, in (d), the right side equals $c1 \oplus c3$, by (2.52); hence, (d) becomes:
$c1 \oplus [c3 \oplus ((c1 \odot (c2 \oplus c3)) \oplus (c2 \odot c3))] = c1 \oplus c3$, i.e. (2.48) holds. □

Lemma 2.2.21 *We have:*
(Z^{10}) + (SU) + (Sass) + (m-La) $\Longrightarrow$ (Z).

Proof. In (Z^{10}) ($x \oplus y = x \Longrightarrow x \oplus y \oplus z = (x \oplus z) \oplus (x \odot (y \oplus z)) \oplus (y \odot z)$), take $Y := 0$, to obtain:
$x \oplus 0 = x \Longrightarrow x \oplus 0 \oplus z = (x \oplus z) \oplus (x \odot (0 \oplus z)) \oplus (0 \odot z)$,
which, by (SU), (m-La), becomes:
$x = x \Longrightarrow x \oplus z = (x \oplus z) \oplus (x \odot z)$, i.e. (Z). □

Corollary 2.2.22 *Let $\mathcal{A} = (A, \oplus, ^-, 0)$ be a right-Z algebra. Then, (Z^{10}) $\Longleftrightarrow$ (Z).*

Proof. By above Proposition 2.2.20 and Lemma 2.2.21. □

• **Consider the following new property (the dual one is omitted)** (see Corollary 7.1.90 (ii)) (m=2):
(Z^{20}) $x \oplus y = x,\ y \oplus z = y \Longrightarrow x \oplus y \oplus z \oplus t = (x \oplus t) \oplus (x \odot (y \oplus t)) \oplus (y \odot (z \oplus t)) \oplus (z \odot t)$.

An alternative notation, suggesting the name (Z^{20}), is:
$x_0 \oplus x_1 = x_0,\ x_1 \oplus x_2 = x_1 \Longrightarrow$
$x_0 \oplus x_1 \oplus x_2 \oplus y_0 = (x_0 \oplus y_0) \oplus (x_0 \odot (x_1 \oplus y_0)) \oplus (x_1 \odot (x_2 \oplus y_0)) \oplus (x_2 \odot y_0)$.

Proposition 2.2.23 *We have:*

$$(Z) + (Scomm) + (Pcomm) + (Sass) \Longrightarrow (Z^{20}).$$

Proof. By PROVER9, Length of proof is 34, in 1.67 seconds.
Suppose we have:

$$c1 \oplus c2 = c1 \tag{2.55}$$

and we have:

$$c2 \oplus c3 = c2. \tag{2.56}$$

We must prove that (since $c1 \oplus c2 \oplus c3 \oplus t = c1 \oplus t$, by (Sass) and (2.55), (2.56)):
$c1 \oplus t = (c1 \oplus t) \oplus (c1 \odot (c2 \oplus t)) \oplus (c2 \odot (c3 \oplus t)) \oplus (c3 \odot t)$,
which, by (Sass), (Scomm), (Pcomm), becomes:
$t \oplus c1 = [(c1 \oplus t) \oplus (c3 \odot t)] \oplus [(c1 \odot (c2 \oplus t)) \oplus (c2 \odot (c3 \oplus t))]$, then:
$t \oplus c1 = (t \oplus c1) \oplus [(t \odot c3) \oplus [(c1 \odot (t \oplus c2)) \oplus (c2 \odot (t \oplus c3))]]$, finally:

$$t \oplus (c1 \oplus [(t \odot c3) \oplus [(c1 \odot (t \oplus c2)) \oplus (c2 \odot (t \oplus c3))]]) = t \oplus c1. \tag{2.57}$$

Indeed, first, from (Z) ($(x \oplus y) \oplus (x \odot y) = x \oplus y$), by (Sass), we obtain:

$$x \oplus (y \oplus (x \odot y)) = x \oplus y. \tag{2.58}$$

Then, in (Sass) ($(x \oplus y) \oplus z = x \oplus (y \oplus z)$), take $Y := y \oplus (x \odot y)$, to obtain:
(a) $(x \oplus (y \oplus (x \odot y))) \oplus z = x \oplus ((y \oplus (x \odot y)) \oplus z)$;

but, in (a), the part $x \oplus (y \oplus (x \odot y))$ equals $x \oplus y$, by (2.58); hence, (a) becomes:
$(x \oplus y) \oplus z = x \oplus ((y \oplus (x \odot y)) \oplus z)$,
which, by (Sass), becomes:

$$x \oplus (y \oplus z) = x \oplus ((y \oplus (x \odot y)) \oplus z). \tag{2.59}$$

Now, in (2.58), take $Y := y \oplus z$, to obtain:
$x \oplus ((y \oplus z) \oplus (x \odot (y \oplus z))) = x \oplus (y \oplus z)$,
which, by (Sass), becomes:

$$x \oplus (y \oplus [z \oplus (x \odot (y \oplus z))]) = x \oplus (y \oplus z). \tag{2.60}$$

Now, in (Sass) ($(x \oplus y) \oplus z = x \oplus (y \oplus z)$), take $X := c1$, $Y := c2$, to obtain:
$(c1 \oplus c2) \oplus z = c1 \oplus (c2 \oplus z)$,
which, by (2.55), becomes:

$$c1 \oplus z = c1 \oplus (c2 \oplus z). \tag{2.61}$$

Similarly, in (Sass) ($(x \oplus y) \oplus z = x \oplus (y \oplus z)$), take $X := c2$, $Y := c3$, to obtain, by (2.56):

$$c2 \oplus z = c2 \oplus (c3 \oplus z). \tag{2.62}$$

Note that we have, by (Sass), (Scomm):

$$x \oplus (y \oplus z) = y \oplus (x \oplus z). \tag{2.63}$$

Now, in (2.63), take $Y := c1$, $Z := c2 \oplus y$, to obtain:
$x \oplus (c1 \oplus (c2 \oplus y)) = c1 \oplus (x \oplus (c2 \oplus y))$,
which, by (2.61), becomes:

$$x \oplus (c1 \oplus y) = c1 \oplus (x \oplus (c2 \oplus y)). \tag{2.64}$$

Similarly, in (2.63), take $Y := c2$, $Z := c3 \oplus y$, to obtain:
$x \oplus (c2 \oplus (c3 \oplus y)) = c2 \oplus (x \oplus (c3 \oplus y))$,
which, by (2.62), becomes:

$$x \oplus (c2 \oplus y) = c2 \oplus (x \oplus (c3 \oplus y)). \tag{2.65}$$

Now, in (2.63) again, take $Y := c1$, $Z := c2$, to obtain:
$x \oplus (c1 \oplus c2) = c1 \oplus (x \oplus c2)$,
which, by (2.55), becomes:

$$x \oplus c1 = c1 \oplus (x \oplus c2). \tag{2.66}$$

Similarly, in (2.63), take $Y := c2$, $Z := c3$, to obtain, by (2.56):

$$x \oplus c2 = c2 \oplus (x \oplus c3). \tag{2.67}$$

Note that $c1 \oplus c2 \oplus c3 = c1 = c1 \oplus c3$, by (Sass) and (2.55), (2.56), hence we have:

$$c1 = c1 \oplus c3. \tag{2.68}$$

Now, in (Sass) ($(x \oplus y) \oplus z = x \oplus (y \oplus z)$), take $Y := c1$, $Z := c3$, to obtain:
$(x \oplus c1) \oplus c3 = x \oplus (c1 \oplus c3)$,
which, by (Scomm), (2.68), becomes:

$$c3 \oplus (x \oplus c1) = x \oplus c1. \tag{2.69}$$

Now, in (2.59), take $Y := c3$, $Z := c1$, to obtain:
$x \oplus (c3 \oplus c1) = x \oplus ((c3 \oplus (x \odot c3)) \oplus c1)$,
which, by (2.68) and (2.69), becomes:
$x \oplus c1 = x \oplus ((x \odot c3) \oplus c1)$,
which, by (Scomm), becomes:

$$x \oplus c1 = x \oplus (c1 \oplus (x \odot c3)). \tag{2.70}$$

Now, in (Sass) ($(x \oplus y) \oplus z = x \oplus (y \oplus z)$), take $Y := c1 \oplus (x \odot c3)$, $Z := y$, to obtain:
$(x \oplus [c1 \oplus (x \odot c3)]) \oplus y = x \oplus ([c1 \oplus (x \odot c3)] \oplus y)$,
which, by (2.70), becomes:
$(x \oplus c1) \oplus y = x \oplus ([c1 \oplus (x \odot c3)] \oplus y)$,
which, by (Sass), becomes:

$$x \oplus (c1 \oplus y) = x \oplus (c1 \oplus [(x \odot c3) \oplus y]). \tag{2.71}$$

Now, in (2.60), take $X := c1$, $Z := c2$, to obtain:
$c1 \oplus (y \oplus [c2 \oplus (c1 \odot (y \oplus c2))]) = c1 \oplus (y \oplus c2)$,
which, by (2.64) and (2.66), becomes:

$$y \oplus (c1 \oplus (c1 \odot (y \oplus c2))) = y \oplus c1. \tag{2.72}$$

Now, in (Sass) ($(x \oplus y) \oplus z = x \oplus (y \oplus z)$), take $Y := c1 \oplus (c1 \odot (x \oplus c2))$, $Z := y$, to obtain:
$(x \oplus [c1 \oplus (c1 \odot (x \oplus c2))]) \oplus y = x \oplus ([c1 \oplus (c1 \odot (x \oplus c2))] \oplus y)$,
which, by (2.72), becomes:
$(x \oplus c1) \oplus y = x \oplus ([c1 \oplus (c1 \odot (x \oplus c2))] \oplus y)$,
which, by (Sass), becomes:

$$x \oplus (c1 \oplus y) = x \oplus (c1 \oplus [(c1 \odot (x \oplus c2)) \oplus y]). \tag{2.73}$$

Now, in (2.60) again, take $X := c2$, $Z := c3$, to obtain:
$c2 \oplus (y \oplus [c3 \oplus (c2 \odot (y \oplus c3))]) = c2 \oplus (y \oplus c3)$,
which, by (2.65) on the left side and by (2.67) on the right side, becomes:

$$y \oplus (c2 \oplus (c2 \odot (y \oplus c3))) = y \oplus c2. \tag{2.74}$$

Now, in (2.64), take $Y := c2 \odot (x \oplus c3)$, to obtain:
$x \oplus (c1 \oplus [c2 \odot (x \oplus c3)]) = c1 \oplus (x \oplus (c2 \oplus [c2 \odot (x \oplus c3)]))$,
which, by (2.74) on the right side, becomes:
$x \oplus (c1 \oplus [c2 \odot (x \oplus c3)]) = c1 \oplus (x \oplus c2)$,
which, by (2.66) on the right side again, becomes:

$$x \oplus (c1 \oplus [c2 \odot (x \oplus c3)]) = x \oplus c1. \tag{2.75}$$

Finally, we shall prove that (2.57) holds.
Indeed, the left side of (2.57) becomes, by (2.71):

$$t \oplus (c1 \oplus [(c1 \odot (t \oplus c2)) \oplus (c2 \odot (t \oplus c3))]) \tag{2.76}$$

which becomes, by (2.73):

$$t \oplus (c1 \oplus (c2 \odot (t \oplus c3))) \tag{2.77}$$

which, by (2.75), equals $t \oplus c1$. □

Lemma 2.2.24 *We have:*

$$(Z^{20}) + (SU) + (Sass) + (m\text{-}La) \Longrightarrow (Z).$$

Proof. In (Z^{20}) $(x \oplus y = x,\ y \oplus z = y \Longrightarrow x \oplus y \oplus z \oplus t = (x \oplus t) \oplus (x \odot (y \oplus t)) \oplus (y \odot (z \oplus t)) \oplus (z \odot t))$, take $Y := 0$, $Z := 0$, to obtain:
$x \oplus 0 = x,\ y \oplus 0 = y \Longrightarrow x \oplus 0 \oplus 0 \oplus t = (x \oplus t) \oplus (x \odot (0 \oplus t)) \oplus (0 \odot (0 \oplus t)) \oplus (z \odot t)$,
which, by (SU), (Sass), (m-La), becomes:
$x = x,\ y = y \Longrightarrow x \oplus t = (x \oplus t) \oplus (x \odot t)$, i.e. (Z). □

Corollary 2.2.25 *Let $\mathcal{A} = (A, \oplus, ^-, 0)$ be a right-Z algebra. Then, $(Z^{20}) \Longleftrightarrow (Z)$.*

Proof. By above Proposition 2.2.23 and Lemma 2.2.24. □

• **One can consider further the properties ($\mathbf{Z^{30}}$), ($\mathbf{Z^{40}}$) etc. Consider finally the following new property (the dual one is omitted)** (see Corollary 7.1.92 (ii)) (m=n):
(Z^{n0}) $x_i \oplus x_{i+1} = x_i,\ i = 0, 1, \ldots, n-1 \Longrightarrow$
$$x_0 \oplus x_1 \oplus \ldots \oplus x_n \oplus y_0 = c_0 \oplus c_1 \oplus \ldots \oplus c_n \oplus c_{n+1},$$
where $c_i = x_{i-1} \odot (x_i \oplus y_0)$, $i = 0, 1, \ldots, n+1$, with $x_{-1} = 1$.

Open problem 2.2.26 *It is an open problem to obtain a direct proof that (Z^{n0}) holds in any right-Z algebra. An indirect proof is given by the proof of Corollary 7.1.92 (ii).*

Lemma 2.2.27 *We have:*

$$(Z^{n0}) + (SU) + (Sass) + (m\text{-}La) \Longrightarrow (Z).$$

Proof. In (Z^{n0}), take $x_i = 0$, for $i = 1, 2, \dots, n$, to obtain, by (SU), (Sass), (m-La):
$x_0 = x_0 \Longrightarrow x_0 \oplus y_0 = (x_0 \oplus y_0) \oplus (x_0 \odot y_0)$, i.e. (Z), since:
$x_0 \oplus x_1 \oplus \dots \oplus x_n \oplus y_0 = x_0 \oplus y_0$ and
$c_0 = x_0 \oplus y_0$, $c_1 = x_0 \odot y_0$, $c_i = 0$, for $i = 2, 3, \dots, n+1$, hence
$c_0 \oplus c_1 \oplus \dots \oplus c_n \oplus c_{n+1} = c_0 \oplus c_1 = (x_0 \oplus y_0) \oplus (x_0 \odot y_0)$. □

Thus, we obtain:

Corollary 2.2.28 *Let $\mathcal{A} = (A, \oplus, ^-, 0)$ be a right-Z algebra. Then, (Z^{n0}) $\Longleftrightarrow$ (Z).*

By Corollaries 2.2.22, 2.2.25, 2.2.28, we obtain a resuming result:

Corollary 2.2.29 *Let $\mathcal{A} = (A, \oplus, ^-, 0)$ be a right-Z algebra. Then,*

$$(Z^{10}) \Longleftrightarrow (Z^{20}) \Longleftrightarrow \dots \Longleftrightarrow (Z^{n0}) \Longleftrightarrow (Z).$$

• **Consider now the following new property (the dual one is omitted)** (see Corollary 7.1.94 (ii)):
(Z^{11}) $x \oplus y = x$, $z \oplus v = z \Longrightarrow$
$x \oplus y \oplus z \oplus v = (x \oplus z) \oplus ((x \oplus v) \odot (y \oplus z)) \oplus (x \odot z \odot (y \oplus v)) \oplus (y \odot v)$.

An alternative notation, suggesting the name (Z^{11}), is:
$x_0 \oplus x_1 = x_0$, $y_0 \oplus y_1 = y_0 \Longrightarrow$
$x_0 \oplus x_1 \oplus y_0 \oplus y_1 = (x_0 \oplus y_0) \oplus ((x_0 \oplus y_1) \odot (x_1 \oplus y_0)) \oplus (x_0 \odot y_0 \odot (x_1 \oplus y_1)) \oplus (x_1 \odot y_1)$.

Proposition 2.2.30 *We have:*

$$(Z) + (Scomm) + (Sass) + (DN) \Longrightarrow (Z^{11}).$$

The proof by PROVER9 (Length of proof is 49, in 42.59 seconds) is omitted. □

Lemma 2.2.31 *We have:*
(Z^{11}) + (SU) + (Sass) + (m-La) $\Longrightarrow$ (Z).

Proof. In (Z^{11}), take $x_1 := 0$, $y_1 = 0$, to obtain:
$x_0 \oplus 0 = x_0$, $y_0 \oplus 0 = y_0 \Longrightarrow$
$x_0 \oplus 0 \oplus y_0 \oplus 0 = (x_0 \oplus y_0) \oplus ((x_0 \oplus 0) \odot (0 \oplus y_0)) \oplus (x_0 \odot y_0 \odot (0 \oplus 0)) \oplus (0 \odot 0)$,
which, by (SU), (m-La), (Sass), becomes:
$x_0 = x_0$, $y_0 = y_0 \Longrightarrow$
$x_0 \oplus y_0 = (x_0 \oplus y_0) \oplus (x_0 \odot y_0)$, i.e. (Z). □

Corollary 2.2.32 *Let $\mathcal{A} = (A, \oplus, ^-, 0)$ be a right-Z algebra. Then, (Z^{11}) $\Longleftrightarrow$ (Z).*

Proof. By above Proposition 2.2.30 and Lemma 2.2.31. □

Note that one could consider similarly the properties ($\mathbf{Z}^{22}$), ($\mathbf{Z}^{33}$), ..., ($\mathbf{Z}^{nn}$) with similar results: their equivalence with (Z).

• Totally ordered Z algebras

By Definitions 2.1.7, a right-Z algebra $\mathcal{A}$ is *totally ordered by* $\leq_m$ if, for all $x, y \in A$, we have:

$$(tot - ord - m) \qquad x \leq_m y \quad or \quad y \leq_m x,$$

i.e. $y \oplus x^- = 1$ or $x \oplus y^- = 1$.

There exist proper right-Z algebras that are totally ordered by $\leq_m$ (hence $\leq_m$ is reflexive), as the following examples prove.

Examples 2.2.33 Proper Z algebras totally ordered by $\leq_m$

Example 1 (the smallest example found by MACE4)

Consider the right-Z algebra $(A = \{0, 2, 3, 1\}, \oplus, ^-, 0)$, with the following tables of $^-$, $\oplus$, $\odot$ and $\vee_m^M$:

x:	0	2	3	1
x^-	1	3	2	0

and

$\oplus$	0	2	3	1
0	0	2	3	1
2	2	2	1	1
3	3	1	1	1
1	1	1	1	1

,

$\odot$	0	2	3	1
0	0	0	0	0
2	0	0	0	2
3	0	0	3	3
1	0	2	3	1

;

$\vee_m^M$	0	2	3	1
0	0	2	3	1
2	2	2	3	1
3	3	1	3	1
1	1	1	1	1

.

It is a proper right-Z algebra, i.e. it does not verify (XX) for $(x, y, z) = (2, 2, 3)$, (YY) for $(x, y, z) = (2, 2, 0)$; it is not an MV algebra for $(x, y) = (2, 3)$: $3 = 2 \vee_m^M 3 \neq 3 \vee_m^M 2 = 1$ (see the above table of $\vee_m^M$).

It is totally ordered by $\leq_m$: $0 \leq_m 2 \leq_m 2 \leq_m 3 \leq_m 3 \leq_m 1$.

Example 2

Consider the algebra $(A = \{0, 2, 3, 4, 1\}, \oplus, ^-, 0)$, with the following tables of $^-$, $\oplus$, $\odot$ and $\vee_m^M$:

x	0	2	3	4	1
x^-	1	2	4	3	0

and

$\oplus$	0	2	3	4	1
0	0	2	3	4	1
2	2	1	2	1	1
3	3	2	3	1	1
4	4	1	1	1	1
1	1	1	1	1	1

,

$\odot$	0	2	3	4	1
0	0	0	0	0	0
2	0	0	0	2	2
3	0	0	0	0	3
4	0	2	0	4	4
1	0	2	3	4	1

;

$\vee_m^M$	0	2	3	4	1
0	0	2	3	4	1
2	2	2	2	4	1
3	3	2	3	4	1
4	4	1	1	4	1
1	1	1	1	1	1

.

It is a proper right-Z algebra, i.e. it does not verify (XX) for $(x, y, z) = (3, 4, 4)$, (YY) for $(x, y, z) = (2, 3, 3)$; it is not an MV algebra for $(x, y) = (4, 3)$: $1 = 4 \vee_m^M 3 \neq 3 \vee_m^M 4 = 4$ (see the above table of $\vee_m^M$).

It is totally ordered by $\leq_m$: see the Hasse-type diagram (Remarks 2.1.12) in Figure 2.1 and recall that $\leq_m$ is reflexive.

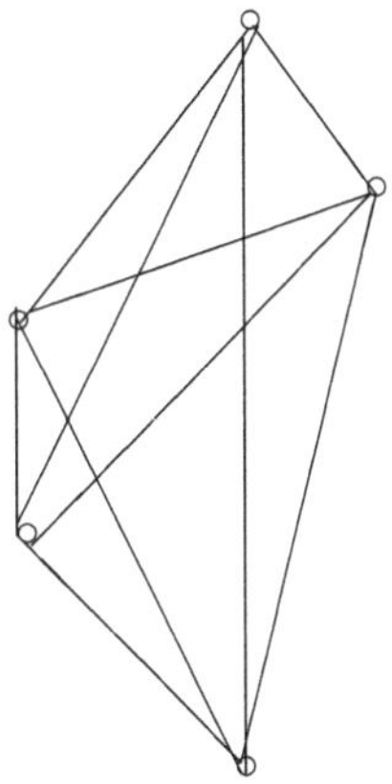

Figure 2.1: The Hasse-type diagram of $(A, \leq_m)$

There exist proper right-Z algebras that are not totally ordered by $\leq_m$, as the following examples prove.

Examples 2.2.34 Proper Z algebras not totally ordered by $\leq_m$

Example 1

x	0	2	3	4	1
x^-	1	2	4	3	0

and

$\oplus$	0	2	3	4	1
0	0	2	3	4	1
2	2	2	1	4	1
3	3	1	1	1	1
4	4	4	1	2	1
1	1	1	1	1	1

,

$\odot$	0	2	3	4	1
0	0	0	0	0	0
2	0	2	3	0	2
3	0	3	2	0	3
4	0	0	0	0	4
1	0	2	3	4	1

;

$\vee_m^M$	0	2	3	4	1
0	0	2	3	4	1
2	2	2	3	1	1
3	3	1	3	4	1
4	4	2	3	4	1
1	1	2	1	1	1

.

It is a proper right-Z algebra, i.e. it does not verify (XX) for $(x, y, z) = (2, 2, 3)$ and (YY) for $(x, y, z) = (2, 2, 3)$; it is not an MV algebra, because $1 = 2 \vee_m^M 4 \neq 4 \vee_m^M 2 = 2$ (see the above table of $\vee_m^M$).
It is not totally ordered by $\leq_m$ (see the Hasse-type diagram (Remarks 2.1.12) in Figure 2.2), because $\leq_m$ is not reflexive: there exists (only) 2 such that $2 \not\leq_m 2$ $(2 \oplus 2 = 2 \neq 1)$.

Example 2
Consider the right-Z algebra $(A = \{0, 2, 3, 4, 1\}, \oplus, ^-, 0)$, with the following tables of $^-$, $\oplus$, $\odot$ and $\vee_m^M$:

x	0	2	3	4	1
x^-	1	2	4	3	0

and

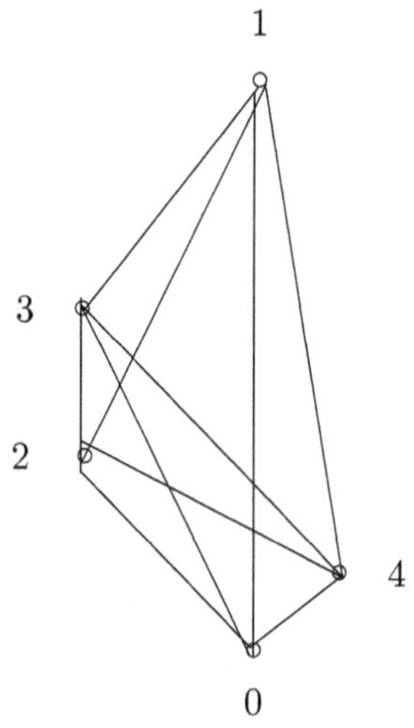

Figure 2.2: The Hasse-type diagram of $(A, \leq_m)$

$\oplus$	0	2	3	4	1
0	0	2	3	4	1
2	2	2	3	3	1
3	3	3	3	3	1
4	4	3	3	4	1
1	1	1	1	1	1

,

$\odot$	0	2	3	4	1
0	0	0	0	0	0
2	0	2	4	4	2
3	0	4	3	4	3
4	0	4	4	4	4
1	0	2	3	4	1

;

$\vee_m^M$	0	2	3	4	1
0	0	2	3	4	1
2	2	2	3	4	1
3	3	3	3	3	1
4	4	3	3	4	1
1	1	2	3	3	1

.

It is a proper right-Z algebra, i.e. it does not verify (XX) for $(x, y, z) = (2, 2, 3)$ and (YY) for $(x, y, z) = (2, 2, 3)$; it is not an MV algebra for $(1, 2)$, because $2 = 1 \vee_m^M 2 \neq 2 \vee_m^M 1 = 1$ (see the above table of $\vee_m^M$).
It is not totally ordered by $\leq_m$, since there exists $(x, y) = (2, 3)$ such that $2 \not\leq_m 3$ and $3 \not\leq_m 2$ ($3 \oplus 2^- = 3 \neq 1$ and $2 \oplus 3^- = 3 \neq 1$).

By Definitions 2.1.7 also, a right-Z algebra $\mathcal{A}$ is *totally ordered by* $\leq_m^M$ if, for all $x, y \in A$, we have:

$$(tot - ord - m - M) \qquad x \leq_m^M y \quad or \quad y \leq_m^M x,$$

i.e. $y \vee_m^M x = y$ or $x \vee_m^M y = x$, where $x \vee_m^M y \stackrel{def.}{=} y \oplus (y^- \odot x)$ and $x \odot y \stackrel{def.}{=} (x^- \oplus y^-)^-$, by (2.1).

Corollary 2.2.35 *If a right-Z algebra $\mathcal{A}$ is totally ordered by $\leq_m^M$, then it is an MV algebra.*

Proof. By Proposition 2.1.9, since $\mathcal{A}$ is an involutive right-m-MEL algebra. □

2.2.4 Properties of XY algebras

We shall present some properties obtained from definition.

Proposition 2.2.36 *Let $\mathcal{A} = (A, \oplus, ^-, 0)$ be an involutive right-m-MEL algebra. Then,*

$$(XX) \Longrightarrow (Z).$$

Proof. In (XX) $((x \oplus y) \odot ((x \odot y) \oplus z) = (y \oplus z) \odot ((y \odot z) \oplus x))$, take $Z := 0$, to obtain:
$(x \oplus y) \odot ((x \odot y) \oplus 0) = (y \oplus 0) \odot ((y \odot 0) \oplus x)$,
which, by (SU), (m-La) (see Proposition 2.1.3), becomes:
$(x \oplus y) \odot (x \odot y) = y \odot x$,
which, by (Pcomm), becomes:
$(x \odot y) \odot (x \oplus y) = x \odot y$, i.e. (Zd);
then, by Proposition 2.2.18, (Z) holds. □

Corollary 2.2.37 *Let $\mathcal{A} = (A, \oplus, ^-, 0)$ be a right-XY algebra. Then, (Z) and (Zd) hold, i.e. $\mathcal{A}$ is a right-Z algebra.*

Proof. By above Propositions 2.2.36 and 2.2.18. □

Remark 2.2.38 *(See Remark 7.1.83)*

We have seen (Proposition 2.2.36 and Corollary 2.2.37) a direct simple proof that (Z) holds in any right-XY algebra. An alternative simple proof can be seen in Corollary 7.1.82 (ii).

Corollary 2.2.39 *Let $\mathcal{A} = (A, \oplus, ^-, 0)$ be a right-XY algebra. Then, (Z^{10}) holds.*

Proof. By Corollary 2.2.37 and Proposition 2.2.20. □

Remark 2.2.40 *(See Remark 7.1.86)*

We have seen (Proposition 2.2.20 and Corollary 2.2.39) a direct complicated proof that (Z^{10}) holds in any right-XY algebra. An alternative simple proof can be seen in Corollary 7.1.85 (ii). More, we have seen that $(Z^{10}) \Longleftrightarrow (Z)$ (Corollary 2.2.22).

Corollary 2.2.41 *Let $\mathcal{A} = (A, \oplus, ^-, 0)$ be a right-XY algebra. Then, (Z^{20}) holds.*

Proof. By Proposition 2.2.23. □

Remark 2.2.42 *(See Remark 7.1.91)*

We have seen (Proposition 2.2.23 and Corollary 2.2.41) a direct more complicated proof that (Z^{20}) holds in any right-XY algebra. An alternative simple proof can be seen in Corollary 7.1.90 (ii). More, we have seen that $(Z^{20}) \Longleftrightarrow (Z)$ (Corollary 2.2.25).

Remark 2.2.43 *(See Remark 7.1.93)*

It is an open problem to give a direct proof that (Z^{n0}) holds in any right-XY algebra. An alternative proof can be seen in Corollary 7.1.92 (ii). More, we have seen that $(Z^{n0}) \Longleftrightarrow (Z)$ (Corollary 2.2.28).

Corollary 2.2.44 *Let $\mathcal{A} = (A, \oplus, ^-, 0)$ be a right-XY algebra. Then, (Z^{11}) holds.*

Proof. By Proposition 2.2.30. □

Remark 2.2.45 *(See Remark 7.1.95)*
We have (Proposition 2.2.30 and Corollary 2.2.44) a direct even more complicated proof that (Z^{11}) holds in any right-XY algebra. An alternative simple proof can be seen in Corollary 7.1.94 (ii). More, we have seen that (Z^{11}) $\Longleftrightarrow$ (Z) (Corollary 2.2.32).

*

Remark 2.2.46 *Consider the property (XX):*
(XX) $(x \oplus y) \odot ((x \odot y) \oplus z) = (y \oplus z) \odot ((y \odot z) \oplus x)$.
If we make on (XX) the following permutation: x, y, z to y, z, x, respectively, then we obtain:
(XX_1) $(y \oplus z) \odot ((y \odot z) \oplus x) = (z \oplus x) \odot ((z \odot x) \oplus y)$.
It follows that we also have:
(XX_2) $(x \oplus y) \odot ((x \odot y) \oplus z) = (z \oplus x) \odot ((z \odot x) \oplus y)$.

Notation 2.2.47 *(See ([1], 6.2))*
We set:

$$\sigma_1(x, y, z) \overset{notation}{=} (x \oplus y) \odot ((x \odot y) \oplus z)$$

and, dually,

$$\sigma_2(x, y, z) \overset{notation}{=} (x \odot y) \oplus ((x \oplus y) \odot z).$$

Remarks 2.2.48
(1) The property (XX) means, by above notations, $\sigma_1(x, y, z) = \sigma_1(y, z, x)$ and, by Remark 2.2.46, we have:

$$\sigma_1(x, y, z) = \sigma_1(y, z, x) = \sigma_1(z, x, y).$$

(2) The dual property (XXd) means, by above notations, $\sigma_2(x, y, z) = \sigma_2(y, z, x)$ and, by a Remark dual to Remark 2.2.46, we have:

$$\sigma_2(x, y, z) = \sigma_2(y, z, x) = \sigma_2(z, x, y).$$

Lemma 2.2.49 *(See ([1], Lemma 6.3))*
Let $\mathcal{A} = (A, \oplus, ^-, 0)$ be a right-XY algebra. For all $i, j \in \{1, 2\}$, every permutation $p : \{1, 2, 3\} \longrightarrow \{1, 2, 3\}$ and all $x_1, x_2, x_3 \in A$, we have:

$$\sigma_i(x_1, x_2, x_3) = \sigma_j(x_{p(1)}, x_{p(2)}, x_{p(3)}),$$

i.e. the terms σ_1, σ_2 are all invariant under permutations of variables, and they coincide.

Proof. Recall that there are $(3! = 6)$ six permutations of the set $\{1, 2, 3\}$; written as tuples, they are (1, 2, 3), (1, 3, 2), (2, 1, 3), (2, 3, 1), (3, 1, 2), (3, 2, 1). This means that there are six permutations of the set $\{x, y, z\}$; written as tuples, they are (x, y, z), (x, z, y), (y, x, z), (y, z, x), (z, x, y), (z, y, x).

(1) By Remarks 2.2.48, we have: $\sigma_1(x, y, z) = \sigma_1(y, z, x) = \sigma_1(z, x, y)$ and by (Scomm), (Pcomm), we obtain:
$(\sigma_1(x, y, z) = \sigma_1(x, z, y)) = (\sigma_1(y, x, z) = \sigma_1(y, z, x)) = (\sigma_1(z, x, y) = \sigma_1(z, y, x))$,
i.e. the term σ_1 is invariant under permutations of variables.

(2) We prove now that $\sigma_1(x, y, z) = \sigma_2(x, y, z)$. Indeed,
in (YY) $(x \oplus y = x \Longrightarrow x \odot (y \oplus z) = y \oplus (x \odot z))$, take $X := x \oplus y$, $Y := z$, $Z := u$, to obtain:
(a) $(x \oplus y) \oplus z = x \oplus y \Longrightarrow (x \oplus y) \odot (z \oplus u) = z \oplus ((x \oplus y) \odot u)$;
now, in (a), take $Z := x \odot y$, $U := z$, to obtain:
(b) $(x \oplus y) \oplus (x \odot y) = x \oplus y \Longrightarrow (x \oplus y) \odot ((x \odot y) \oplus z) = (x \odot y) \oplus ((x \oplus y) \odot z)$;
but, in (b), the term $(x \oplus y) \oplus (x \odot y)$ equals $x \oplus y$, by (Z); hence, (b) becomes:
(b') $x \oplus y = x \oplus y \Longrightarrow (x \oplus y) \odot ((x \odot y) \oplus z) = (x \odot y) \oplus ((x \oplus y) \odot z)$,
hence we obtain:
$(x \oplus y) \odot ((x \odot y) \oplus z) = (x \odot y) \oplus ((x \oplus y) \odot z)$,
which means $\sigma_1(x, y, z) = \sigma_2(x, y, z)$.

(3) By above (1) and (2), it follows that the term σ_2 is invariant under permutations of variables too. Thus, the terms σ_1, σ_2 are invariant under permutations of variables, and they coincide. □

Remark 2.2.50 *(See [1])*
By above Lemma 2.2.49, the terms σ_1, σ_2 are invariant under permutations of variables, and they coincide. For all $x, y, z \in A$, we shall denote by $\sigma(x, y, z)$ the common above value.

Corollary 2.2.51 *Let $\mathcal{A} = (A, \oplus, ^-, 0)$ be a right-XY algebra. Then,*

$$(XX) \Longleftrightarrow (XXd).$$

Proof. By Remarks 2.2.48 and by Lemma 2.2.49. □

Proposition 2.2.52

$$(DN) \Longrightarrow ((YY) \Leftrightarrow (YYd)).$$

Proof. (YY) is $x \oplus y = x \Longrightarrow x \odot (y \oplus z) = y \oplus (x \odot z)$ and
(YYd) is $\quad x \odot y = x \Longrightarrow x \oplus (y \odot z) = y \odot (x \oplus z)$
$\Longleftrightarrow \quad (x^- \oplus y^-)^- = x \Longrightarrow x \oplus (y \odot z) = y \odot (x \oplus z)$
$\Longleftrightarrow \quad x^- \oplus y^- = x^- \Longrightarrow x \oplus (y \odot z) = y \odot (x \oplus z)$, by (DN).

Now, in (YY), take $X := x^-$, $Y := y^-$, $Z := z^-$, to obtain:
(a) $x^- \oplus y^- = x^- \Longrightarrow x^- \odot (y^- \oplus z^-) = y^- \oplus (x^- \odot z^-)$;
but, in (a), the part $x^- \odot (y^- \oplus z^-) = y^- \oplus (x^- \odot z^-)$ implies
$(x^- \odot (y^- \oplus z^-))^- = (y^- \oplus (x^- \odot z^-))^-$,
hence (a) becomes:

(b) $x^- \oplus y^- = x^- \Longrightarrow (x^- \odot (y^- \oplus z^-))^- = (y^- \oplus (x^- \odot z^-))^-$,
which becomes, by (DN):
$(x \odot y)^- = x^- \Longrightarrow x \oplus (y^- \oplus z^-)^- = y \odot (x^- \odot z^-)^-$,
which becomes, by (DN):
$x \odot y = x \Longrightarrow x \oplus (y \odot z) = y \odot (x \oplus z)$, i.e. (YYd).

Conversely, in (YYd) ($x^- \oplus y^- = x^- \Longrightarrow x \oplus (y \odot z) = y \odot (x \oplus z)$), take $X := x^-$, $Y := y^-$, $Z := z^-$, to obtain:
(a') $x^= \oplus y^= = x^= \Longrightarrow x^- \oplus (y^- \odot z^-) = y^- \odot (x^- \oplus z^-)$;
but, in (a'), the part $x^- \oplus (y^- \odot z^-) = y^- \odot (x^- \oplus z^-)$ implies
$(x^- \oplus (y^- \odot z^-))^- = (y^- \odot (x^- \oplus z^-))^-$,
hence (a') becomes, by (DN):
(b') $x \oplus y = x \Longrightarrow (x^- \oplus (y^- \odot z^-))^- = (y^- \odot (x^- \oplus z^-))^-$,
which becomes, by (DN):
$x \oplus y = x \Longrightarrow x \odot (y^- \odot z^-)^- = y \oplus (x^- \oplus z^-)^-$,
which becomes:
$x \oplus y = x \Longrightarrow x \odot (y \oplus z) = y \oplus (x \odot z)$, i.e. (YY). □

Then, we can prove the following result.

Proposition 2.2.53 *If $\mathcal{A} = (A, \oplus, ^-, 0)$ is a right-XY algebra, then $(A, \odot, ^-, 1)$ is a left-XY algebra, where $x \odot y = (x^- \oplus y^-)^-$, $1 = 0^-$.*

Proof. By Proposition 2.1.3, Corollary 2.2.51, Proposition 2.2.52. □

Proposition 2.2.54 *For all $x, y \in A$,*

$$x \oplus y = x \iff x \odot y = y.$$

Proof. Suppose $a \oplus b = a$; we must prove that $a \odot b = b$.
Indeed, from (YY) ($x \oplus y = x \Longrightarrow x \odot (y \oplus z) = y \oplus (x \odot z)$), since $a \oplus b = a$, we obtain:
$a \odot (b \oplus z) = b \oplus (a \odot z)$,
which, for $z := 0$, becomes:
$a \odot (b \oplus 0) = b \oplus (a \odot 0)$,
which, by (SU), (m-La), becomes:
$a \odot b = b$.

Conversely, suppose $a \odot b = b$, i.e. $(a^- \oplus b^-)^- = b$, hence $a^- \oplus b^- = b^-$, by (DN), hence $b^- \oplus a^- = b^-$, by (Scomm); we must prove that $a \oplus b = a$.
Indeed, from (YY) ($x \oplus y = x \Longrightarrow x \odot (y \oplus z) = y \oplus (x \odot z)$), since $b^- \oplus a^- = b^-$, we obtain:
$b^- \odot (a^- \oplus z) = a^- \oplus (b^- \odot z)$,
which, for $z := 0$, becomes:
$b^- \odot (a^- \oplus 0) = a^- \oplus (b^- \odot 0)$,
which, by (SU), (m-La), becomes:
$b^- \odot a^- = a^-$,
which, by definition of $\oplus$ from $\odot$, becomes:

$(b^{=} \oplus a^{=})^{-} = a^{-}$, which, by (DN), becomes:
$(b \oplus a)^{-} = a^{-}$,
which, by (DN) again, becomes:
$b \oplus a = a$, hence $a \oplus b = a$, by (Scomm). □

Proposition 2.2.55 *For all $x, y \in A$,*

$$x \oplus y = x \iff x^{-} \oplus y^{-} = y^{-}.$$

Proof. By Proposition 2.2.54, since $x \odot y = y \iff (x^{-} \oplus y^{-})^{-} = y \iff x^{-} \oplus y^{-} = y^{-}$, by (DN). □

Proposition 2.2.56 *Let $\mathcal{A} = (A, \oplus, ^{-}, 0)$ be a right-XY algebra. For all $x \in A$,*

$$0 \leq_m x \leq_m 1 \quad and \quad 0 \leq_m^M x.$$

Proof. By Proposition 2.1.5. □

Note that, in any XY algebra, $\leq_m$ is not reflexive, antisymmetric or transitive, in general.

2.2.5 Totally ordered proper XY algebras do not exist

We shall prove here that totally ordered proper XY algebras do not exist: they are MV algebras.

By Definitions 2.1.7, a right-XY algebra $\mathcal{A}$ is *totally ordered by* $\leq_m$ if, for all $x, y \in A$, we have:

$$(tot - ord - m) \qquad x \leq_m y \quad or \quad y \leq_m x,$$

i.e. $y \oplus x^{-} = 1$ or $x \oplus y^{-} = 1$.

Note that if a right-XY algebra $\mathcal{A}$ is *totally ordered by* $\leq_m$, then $x \oplus x^{-} = 1$, for all $x \in A$, i.e. (m-Re^R) holds. Hence, we immediately obtain the following result.

Proposition 2.2.57 *If a right-XY algebra $\mathcal{A}$ is totally ordered by $\leq_m$, then it is an MV algebra.*

Proof. By Corollary 2.2.67. □

By Definitions 2.1.7 also, a right-XY algebra $\mathcal{A}$ is *totally ordered by* $\leq_m^M$ if, for all $x, y \in A$, we have:

$$(tot - ord - m - M) \qquad x \leq_m^M y \quad or \quad y \leq_m^M x,$$

i.e. $y \vee_m^M x = y$ or $x \vee_m^M y = x$, where $x \vee_m^M y \overset{def.}{=} y \oplus (y^{-} \odot x)$ and $x \odot y \overset{def.}{=} (x^{-} \oplus y^{-})^{-}$, by (2.1).

Corollary 2.2.58 *If a right-XY algebra $\mathcal{A}$ is totally ordered by $\leq_m^M$, then it is an MV algebra.*

Proof. By Proposition 2.1.9. □

We conclude, by Proposition 2.2.57 and Corollary 2.2.58, that all proper XY algebras are not totally ordered.

2.2.6 Connections between Z algebras, XY algebras and MV algebras

We know now that any XY algebra is an Z algebra (Corollary 2.2.37) and that any Z algebra verifying (XX) and (YY) is an XY algebra. We write:

$$\mathbf{XY}^R \subseteq \mathbf{Z}^R, \qquad \mathbf{Z}^R + (\mathrm{XX}) + (\mathrm{YY}) = \mathbf{XY}^R.$$

The goal of this subsection, reached in Corollary 2.2.68, is to prove that the XY algebra is an ancestor of the MV algebra.

• **First, we shall prove, in Theorem 2.2.63 below, that any MV algebra is an XY algebra.**
Indeed, any right-MV algebra is an involutive right-m-MEL algebra, by definition, and we shall see next that it verifies the properties (Z), (XX), (YY).

Proposition 2.2.59 *Let $\mathcal{A} = (A, \oplus, ^-, 0)$ be a right-MV algebra. Then, the property (Z) holds, where:*
(Z) $\quad (x \oplus y) \oplus (x \odot y) = x \oplus y.$

Proof. (Based on a proof by PROVER 9, Length of proof was 25; the length of the expanded proof was 49.)

Since $\mathcal{A}$ is an MV algebra, then (SU), (Scomm), (Sass), (Neg0-1), (m-La^R), (DN), ($\vee_m$-comm) and (Neg1-0), (m-Re^R) hold.

First, since $x \oplus (y \oplus z) \overset{(Sass)}{=} (x \oplus y) \oplus z \overset{(Scomm)}{=} (y \oplus x) \oplus z \overset{(Sass)}{=} y \oplus (x \oplus z)$, it follows that we have:

$$x \oplus (y \oplus z) = y \oplus (x \oplus z). \tag{2.78}$$

Then, since $x \oplus (y \oplus x^-) \overset{(2.78)}{=} y \oplus (x \oplus x^-) \overset{(m-Re^R)}{=} y \oplus 1 \overset{(m-La^R)}{=} 1$, it follows that we have:

$$x \oplus (y \oplus x^-) = 1. \tag{2.79}$$

Note that ($\vee_m$-comm) ($x \vee_m^M y = y \vee_m^M x$) means that $y \oplus (y \oplus x^-)^- = x \oplus (x \oplus y^-)^-$, i.e. we have:

$$x \oplus (x \oplus y^-)^- = y \oplus (y \oplus x^-)^-. \tag{2.80}$$

Then, from (2.80), by (Scomm), we obtain:

$$x \oplus (y^- \oplus x)^- = y \oplus (y \oplus x^-)^-. \tag{2.81}$$

Now, in (2.80), take $Y := x \oplus y^-$, to obtain:
$x \oplus (x \oplus (x \oplus y^-)^-)^- = (x \oplus y^-) \oplus ((x \oplus y^-) \oplus x^-)^-$,
which, by (2.81), becomes:
$x \oplus (y \oplus (x^- \oplus y)^-)^- = (x \oplus y^-) \oplus ((x \oplus y^-) \oplus x^-)^-$,

which, by (Sass), becomes:
$x \oplus (y \oplus (x^- \oplus y)^-)^- = (x \oplus y^-) \oplus (x \oplus (y^- \oplus x^-))^-$,
which, by (2.79), becomes:
$x \oplus (y \oplus (x^- \oplus y)^-)^- = (x \oplus y^-) \oplus 1^-$,
which, by (Neg1-0), (SU), becomes:

$$x \oplus (y \oplus (x^- \oplus y)^-)^- = x \oplus y^-. \tag{2.82}$$

Then, in (2.78), take $Y := z$, $Z := (y^- \oplus z)^-$, to obtain:
$x \oplus (z \oplus (y^- \oplus z)^-) = z \oplus (x \oplus (y^- \oplus z)^-)$,
which, by (2.81), becomes:
$x \oplus (y \oplus (y \oplus z^-)^-) = z \oplus (x \oplus (y^- \oplus z)^-)$, i.e.

$$x \oplus (y \oplus (z^- \oplus x)^-) = y \oplus (z \oplus (z \oplus x^-)^-). \tag{2.83}$$

Then, in (2.83), take $X := (x^- \oplus y^-)^-$, $Y := x$, $Z := y$, to obtain:
$(x^- \oplus y^-)^- \oplus (x \oplus (y^- \oplus (x^- \oplus y^-)^-)^-) = x \oplus (y \oplus (y \oplus (x^- \oplus y^-)^=)^-)$,
which, by (2.82) for $Y := y^-$, becomes:
$(x^- \oplus y^-)^- \oplus (x \oplus y^=) = x \oplus (y \oplus (y \oplus (x^- \oplus y^-)^=)^-)$,
which, by (DN), becomes:
$(x^- \oplus y^-)^- \oplus (x \oplus y) = x \oplus (y \oplus (y \oplus (x^- \oplus y^-))^-)$,
which, by (2.79), becomes:
$(x^- \oplus y^-)^- \oplus (x \oplus y) = x \oplus (y \oplus 1^-)$,
which, by (Neg1-0) and (SU), becomes:
$(x^- \oplus y^-)^- \oplus (x \oplus y) = x \oplus y$,
which, by (Scomm) and the definition of $\odot$, becomes:
$(x \oplus y) \oplus (x \odot y) = x \oplus y$, i.e. (Z) holds. □

Proposition 2.2.60 *Let $\mathcal{A} = (A, \oplus, ^-, 0)$ be a right-MV algebra. Then, the property (XX) holds, where:*
(XX) $\quad (x \oplus y) \odot ((x \odot y) \oplus z) = (y \oplus z) \odot ((y \odot z) \oplus x)$.

Proof 1. By ([10], Proposition 1.6.2 (1.16)), the dual property (XXd) holds. Hence, (XX) holds, by Corollary 2.2.51.

Proof 2. A direct proof by PROVER9 on Linux Operating System, made by Michael Kinyon for me, is very long (the Length of the proof was 152, the length of the expanded proof was 427) and is omitted.

Proof 3. A direct 'hybrid' proof, based on a proof by PROVER9 (Length of the proof was 92, the length of the expanded proof was 280), which uses the above first equivalent definition of MV algebras, is presented in **Appendix C**. □

Corollary 2.2.61 *Let $\mathcal{A} = (A, \oplus, ^-, 0)$ be a right-MV algebra. Then,*

$$(XX) \implies (Z).$$

Proof. By Proposition 2.2.36. □

Proposition 2.2.62 *Let $\mathcal{A} = (A, \oplus, ^-, 0)$ be a right-MV algebra. Then, the property (YY) holds, where:*
(YY) $\quad x \oplus y = x$ *implies* $x \odot (y \oplus z) = y \oplus (x \odot z)$.

Proof. (Based on a proof by PROVER9, Length of proof was 47.)
Let $(A, \oplus, ^-, 0)$ be an MV algebra, hence (SU), (Scomm), (Sass), (m-La^R), (Neg0-1), (DN), ($\vee_m$-comm) and also (Neg1-0), (m-Re^R) hold. Since $x \vee_m^M y \stackrel{def.}{=} y \oplus (y \oplus x^-)^-$, then ($\vee_m$-comm) means:

(2.84)
$$y \oplus (y \oplus x^-)^- = x \oplus (x \oplus y^-)^-.$$

We must prove that (YY) holds, i.e. $x \oplus y = x \Longrightarrow x \odot (y \oplus z) = y \oplus (x \odot z)$; since $x \odot y \stackrel{def.}{=} (x^- \oplus y^-)^-$, it follows that we must prove:
(YY) $x \oplus y = x \Longrightarrow (x^- \oplus (y \oplus z)^-)^- = y \oplus (x^- \oplus z^-)^-$.
Suppose we have:

(2.85)
$$c1 \oplus c2 = c1;$$

we must prove:

(2.86)
$$(c1^- \oplus (c2 \oplus c3)^-)^- = c2 \oplus (c1^- \oplus c3^-)^-.$$

Indeed, first, in (Sass) ($(x \oplus y) \oplus z = x \oplus (y \oplus z)$), take first $X := c1$, $Y := c2$ to obtain, by (2.85):

(2.87)
$$c1 \oplus z = c1 \oplus (c2 \oplus z),$$

then, in (Sass) also, take $Y := c1$, $Z := c2$, to obtain, by (2.85):

(2.88)
$$(x \oplus c1) \oplus c2 = x \oplus c1.$$

Now, in (2.84), take $Y := x \oplus y^-$ to obtain:
(a) $(x \oplus y^-) \oplus ((x \oplus y^-) \oplus x^-)^- = x \oplus (x \oplus (x \oplus y^-)^-)^-$;
but, in (a), the part on the left side $(x \oplus y^-) \oplus x^-$ equals 1, by (Sass), (Scomm), (m-Re^R), (m-La^R), and the part on the right side $x \oplus (x \oplus y^-)^-$ equals $y \oplus (y \oplus x^-)^-$, by (2.84); hence, (a) becomes:
$(x \oplus y^-) \oplus 1^- = x \oplus (y \oplus (y \oplus x^-)^-)^-$,
which, by (Neg1-0), (SU), becomes:

(2.89)
$$x \oplus (y \oplus (y \oplus x^-)^-)^- = x \oplus y^-.$$

Now, in (2.89), take $X := x^-$, to obtain, by (DN):
(b) $x^- \oplus (y \oplus (y \oplus x)^-)^- = x^- \oplus y^-$;
then, in (b), take $X := c2 \oplus x$, $Y := c1$, to obtain:
$(c2 \oplus x)^- \oplus (c1 \oplus (c1 \oplus (c2 \oplus x))^-)^- = (c2 \oplus x)^- \oplus c1^-$,
which, by (2.87), becomes:

(2.90)
$$(c2 \oplus x)^- \oplus (c1 \oplus (c1 \oplus x)^-)^- = (c2 \oplus x)^- \oplus c1^-.$$

Now, in (Sass) also, take $Z := ((x \oplus y) \oplus z^-)^-$, to obtain:
$(x \oplus y) \oplus ((x \oplus y) \oplus z^-)^- = x \oplus (y \oplus ((x \oplus y) \oplus z^-)^-)$,
which, by (2.84) on left side, becomes:
(c) $z \oplus (z \oplus (x \oplus y)^-)^- = x \oplus (y \oplus ((x \oplus y) \oplus z^-)^-)$;
then, in (c) take $Z := x \oplus (y \oplus z^-)$, to obtain:
(c') $[x \oplus (y \oplus z^-)] \oplus ([x \oplus (y \oplus z^-)] \oplus (x \oplus y)^-)^- = x \oplus (y \oplus ((x \oplus y) \oplus [x \oplus (y \oplus z^-)]^-)^-)$;
but, in (c'), the part $(x \oplus y) \oplus [x \oplus (y \oplus z^-)]^-$ equals $z \oplus (z \oplus (x \oplus y)^-)^-$, by (c); hence, (c') becomes:
(c") $[x \oplus (y \oplus z^-)] \oplus ([x \oplus (y \oplus z^-)] \oplus (x \oplus y)^-)^- = x \oplus (y \oplus (z \oplus (z \oplus (x \oplus y)^-)^-)^-)$;
but, in (c"), the part $[x \oplus (y \oplus z^-)] \oplus (x \oplus y)^-$ equals 1, by (Sass), (Scomm), (m-Re^R), (m-La^R); hence, (c") becomes:
$[x \oplus (y \oplus z^-)] \oplus 1^- = x \oplus (y \oplus (z \oplus (z \oplus (x \oplus y)^-)^-)^-)$,
which, by (Neg1-0), (SU), becomes:

$$x \oplus (y \oplus z^-) = x \oplus (y \oplus (z \oplus (z \oplus (x \oplus y)^-)^-)^-). \tag{2.91}$$

Now, in (2.84), take $X := x \oplus y$, to obtain:
(d) $y \oplus (y \oplus (x \oplus y)^-)^- = (x \oplus y) \oplus ((x \oplus y) \oplus y^-)^-$;
but, in (d), the part $(x \oplus y) \oplus y^-$ equals 1, by (Sass), (m-Re^R), (m-La^R); hence, (d) becomes, by (Neg1-0), (SU):

$$y \oplus (y \oplus (x \oplus y)^-)^- = x \oplus y. \tag{2.92}$$

Now, in (2.84) again, take $X := (x^- \oplus y^-)^-$, to obtain, by (DN):
(e) $y \oplus (y \oplus (x^- \oplus y^-))^- = (x^- \oplus y^-)^- \oplus ((x^- \oplus y^-)^- \oplus y^-)^-$;
but, in (e), the part on left side $y \oplus (x^- \oplus y^-)$ equals 1, by (Scomm), (Sass), (m-Re^R), (m-La^R), and the part on right side $(x^- \oplus y^-)^- \oplus y^-$ equals $x \oplus (x \oplus y^{=})^-$, by (2.84); hence, (e) becomes, by (Neg1-0), (SU) and (DN):

$$(x^- \oplus y^-)^- \oplus (x \oplus (x \oplus y)^-)^- = y. \tag{2.93}$$

Then, in (2.93), interchange x with y to obtain:
$(y^- \oplus x^-)^- \oplus (y \oplus (y \oplus x)^-)^- = x$,
which, by taking $X := x^-$, becomes, by (DN):

$$(y^- \oplus x)^- \oplus (y \oplus (y \oplus x^-)^-)^- = x^-. \tag{2.94}$$

Now, in (2.93), take $Y := c2$, $X := y \oplus c1$, to obtain:
(f) $((y \oplus c1)^- \oplus c2^-)^- \oplus ((y \oplus c1) \oplus ((y \oplus c1) \oplus c2)^-)^- = c2$;
but, in (f), the part $(y \oplus c1) \oplus c2$ equals $y \oplus c1$, by (2.88); hence, (f) becomes:
$((y \oplus c1)^- \oplus c2^-)^- \oplus ((y \oplus c1) \oplus (y \oplus c1)^-)^- = c2$,
which, by (m-Re^R), (Neg1-0), (SU), becomes:
$((y \oplus c1)^- \oplus c2^-)^- = c2$,
which, by (DN), becomes:
(f ') $(y \oplus c1)^- \oplus c2^- = c2^-$;
then, in (f '), take $Y := (c1 \oplus x^-)^-$, to obtain:
$((c1 \oplus x^-)^- \oplus c1)^- \oplus c2^- = c2^-$,
which, by (2.84), becomes:

$$(x \oplus (x \oplus c1^-)^-)^- \oplus c2^- = c2^-. \tag{2.95}$$

Now, in (2.91), take $Y := (c1^- \oplus x)^-$, $Z := c2^-$, to obtain, by (DN):
(h) $x \oplus ((c1^- \oplus x)^- \oplus c2) = x \oplus ((c1^- \oplus x)^- \oplus (c2^- \oplus (c2^- \oplus (x \oplus (c1^- \oplus x)^-)^-)^-)^-)$;
but, in (h), the part $c2^- \oplus (x \oplus (c1^- \oplus x)^-)^-$ equals $c2^-$, by (2.95); hence, (h) becomes, by (DN):
$x \oplus ((c1^- \oplus x)^- \oplus c2) = x \oplus ((c1^- \oplus x)^- \oplus (c2^- \oplus c2)^-)$,
which, by (m-Re^R), (Neg1-0), (SU), becomes:

$$x \oplus ((c1^- \oplus x)^- \oplus c2) = x \oplus (c1^- \oplus x)^-. \tag{2.96}$$

Now, in (2.89), take $Y := c2 \oplus (c1^- \oplus x^-)^-$, to obtain:
(i) $x \oplus ([c2 \oplus (c1^- \oplus x^-)^-] \oplus ([c2 \oplus (c1^- \oplus x^-)^-] \oplus x^-)^-)^- = x \oplus [c2 \oplus (c1^- \oplus x^-)^-]^-$;
but, in (i), the part $[c2 \oplus (c1^- \oplus x^-)^-] \oplus x^-$ equals $x^- \oplus (c1^- \oplus x^-)^-$, by (2.96), which equals $c1 \oplus (c1 \oplus x^=)^-$, by (2.84); hence, (i) becomes, by (DN):
$x \oplus ([c2 \oplus (c1^- \oplus x^-)^-] \oplus (c1 \oplus (c1 \oplus x)^-)^-)^- = x \oplus [c2 \oplus (c1^- \oplus x^-)^-]^-$,
which, by (Sass), becomes:
(i') $x \oplus (c2 \oplus [(c1^- \oplus x^-)^- \oplus (c1 \oplus (c1 \oplus x)^-)^-])^- = x \oplus [c2 \oplus (c1^- \oplus x^-)^-]^-$;
but, in (i'), the part $(c1^- \oplus x^-)^- \oplus (c1 \oplus (c1 \oplus x)^-)^-$ equals x, by (2.93); hence, (i') becomes:

$$x \oplus (c2 \oplus x)^- = x \oplus [c2 \oplus (c1^- \oplus x^-)^-]^-. \tag{2.97}$$

Finally, in (2.94), take $X := c2 \oplus (c1^- \oplus x^-)^-$, $Y := x$, to obtain:
(j) $(x^- \oplus [c2 \oplus (c1^- \oplus x^-)^-])^- \oplus (x \oplus (x \oplus [c2 \oplus (c1^- \oplus x^-)^-]^-)^-)^- = [c2 \oplus (c1^- \oplus x^-)^-]^-$;
but, in (j), the part $x^- \oplus [c2 \oplus (c1^- \oplus x^-)^-]$ equals $x^- \oplus (c1^- \oplus x^-)^-$, by (2.96), which equals $c1 \oplus (c1 \oplus x^=)^-$, by (2.84); hence, (j) becomes, by (DN):
(j') $(c1 \oplus (c1 \oplus x)^-)^- \oplus (x \oplus (x \oplus [c2 \oplus (c1^- \oplus x^-)^-]^-)^-)^- = [c2 \oplus (c1^- \oplus x^-)^-]^-$;
but, in (j'), the part $x \oplus [c2 \oplus (c1^- \oplus x^-)^-]^-$ equals $x \oplus (c2 \oplus x)^-$, by 2.97); hence, (j') becomes: 120
(j") $(c1 \oplus (c1 \oplus x)^-)^- \oplus (x \oplus (x \oplus (c2 \oplus x)^-)^-)^- = [c2 \oplus (c1^- \oplus x^-)^-]^-$;
but, in (j"), the part $x \oplus (x \oplus (c2 \oplus x)^-)^-$ equals $c2 \oplus x$, by (2.92); hence, (j") becomes:
$(c1 \oplus (c1 \oplus x)^-)^- \oplus (c2 \oplus x)^- = [c2 \oplus (c1^- \oplus x^-)^-]^-$,
which, by (Scomm), becomes:
(j"') $(c2 \oplus x)^- \oplus (c1 \oplus (c1 \oplus x)^-)^- = [c2 \oplus (c1^- \oplus x^-)^-]^-$;
but, in (j"'), the left part equals $(c2 \oplus x)^- \oplus c1^-$, by (2.90); hence, (j"') becomes:
$(c2 \oplus x)^- \oplus c1^- = [c2 \oplus (c1^- \oplus x^-)^-]^-$,
which, by (Scomm) and (DN), becomes:
$(c1^- \oplus (c2 \oplus x)^-)^- = c2 \oplus (c1^- \oplus x^-)^-$,
which, for $X := c3$, becomes:
$(c1^- \oplus (c2 \oplus c3)^-)^- = c2 \oplus (c1^- \oplus c3^-)^-$, i.e. (2.86) holds. □

Theorem 2.2.63 *Any right-MV algebra is a cancellative right-XY algebra.*

Proof. Let $\mathcal{A} = (A, \oplus, ^-, 0)$ be a right-MV algebra. Then, it is cancellative. Then, $\mathcal{A}$ is an involutive right-m-MEL algebra, by Remarks 2.2.2. It verifies (XX) and (YY), by Propositions 2.2.60 and 2.2.62. Hence, $\mathcal{A}$ is a cancellative right-XY algebra. □

We write:

$$\mathbf{MV}^R \subseteq \mathbf{CXY}^R.$$

Consider the following property of compatibility of $\leq_m$: for all $x, y, z \in A$,
(Cp) (compatibility) $x \leq_m y \Longrightarrow x \oplus z \leq_m y \oplus z$.

While any MV algebra verifies this property (Cp), an XY algebra does not verify it, as the following example, found by MACE4, shows.

Example 2.2.64 Consider the right-XY algebra $\mathcal{A} = (A, \oplus, ^-, 0)$, $1 \overset{def.}{=} 0^-$, with the following tables (the non-commutative table of $\vee_m^M$ shows that $\mathcal{A}$ is not an MV algebra):

x:	0	2	3	1
x^-	1	3	2	0

and

$\oplus$	0	2	3	1
0	0	2	3	1
2	2	0	3	1
3	3	3	1	1
1	1	1	1	1

,

$\odot$	0	2	3	1
0	0	0	0	0
2	0	0	2	2
3	0	2	1	3
1	0	2	3	1

,

$\vee_m^M$	0	2	3	1
0	0	2	3	1
2	2	0	3	1
3	3	1	3	1
1	1	3	3	1

.

For $x = 0$, $y = 2$, $z = 2$, we have: $x \leq_m y$, since $y \oplus x^- = 1$ $(2 \oplus 0^- = 1)$, but $x \oplus z \not\leq_m y \oplus z$, since $(y \oplus z) \oplus (x \oplus z)^- \neq 1$ $((2 \oplus 2) \oplus (0 \oplus 2)^- = 3 \neq 1)$. Hence, the property (Cp) is not verified.

• **Second, we shall prove, in Corollary 2.2.68 below, that right-XY algebras verifying (m-Re^R) coincide with right-MV algebras.**

Indeed, we prove first the following result.

Proposition 2.2.65 *In any involutive right-m-MEL algebra, we have:*

$$(XX) + (m - Re^R) \Longrightarrow (\vee_m - comm).$$

Proof. (Based on the proof by PROVER9, Length of proof was 39, in 15.19 seconds.)

We have to prove that $x \vee_m^M y = y \vee_m^M x$, i.e. $y \oplus (y^- \odot x)^- = x \oplus (x^- \odot y)^-$; since $x \odot y = (x^- \oplus y^-)^-$, it follows that we must prove that:

$$y \oplus (y \oplus x^-)^- = x \oplus (x \oplus y^-)^-. \tag{2.98}$$

The property (XX) is used in the proof both under the initial form:

$$(x \oplus y) \odot ((x \odot y) \oplus z) = (y \oplus z) \odot ((y \odot z) \oplus x) \tag{2.99}$$

and also under the equivalent form (see Remark 2.2.46):

$$(x \oplus y) \odot ((x \odot y) \oplus z) = (z \oplus x) \odot ((z \odot x) \oplus y). \tag{2.100}$$

But, note that $(x \oplus y) \odot ((x \odot y) \oplus z)$ becomes, since $x \odot y = (x^- \oplus y^-)^-$, $[(x \oplus y)^- \oplus ((x^- \oplus y^-)^- \oplus z)^-]^-$; hence, (2.99) becomes:

$$(2.101) \quad [(x \oplus y)^- \oplus ((x^- \oplus y^-)^- \oplus z)^-]^- = [(y \oplus z)^- \oplus ((y^- \oplus z^-)^- \oplus x)^-]^-$$

and (2.100) becomes:

$$(2.102) \quad [(x \oplus y)^- \oplus ((x^- \oplus y^-)^- \oplus z)^-]^- = [(z \oplus x)^- \oplus ((z^- \oplus x^-)^- \oplus y)^-]^-.$$

First, from (2.101), by (Scomm), we obtain:

$$(2.103) \quad [(x \oplus y)^- \oplus ((y^- \oplus x^-)^- \oplus z)^-]^- = [(z \oplus y)^- \oplus ((z^- \oplus y^-)^- \oplus x)^-]^-.$$

Then, from (2.102), by (DN), we obtain:

$$(2.104) \quad (x \oplus y)^- \oplus ((x^- \oplus y^-)^- \oplus z)^- = (z \oplus x)^- \oplus ((z^- \oplus x^-)^- \oplus y)^-.$$

Now, from (2.102), by replacing x by x^- and using (DN), we obtain:

$$(2.105) \quad [(x^- \oplus y)^- \oplus ((x \oplus y^-)^- \oplus z)^-]^- = [(z \oplus x^-)^- \oplus ((z^- \oplus x)^- \oplus y)^-]^-.$$

Now, we prove:

$$(2.106) \quad [(x \oplus y)^- \oplus (x \oplus (x \oplus y)^-)^-]^- = x.$$

Indeed, in (2.103), take $Y := x^-$, $Z := y$, to obtain:
(a) $[(x \oplus x^-)^- \oplus ((x^= \oplus x^-)^- \oplus y)^-]^- = [(y \oplus x^-)^- \oplus ((y^- \oplus x^=)^- \oplus x)^-]^-$,
which, by (m-Re^R), (Neg1-0), (SU), (DN), becomes:
$y = [(y \oplus x^-)^- \oplus ((y^- \oplus x)^- \oplus x)^-]^-$,
which, by (Scomm), becomes:
$y = [(y \oplus x^-)^- \oplus (x \oplus (y^- \oplus x)^-)^-]^-$,
which, by interchanging x with y, becomes:
(a') $[(x \oplus y^-)^- \oplus (y \oplus (x^- \oplus y)^-)^-]^- = x$.
Now, in (a'), take $Y := x^= \oplus y$, to obtain:
$[(x \oplus (x^= \oplus y)^-)^- \oplus ((x^= \oplus y) \oplus (x^- \oplus (x^= \oplus y))^-)^-]^- = x$,
which, by (Sass), (m-Re^R), (m-La^R), (Neg1-0), (SU), (DN), becomes:
$[(x \oplus (x \oplus y)^-)^- \oplus (x \oplus y)^-]^- = x$,
which, by (Scomm), becomes:
$[(x \oplus y)^- \oplus (x \oplus (x \oplus y)^-)^-]^- = x$, i.e. (2.106) holds.

Now, from (2.106), by (DN), we obtain:

$$(2.107) \quad (x \oplus y)^- \oplus (x \oplus (x \oplus y)^-)^- = x^-.$$

Now, we prove:

$$(2.108) \quad x \oplus (y \oplus (y \oplus x^-)^-)^- = y^- \oplus x.$$

Indeed, in (2.105), take $Z := (x \oplus y^-)^=$, to obtain:
$[(x^- \oplus y)^- \oplus ((x \oplus y^-)^- \oplus (x \oplus y^-)^=)^-]^-$

$= [((x \oplus y^-)^= \oplus x^-)^- \oplus ((((x \oplus y^-)^=)^- \oplus x)^- \oplus y)^-]^-$,
which, by (m-Re^R), (Neg1-0), (SU), (DN), becomes:
$x^- \oplus y = [((x \oplus y^-) \oplus x^-)^- \oplus (((x \oplus y^-)^- \oplus x)^- \oplus y)^-]^-$,
which, by (Sass), (Scomm), (m-Re^R), (m-La^R), Neg1-0), (SU), becomes:
$x^- \oplus y = [(((x \oplus y^-)^- \oplus x)^- \oplus y)^-]^-$,
which, by (DN), becomes:
$x^- \oplus y = ((x \oplus y^-)^- \oplus x)^- \oplus y$,
which, by (Scomm), becomes:
$x^- \oplus y = y \oplus (x \oplus (x \oplus y^-)^-)^-$, i.e. (2.108) holds.

Now, we prove:

$$(2.109) \qquad (x^- \oplus y)^- \oplus (x \oplus (x \oplus y^-)^-)^- = y^-.$$

Indeed, in (2.104), take $X := x^-$, $Z := x$, to obtain:
$(x^- \oplus y)^- \oplus ((x^= \oplus y^-)^- \oplus x)^- = (x \oplus x^-)^- \oplus ((x^- \oplus x^=)^- \oplus y)^-$,
which, by (m-Re^R), (Neg1-0), (SU), (DN), becomes:
$(x^- \oplus y)^- \oplus ((x \oplus y^-)^- \oplus x)^- = y^-$,
which, by (Scomm), becomes:
$(x^- \oplus y)^- \oplus (x \oplus (x \oplus y^-)^-)^- = y^-$, i.e. (2.109) holds.

Finally, we prove (2.98).
Indeed, in (2.107), take $X := (x \oplus (x \oplus y^-)^-)^-$, to obtain:
$((x \oplus (x \oplus y^-)^-)^- \oplus y)^- \oplus ((x \oplus (x \oplus y^-)^-)^- \oplus ((x \oplus (x \oplus y^-)^-)^- \oplus y)^-)^- = (x \oplus (x \oplus y^-)^-)^=$,
which, by (DN), becomes:
(b) $((x \oplus (x \oplus y^-)^-)^- \oplus y)^- \oplus ((x \oplus (x \oplus y^-)^-)^- \oplus ((x \oplus (x \oplus y^-)^-)^- \oplus y)^-)^- = x \oplus (x \oplus y^-)^-$;
but, the parts $(x \oplus (x \oplus y^-)^-)^- \oplus y$ equal $x^- \oplus y$, by (2.108); hence, (b) becomes:
$(x^- \oplus y)^- \oplus ((x \oplus (x \oplus y^-)^-)^- \oplus (x^- \oplus y)^-)^- = x \oplus (x \oplus y^-)^-$,
which, by (Scomm), becomes:
(b') $(x^- \oplus y)^- \oplus ((x^- \oplus y)^- \oplus (x \oplus (x \oplus y^-)^-)^-)^- = x \oplus (x \oplus y^-)^-$;
but, the part $(x^- \oplus y)^- \oplus (x \oplus (x \oplus y^-)^-)^-$ becomes y^-, by (2.109); hence, (b') becomes:
$(x^- \oplus y)^- \oplus y^= = x \oplus (x \oplus y^-)^-$,
which, by (DN), (Scomm) becomes:
$y \oplus (y \oplus x^-)^- = x \oplus (x \oplus y^-)^-$, i.e. (2.98) holds. □

Note that the above Proposition says that any involutive right-m-BE algebra verifying (XX) is a right-MV algebra.

Corollary 2.2.66 *In any involutive right-m-MEL algebra, we have:*

$$(XX) + (m - Re^R) \Longleftrightarrow (\vee_m - comm).$$

Proof. By Proposition 2.2.65 and since any right-MV algebra is an involutive right-m-MEL algebra verifying (m-Re^R) and (XX). □

Note that the above Corollary 2.2.66 says that involutive right-m-BE algebras verifying (XX) coincide with right-MV algebras, i.e.

$$\mathbf{m-BE}^R_{(DN)} + (XX) = \mathbf{MV}^R.$$

Consequently, we also have:

$$\mathbf{m-pre-BCK}^R_{(DN)} + (XX) = \mathbf{MV}^R,$$

$$\mathbf{m-aBE}^R_{(DN)} + (XX) = \mathbf{MV}^R,$$

$$\mathbf{m-BCK}^R + (XX) = \mathbf{m-BCK}^R_{(DN)} + (XX) = \mathbf{MV}^R.$$

Corollary 2.2.67 *In any right-XY algebra, we have:*

$$(m - Re^R) \Longrightarrow (\vee_m - comm).$$

Proof. By Proposition 2.2.65, since any right-XY algebra is an involutive right-m-MEL algebra verifying (XX). □

Note that the above Corollary 2.2.67 says that any right-XY algebra verifying (m-Re^R) is a right-MV algebra.

Corollary 2.2.68 *In any right-XY algebra, we have:*

$$(m - Re^R) \Longleftrightarrow (\vee_m - comm).$$

Proof. By Corollary 2.2.67 and since any right-MV algebra is a right-XY algebra verifying (m-Re^R). □

Note that the above Corollary 2.2.68 says that right-XY algebras verifying (m-Re^R) coincide with right-MV algebras, i.e.

$$\mathbf{XY}^R + (m - Re^R) = \mathbf{MV}^R.$$

Thus, we have the connections from the following Figure 2.3 (see Remark 2.1.13).

2.2.7 Examples of proper XY algebras, not cancellative and cancellative

Examples 2.2.69 Not cancellative proper right-XY algebras (i.e. that are not right-MV algebras)

- **Example 1 (the smallest found by MACE4)**

x	0	2	3	4	5	6	1
x^-	1	3	2	5	4	6	0

and

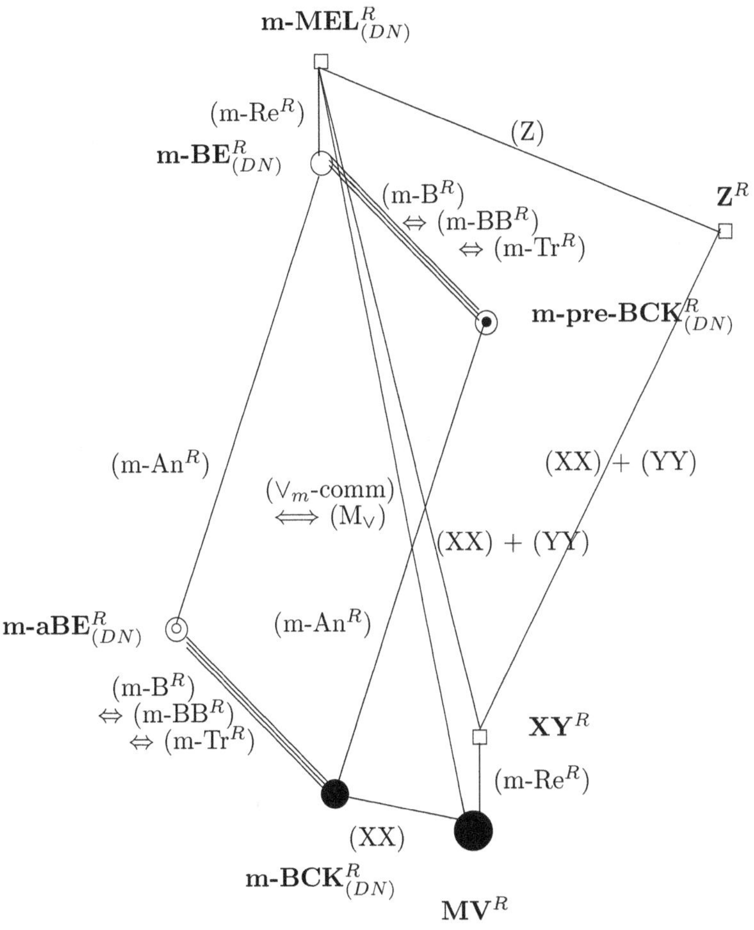

Figure 2.3: The connections between $\mathbf{m\text{-}MEL}^R_{(DN)}$, $\mathbf{m\text{-}BCK}^R_{(DN)}$, $\mathbf{Z}^R$, $\mathbf{XY}^R$ and $\mathbf{MV}^R$

$\oplus$	0	2	3	4	5	6	1
0	0	2	3	4	5	6	1
2	2	6	6	6	4	4	1
3	3	6	6	6	4	4	1
4	4	6	6	6	4	4	1
5	5	4	4	4	6	6	1
6	6	4	4	4	6	6	1
1	1	1	1	1	1	1	1

,

$\odot$	0	2	3	4	5	6	1
0	0	0	0	0	0	0	0
2	0	6	6	5	6	5	2
3	0	6	6	5	6	5	3
4	0	5	5	6	5	6	4
5	0	6	6	5	6	5	5
6	0	5	5	6	5	6	6
1	0	2	3	4	5	6	1

.

The property ($\vee_m$-comm) ($x \vee_m^M y = y \vee_m^M x$) is not verified for $(x, y) = (2, 5)$: $6 = 2 \vee_m^M 5 \neq 5 \vee_m^M 2 = 4$. The property (C) is not verified for $x = 2$, $y = 4$, $z = 3$.

• **Example 2**

x	0	2	3	4	5	6	7	1
x^-	1	3	2	4	6	5	7	0

and

$\oplus$	0	2	3	4	5	6	7	1
0	0	2	3	4	5	6	7	1
2	2	7	7	7	6	7	6	1
3	3	7	7	7	6	7	6	1
4	4	7	7	7	6	7	6	1
5	5	6	6	6	7	6	7	1
6	6	7	7	7	6	7	6	1
7	7	6	6	6	7	6	7	1
1	1	1	1	1	1	1	1	1

,

$\odot$	0	2	3	4	5	6	7	1
0	0	0	0	0	0	0	0	0
1	0	2	3	4	5	6	7	1
2	0	7	7	7	7	5	5	2
3	0	7	7	7	7	5	5	3
4	0	7	7	7	7	5	5	4
5	0	7	7	7	7	5	5	5
6	0	5	5	5	5	7	7	6
7	0	5	5	5	5	7	7	7

.

The property ($\vee_m$-comm) is not verified for $(x, y) = (3, 5)$: $7 = 3 \vee_m^M 5 \neq 5 \vee_m^M 3 = 6$. The property (C) is not verified for $x = 2$, $y = 5$, $z = 4$.

• **Example 3**

x	0	2	3	4	5	6	7	8	1
x^-	1	3	2	4	6	5	8	7	0

and

$\oplus$	0	2	3	4	5	6	7	8	1
0	0	2	3	4	5	6	7	8	1
2	2	4	4	5	4	5	2	8	1
3	3	4	4	5	4	5	3	8	1
4	4	5	5	4	5	4	4	8	1
5	5	4	4	5	4	5	5	8	1
6	6	5	5	4	5	4	6	8	1
7	7	2	3	4	5	6	0	8	1
8	8	8	8	8	8	8	8	1	1
1	1	1	1	1	1	1	1	1	1

,

$\odot$	0	2	3	4	5	6	7	8	1
0	0	0	0	0	0	0	0	0	0
2	0	4	4	6	6	4	7	2	2
3	0	4	4	6	6	4	7	3	3
4	0	6	6	4	4	6	7	4	4
5	0	6	6	4	4	6	7	5	5
6	0	4	4	6	6	4	7	6	6
7	0	7	7	7	7	7	0	7	7
8	0	2	3	4	5	6	7	1	8
1	0	2	3	4	5	6	7	8	1

.

The property ($\vee_m$-comm) is not verified for $(x, y) = (4, 8)$: $8 = 4 \vee_m^M 8 \neq 8 \vee_m^M 4 = 4$. The property (C) is not verified for $x = 2$, $y = 3$, $z = 3$.

Examples 2.2.70 Cancellative proper right-XY algebras (i.e. that are not right-MV algebras)

- **Example 1**

x	0	2	1
x^-	1	2	0

and

$\oplus$	0	2	1
0	0	2	1
2	2	2	1
1	1	1	1

,

$\odot$	0	2	1
0	0	0	0
2	0	2	2
1	0	2	1

;

$\vee_m^M$	0	2	1
0	0	2	1
2	2	2	1
1	1	2	1

.

The property ($\vee_m$-comm) is not verified for $x = 1$, $y = 2$.

- **Example 2**

x	0	2	3	1
x^-	1	2	3	0

and

$\oplus$	0	2	3	1
0	0	2	3	1
2	2	2	1	1
3	3	1	3	1
1	1	1	1	1

,

$\odot$	0	2	3	1
0	0	0	0	0
2	0	2	0	2
3	0	0	3	3
1	0	2	3	1

;

$\vee_m^M$	0	2	3	1
0	0	2	3	1
2	2	2	3	1
3	3	2	3	1
1	1	2	3	1

.

The property ($\vee_m$-comm) is not verified for $x = 2$, $y = 3$.

- **Example 3**

x	0	2	3	1
x^-	1	3	2	0

and

$\oplus$	0	2	3	1
0	0	2	3	1
2	2	1	2	1
3	3	2	0	1
1	1	1	1	1

,

$\odot$	0	2	3	1
0	0	0	0	0
2	0	1	3	2
3	0	3	0	3
1	0	2	3	1

;

$\vee_m^M$	0	2	3	1
0	0	2	3	1
2	2	2	1	1
3	3	2	0	1
1	1	2	2	1

.

The property ($\vee_m$-comm) is not verified for $x = 2$, $y = 1$.

- **Example 4**

x	0	2	3	4	5	1
x^-	1	2	3	5	4	0

and

$\oplus$	0	2	3	4	5	1
0	0	2	3	4	5	1
2	2	1	4	1	4	1
3	3	4	3	4	3	1
4	4	1	4	1	4	1
5	5	4	3	4	3	1
1	1	1	1	1	1	1

,

$\odot$	0	2	3	4	5	1
0	0	0	0	0	0	0
2	0	0	5	5	0	2
3	0	5	3	3	5	3
4	0	5	3	3	5	4
5	0	0	5	5	0	5
1	0	2	3	4	5	1

.

The property ($\vee_m$-comm) is not verified for $x = 1$, $y = 3$: $3 = 1 \vee_m^M 3 \neq 3 \vee_m^M 1 = 1$.

- **Example 5**

x	0	2	3	4	5	6	7	1
x^-	1	3	2	5	4	7	6	0

and

$\oplus$	0	2	3	4	5	6	7	1
0	0	2	3	4	5	6	7	1
2	2	2	1	4	6	6	4	1
3	3	1	3	1	3	1	3	1
4	4	4	1	2	6	6	2	1
5	5	6	3	6	3	1	5	1
6	6	6	1	6	1	1	6	1
7	7	4	3	2	5	6	0	1
1	1	1	1	1	1	1	1	1

,

$\odot$	0	2	3	4	5	6	7	1
0	0	0	0	0	0	0	0	0
2	0	2	0	2	0	2	0	2
3	0	0	3	7	5	5	7	3
4	0	2	7	2	7	4	0	4
5	0	0	5	7	3	3	7	5
6	0	2	5	4	3	1	7	6
7	0	0	7	0	7	7	0	7
1	0	2	3	4	5	6	7	1

.

The property ($\vee_m$-comm) is not verified for $x = 3$, $y = 4$: $6 = 3 \vee_m^M 4 \neq 4 \vee_m^M 3 = 1$.

- **Example 6**

x	0	2	3	4	5	6	7	1
x^-	1	2	3	5	4	7	6	0

and

$\oplus$	0	2	3	4	5	6	7	1
0	0	2	3	4	5	6	7	1
2	2	1	4	1	6	1	4	1
3	3	4	1	1	7	4	1	1
4	4	1	1	1	4	1	1	1
5	5	6	7	4	0	2	3	1
6	6	1	4	1	2	1	4	1
7	7	4	1	1	3	4	1	1
1	1	1	1	1	1	1	1	1

,

$\odot$	0	2	3	4	5	6	7	1
0	0	0	0	0	0	0	0	0
2	0	0	5	7	0	5	0	2
3	0	5	0	6	0	0	5	3
4	0	7	6	1	5	3	2	4
5	0	0	0	5	0	0	0	5
6	0	5	0	3	0	0	5	6
7	0	0	5	2	0	5	0	7
1	0	2	3	4	5	6	7	1

.

The property ($\vee_m$-comm) is not verified for $x = 1$, $y = 4$: $4 = 1 \vee_m^M 4 \neq 4 \vee_m^M 1 = 1$.

- **Example 7**

x	0	2	3	4	5	6	7	8	1
x^-	1	4	5	2	3	7	6	8	0

and

$\oplus$	0	2	3	4	5	6	7	8	1
0	0	2	3	4	5	6	7	8	1
2	2	2	2	1	1	7	7	7	1
3	3	2	2	5	1	8	7	7	1
4	4	1	5	4	5	4	1	5	1
5	5	1	1	5	1	5	1	1	1
6	6	7	8	4	5	6	7	8	1
7	7	7	7	1	1	7	7	7	1
8	8	7	7	5	1	8	7	7	1
1	1	1	1	1	1	1	1	1	1

,

$\odot$	0	2	3	4	5	6	7	8	1
0	0	0	0	0	0	0	0	0	0
2	0	2	3	0	3	0	2	3	2
3	0	3	0	0	0	0	3	0	3
4	0	0	0	4	4	6	6	6	4
5	0	3	0	4	4	6	8	6	5
6	0	0	0	6	6	6	6	6	6
7	0	2	3	6	8	6	7	8	7
8	0	3	0	6	6	6	8	6	8
1	0	2	3	4	5	6	7	8	1

.

The property ($\vee_m$-comm) is not verified for $x = 1$, $y = 6$: $7 = 1 \vee_m^M 6 \neq 6 \vee_m^M 1 = 1$.

2.3 Homomorphisms of XY algebras

We introduce now the following definition (the dual one is omitted).

Definition 2.3.1 Let $\mathcal{A} = (A, \oplus, ^-, 0_A)$, with $1_A \stackrel{def.}{=} 0_A^-$, and $\mathcal{B} = (B, \oplus, ^-, 0_B)$, with $1_B \stackrel{def.}{=} 0_B^-$, be two right-XY algebras. A function $f : A \longrightarrow B$ is said to be a *right-XY algebras homomorphism* if, for all $x, y \in A$,
(H1) $f(x \oplus y) = f(x) \oplus f(y)$,
(H2) $f(x^-) = f(x)^-$,
(H3) $f(0_A) = 0_B$.

Corollary 2.3.2 *If $f : A \longrightarrow B$ is a right-XY algebras homomorphism, then we have: for all $x, y \in A$,*
(H1') $f(x \odot y) = f(x) \odot f(y)$,
(H4) if $x \leq_m y$, then $f(x) \leq_m f(y)$,
(H4') if $x \leq_m^M y$, then $f(x) \leq_m^M f(y)$,
(H5) $f(1_A) = 1_B$.

Proof. (H1'): $f(x \odot y) = f((x^- \oplus y^-)^-) \stackrel{(H2)}{=} f(x^- \oplus y^-)^-$ $\stackrel{(H1)}{=} (f(x^-) \oplus f(y^-))^- \stackrel{(H2)}{=} (f(x)^- \oplus f(y)^-)^- = f(x) \odot f(y)$.

(H4): if $x \leq_m y$, i.e. $y \oplus x^- = 1_A$, then $f(y \oplus x^-) = f(1_A)$; hence, $f(y) \oplus f(x)^- = 1_B$, by (H1), (H2), (H5); thus, $f(x) \leq_m f(y)$.

(H4'): if $x \leq_m^M y$, i.e. $x \oplus (x \oplus y^-)^- = y$, then $f(x \oplus (x \oplus y^-)^-) = f(y)$; hence, $f(x) \oplus (f(x) \oplus f(y)^-)^- = f(y)$, by (H1), (H2); thus, $f(x) \leq_m^M f(y)$.

(H5): $f(1_A) = f(0_A^-) = f(0_A)^- = 0_B^- = 1_B$, by (H2), (H3). □

Denote by $\mathcal{C}(\mathbf{XY}^R)$ the category whose objects are the right-XY algebras and morphisms are the right-XY algebras homomorphisms.

Chapter 3

Strong m_0-ME structures. Strong m_0-ME positive cones

Starting from the framework centered on BCK algebras, we have introduced the m-BCK algebras ('m' coming from 'magma') and other more general algebras, thus obtaining an analogous framework centered on m-BCK algebras, such that, for example, the involutive BCK algebras are definitionally equivalent to (involutive) m-BCK algebras. The m_0-ME algebras, the commutative groups and the m-MEL algebras, the MV algebras, the Boolean algebras belong to this new framework, presented in the paper [20], in 2020, and in the monograph [21], in 2022 (see **Appendix A**).

In this chapter, we introduce and study, in this new framework, the strong m_0-ME structures and the strong m_0-ME positive cones as the 'ancestors' of commutative lattice-ordered groups with strong unit and their positive cones, respectively.

3.1 (Involutive) m_0-ME algebras vs. commutative groups

In this section, we recall from the framework centered on m-BCK algebras (see **Appendix A**) the m_0-ME algebras and the commutative groups, The 'ancestor' of lattice-ordered commutative l-groups with strong unit, called *strong m_0-ME structures*, introduced in the second section, will belong to this framework too.

Recall the following well-known definition (the dual one is omitted).

A *commutative (right, i.e. additive) group* is an algebra $\mathcal{G} = (G, +, -, 0_G)$ such that:

(j) $(G, +, 0_G)$ is a commutative monoid, i.e. for all $x, y, z \in G$,

(SU) $\quad x + 0_G = x = 0_G + x$,

(Scomm) $\quad x + y = y + x$,

(Sass) $\quad (x + y) + z = x + (y + z)$;

(jj) for each $x \in G$, we have:
(m_0-Re) (inverse) $x + (-x) = 0_G$.

Denote by **c. group** the class of all commutative groups.

Note that, in the sequel, $x - y$ means $x + (-y)$, or $x + -y$.

Note that in any commutative group $\mathcal{G} = (G, +, -, 0_G)$, we have: for all $x, y \in G$,
(N0) $-0_G = 0_G$,
(DN) $-(-x) = x$,
(NegS) $-(x + y) = (-y) + (-x)$.

Remark 3.1.1 *(See ([21], Remarks 5.4.7))*
Since the commutative group $\mathcal{G} = (G, +, -, 0_G)$ is an involutive algebra, by (DN), it follows that we can define the dual operation, $\cdot$, by (see (2.1)): for all $x, y \in G$,

$$x \cdot y \overset{def.}{=} -(-x + -y).$$

Then, $x \cdot y = - - y + - - x \overset{(DN)}{=} y + x \overset{(Scomm)}{=} x + y$, *i.e. they coincide (+ is selfdual).*

In any commutative group $\mathcal{G} = (G, +, -, 0_G)$, the cancellation property (C) holds: for all $x, y, z \in G$,
(C) $z + x = z + y \Longrightarrow x = y$.
Note that, by above Remark, the cancellation property (C) for commutative groups coincides with the usual cancellation property (C) in MV algebras and XY algebras.

We introduce the following definition.

Definition 3.1.2 A *strong commutative monoid*, or a *strong monoid* for short, is an algebra $\mathcal{M} = (M, +, 0_M, u_M)$ of type $(2, 0, 0)$ such that:
(i) $(M, +, 0_M)$ is a commutative monoid,
(ii) $u_M \neq 0_M$ is an element of M verifying the following particular cancellation property: for all $x, y \in M$,

$$(C_u^{1R}) \qquad x + u_M = y + u_M \implies x = y.$$

Here are two examples of strong monoids.

Examples 3.1.3
Consider the two strong monoids with the following tables of $+$:

$+$:	0_M	u_M	2	3
0_M	0_M	u_M	2	3
u_M	u_M	0_M	3	2
2	2	3	2	3
3	3	2	3	2

;

$+$:	0_M	u_M	2	3	4
0_M	0_M	u_M	2	3	4
u_M	u_M	2	3	0_M	4
2	2	3	0_M	u_M	4
3	3	0_M	u_M	2	4
4	4	4	4	4	4

.

Notation 3.1.4 *Let* $\mathbf{N}$[1] *be the set of all positive integers, i.e.* $\mathbf{N} = \{0, 1, 2, \ldots\}$.

Let $\mathcal{M} = (M, +, 0_M, u_M)$ *be a strong commutative monoid. Let* $0u_M = 0_M$ *and, for each integer* $n \geq 0$ *(i.e.* $n \in \mathbf{N}$*), let* $(n+1)u_M = nu_M + u_M$.
Hence, $nu_M = \underbrace{u_M + u_M + \ldots + u_M}_{n \text{ times}} \in M$.

Corollary 3.1.5 *Let* $\mathcal{M} = (M, +, 0_M, u_M)$ *be a strong commutative monoid. Then, the following particular cancellation property holds: for all* $x, y \in M$ *and for each integer* $n \geq 1$,

$$(C_u^R) \qquad x + nu_M = y + nu_M \implies x = y.$$

Proof. We prove by induction on $n \geq 1$.
- For $n = 1$, (C_u^R) holds by (C_u^{1R}).
- Suppose that (C_u^R) holds for n.
- We prove that (C_u^R) holds for $n+1$. Indeed,

$x + (n+1)u_M = y + (n+1)u_M$
$\iff x + (nu_M + u_M) = y + (nu_M + u_M)$
$\overset{(Sass)}{\iff} (x + nu_M) + u_M = (y + nu_M) + u_M$
$\overset{(C_u^{1R})}{\implies} x + nu_M = y + nu_M$
$\implies x = y$, by the inductive hypothesis. □

Note that Corollary 3.1.5 (hence (C_u^R)) holds for each $n \in \mathbf{N}$.

*

Recall the following definitions from ([21], Chapter 7):

Definitions 3.1.6 (See [21], subsection 7.1.1)

(1) A *m_0-ME algebra* is an algebra $\mathcal{M} = (M, +, -, 0_M)$ of type $(2, 1, 0)$ verifying the following axioms: for all $x, y, z \in M$,

(SU) $x + 0_M = x = 0_M + x$,
(Scomm) $x + y = y + x$,
(Sass) $(x + y) + z = x + (y + z)$,
(N0) $-0_M = 0_M$,

i.e. it is an additive monoid verifying (N0).

(1') Dually, a *m_1-ME algebra* is an algebra $\mathcal{M} = (M, \cdot, ^{-1}, 1_M)$ of type $(2, 1, 0)$ verifying the following axioms: for all $x, y, z \in M$,

(PU) $x \cdot 1_M = x = 1_M \cdot x$,
(Pcomm) $x \cdot y = y \cdot x$,
(Pass) $(x \cdot y) \cdot z = x \cdot (y \cdot z)$,
(N1) ${1_M}^{-1} = 1_M$,

i.e. it is a multiplicative monoid verifying (N1).

[1] Our LaTeX has no a command for the usual notation of this set.

Denote by m_0-**ME** the class of m_0-ME algebras and by m_1-**ME** the class of m_1-ME algebras.

Remark 3.1.7 *Any commutative group is a m_0-ME algebra.*

Define, in a m_0-ME algebra $\mathcal{M}$ a basic internal (i.e. defined by using the operations on M) binary relation (see [21]) $\leq_{m_0}$ by: for all $x, y \in M$,

$$x \leq_{m_0} y \stackrel{def.}{\Longleftrightarrow} y + (-x) = 0_M.$$

We shall denote:
- the *reflexivity* of $\leq_{m_0}$, and hence of $\mathcal{M}$, by (m_0-Re),
- the *antisymmetry* of $\leq_{m_0}$, and hence of $\mathcal{M}$, by (m_0-An) and
- the *transitivity* of $\leq_{m_0}$, and hence of $\mathcal{M}$, by (m_0-Tr),

hence we have the following properties:

(m_0-Re) $x + (-x) = 0_M$,
(m_0-Re') $x \leq_{m_0} x$;

(m_0-An) $y + (-x) = 0_M$ and $x + (-y) = 0_M$ imply $x = y$,
(m_0-An') $x \leq_{m_0} y$ and $y \leq_{m_0} x$ imply $x = y$;

(m_0-Tr) $y + (-x) = 0_M$ and $z + (-y) = 0_M$ imply $z + (-x) = 0_M$,
(m_0-Tr') $x \leq_{m_0} y$ and $y \leq_{m_0} z$ imply $x \leq_m z$.

Dually, define, in a m_1-ME algebra $\mathcal{M}$ a basic internal binary relation (see [21]) $\geq_{m_1}$ by: for all $x, y \in M$,

$$x \geq_{m_1} y \stackrel{def.}{\Longleftrightarrow} y \cdot x^{-1} = 1_M.$$

Remark 3.1.8 *Any commutative group is a m_0-ME algebra verifying (m_0-Re).*

Definitions 3.1.9

(1) An *involutive m_0-ME algebra* is a m_0-ME algebra $\mathcal{M} = (M, +, -, 0_M)$ verifying the following axiom: for all $x \in M$,
(DN) $-(-x) = x$, or $= x = x$.

(1') Dually, an *involutive m_1-ME algebra* is a m_1-ME algebra $\mathcal{M} = (M, \cdot, ^{-1}, 1_M)$ verifying the axiom: for all $x \in M$,
(DN) $(x^{-1})^{-1} = x$.

Denote by m_0-$\mathbf{ME}_{(DN)}$ the class of involutive m_0-ME algebras and by m_1-$\mathbf{ME}_{(DN)}$ the class of involutive m_1-ME algebras.

Remark 3.1.10 *Any commutative group is an involutive m_0-ME algebra verifying (m_0-Re).*

Let $\mathcal{M} = (M, +, -, 0_M)$ be an involutive m$_0$-ME algebra. Let us introduce a new operation, $\cdot$, defined by: for all $x, y \in M$,

$$x \cdot y \overset{def.}{=} -((-x) + (-y)). \tag{3.1}$$

Then, we have the following result.

Proposition 3.1.11 *Let $\mathcal{M} = (M, +, -, 0_M)$ be an involutive m$_0$-ME algebra. Then, the algebra $(M, \cdot, ^{-1} \overset{def.}{=} -, 1_M \overset{def.}{=} 0_M)$ is an involutive m$_1$-ME algebra.*

Proof.

(PU): $x \cdot 1_M = x \cdot 0_M = -((-x)+(-0_M)) \overset{(N0)}{=} -((-x)+0_M) \overset{(SU)}{=} -(-x) \overset{(DN)}{=} x$.

(Pcomm): $x \cdot y = -((-x) + (-y)) \overset{(Scomm)}{=} -((-y) + (-x)) = y \cdot x$,

(Pass): $(x \cdot y) \cdot z = -((-x) + (-y)) \cdot z = -[= ((-x) + (-y)) + (-z)]$ $\overset{(DN)}{=} -[((-x) + (-y)) + (-z)] \overset{(Sass)}{=} -[(-x) + ((-y) + (-z))]$ $\overset{(DN)}{=} -[(-x)+ = ((-y) + (-z))] = x \cdot -((-y) + (-z)) = x \cdot (y \cdot z)$.

(N1): $(1_M)^{-1} = -(0_M) \overset{(N0)}{=} 0_M = 1_M$.

(DN): $(x^{-1})^{-1} = -(-x) \overset{(DN)}{=} x$. □

Note that we shall often write $x - y$ instead of $x + (-y)$ in the sequel.

Proposition 3.1.12 *(See ([21], Remarks 7.1.1 (4)))*
Any (involutive) m$_0$-ME algebra verifying (m$_0$-Re) (i.e. $x + (-x) = 0_M$) is a commutative group.

Proof. Obviously, by above definition of the commutative group. □

Hence, the (involutive) m$_0$-ME algebras verifying (m$_0$-Re) coincide with the commutative groups.

We write (see the connections in **Appendix A**):

$$\mathbf{m_0 - ME} \ + \ (m_0 - Re) \ = \ \mathbf{c.\,group}.$$

$$\mathbf{m_0 - ME_{(DN)}} \ + \ (m_0 - Re) \ = \ \mathbf{c.\,group}.$$

Note that, since in this work all the algebras/structures are *commutative*, we shall freely omit in the sequel the adjective 'commutative'.

3.2 Strong m$_0$-ME structures vs. unital l-groups

Recall the following definitions (the dual ones are omitted).

Definitions 3.2.1

(1) A *partially ordered (right-) group*, or a *(right-) po-group* for short, is a structure $\mathcal{G} = (G, \leq, +, -, 0_G)$ such that:

(j) $(G, +, -, 0_G)$ is a commutative group,
(jj) the binary relation $\leq$ is an external (i.e. not defined by the operations on G) order relation (i.e. it is reflexive, antisymmetric and transitive) on G, which is compatible with the sum $+$, i.e. for all $x, y, z \in G$, we have:
(Cp) (compatibility) $x \leq y \Longrightarrow x + z \leq y + z$,
(hence, $x \leq y \Longleftrightarrow x + z \leq y + z$.)

(2) A *lattice ordered group*, or an *l-group* for short, is an algebra $\mathcal{G} = (G, \vee, \wedge, +, -, 0_G)$ such that the external order relation $\leq$ is a lattice order (hence, $(G, \vee, \wedge)$ is a lattice), where $x \leq y \Longleftrightarrow x \vee y = y \Longleftrightarrow x \wedge y = x$.

(3) An *l-group with strong unit*, or a *unital l-group* for short, is an algebra

$$\mathcal{G}_u = (G, \vee, \wedge, +, -, 0_G, u = u_G)$$

such that:
(j) $(G, \vee, \wedge, +, -, 0_G)$ is an l-group,
(jj) u_G is an element of G verifiying:
(g0) $0_G < u_G$ and
(g1) for each $x \in G$, there exists $n_x \in \mathbf{N}$ such that $\mid x \mid \stackrel{def.}{=} x \vee -x \leq n_x u_G$.

Note that the *positive cone* $G^+ = \{x \in G \mid x \geq 0_G\}$ of a po-group/l-group $\mathcal{G}$ verifies also the cancellation property (C).

Recall also the following definitions.

Definition 3.2.2 A poset (partially ordered set) $\mathcal{A} = (A, \leq)$ will be said to be an *Ore lattice*, if for each two elements $x, y \in A$, there exist $\inf(x, y)$ and $\sup(x, y)$.

Definition 3.2.3 An algebra $\mathcal{A} = (A, \wedge, \vee)$ or, dually, $\mathcal{A} = (A, \vee, \wedge)$, of type $(2, 2)$, will be said to be a *Dedekind lattice*, if the following properties hold: for all $x, y, z \in A$,

(Wid)	(idempotency of $\wedge$)	$x \wedge x = x$,
(Wcomm)	(commutativity of $\wedge$)	$x \wedge y = y \wedge x$,
(Wass)	(associativity of $\wedge$)	$x \wedge (y \wedge z) = (x \wedge y) \wedge z$,
(Wabs)	(absorbtion of $\wedge$ over $\vee$)	$x \wedge (x \vee y) = x$, and also
(Vid)	(idempotency of $\vee$)	$x \vee x = x$,
(Vcomm)	(commutativity of $\vee$)	$x \vee y = y \vee x$,
(Vass)	(associativity of $\vee$)	$(x \vee y) \vee z = x \vee (y \vee z)$,
(Vabs)	(absorbtion of $\vee$ over $\wedge$)	$x \vee (x \wedge y) = x$,

where 'W' comes from 'wedge' (the LaTeX command for the meet, $\wedge$), 'V' comes from 'vee' (the LaTeX command for the join, $\vee$).

Recall finally that the two kinds of lattices are *definitionally equivalent*:

Theorem 3.2.4

(1) Let $\mathcal{A} = (A, \leq)$ be an Ore lattice. Define $\delta(\mathcal{A}) \stackrel{def.}{=} (A, \wedge, \vee)$, where for all $x, y \in A$:

$$x \wedge y \stackrel{def.}{=} \inf(x,y) \ \ and \ \ x \vee y \stackrel{def.}{=} \sup(x,y).$$

Then, $\delta(\mathcal{A})$ is a Dedekind lattice.

(1') Let $\mathcal{A} = (A, \wedge, \vee)$ be a Dedekind lattice. Define $\omega(\mathcal{A}) \stackrel{def.}{=} (A, \leq)$, where for all $x, y \in A$:
$x \leq y \stackrel{def.}{\Longleftrightarrow} x \wedge y = x$ or, equivalently, $x \leq y \stackrel{def.}{\Longleftrightarrow} x \vee y = y$.
Then, $\omega(\mathcal{A})$ is an Ore lattice.

(2) The two mappingss, δ and ω, are mutually inverse.

Recall that, in any l-group $(G, \vee, \wedge, +, -, 0_G)$, the following additional properties hold: for all $x, y, z \in G$,
(g2) $z + (x \vee y) = (z + x) \vee (z + y)$,
(g3) $z + (x \wedge y) = (z + x) \wedge (z + y)$;
(g4) $-(x \vee y) = (-x) \wedge (-y)$ (De Morgan law 1),
(g5) $-(x \wedge y) = (-x) \vee (-y)$ (De Morgan law 2);
(g6) the lattice $(G, \vee, \wedge)$ is distributive, i.e. the following two equivalent properties hold:
(Wdis) (distributivity of $\vee$ over $\wedge$) $z \wedge (x \vee y) = (z \wedge x) \vee (z \wedge y)$,
(Vdis) (distributivity of $\wedge$ over $\vee$) $z \vee (x \wedge y) = (z \vee x) \wedge (z \vee y)$,
(g7) $x \leq y \Longleftrightarrow -y \leq -x$.

Note that we shall freely write $x \leq y$ or $y \geq x$ in the sequel.

*

We introduce now the following new notions.

Definitions 3.2.5

(1) A *m_0-ME structure* is a structure $\mathcal{M} = (M, \leq_m, +, ^{\smile}, 0_M)$ such that:
(i) $(M, +, ^{\smile}, 0_M)$ is an involutive m_0-ME algebra,
(ii) $\leq_m$ is an external binary relation on M such that $0_M \leq_m 0_M$ ('m' comes from 'magma').

(1') Dually, a *m_1-ME structure* is a structure $\mathcal{M} = (M, \leq_m, \cdot, ^{-1}, 1_M)$ such that:
(i') $(M, \cdot, ^{-1}, 1_M)$ is an involutive m_1-ME algebra,
(ii') $\leq_m$ is an external binary relation on M such that $1_M \leq_m 1_M$ ('m' comes from 'magma').

Note that we shall freely write $x \leq_m y$ or $y \geq_m x$ in the sequel.

Definitions 3.2.6

(1) A *strong m_0-ME structure* is a structure

$$\mathcal{M}_u = (M, \leq_m, +, ^{\smile}, 0_M, u = u_M)$$

such that:
(i) $(M, \leq_m, +, ^{\smile}, 0_M)$ is a m_0-ME structure,
(ii) u_M is an element of M verifying:

($s0^R$) $0_M <_m u$;
($s1^R$) for each $x \in M$, there exists $n_x \in \mathbf{N}$ such that $x, \smile x \leq_m n_x u$;
($s2^R$) (i) for each $x \geq_m 0_M$, there exists $n_x \in \mathbf{N}$, and there exist the unique elements of M:
$0_M \leq_m x_1, x_2, \ldots, x_{n_x} \leq_m u$ (hence $0_M \leq_m \smile x_1, \smile x_2, \ldots, \smile x_{n_x} \leq_m u$),
s.t. $x = x_1 + \ldots + x_{n_x}$ ($\leq_m n_x u$), $\smile x = \smile x_1 \smile \ldots \smile x_{n_x}$ ($\leq_m n_x u$),
(ii) if $x = x_1 + \ldots + x_n \geq_m 0_M$ and $y = y_1 + \ldots + y_n \geq_m 0_M$,
then $x = y \Longleftrightarrow x_i = y_i$, $i = 1, \ldots, n$,
(iii) if $x = x_1 + \ldots + x_n \geq_m 0_M$ and $y = y_1 + \ldots + y_n \geq_m 0_M$,
then $x \leq_m y \Longleftrightarrow x_i \leq_m y_i$, $i = 1, \ldots, n$;
($s3^R$) for $0_M \leq_m x, y \leq_m u$ (i.e. $x, y \in [0_M, u]$),
(i) there exist $\inf_m(u, x + y)$, $\smile \inf_m(u, \smile x \smile y)$ and belong to $[0_M, u]$,
(ii) $\inf_m(u, x + u) = u$, $\smile \inf_m(u, \smile x \smile 0_M) = 0_M$, $\smile \inf_m(u, \smile x \smile u) = x$,
(iii) $x \leq_m y$ ($x, y \neq 0_M$) $\Longleftrightarrow u \leq_m y - x$;
($s4^R$) for $0_M \leq_m x, y, z \leq_m u$,
(i) there exist and are equal $\inf_m(u, x + \inf_m(u, y + z)) = \inf_m(u, x + (y + z))$,
(ii) $\inf_m(u, \smile \inf_m(u, \smile x \smile y)) = \smile \inf_m(u, \smile x \smile y)$;
($s5^R$) for $0_M \leq_m x_0, x_1, \ldots, x_n$, $y_0 \leq_m u$,
(V^{00}) $x_0 + y_0 = \inf_m(u, x_0 + y_0) \smile \inf_m(u, \smile x_0 \smile y_0)$,
(V^{10}) if $\inf_m(u, x_0 + x_1) = x_0$, then $x_0 + x_1 + y_0 = \inf_m(u, x_0 + y_0)$
$\smile \inf_m(u, \smile x_0 \smile \inf_m(u, x_1 + y_0)) \smile \inf_m(u, \smile x_1 \smile y_0)$,
(V^{n0}) if $\inf_m(u, x_i + x_{i+1}) = x_i$, for $i = 0, 1, \ldots, n-1$, then
$\sum_{i=0}^{n} x_i + y_0 = \sum_{i=0}^{n+1} c_i$, where
$c_i = \smile \inf_m(u, \smile x_{i-1} \smile \inf_m(u, x_i + y_0))$, with $x_{-1} = u$, $x_{n+1} = 0_M$;
(XX_u^R) for $0_M \leq_m x, y, z \leq_m u$, there exist and are equal
$\smile \inf_m(u, \smile \inf_m(u, x + y) \smile \inf_m(u, z \smile \inf_m(u, \smile x \smile y)))$
$= \smile \inf_m(u, \smile \inf_m(u, y + z) \smile \inf_m(u, x \smile \inf_m(u, \smile y \smile z)))$;
(YY_u^R) for $0_M \leq_m x, y, z \leq_m u$,
if $\inf_m(u, x + y) = x$, then there exist and are equal
$\smile \inf_m(u, \smile x \smile \inf_m(u, y + z)) = \inf_m(u, y \smile \inf_m(u, \smile x \smile z))$;
(C_u^{1R}) $x + u = y + u \Longrightarrow x = y$, for all $x, y \in M$;
(C_{pu}^{1R}) $x \leq_m y \Longleftrightarrow x + u \leq_m y + u$, for all $x, y \in M$;
(m-Tr_u^R) $x \leq_m nu$ and $nu \leq_m ru \Longrightarrow x \leq_m ru$, for all $x \in M$, $n, r \in \mathbf{N}$;
(Nu^{1R}) $\smile u = u$;
(NNu^{1R}) $\smile (\smile u) = u$;
(S^{1R}) $\smile (x + u) = \smile x + u$, for each $x \in M$,
where $\smile x \smile y$ means $\smile x + \smile y$.

(1') Dually, a *strong* m_1*-ME structure* is a structure

$$\mathcal{M}_u = (M, \leq_m, \cdot, ^{-1}, 1_M, u = u_M)$$

such that:
(i') $(M, \leq_m, \cdot, ^{-1}, 1_M)$ is a m_1-ME structure,
(ii') u_M is an element of M verifying:
(s0) $1_M <_m u_M$;

(s1) for each $x \in M$, there exists $n_x \in \mathbf{N}$ such that $x, x^{-1} \leq_m n_x u$;
(s2) (i) for each $x \geq_m 0_M$, there exists $n_x \in \mathbf{N}$
and there exist the unique elements of M
$1_M \leq_m x_1, x_2, \ldots, x_{n_x} \leq_m u$ (hence $1_M \leq_m x_1^{-1}, x_2^{-1}, \ldots, x_{n_x}^{-1} \leq_m u$),
s.t. $x = x_1 \cdot \ldots \cdot x_{x_n}$ ($\leq_m n_x u$), $x^{-1} = x_1^{-1} \cdot \ldots \cdot x_{x_n}^{-1}$ ($\leq_m n_x u$),
(ii) if $x = x_1 \cdot \ldots \cdot x_n \geq_m 0_M$ and $y = y_1 \cdot \ldots \cdot y_n \geq_m 0_M$,
then $x = y \Longleftrightarrow x_i = y_i$, $i = 1, \ldots, n$,
(iii) if $x = x_1 \cdot \ldots \cdot x_n \geq_m 0_M$ and $y = y_1 \cdot \ldots \cdot y_n \geq_m 0_M$,
then $x \leq_m y \Longleftrightarrow x_i \leq_m y_i$, $i = 1, \ldots, n$;
(s3) for $1_M \leq_m x, y \leq_m u$ (i.e. $x, y \in [1_M, u]$),
(i) there exist $\sup_m(u, x \cdot y)$, $\sup_m(u, x^{-1} \cdot y^{-1})^{-1}$ and belong to $[1_M, u]$,
(ii) $\sup_m(u, x \cdot 0_M) = 0_M$, $\sup_m(u, x^{-1} \cdot u^{-1})^{-1} = u$, $\sup_m(u, x^{-1} \cdot 0_M^{-1})^{-1} = x$,
(iii) $x \leq_m y$ $(x, y \neq 1_M) \Longleftrightarrow u \leq_m y \cdot x^{-1}$;
(s4) for $1_M \leq_m x, y, z \leq_m u$,
(i) there exist and are equal $\sup_m(u, x \cdot \sup_m(u, y \cdot z)) = \sup_m(u, x \cdot (y \cdot z))$,
(ii) $\sup_m(u, \sup_m(u, x^{-1} \cdot y^{-1})^{-1}) = \sup_m(u, x^{-1} \cdot y^{-1})^{-1}$;
(s5) for $1_M \leq_m x_0, x_1, \ldots, x_n, y_0 \leq_m u$,
(V^{00}) $x_0 \cdot y_0 = \sup_m(u, x_0 \cdot y_0) \cdot \sup_m(u, x_0^{-1} \cdot y_0^{-1})^{-1}$,
(V^{10}) if $\sup_m(u, x_0 \cdot x_1) = x_0$, then $x_0 \cdot x_1 \cdot y_0 = \sup_m(u, x_0 \cdot y_0)$
$\cdot \sup_m(u, x_0^{-1} \cdot \sup_m(u, x_1 \cdot y_0)^{-1})^{-1} \cdot \sup_m(u, x_1^{-1} \cdot y_0^{-1})^{-1}$,
(V^{n0}) if $\sup_m(u, x_i \cdot x_{i+1}) = x_i$, for $i = 0, \ldots, n-1$, then
$\prod_{i=0}^{n} x_i \cdot y_0 = \prod_{i=0}^{n+1} c_i$, where
$c_i = \sup_m(u, x_{i-1}^{-1} \cdot \sup_m(u, x_i \cdot y_0)^{-1})^{-1}$, with $x_{-1} = u$, $x_{n+1} = 0_M$;
(XX_u) for $1_M \leq_m x, y, z \leq_m u$, there exist and are equal
$\sup_m(u, \sup_m(u, x \cdot y)^{-1} \cdot \sup_m(u, z \cdot \sup_m(u, x^{-1} \cdot y^{-1})^{-1})^{-1})^{-1}$
$= \sup_m(u, \sup_m(u, y \cdot z)^{-1} \cdot \sup_m(u, x \cdot \sup_m(u, y^{-1} \cdot z^{-1})^{-1})^{-1})^{-1}$;
(YY_u) for $1_M \leq_m x, y, z \leq_m u$,
if $\sup_m(u, x \cdot y) = x$, then there exist and are equal
$\sup_m(u, x^{-1} \cdot \sup_m(u, y \cdot z)^{-1})^{-1} = \sup_m(u, y \cdot \sup_m(u, x^{-1} \cdot z^{-1})^{-1})$;
(C_u^1) $x \cdot u = y \cdot u \Longrightarrow x = y$, for all $x, y \in M$;
(C_{pu}^1) $x \leq_m y \Longleftrightarrow x \cdot u \leq_m y \cdot u$, for all $x, y \in M$;
(m-Tr_u) $x \leq_m nu$ and $nu \leq_m ru \Longrightarrow x \leq_m ru$, for each $x \in M$, $n, r \in \mathbf{N}$;
(Nu^1) $u^{-1} = u$;
(NNu^1) $\smile (u^{-1})^{-1} = u$;
(S^1) $(x \cdot u)^{-1} = x^{-1} \cdot u$, for each $x \in M$.

Denote by **s-m$_0$-ME** the class of all strong m$_0$-ME structures and by **s-m$_1$-ME** the class of all strong m$_1$-ME structures.

Remark 3.2.7 *The axioms of strong m$_0$-ME structures are verified by the particular example $\mathbf{T_m}(M^+)$ (see next Chapter 5).*

Note that the properties (s3^R) - (s5^R), (XX$_u^R$), (YY$_u^R$) refer to the unit interval $[0_M, u_M]$.

Corollary 3.2.8 *(See Proposition 2.2.36)*

Let $\mathcal{M}_u = (M, \leq_m, +, \smile, 0_M, u = u_M)$ be a strong m_0-ME structure. For $0_M \leq_m x, y, z \leq_m u$, we have:

$$(XX_u^R) \implies (Z_u^R),$$

where:
(Z_u^R) $\inf_m(u, x+y) = \inf_m(u, \inf_m(u, x+y) \smile \inf_m(u, \smile x \smile y))$.

Proof. In (XX_u), i.e. in the equality:
$\smile \inf_m(u, \smile \inf_m(u, x+y) \smile \inf_m(u, z \smile \inf_m(u, \smile x \smile y)))$
$= \smile \inf_m(u, \smile \inf_m(u, y+z) \smile \inf_m(u, x \smile \inf_m(u, \smile y \smile z)))$,
take $Z := 0_M$, to obtain:
$\smile \inf_m(u, \smile \inf_m(u, x+y) \smile \inf_m(u, 0_M \smile \inf_m(u, \smile x \smile y)))$
$= \smile \inf_m(u, \smile \inf_m(u, y+0_M) \smile \inf_m(u, x \smile \inf_m(u, \smile y \smile 0_M)))$,
which, by (DN), (SU), becomes:
$\inf_m(u, \smile \inf_m(u, x+y) \smile \inf_m(u, \smile \inf_m(u, \smile x \smile y)))$
$= \inf_m(u, \smile \inf_m(u, y) \smile \inf_m(u, x + \smile \inf_m(u, \smile y \smile 0_M)))$,
which, by ($s3^R$) (ii), ($s4^R$) (ii), becomes:
$\inf_m(u, \smile \inf_m(u, x+y) \smile (\smile \inf_m(u, \smile x \smile y)))$
$= \inf_m(u, \smile \inf_m(u, y) \smile \inf_m(u, x + 0_M))$,
which, by (DN), (SU), becomes:
$\inf_m(u, \smile \inf_m(u, x+y) + \inf_m(u, \smile x \smile y))$
$= \inf_m(u, \smile y \smile \inf_m(u, x))$,
hence:
(a) $\inf_m(u, \smile \inf_m(u, x+y) + \inf_m(u, \smile x \smile y)) = \inf_m(u, \smile y \smile x)$;
now, in (a), take $X := \smile x \in [0_M, u]$, $Y := \smile y \in [0_M, u]$, to obtain, by (DN):
$\inf_m(u, \smile \inf_m(u, \smile x + \smile y) + \inf_m(u, x+y)) = \inf_m(u, y+x)$,
which, by (Scomm), becomes:
$\inf_m(u, \inf_m(u, x+y) \smile \inf_m(u, \smile x \smile y)) = \inf_m(u, x+y)$, i.e. ($Z_u^R$). □

Corollary 3.2.9
Let $\mathcal{M}_u = (M, \leq_m, +, \smile, 0_M, u = u_M)$ be a strong m_0-ME structure. For $0_M \leq_m x, y \leq_m u$, we have (see property ($s5^R$)):

$$(V^{00}) \implies (Z_u^R).$$

Proof. (V^{00}), i.e. $x + y = \inf_m(u, x+y) \smile \inf_m(u, \smile x \smile y)$
obviously implies:
(Z_u^R) $\inf_m(u, x+y) = \inf_m(u, \inf_m(u, x+y) \smile \inf_m(u, \smile x \smile y))$. □

Corollary 3.2.10
Let $\mathcal{M}_u = (M, \leq_m, +, \smile, 0_M, u = u_M)$ be a strong m_0-ME structure. For $0_M \leq_m x_0, x_1, y_0 \leq_m u$, we have (see property ($s5^R$)):

$$(V^{10}) \implies (V^{00}).$$

Proof. Recall that (V^{00}) is $x_0 + y_0 = \inf_m(u, x_0 + y_0) \smile \inf_m(u, \smile x_0 \smile y_0)$ and

(V^{10}) is: if $\inf_m(u, x_0 + x_1) = x_0$, then $x_0 + x_1 + y_0 = \inf_m(u, x_0 + y_0)$
$\smile \inf_m(u, \smile x_0 \smile \inf_m(u, x_1 + y_0)) \smile \inf_m(u, \smile x_1 \smile y_0)$.

Then, take $x_1 = 0_M$ in (V^{10}), to obtain, by (SU):
if $\inf_m(u, x_0) = x_0$ (that is true), then $x_0 + y_0 =$
$\inf_m(u, x_0 + y_0) \smile \inf_m(u, \smile x_0 \smile \inf_m(u, y_0)) \smile \inf_m(u, \smile 0_M \smile y_0)$,
which, by ($s3^R$) (ii), becomes:
$x_0 + y_0 = \inf_m(u, x_0 + y_0) \smile \inf_m(u, \smile x_0 \smile y_0) + 0_M$,
which, by (SU), becomes:
$x_0 + y_0 = \inf_m(u, x_0 + y_0) \smile \inf_m(u, \smile x_0 \smile y_0)$, i.e. ($V^{00}$) holds. □

Note that, if $\mathcal{M}_u = (M, \leq_m, +, \smile, 0_M, u_M)$ is a strong m_0-ME structure, then $(M, +, 0_M, u_M)$ is a strong monoid. Hence, (C_u^R) holds, by Corollary 3.1.5.

Corollary 3.2.11 *Let $\mathcal{M}_u = (M, \leq_m, +, \smile, 0_M, u = u_M)$ be a strong m_0-ME structure. We have the following particular compatibility property: for all $x, y \in M$ and each integer $n \geq 1$,*
(C_{pu}^R) $\quad x \leq_m y \iff x + nu \leq_m y + nu.$

Proof. We prove by induction on $n \geq 1$.
- For $n = 1$, (C_{pu}^R) holds by (C_{pu}^{1R}).
- Suppose that (C_{pu}^R) holds for n, i.e. $x \leq_m y \iff x + nu \leq_m y + nu$.
- We shall prove that (C_{pu}^R) holds for $n+1$, i.e.

$x \leq_m y \iff x + (n+1)u \leq_m y + (n+1)u$. Indeed,
$x \leq_m y \iff x + nu \leq_m y + nu$, by the inductive hypothesis
$\overset{(C_{pu}^{1R})}{\iff} (x + nu) + u \leq_m (y + nu) + u$
$\overset{(Sass)}{\iff} x + (nu + u) \leq_m y + (nu + u)$
$\iff x + (n+1)u \leq_m y + (n+1)u$. □

Remark 3.2.12

$$(S^{1R}) \implies (Nu^{1R}).$$

Indeed, $\smile u \overset{(SU)}{=} \smile (0_M + u) \overset{(S^{1R})}{=} \smile 0_M + u \overset{(N0)}{=} 0_M + u \overset{(SU)}{=} u$.

Corollary 3.2.13 *Let $\mathcal{M}_u = (M, \leq_m, +, \smile, 0_M, u = u_M)$ be a strong m_0-ME structure. We have: for each integer $n \geq 1$,*
(S^R) $\quad \smile (x + nu) = \smile x + nu.$

Proof. We prove by induction on $n \geq 1$.
- For $n = 1$, (S^R) holds by (S^{1R}).
- Suppose that (S^R) holds for n.
- We shall prove that (S^R) holds for $n+1$. Indeed,

$\smile (x+(n+1)u) = \smile (x+(nu+u)) \overset{(Sass)}{=} \smile ((x+nu)+u) \overset{(S^{1R})}{=} \smile (x+nu)+u \overset{ind.\ hyp.}{=}$
$(\smile x + nu) + u \overset{(Sass)}{=} \smile x + (nu + u) = \smile x + (n+1)u$. □

Corollary 3.2.14 *Let $\mathcal{M}_u = (M, \leq_m, +, \smile, 0_M, u = u_M)$ be a strong m_0-ME structure. We have: for each integer $n \geq 1$,*
(Nu^R) $\quad \smile nu = nu.$

Proof. By above (S^R), $\smile nu = \smile (0_M + nu) = \smile 0_M + nu = 0_M + nu = nu.$ □

Corollary 3.2.15 *Let $\mathcal{M}_u = (M, \leq_m, +, \smile, 0_M, u = u_M)$ be a strong m_0-ME structure. We have: for each integer $n \geq 1$,*
(NNu^R) $\quad \smile (\smile nu) = nu.$

Proof. We prove by induction on $n \geq 1$.
- For $n = 1$, (NNu^R) holds by (NNu^{1R}).
- Suppose that (NNu^R) holds for n.
- We shall prove that (NNu^R) holds for $n+1$. Indeed,

$\smile (\smile (n+1)u) = \smile (\smile (nu + u)) \overset{(s2^R)(i)}{=} \smile (\smile nu \smile u)$
$\overset{(s2^R)(i)}{=} \smile (\smile nu) \smile (\smile u) \overset{(NNu^{1R})}{=} \smile (\smile nu) + u = nu + u = (n+1)u$, by the inductive hypothesis. □

Remarks 3.2.16 *(The dual ones are omitted)*

(i) Concerning the property ($s1^R$), note that for $0_M \in M$, there exists $0 \in \mathbf{N}$ such that $0_M \leq_m 0u_M = 0_M$, by Notation 3.1.4 and by Definitions 3.2.5.

(ii) The strong m_0-ME structure $\mathcal{M}_u$ does not verify the cancellation property (C) for any element, but it verifies the particular cancellation property (C_u^{1R}), hence it verifies the particular cancellation property (C_u^R), by Corollary 3.1.5.

(iii) The external binary relation $\leq_m$ does not verify the compatibility property: for all $x, y, z \in M$,
(Cp) $x \leq_m y \Longrightarrow x + z \leq_m y + z$,
but it verifies the particular stronger compatibility property (C_{pu}^{1R}) (with $\Longleftrightarrow$), hence it verifies the particular stronger compatibility property (C_{pu}^R), by Corollary 3.2.11.

(iv) The external binary relation $\leq_m$ is not reflexive, antisymmetric or transitive, i.e. it does not verify: for all $x, y, z \in M$,
(m-Re^R) $x \leq_m x$,
(m-An^R) $x \leq_m y$ and $y \leq_m x$ imply $x = y$,
(m-Tr^R) $x \leq_m y$ and $y \leq_m z$ imply $x \leq_m z$,
but it verifies the particular transitive property (m-Tr_u^R).

(v) The external binary relation $\leq_m$ is not a lattice order, but there exists particular infimum w.r.t. to $\leq_m$, denoted $\inf_m$, involved in the properties ($s3^R$), ($s4^R$), ($s5^R$), (XX_u^R), (YY_u^R).

Corollary 3.2.17 *We have:*
(i) $u_M \leq_m u_M$,
(ii) $0_M \leq_m nu_M$, for each $n \in \mathbf{N}$,
(iii) $n < m' = n + r$ implies $nu_M \leq_m m'u_M$.

Proof. (i): $0_M \leq_m 0_M$ (that is true) is equivalent, by (C_{pu}^{1R}), with $0_M + u_M \leq_m 0_M + u_M$, i.e. with $u_M \leq_m u_M$, by (SU).

(ii): By induction on $n \geq 0$.

• For $n = 0$, $0_M \leq_m 0u_M = 0_M$, by Notation 3.1.4, that is true by Definition 3.2.5.

• Suppose that $0_M \leq_m nu_M$.

• We prove that $0_M \leq_m (n+1)u_M$. Indeed,

$$\begin{array}{ll} & 0_M <_m u_M, \text{ by } (\mathrm{s}0^R), \\ \Longleftrightarrow & 0_M + nu_M \leq_m u_M + nu_M, \text{ by above } (\mathrm{C}^R_{pu}), \\ \Longleftrightarrow & nu_M \leq_m u_M + nu_M, \text{ by (SU)}, \\ \Longleftrightarrow & nu_M \leq_m (n+1)u_M. \end{array}$$

By induction hypothesis, we have $0_M \leq_m nu_M$; since we also have $nu_M \leq_m (n+1)u_M$, it follows, by (m-Tr^R_u), that $0_M \leq_m (n+1)u_M$.

(iii): By above (ii), $0_M \leq_m ru_M$ is true, which is equivalent, by (C^R_{pu}), with $0_M + nu_M \leq_m ru_M + nu_M$, i.e. with $nu_M \leq_m (r+n)u_M$, by (SU), i.e. with $nu_M \leq_m m'u_M$. □

3.3 Strong m_0-ME positive cones vs. positive-unital l-groups

Let $\mathcal{G}_u = (G, \vee, \wedge, +, -, 0_G, u = u_G)$ be a unital commutative l-group, with $x \leq y \Longleftrightarrow x \vee y = y \Longleftrightarrow x \wedge y = x$ and $0_G < u \in G$, and let $G^+ = \{x \in G \mid x \geq 0_G\}$ be its positive cone. Then, G^+ is closed under $\vee$, $\wedge$ and $+$, 0_G, u, but it is not closed under $-$.

Recall (1.3) from Chapter 1 that we can endow G^+ with a negation $\smile$ defined by: for all $x \in G^+$,

$$\smile x \stackrel{def.}{=} \begin{cases} u - x, & \text{if} \quad u - x \geq 0_G, \\ x - u, & \text{if} \quad u - x < 0_G, \quad x \neq 0_G, \\ u \wedge x, & \quad\quad \text{otherwise}, \end{cases} \quad \smile 0_G \stackrel{def.}{=} 0_G,$$

$$\smile (x+u) \stackrel{def.}{=} \smile x.$$

Note that $x \in [0_G, u] \Longleftrightarrow \smile x \in [0_G, u]$; $\smile(\smile x) = x$ and $x + \smile x = u$, for all $x \in [0_G, u]$; $\smile u = 0_G$.

Then,

$$\mathcal{G}^+_u = (G^+, \vee, \wedge, +, \smile, 0_G, u = u_G)$$

is the associated *positive-unital commutative l-group* (Definition 1.4.1) by the functor $\mathbf{c}^+$.

*

We introduce now the following new notions.

Definitions 3.3.1

(1) A *strong m_0-ME positive cone* is a structure

$$\mathcal{M}^+_u = (M^+, \leq_m, +, \smile, 0_{M^+}, u = u_{M^+})$$

such that:
(i) $(M^+, \leq_m, +, \smile, 0_{M^+})$ is a m_0-ME structure,
(ii) $0_{M^+} <_m u$ is an element of M^+ verifying:
(s0^R) $0_{M^+} \leq_m x$, for each $x \in M^+$;
(s1^R) for all $x \in M^+$, there exists $n_x \in \mathbf{N}$ such that $x, \smile x \leq_m n_x u$;
(s2^{+R}) (i) for all $x \in M^+$, there exists $n_x \in \mathbf{N}$
and there exist the unique elements of M^+:
$x_1, x_2, \ldots, x_{n_x} \leq_m u$ (hence $0_{M^+} \leq_m \smile x_1, \smile x_2, \ldots, \smile x_{n_x} \leq_m u$)
s.t. $x = x_1 + \ldots + x_{n_x}$ $(\leq_m n_x u)$, $\smile x = \smile x_1 \smile \ldots \smile x_{n_x}$ $(\leq_m n_x u)$,
(ii) if $x = x_1 + \ldots + x_n$ and $y = y_1 + \ldots + y_n$,
then $x = y \Longleftrightarrow x_i = y_i$, $i = 1, \ldots, n$,
(iii) if $x = x_1 + \ldots + x_n$ and $y = y_1 + \ldots + y_n$,
then $x \leq_m y \Longleftrightarrow x_i \leq_m y_i$, $i = 1, \ldots, n$;
(s3^R) for $0_{M^+} \leq_m x, y \leq_m u = u_{M^+}$ (i.e. $x, y \in [0_{M^+}, u]$),
(i) there exist $\inf_m(u, x + y)$, $\smile \inf_m(u, \smile x \smile y)$ and belong to $[0_{M^+}, u]$,
(ii) $\inf_m(u, x+u) = u$, $\smile \inf_m(u, \smile x \smile 0_{M^+}) = 0_{M^+}$, $\smile \inf_m(u, \smile x \smile u) = x$,
(iii) $x \leq_m y$ $(x, y \neq 0_{M^+}) \Longleftrightarrow u \leq_m y \smile x$;
(s4^R) for $0_{M^+} \leq_m x, y, z \leq_m u = u_{M^+}$,
(i) there exist and are equal $\inf_m(u, x + \inf_m(u, y + z)) = \inf_m(u, x + (y + z))$,
(ii) $\inf_m(u, \smile \inf_m(u, \smile x \smile y)) = \smile \inf_m(u, \smile x \smile y)$;
(s5^R) for $0_{M^+} \leq_m x_0, x_1, \ldots, x_n$, $y_0 \leq_m u = u_{M^+}$,
(V^{00}) $x_0 + y_0 = \inf_m(u, x_0 + y_0) \smile \inf_m(u, \smile x_0 \smile y_0)$,
(V^{10}) if $\inf_m(u, x_0 + x_1) = x_0$, then $x_0 + x_1 + y_0 = \inf_m(u, x_0 + y_0)$
$\smile \inf_m(u, \smile x_0 \smile \inf_m(u, x_1 + y_0)) \smile \inf_m(u, \smile x_1 \smile y_0)$,
(V^{n0}) if $\inf_m(u, x_i + x_{i+1}) = x_i$, for $i = 0, \ldots, n-1$, then
$\sum_{i=0}^{n} x_i + y_0 = \sum_{i=0}^{n+1} c_i$, where
$c_i = \smile \inf_m(u, \smile x_{i-1} \smile \inf_m(u, x_i + y_0))$, with $x_{-1} = u$, $x_{n+1} = 0_{M^+}$;
(XX_u^R) for $0_{M^+} \leq_m x, y, z \leq_m u = u_{M^+}$, there exist and are equal
$\smile \inf_m(u, \smile \inf_m(u, x + y) \smile \inf_m(u, z \smile \inf_m(u, \smile x \smile y)))$
$= \smile \inf_m(u, \smile \inf_m(u, y + z) \smile \inf_m(u, x \smile \inf_m(u, \smile y \smile z)))$;
(YY_u^R) for $0_{M^+} \leq_m x, y, z \leq_m u = u_{M^+}$,
if $\inf_m(u, x + y) = x$, then there exist and are equal
$\smile \inf_m(u, \smile x \smile \inf_m(u, y + z)) = \inf_m(u, y \smile \inf_m(u, \smile x \smile z))$;
(C_u^{1R}) $x + u = y + u \Longrightarrow x = y$, for all $x, y \in M^+$;
(C_{pu}^{1R}) $x \leq_m y \Longleftrightarrow x + u \leq_m y + u$, for all $x, y \in M^+$;
(m-Tr_u^R) $x \leq_m nu$ and $nu \leq_m ru \Longrightarrow x \leq_m ru$, for each $x \in M^+$, $n, r \in \mathbf{N}$;
(Nu^{+1R}) $\smile u = 0_{M^+}$;
(NNu^{1R}) $\smile(\smile u) = u$;
(S^{+1R}) $\smile(x + u) = \smile x \smile u = \smile x$, for each $x \in M^+$,
where $\smile x \smile y$ means $\smile x + \smile y$.

(1') Dually, a *strong m_1-ME positive cone* is a structure

$$\mathcal{M}_u^+ = (M^+, \leq_m, \cdot, ^{-1}, 1_{M^+}, u = u_{M^+})$$

such that:

(i') $(M^+, \leq_m, \cdot, ^{-1}, 1_{M^+})$ is a m_1-ME structure,
(ii') $1_{M^+} <_m u_{M^+}$ is an element of M^+ verifying:
(s0) $1_{M^+} \leq_m x$, for each $x \in M^+$;
(s1) for each $x \in M^+$, there exists $n_x \in \mathbf{N}$ such that $x, x^{-1} \leq_m n_x u$;
(s2$^+$) (i) there exists $n_x \in \mathbf{N}$
and there exist the unique elements of M^+:
$x_1, x_2, \ldots, x_{n_x} \leq_m u$ (hence $1_{M^+} \leq_m x_1^{-1}, x_2^{-1}, \ldots, x_{n_x}^{-1} \leq_m u$)
s.t. $x = x_1 \cdot \ldots \cdot x_{n_x}$ ($\leq_m n_x u$), $x^{-1} = x_1^{-1} \cdot \ldots \cdot x_{n_x}^{-1}$ ($\leq_m n_x u$),
(ii) if $x = x_1 \cdot \ldots \cdot x_n$ and $y = y_1 \cdot \ldots \cdot y_n$,
then $x = y \Longleftrightarrow x_i = y_i$, $i = 1, \ldots, n$,
(iii) if $x = x_1 \cdot \ldots \cdot x_n$ and $y = y_1 \cdot \ldots \cdot y_n$,
then $x \leq_m y \Longleftrightarrow x_i \leq_m y_i$, $i = 1, \ldots, n$;
(s3) for $1_{M^+} \leq_m x, y \leq_m u$ (i.e. $x, y \in [1_{M^+}, u]$),
(i) there exist $\sup_m(u, x \cdot y)$, $\sup_m(u, x^{-1} \cdot y^{-1})^{-1}$ and belong to $[1_{M^+}, u]$,
(ii) $\sup_m(u, x \cdot 0_{M^+}) = 0_{M^+}$, $\sup_m(u, x^{-1} \cdot u^{-1})^{-1} = u$, $\sup_m(u, x^{-1} \cdot 0_{M^+}^{-1})^{-1} = x$,
(iii) $x \leq_m y$ $(x, y \neq 1_{M^+}) \Longleftrightarrow u \leq_m y \cdot x^{-1}$;
(s4) for $1_{M^+} \leq_m x, y, z \leq_m u$,
(i) there exist and are equal $\sup_m(u, x \cdot \sup_m(u, y \cdot z)) = \sup_m(u, x \cdot (y \cdot z))$,
(ii) $\sup_m(u, \sup_m(u, x^{-1} \cdot y^{-1})^{-1}) = \sup_m(u, x^{-1} \cdot y^{-1})^{-1}$;
(s5) for $1_{M^+} \leq_m x_0, x_1, \ldots, x_n$, $y_0 \leq_m u$,
(V^{00}) $x_0 \cdot y_0 = \sup_m(u, x_0 \cdot y_0) \cdot \sup_m(u, x_0^{-1} \cdot y_0^{-1})^{-1}$,
(V^{10}) if $\sup_m(u, x_0 \cdot x_1) = x_0$, then $x_0 \cdot x_1 \cdot y_0 = \sup_m(u, x_0 \cdot y_0)$
$\cdot \sup_m(u, x_0^{-1} \cdot \sup_m(u, x_1 \cdot y_0)^{-1})^{-1} \cdot \sup_m(u, x_1^{-1} \cdot y_0^{-1})^{-1}$,
(V^{n0}) if $\sup_m(u, x_i \cdot x_{i+1}) = x_i$, for $i = 0, \ldots, n-1$, then
$\prod_{i=0}^{n} x_i \cdot y_0 = \prod_{i=0}^{n+1} c_i$, where
$c_i = \sup_m(u, x_{i-1}^{-1} \cdot \sup_m(u, x_i \cdot y_0)^{-1})^{-1}$, with $x_{-1} = u$, $x_{n+1} = 0_{M^+}$;
(XX_u) for $1_{M^+} \leq_m x, y, z \leq_m u$, there exist and are equal
$\sup_m(u, \sup_m(u, x \cdot y)^{-1} \cdot \sup_m(u, z \cdot \sup_m(u, x^{-1} \cdot y^{-1})^{-1})^{-1}))^{-1}$
$= \sup_m(u, \sup_m(u, y \cdot z)^{-1} \cdot \sup_m(u, x \cdot \sup_m(u, y^{-1} \cdot z^{-1})^{-1})^{-1})^{-1}$;
(YY_u) for $1_{M^+} \leq_m x, y, z \leq_m u$,
if $\sup_m(u, x \cdot y) = x$, then there exist and are equal
$\sup_m(u, x^{-1} \cdot \sup_m(u, y \cdot z)^{-1})^{-1} = \sup_m(u, y \cdot \sup_m(u, x^{-1} \cdot z^{-1})^{-1})$;
(C_u^1) $x \cdot u = y \cdot u \Longrightarrow x = y$, for all $x, y \in M^+$;
(C_{pu}^1) $x \leq_m y \Longleftrightarrow x \cdot u \leq_m y \cdot u$, for all $x, y \in M^+$;
(m-Tr_u) $x \leq_m nu$ and $nu \leq_m ru \Longrightarrow x \leq_m ru$, for each $x \in M^+$, $n, r \in \mathbf{N}$;
(Nu^{+1}) $(u)^{-1} = 1_{M^+}$;
(NNu^1) $((u)^{-1})^{-1} = u$;
(S^{+1}) $(x \cdot u)^{-1} = x^{-1} \cdot (u)^{-1} = x^{-1}$, for each $x \in M^+$.

Denote by **s-m$_0$-ME$^+$** the class of all strong m_0-ME positive cones and by **s-m$_1$-ME$^+$** the class of all strong m_1-ME positive cones.

Remark 3.3.2 *The axioms of strong m_0-ME positive cones are verified in the particular example $G^+(A)$ (see next Chapter 7).*

Note that the properties ($s3^R$) - ($s5^R$), (XX_u^R), (YY_u^R) refer to the unit interval $[0_{M^+}, u_{M^+}]$.

Corollary 3.3.3 *(See Proposition 2.2.36)*
Let $\mathcal{M}_u^+ = (M^+, \leq_m, +, \smile, 0_{M^+}, u = u_{M^+})$ *be a strong* m_0*-ME positive cone. For* $0_{M^+} \leq_m x, y, z \leq_m u$*, we have:*

$$(XX_u^R) \implies (Z_u^R),$$

where: for all $x, y \in M^+$*,*
(Z_u^R) $\inf_m(u, x+y) = \inf_m(u, \inf_m(u, x+y) \smile \inf_m(u, \smile x \smile y))$.

Proof. Similarly to the proof of Corollary 3.2.8. □

Corollary 3.3.4
Let $\mathcal{M}_u^+ = (M^+, \leq_m, +, \smile, 0_{M^+}, u = u_{M^+})$ *be a strong* m_0*-ME positive cone. For* $0_{M^+} \leq_m x, y \leq_m u$*, we have (see property* $(s5^R)$*):*

$$(V^{00}) \implies (Z_u^R).$$

Proof. Similarly to the proof of Corollary 3.2.9. □

Corollary 3.3.5
Let $\mathcal{M}_u^+ = (M^+, \leq_m, +, \smile, 0_{M^+}, u = u_{M^+})$ *be a strong* m_0*-ME positive cone. For* $0_{M^+} \leq_m x_0, x_1, y_0 \leq_m u$*, we have (see property* $(s5^R)$*):*

$$(V^{10}) \implies (V^{00}).$$

Proof. Similarly to the proof of Corollary 3.2.10. □

More, we have here a result that is not true for strong m_0-ME structures:

Corollary 3.3.6
Let $\mathcal{M}_u^+ = (M^+, \leq_m, +, \smile, 0_{M^+}, u = u_{M^+})$ *be a strong* m_0*-ME positive cone. For* $0_{M^+} \leq_m x_0, x_1, \ldots, x_n, y_0 \leq_m u$*, we have (see property* $(s5^R)$*):*

$$(V^{n0}) \implies (V^{00}).$$

Proof. Recall that (V^{00}) is $x_0 + y_0 = \inf_m(u, x_0 + y_0) \smile \inf_m(u, \smile x_0 \smile y_0)$ and
(V^{n0}) is: if $\inf_m(u, x_i + x_{i+1}) = x_1$, for $i = 0, \ldots, n-1$, then
$\sum_{i=0}^{n} x_i + y_0 = \sum_{i=0}^{n+1} c_i$, where
$c_i = \smile \inf_m(u, \smile x_{i-1} \smile \inf_m(u, x_i + y_0))$, with $x_{-1} = u$, $x_{n+1} = 0_{M^+}$.
Then, take $x_1 = \ldots = x_n = 0_{M^+}$ in (V^{n0}), to obtain:
if $\inf_m(u, x_0) = x_0$ (that is true), then $x_0 + y_0 = c_0 + c_1$, where
$c_0 = \smile \inf_m(u, \smile x_{-1} \smile \inf_m(u, x_0 + y_0))$
$= \smile \inf_m(u, -u \smile \inf_m(u, x_0 + y_0))$
$= \smile \inf_m(u, 0_{M^+} \smile \inf_m(u, x_0 + y_0))$, by (Nu^{+1R}),
$= \smile \inf_m(u, \smile \inf_m(u, x_0 + y_0))$, by (SU),
$= \smile \inf_m(u, \smile \inf_m(u, \smile(\smile x_0) \smile (\smile y_0)))$, by (DN),

$=\smile(\smile \inf_m(u,\smile(\smile x_0)\smile(\smile y_0)))$, by ($s4^R$) (ii),
$= \inf_m(u, x_0 + y_0)$, by (DN),
$c_1 =\smile \inf_m(u,\smile x_0 \smile \inf_m(u, y_0))$
$=\smile \inf_m(u,\smile x_0 \smile y_0)$,
$c_2 =\smile \inf_m(u,\smile x_1 \smile \inf_m(u, x_2 + y_0))$
$=\smile \inf_m(u,\smile 0_{M^+} \smile \inf_m(u, y_0))$
$=\smile \inf_m(u,\smile 0_{M^+} \smile y_0)$
$= 0_{M^+}$, by ($s3^R$) (ii).
Hence, $x_0 + y_0 = \inf_m(u, x_0 + y_0)+\smile \inf_m(u,\smile x_0 \smile y_0)$, i.e. ($V^{00}$) holds. □

Open problems 3.3.7 *Concerning the property ($s5^R$) in a strong m_0-ME positive cone:*

(1) Does (V^{00}) imply (V^{10})? (see Proposition 3.4.16).

(2) Does (V^{00}) imply (V^{n0})?

Is there a connection between the properties (Z^{00}), (Z^{10}), ... , (Z^{n0}) in a right-XY algebra (they are equivalent, see Remarks 7.1.93) and the properties (V^{00}), (V^{10}), ... , (V^{n0}) from ($s5^R$) in a strong m_0-ME positive cone?

Note that, if $\mathcal{M}_u^+ = (M^+, \leq_m, +, \smile, 0_{M^+}, u = u_{M^+})$ is a strong m_0-ME positive cone, then $(M^+, +, 0_{M^+}, u)$ is a strong commutative monoid. Hence, by Corollary 3.1.5, the property (C_u^R) holds.

Corollary 3.3.8 *Let $\mathcal{M}_u^+ = (M^+, \leq_m, +, \smile, 0_{M^+}, u = u_{M^+})$ be a strong m_0-ME positive cone. We have the following particular compatibility property: for all $x, y \in M^+$ and each integer $n \geq 1$,*
(C_{pu}^R) $\quad x + nu \leq_m y + nu \Longrightarrow x \leq_m y.$

Proof. Similarly to the proof of Corollary 3.2.11. □

Remark 3.3.9

$$(S^{+1R}) \Longrightarrow (Nu^{+1R}).$$

Indeed, $\smile u \overset{(SU)}{=} \smile (0_{M^+} + u) \overset{(S^{+1R})}{=} \smile 0_{M^+} \overset{(N0)}{=} 0_{M^+}.$

Corollary 3.3.10 *Let $\mathcal{M}_u = (M^+, \leq_m, +, \smile, 0_{M^+}, u = u_{M^+})$ be a strong m_0-ME positive cone. We have: for each integer $n \geq 1$,*
(S^{+R}) $\quad \smile (x + nu) =\smile x \smile nu =\smile x.$

Proof. We prove by induction on $n \geq 1$.
• For $n = 1$, (S^{+R}) holds by (S^{+1R}).
• Suppose that (S^{+R}) holds for n.
• We shall prove that (S^{+R}) holds for $n + 1$. Indeed,

$\smile (x + (n + 1)u) =\smile (x + (nu + u)) \overset{(Sass)}{=} \smile ((x + nu) + u$
$\overset{(S^{+1R})}{=} \smile (x + nu) \overset{ind.\ hyp.}{=} \smile x.$ □

Corollary 3.3.11 *Let* $\mathcal{M}_u^+ = (M^+, \leq_m, +, \smile, 0_{M^+}, u = u_{M^+})$ *be a strong* m_0*-ME positive cone. We have: for each integer* $n \geq 1$,
(Nu^{+R}*)* $\qquad \smile nu = 0_{M^+}$.

Proof. By (S^{+R}), $\smile nu = \smile (0_{M^+} + nu) = \smile 0_{M^+} \overset{(N0)}{=} 0_{M^+}$. □

Corollary 3.3.12 *Let* $\mathcal{M}_u^+ = (M^+, \leq_m, +, \smile, 0_{M^+}, u = u_{M^+})$ *be a strong* m_0*-ME positive cone. We have: for each integer* $n \geq 1$,
(NNu^R*)* $\qquad \smile(\smile nu) = nu$.

Proof. Similarly to the proof of Corollary 3.2.15. □

Remark 3.3.13 *(The dual one is omitted)*
The properties (s2^{+R}*), (Nu*$^{+1R}$*) and (S*$^{+1R}$*) of a strong* m_0*-ME positive cone differ from the properties (s2*^{R}*), (Nu*^{1R}*) and (S*^{1R}*), respectively, of a strong* m_0*-ME structure.*

Corollary 3.3.14 *We have:*
(i) $u_{M^+} \leq_m u_{M^+}$,
(ii) $0_{M^+} \leq_m nu_{M^+}$, *for each* $n \in \mathbf{N}$,
(iii) $n < m' = n + r$ *implies* $nu_{M^+} \leq_m m'u_{M^+}$.

Proof. Similarly to the proof of Corollary 3.2.17. □

Corollary 3.3.15 *For* $0_{M^+} \leq_m x_0, x_1, \ldots, x_{n-1}, x_n \leq_m u = u_{M^+}$, *there exist*
$\inf_m(u, x_n + \inf_m(u, x_0 + x_1 + \ldots + x_{n-1}))$
$= \inf_m(u, x_n + (x_0 + x_1 + \ldots + x_{n-1}))$.

Proof. By induction on $n \geq 2$.

• The statement is true for $n = 2$, by (s4^R):
For $0_{M^+} \leq_m x_0, x_1, x_2 \leq_m u$, there exist
$\inf_m(u, x_2 + \inf_m(u, x_0 + x_1)) = \inf_m(u, x_2 + (x_0 + x_1))$.

• Suppose that the statement is true for $n - 1$,
i.e. for $0_{M^+} \leq_m x_0, x_1, \ldots, x_{n-2}, x_{n-1} \leq_m u$, there exist
$\inf_m(u, x_{n-1} + \inf_m(u, x_0 + x_1 + \ldots + x_{n-2}))$
$= \inf_m(u, x_{n-1} + (x_0 + x_1 + \ldots + x_{n-2}))$.

• We shall prove that the statement is true for n. Indeed,
$\inf_m(u, x_n + \inf_m(u, x_0 + x_1 + \ldots + x_{n-1}))$
equals, by inductive hypothesis for $n - 1$,
$= \inf_m(u, x_n + \inf_m(u, x_{n-1} + \inf_m(u, x_0 + x_1 + \ldots + x_{n-2}))$,
which equals, by inductive hypothesis for $n = 2$,
$= \inf_m(u, x_n + (x_{n-1} + \inf_m(u, x_0 + x_1 + \ldots + x_{n-2})))$,
which equals, by (Sass) in M^+,
$= \inf_m(u, (x_n + x_{n-1}) + \inf_m(u, x_0 + x_1 + \ldots + x_{n-2}))$,
which equals, by inductive hypothesis for $n - 1$,
$= \inf_m(u, (x_n + x_{n-1}) + (x_0 + x_1 + \ldots + x_{n-2}))$,
which equals, by (Sass), (Scomm) in M^+,
$= \inf_m(u, x_n + (x_0 + x_1 + \ldots + x_{n-1}))$. □

3.4 Connections

We show the connections existing between the strong m_0-ME structure and the commutative l-group with strong unit and between the strong m_0-ME positive cone and the positive-unital commutative l-group.

*

• First, we shall prove that any commutative l-group is a m_0-ME structure.

Proposition 3.4.1 *Let $\mathcal{G} = (G, \vee, \wedge, +, -, 0_G)$ be a commutative l-group, with $x \leq y \iff x \wedge y = x \iff x \vee y = y$. Then, $(G, \leq_m = \leq, +, -, 0_G)$ is a m_0-ME structure.*

Proof. Since $(G, +, -, 0_G)$ is a commutative group, then (SU), (Scomm), (Sass), (N0), (DN) hold. It follows that $(G, +, -, 0_G)$ is an involutive m_0-ME algebra.

Since $\leq$ is a lattice order relation, then it is reflexive, hence $0_G \leq 0_G$. It follows that $(G, \leq, +, -, 0_G)$ is a m_0-ME structure. □

• Now, we shall analyse the properties ($s0^R$), ($s1^R$), ($s2^R$), ($s3^R$), ($s4^R$), (XX_u^R), (YY_u^R) in a commutative l-group with strong unit.

Let $\mathcal{G}_u = (G, \vee, \wedge, +, -, 0_G, u = u_G)$ be a commutative l-group with strong unit $u = u_G$, with $x \leq y \iff x \wedge y = x \iff x \vee y = y$.
Then, the fact that u_G is a strong unit (axioms (g0) and (g1) in Definitions 3.2.1) and the elements a, b, c of G verify: $0_G \leq a, b, c \leq u_G$ is expressed by the following conditions (α):

(α)	$0_G \wedge u_G = 0_G$	(i.e. $0_G \leq u_G$) (g0),
	$\forall x \in G, \exists n_x \in \mathbf{N}, (x \vee -x) \vee n_x u_G = n_x u_G$	(i.e. $x \vee -x \leq n_x u_G$) (g1);
	$0_G \wedge a = 0_G$	(i.e. $0_G \leq a$),
	$0_G \wedge b = 0_G$	(i.e. $0_G \leq b$),
	$0_G \wedge c = 0_G$	(i.e. $0_G \leq c$);
	$a \wedge u_G = a$	(i.e. $a \leq u_G$),
	$b \wedge u_G = b$	(i.e. $b \leq u_G$),
	$c \wedge u_G = c$	(i.e. $c \leq u_G$);
	$(a \vee -a) \vee u_G = u_G$	(i.e. $a \vee -a \leq u_G$),
	$(b \vee -b) \vee u_G = u_G$	(i.e. $b \vee -b \leq u_G$),
	$(c \vee -c) \vee u_G = u_G$	(i.e. $c \vee -c \leq u_G$).

Proposition 3.4.2 *Let $\mathcal{G}_u = (G, \vee, \wedge, +, -, 0_G, u = u_G)$ be a commutative l-group with strong unit $u = u_G$, with $x \leq y \iff x \wedge y = x \iff x \vee y = y$. Then, the following properties hold:*
($s0^R$) $0_G < u_G$;
($s1^R$) For each $x \in G$, there exists $n_x \in \mathbf{N}$, such that $x, -x \leq n_x u_G$;
($s2^R$) (i) For each $0_G \leq x \in G$, there exists $n_x \in \mathbf{N}$ and there exist the unique elements of G:

$0_G \leq x_1, x_2, \ldots, x_{n_x} \leq u_G$ (hence $0_G \leq -x_1, -x_2, \ldots, -x_{n_x} \leq u_G$), such that $x = x_1 + x_2 + \ldots + x_{n_x} (\leq n_x u_G)$, $-x = -x_1 - x_2 - \ldots - x_{n_x} (\leq n_x u_G)$;

(ii) if $0_G \leq x = x_1 + x_2 + \ldots + x_n \in G$ and $0_G \leq y = y_1 + y_2 + \ldots + y_n \in G$, then $x = y \iff x_i = y_i$, for all $i = 1, 2, \ldots, n$;

(iii) if $0_G \leq x = x_1 + x_2 + \ldots + x_n \in G$ and $0_G \leq y = y_1 + y_2 + \ldots + y_n \in G$, then $x \leq y \iff x_i \leq y_i$, for all $i = 1, 2, \ldots, n$.

Proof. ($s0^R$): follows by (g0).

($s1^R$): follows by (g1), since x, $-x \leq x \vee -x \leq n_x u_G$ and since $\leq$ is transitive.

($s2^R$) (i): follows from the proof of ([10], Lemma 7.1.3), where for each integer $k \geq 1$, x_k is inductively defined by:
$x_1 = x \wedge u_G$ and $x_{k+1} = (x - x_1 - \ldots - x_k) \wedge u_G$.

($s2^R$) (ii): follows from the proof of ([10], Lemma 7.1.3).

($s2^R$) (iii): follows as a consequence of the proof of ([10], Lemma 7.1.3). □

Remark 3.4.3 *Based on (1.3), it follows that, if we want to find if the commutative l-group with strong unit u, $(G, \vee, \wedge, +, -, 0_G, u = u_G)$, verifies the properties ($s3^R$), ($s4^R$), ($s5^R$), (XX^R_u), (YY^R_u) and (Nu^{1R}), (NNu^{1R}), (S^{1R}) from Definitions 3.2.6 of strong m_0-ME structures, then we have to replace "$\smile x$" by "$u_G - x$" in each of these properties.*

Proposition 3.4.4 *Let $\mathcal{G}_u = (G, \vee, \wedge, +, -, 0_G, u = u_G)$ be a commutative l-group with strong unit $u = u_G$, with $x \leq y \iff x \wedge y = x \iff x \vee y = y$. Then, the following properties hold (see Remark 3.4.3):*
($s3^R$) For $0_G \leq a, b \leq u_G$,
(i) there exist $\inf(u_G, a + b)$, $u_G - \inf(u_G, (u_G - a) + (u_G - b))$ and belong to $[0_G, u_G]$,
(ii) $u_G \wedge (a + u_G) = u_G$,
$u_G - (u_G \wedge ((u_G - a) + (u_G - 0_G))) = 0_G$, $u_G - u_G \wedge ((u_G - a) + (u_G - u_G)) = a$,
(iii) $a \leq b$ $(a, b \neq 0_G) \iff u_G \leq b + (u_G - a)$.

Proof. ($s3^R$) (i): Obviously, they exist because there exists $\inf(x, y) = x \wedge y$ for any $x, y \in G$. We prove now that they belong to $[0 = 0_G, u = u_G]$.

- $u \wedge (a+b) \leq u \iff (u \wedge (a+b)) \vee u = u$, that is true, by (Vcomm) and (Vabs).
- $0 \leq u \wedge (a + b) \iff 0 \wedge (u \wedge (a + b)) = 0$.

Indeed, by a proof by PROVER9 of Length 17 (Length of expanded proof, 24) (proof presented in **Appendix C** together with the program, where u is denoted by 1), we have:
First, in (Wass) ($(x \wedge y) \wedge z = x \wedge (y \wedge z)$), take $X := 0$, $Y := u$, to obtain:
$(0 \wedge u) \wedge z = 0 \wedge (u \wedge z)$,
which, since $0 \leq u$, becomes:
$0 \wedge z = 0 \wedge (u \wedge z)$, i.e.

$$0 \wedge (u \wedge x) = 0 \wedge x. \tag{3.2}$$

Then, in (Wass) again, take $X := 0$, $Y := a$, to obtain:
$(0 \wedge a) \wedge z = 0 \wedge (a \wedge z)$,

which, since $0 \leq a$, becomes:
$0 \wedge z = 0 \wedge (a \wedge z)$, i.e.

$$0 \wedge (a \wedge x) = 0 \wedge x. \tag{3.3}$$

Now, (Cp) ($x \leq y \Longleftrightarrow x+z \leq y+z$) is equivalent to $x \wedge y = x \Longleftrightarrow (x+z) \wedge (y+z) = x + z$, which, for $X := 0$, $Y := b$, becomes:
$0 \wedge b = 0 \Longleftrightarrow (0 + z) \wedge (b + z) = 0 + z$,
which, since $0 \leq b$, becomes:
$(0 + z) \wedge (b + z) = 0 + z$,
which, by (SU) and (Scomm), becomes:

$$x \wedge (x + b) = x. \tag{3.4}$$

Then, in (3.3), take $X := a + b$, to obtain:
$0 \wedge (a \wedge (a + b)) = 0 \wedge (a + b)$,
which, by (3.4) for $X := a$, becomes:
$0 \wedge a = 0 \wedge (a + b)$, i.e.

$$0 \wedge (a + b) = 0. \tag{3.5}$$

Finally, $0 \wedge (u \wedge (a + b)) \overset{(3.2)}{=} 0 \wedge (a + b) \overset{(3.5)}{=} 0$.

• $u - (u \wedge ((u - a) + (u - b))) \leq u \Longleftrightarrow (u + -(u \wedge ((u + -a) + (u + -b)))) \vee u = u$.
The proof by Prover9, of Length 94, is omitted.

• $0 \leq u - (u \wedge ((u - a) + (u - b))) \Longleftrightarrow 0 \wedge (u + -(u \wedge ((u + -a) + (u + -b)))) = 0$.
Indeed, by a proof of Prover9 of Length 23, we have:
First, in (Wass) ($(x \wedge y) \wedge z = x \wedge (y \wedge z)$), take $Y := x$, to obtain:
$(x \wedge x) \wedge z = x \wedge (x \wedge z)$,
which, by (Wid), becomes:
$x \wedge z = x \wedge (x \wedge z)$, i.e.

$$x \wedge (x \wedge y) = x \wedge y. \tag{3.6}$$

Then, in (g3) ($(x + y) \wedge (x + z) = x + (y \wedge z)$), take $Y := 0$, to obtain:
$(x + 0) \wedge (x + z) = x + (0 \wedge z)$,
which, by (SU), becomes:

$$x \wedge (x + y) = x + (0 \wedge y). \tag{3.7}$$

Then, note that (Cp) ($x \leq y \Longleftrightarrow x + z \leq y + z$) is equivalent to $x \wedge y = x \Longleftrightarrow (x + z) \wedge (y + z) = x + z$
and note that, obviously, we have:

$$x + (y + z) = y + (x + z). \tag{3.8}$$

Finally, $0 \wedge (u + -(u \wedge ((u + -a) + (u + -b)))) = 0$,
$\Longleftrightarrow 0 \wedge (u + -(u \wedge (u + (-a + (u + -b))))) = 0$, by (Sass)
$\Longleftrightarrow 0 \wedge (u + -(u \wedge (u + (u + (-a + -b))))) = 0$, by (3.8),

$\Longleftrightarrow 0 \wedge (u + -[u + (0 \wedge (u + (-a + -b)))]) = 0$, by (3.7)
$\Longleftrightarrow (0 + x) \wedge ((u + -[u + (0 \wedge (u + (-a + -b)))]) + x) = 0 + x$, by (Cp),
$\Longleftrightarrow x \wedge ((u + -[u + (0 \wedge (u + (-a + -b)))]) + x) = x$, by (SU),
$\Longleftrightarrow x \wedge (u + (-[u + (0 \wedge (u + (-a + -b)))] + x)) = x$, by (Sass),
$\Longleftrightarrow x \wedge (u + (x + -[u + (0 \wedge (u + (-a + -b)))])) = x$, by (Scomm),
$\Longleftrightarrow [u + (0 \wedge (u + (-a + -b)))] \wedge (u + ([u + (0 \wedge (u + (-a + -b)))] + -[u + (0 \wedge (u + (-a + -b)))])) = u + (0 \wedge (u + (-a + -b)))$, for $X := u + (0 \wedge (u + (-a + -b)))$,
$\Longleftrightarrow [u + (0 \wedge (u + (-a + -b)))] \wedge (u + 0) = u + (0 \wedge (u + (-a + -b)))$, by ($m_0$-Re),
$\Longleftrightarrow [u + (0 \wedge (u + (-a + -b)))] \wedge u = u + (0 \wedge (u + (-a + -b)))$, by (SU),
$\Longleftrightarrow u \wedge [u + (0 \wedge (u + (-a + -b)))] = u + (0 \wedge (u + (-a + -b)))$, by (Wcomm),
$\Longleftrightarrow u + (0 \wedge (0 \wedge (u + (-a + -b)))) = u + (0 \wedge (u + (-a + -b)))$, by (3.7) for $X := u$, $Y := 0 \wedge (u + (-a + -b))$,
$\Longleftrightarrow u + (0 \wedge (u + (-a + -b))) = u + (0 \wedge (u + (-a + -b)))$, by (3.6), that is true.

($s3^R$) (ii): • $u_G \wedge (a + u_G) = u_G$.

Indeed, in (g3) $(z + (x \wedge y) = (z + x) \wedge (z + y))$, take $X := 0_G$, $Y := a$, $Z := u_G$, to obtain:
$u_G + (0_G \wedge a) = (u_G + 0_G) \wedge (u_G + a)$,
which, by (SU), (Scomm) becomes:

$$u_G + (0_G \wedge a) = u_G \wedge (a + u_G); \tag{3.9}$$

now, since $0_G \leq a$, we obtain that $0_G \wedge a = 0_G$; hence, (3.9) becomes:
$u_G + 0_G = u_G \wedge (a + u_G)$,
which, by (SU) again, becomes:
$u_G \wedge (a + u_G) = u_G$. This proof is complete.

• $u_G - (u_G \wedge ((u_G - a) + (u_G - 0_G))) = 0_G$.

Indeed, (based on a proof by PROVER9 of Length 33):

First, in (g3) $((x + y) \wedge (x + z) = x + (y \wedge z))$, take $Y := 0_G$ to obtain, by (SU):

$$x \wedge (x + z) = x + (0_G \wedge z). \tag{3.10}$$

Now, in (3.10), take $Z := -x + y$, to obtain:
(*) $x \wedge (x + (-x + y)) = x + (0_G \wedge (-x + y))$;
but, in (*), the part $x + (-x + y)$ equals y, by (Sass), (m_0-Re), (SU); hence, (*) becomes:

$$x \wedge y = x + (0_G \wedge (-x + y)). \tag{3.11}$$

Then, in (3.11), take $X := -x$, to obtain, by (DN), (Scomm):

$$-x \wedge y = -x + (0_G \wedge (y + x)). \tag{3.12}$$

Now, in (g3) $((x + y) \wedge (x + z) = x + (y \wedge z))$ again, take $Z := -x$, to obtain:
$(x + y) \wedge (x + -x) = x + (y \wedge -x)$,
which, by (m_0-Re), becomes:
$(x + y) \wedge 0_G = x + (y \wedge -x)$,
which, by (Wcomm) $(x \wedge y = y \wedge x)$, becomes:

$$0_G \wedge (x + y) = x + (y \wedge -x). \tag{3.13}$$

Now, from (Cp) ($x \leq y \Longleftrightarrow x + z \leq y + z$), hence, equivalently, from:
$x \wedge y = x \Longleftrightarrow (x + z) \wedge (y + z) = x + z$,
which, by (Scomm), becomes:
$x \wedge y = x \Longleftrightarrow (z + x) \wedge (z + y) = x + z$,
by (g3), we obtain:

$$x \wedge y = x \Longleftrightarrow z + (x \wedge y) = x + z. \tag{3.14}$$

Now, since $a \leq u_G$, i.e. $a \wedge u_G = a$, we have:

$$u_G \wedge a = a. \tag{3.15}$$

Now, in (3.11), take $X := u_G$, $Y := a$, to obtain:
$u_G \wedge a = u_G + (0_G \wedge (-u_G + a))$,
which, by (3.15), becomes:
$a = u_G + (0_G \wedge (-u_G + a))$,
which, by (Scomm), becomes:
$a = u_G + (0_G \wedge (a + -u_G))$,
which, by (3.13), becomes:
$a = u_G + (a + (-u_G \wedge -a))$,
which, by (3.12), becomes:
$a = u_G + (a + (-u_G + (0_G \wedge (-a + u_G))))$,
which, since $u_G + -u_G = 0_G$, becomes:
$a = a + (0_G \wedge (-a + u_G))$,
which, by (Scomm), becomes:
$a + (0_G \wedge (u_G + -a)) = a$,
which, by (3.14), becomes, equivalently:

$$0_G \wedge (u_G + -a) = 0_G. \tag{3.16}$$

Finally, $u_G - (u_G \wedge ((u_G - a) + (u_G - 0_G))))$

$$\begin{aligned}
&= u_G + -(u_G \wedge ((u_G + -a) + (u_G + -0_G)))\\
&\overset{(N0)}{=} u_G + -(u_G \wedge ((u_G + -a) + (u_G + 0_G)))\\
&\overset{(SU)}{=} u_G + -(u_G \wedge ((u_G + -a) + u_G))\\
&\overset{(Scomm)}{=} u_G + -(u_G \wedge (u_G + (u_G + -a)))\\
&\overset{(3.10)}{=} u_G + -(u_G + (0_G \wedge (u_G + -a)))\\
&\overset{(NegS)}{=} u_G + (-(0_G \wedge (u_G + -a)) + -u_G)\\
&\overset{(Scomm)}{=} u_G + (-u_G + -(0_G \wedge (u_G + -a)))\\
&\overset{(Sass),(m_0-Re),(SU)}{=} -(0_G \wedge (u_G + -a))\\
&\overset{(3.16)}{=} -0_G\\
&\overset{(N0)}{=} 0_G.
\end{aligned}$$

This proof is complete.

- $u_G - u_G \wedge ((u_G - a) + (u_G - u_G)) = a$.

Indeed, (based on a proof by PROVER9 of Length 21):

First, in (g3) ($(x+y)\wedge(x+z) = x+(y\wedge z)$), take $Y := 0_G$, to obtain:
$(x+0_G)\wedge(x+z) = x+(0_G\wedge z)$,
which, by (SU), becomes:

$$x\wedge(x+z) = x+(0_G\wedge z). \tag{3.17}$$

Then, since $0_G \leq a$, we have:

$$0_G\wedge a = 0_G. \tag{3.18}$$

Now, in (Cp) ($x\leq y \Longleftrightarrow x+z\leq y+z$), hence, equivalently, in:
$x\wedge y = x \Longleftrightarrow (x+z)\wedge(y+z) = x+z$,
take $X := 0_G$, $Y := a$, to obtain:
$0_G\wedge a = 0_G \Longleftrightarrow (0_G+z)\wedge(a+z) = 0_G+z$,
which, by (SU), becomes:
$0_G\wedge a = 0_G \Longleftrightarrow z\wedge(a+z) = z$,
which, by (3.18), implies:
$z\wedge(a+z) = z$,
which, for $Z := -a$, becomes:
$-a\wedge(a+-a) = -a$,
which, by (m_0-Re), becomes:

$$-a\wedge 0_G = -a. \tag{3.19}$$

Finally, $u_G - u_G\wedge((u_G-a)+(u_G-u_G))$

$$\begin{aligned}
&= u_G + -u_G\wedge((u_G+-a)+(u_G+-u_G))\\
&\overset{(m_0-Re)}{=} u_G + -u_G\wedge((u_G+-a)+0_G)\\
&\overset{(SU)}{=} u_G + -u_G\wedge(u_G+-a)\\
&\overset{(3.17)}{=} u_G + -(u_G+(0_G\wedge -a))\\
&\overset{(NegS)}{=} u_G + (-(0_G\wedge -a)+-u_G))\\
&\overset{(Scomm)}{=} u_G + (-u_G+-(0_G\wedge -a))\\
&\overset{(Sass),(m_0-Re),(SU)}{=} -(0_G\wedge -a)\\
&\overset{(3.19)}{=} -(-a)\\
&\overset{(DN)}{=} a.
\end{aligned}$$

This proof is complete.

($s3^R$) (iii): $u_G \leq b+(u_G-a) \Longleftrightarrow 0_G\leq b-a$ and, obviously, in G, $a\leq b$ ($a,b\neq 0_G$) $\Longleftrightarrow 0_G\leq b-a$. $\square$

Proposition 3.4.5 *Let $\mathcal{G}_u = (G,\vee,\wedge,+,-,0_G,u=u_G)$ be a commutative l-group with strong unit $u=u_G$, with $x\leq y \Longleftrightarrow x\wedge y = x \Longleftrightarrow x\vee y = y$. Then, the following property holds:*
($s4^R$) For $0_G\leq a,b,c\leq u_G$,
(i) there exist and are equal

$$\inf(u_G, a+\inf(u_G, b+c)) = \inf(u_G, a+(b+c)).$$

Proof. (i): (Michael Kinyon) We have $0_G \leq a$, i.e. $0_G \wedge a = 0_G$.
In (g3) $(z + (x \wedge y) = (z + x) \wedge (z + y))$, take $x := 0_G$ and $y := a$, to obtain:
$z + (0_G \wedge a) = (z + 0_G) \wedge (z + a)$,
which becomes, by the above hypothesis and (SU):
$z + 0_G = z \wedge (z + a)$,
which becomes, by (SU) again:
(*) $z = z \wedge (z + a)$.
Then, $\inf(u_G, a + \inf(u_G, b + c)) = u_G \wedge (a + (u_G \wedge (b + c)))$
$\stackrel{(g3)}{=} u_G \wedge ((a + u_G) \wedge (a + (b + c)))$
$\stackrel{(Sass),(Scomm)}{=} (u_G \wedge (u_G + a)) \wedge (a + (b + c))$
$\stackrel{(*)}{=} u_G \wedge (a + (b + c)) = \inf(u_G, a + (b + c))$. □

Proposition 3.4.6 *Let $\mathcal{G}_u = (G, \vee, \wedge, +, -, 0_G, u = u_G)$ be a commutative l-group with strong unit $u = u_G$, with $x \leq y \Longleftrightarrow x \wedge y = x \Longleftrightarrow x \vee y = y$. Then, the following property holds (see Remark 3.4.3):*
($s4^R$) For $0_G \leq a, b \leq u_G$,
(ii) $\inf(u_G, u_G + -\inf(u_G, (u_G + -a) + (u_G + -b)))$
$= u_G + -\inf(u_G, (u_G + -a) + (u_G + -b))$, *i.e.*

$$u_G \wedge (u_G + -(u_G \wedge ((u_G + -a) + (u_G + -b)))) = u_G + -(u_G \wedge ((u_G + -a) + (u_G + -b))).$$

Proof. (ii): By PROVER9, Length of proof was 99, in 18.02 seconds; the renumbered expanded proof had the Length 344. See the 'hybrid' proof in **Appendix C.** □

Proposition 3.4.7 *(See Proposition 2.2.60)*
Let $\mathcal{G}_u = (G, \vee, \wedge, +, -, 0_G, u = u_G)$ be a commutative l-group with strong unit $u = u_G$, with $x \leq y \Longleftrightarrow x \wedge y = x \Longleftrightarrow x \vee y = y$. Then, the following property holds (see Remark 3.4.3):
(XX_u^R) for $0_G \leq a, b, c \leq u_G$,
$u_G + -\inf(u_G, [(u_G + -\inf(u_G, a + b)) + (u_G + -\inf(u_G, [c + (u_G + -\inf(u_G, (u_G + -a) + (u_G + -b)))]))]) =$
$u_G + -\inf(u_G, [(u_G + -\inf(u_G, b + c)) + (u_G + -\inf(u_G, [a + (u_G + -\inf(u_G, (u_G + -b) + (u_G + -c)))]))])$, *i.e.*

$$u_G + -(u_G \wedge [(u_G + -(u_G \wedge (a + b))) + (u_G + -(u_G \wedge [c + (u_G + -(u_G \wedge ((u_G + -a) + (u_G + -b))))]))]) =$$
$$u_G + -(u_G \wedge [(u_G + -(u_G \wedge (b + c))) + (u_G + -(u_G \wedge [a + (u_G + -(u_G \wedge ((u_G + -b) + (u_G + -c))))]))]).$$

Proof. (By PROVER9, Length of proof was 79, in 2.33 seconds; the renumbered expanded proof had the Length 358.) See the 'hybrid' proof in **Appendix C**. □

Corollary 3.4.8 *(See Corollary 2.2.61)*
Let $\mathcal{G}_u = (G, \vee, \wedge, +, -, 0_G, u = u_G)$ be a commutative l-group with strong unit $u = u_G$, with $x \leq y \Longleftrightarrow x \wedge y = x \Longleftrightarrow x \vee y = y$. Then, the following property

holds (see Remark 3.4.3): for all $0_G \leq a, b, c \leq u_G$,

$$(XX_u^R) \implies (Z_u^R),$$

where:
(Z_u^R) $u_G \wedge (a+b) = u_G \wedge ((u_G \wedge (a+b)) + (u_G + -(u_G \wedge ((u_G + -a) + (u_G + -b)))))$.

Proof.

In order to simplify the writting (less space on a line), we shall replace 0_G by 0 and u_G by 1.

Hence, we have to prove that, for all $0 \leq a, b, c \leq 1$: $(XX_u^R) \implies (Z_u^R)$, where: (XX_u^R) is:
$1+-(1\wedge[(1+-(1\wedge(a+b)))+(1+-(1\wedge[c+(1+-(1\wedge((1+-a)+(1+-b))))]))]) =$
$1+-(1\wedge[(1+-(1\wedge(b+c)))+(1+-(1\wedge[a+(1+-(1\wedge((1+-b)+(1+-c))))]))])$
and
(Z_u^R) is:
$1 \wedge (a+b) = 1 \wedge [(1 \wedge (a+b)) + (1 + -(1 \wedge ((1+-a) + (1+-b))))]$.

Indeed, take $c = 0$ in (XX_u), to obtain, by (SU):
(*) $1+-(1\wedge[(1+-(1\wedge(a+b)))+(1+-(1\wedge[1+-(1\wedge((1+-a)+(1+-b)))]))]) =$
$1+-(1\wedge[(1+-(1\wedge b))+(1+-(1\wedge[a+(1+-(1\wedge((1+-b)+(1+-0))))]))])$;
but, in (*), the part $1 \wedge [1 + -(1 \wedge ((1+-a) + (1+-b)))]$ on the left side equals $1 + -(1 \wedge ((1+-a) + (1+-b)))$, by $(\mathrm{s}4^R)$ (ii) (Proposition 3.4.6), and the part $1+-(1\wedge((1+-b)+(1+-0)))$ on the right side equals 0, by $(\mathrm{s}3^R)$ (ii) (Proposition 3.4.4); then, (*) becomes, by (SU):
$1 + -(1 \wedge [(1 + -(1 \wedge (a+b))) + (1 + -(1 + -(1 \wedge ((1+-a) + (1+-b)))))]) =$
$1 + -(1 \wedge [(1 + -(1 \wedge b)) + (1 + -(1 \wedge a))])$,
which, since $b \leq 1$ (hence, $1 \wedge b = b$) and $a \leq 1$ (hence, $1 \wedge a = a$), becomes:
$1 + -(1 \wedge [(1 + -(1 \wedge (a+b))) + (1 + -(1 + -(1 \wedge ((1+-a) + (1+-b)))))]) =$
$1 + -(1 \wedge [(1+-b) + (1+-a)])$,
which, by (NegS) on the left side and by (Scomm) on the right side, becomes:
$1 + -(1 \wedge [(1 + -(1 \wedge (a+b))) + (1 + (-(-(1 \wedge ((1+-a) + (1+-b)))) + -1))]) =$
$1 + -(1 \wedge [(1+-a) + (1+-b)])$,
which, by (DN) becomes:
$1 + -(1 \wedge [(1 + -(1 \wedge (a+b))) + (1 + ((1 \wedge ((1+-a) + (1+-b))) + -1))]) =$
$1 + -(1 \wedge [(1+-a) + (1+-b)])$,
which, by (Scomm), becomes:
$1 + -(1 \wedge [(1 + -(1 \wedge (a+b))) + (1 + (-1 + (1 \wedge ((1+-a) + (1+-b)))))]) =$
$1 + -(1 \wedge [(1+-a) + (1+-b)])$,
which, by (Sass), becomes:
$1 + -(1 \wedge [(1 + -(1 \wedge (a+b))) + ((1+-1) + (1 \wedge ((1+-a) + (1+-b))))]) =$
$1 + -(1 \wedge [(1+-a) + (1+-b)])$,
which, by (m_0-Re) and (SU), becomes:
(**) $1 + -(1 \wedge [(1 + -(1 \wedge (a+b))) + (1 \wedge ((1+-a) + (1+-b)))]) =$
$1 + -(1 \wedge [(1+-a) + (1+-b)])$.

Now, put in (**) $A := 1 + -a$; it follows that:

$$A+a = (1+-a)+a \overset{(Sass)}{=} 1+(-a+a) \overset{(Scomm)}{=} 1+(a+-a) \overset{(m_0-Re))}{=} 1+0 \overset{(SU)}{=} 1,$$

hence, $a + A = 1$, hence $a = 1 + -A$; since $0 \leq a$, it follows that $-a \leq 0$, hence $A = 1 + -a \leq 1 + 0 = 1$ and since $a \leq 1$, it follows that $0 = a + -a \leq 1 + -a = A$; hence, $0 \leq A \leq 1$.

Similarly, put in (**) $B := 1 + -b$; it follows $b = 1 + -B$ and $0 \leq B \leq 1$.

Then, (**) becomes:
$1 + -(1 \wedge [(1 + -(1 \wedge ((1 + -A) + (1 + -B)))) + (1 \wedge (A + B))]) =$
$1 + -(1 \wedge [A + B])$,
which, by (C), (DN) and (Scomm), becomes:
$1 \wedge [(1 \wedge (A + B)) + (1 + -(1 \wedge ((1 + -A) + (1 + -B))))] =$
$1 \wedge [A + B]$, i.e. (Z_u^R) holds. □

A direct proof that (Z_u^R) holds follows.

Proposition 3.4.9 *(See Proposition 2.2.59)*
Let $\mathcal{G}_u = (G, \vee, \wedge, +, -, 0_G, u = u_G)$ be a commutative l-group with strong unit $u = u_G$, with $x \leq y \Longleftrightarrow x \wedge y = x \Longleftrightarrow x \vee y = y$. Then, for all $0_G \leq a, b \leq u_G$, the property (Z_u^R) holds (see Remark 3.4.3), where:
(Z_u^R) $u_G \wedge (a+b) = u_G \wedge [(u_G \wedge (a+b)) + (u_G + -(u_G \wedge ((u_G + -a) + (u_G + -b))))]$.

Proof. (By PROVER9, Length of proof was 25, in 0.67 seconds; the expanded proof had the Length 49.)

In order to simplify the writting (less space on a line), we shall replace 0_G by 0 and u_G by 1.

Hence, we have to prove that, for all $0 \leq a, b \leq 1$:

$$1 \wedge (a + b) = 1 \wedge [(1 \wedge (a + b)) + (1 + -(1 \wedge ((1 + -a) + (1 + -b))))]. \tag{3.20}$$

There are not particular hypotheses used by PROVER9 (among the given ones - see (α)).

The following properties are obvious, by (SU), (Scomm), (Sass), (m_0-Re), (NegS) and (Wcomm) ($x \wedge y = y \wedge x$):

$$x + (y + z) = y + (x + z), \tag{3.21}$$

$$x + (-x + y) = y, \tag{3.22}$$

$$-x + (y + x) = y, \tag{3.23}$$

$$x + (y + (z + -x)) = y + z. \tag{3.24}$$

Now, in (g3) ($(x + y) \wedge (x + z) = x + (y \wedge z)$), take first $Y := 0$, to obtain, by (SU):

$$x \wedge (x + z) = x + (0 \wedge z). \tag{3.25}$$

Then, in (g3) again, take $Z := -x$, to obtain, by (m_0-Re),
$(x+y) \wedge 0 = x + (y \wedge -x)$,
which, by (Wcomm), becomes:

$$0 \wedge (x+y) = x + (y \wedge -x). \tag{3.26}$$

Now, in (3.23), take $X := 0 \wedge x$, to obtain:
$-(0 \wedge x) + (y + (0 \wedge x)) = y$,
which, by (3.25), becomes:
(*) $-(0 \wedge x) + (y \wedge (y+x)) = y$;
then, in (NegS) ($-(x+y) = -y + -x$), take $X := -(0 \wedge y)$ and $Y := x \wedge (x+y)$, to obtain, by (DN):
$-(-(0 \wedge y) + (x \wedge (x+y))) = -(x \wedge (x+y)) + (0 \wedge y)$,
which, by (*), becomes:
$-x = -(x \wedge (x+y)) + (0 \wedge y)$,
hence:

$$-(x \wedge (x+y)) + (0 \wedge y) = -x. \tag{3.27}$$

Finally, in (3.22), take $Y := y \wedge -(-x)$, to obtain, by (DN):
$x + (-x + (y \wedge -(-x))) = y \wedge x$,
which, by (3.26), becomes:
$x + (0 \wedge (-x+y)) = y \wedge x$,
hence:

$$x \wedge y = y + (0 \wedge (-y+x)). \tag{3.28}$$

Now, we are ready to prove (3.20).
Indeed, the following equivalent identities hold (copying the lines from the renumbered extended PROVER9 proof):

11 $\quad 1 \wedge ((1 \wedge (a+b)) + (1 + -(1 \wedge ((1 + -a) + (1 + -b))))) = 1 \wedge (a+b)$
12 $\iff 1 \wedge ((1 \wedge (b+a)) + (1 + -(1 \wedge ((1 + -a) + (1 + -b))))) = 1 \wedge (a+b)$, by (Scomm),
13 $\iff 1 \wedge ((1 \wedge (b+a)) + (1 + -(1 \wedge ((1 + -b) + (1 + -a))))) = 1 \wedge (a+b)$, by (Scomm),
14 $\iff 1 \wedge ((1 \wedge (b+a)) + (1 + -(1 \wedge (1 + (-b + (1 + -a)))))) = 1 \wedge (a+b)$, by (Sass),
15 $\iff 1 \wedge ((1 \wedge (b+a)) + (1 + -(1 \wedge (1 + (-b + (1 + -a)))))) = 1 \wedge (b+a)$, by (Scomm),

19 $\iff 1 \wedge ((1 \wedge (b+a)) + (1 + -(1 \wedge (1 + (1 + (-b + -a)))))) = 1 \wedge (b+a)$, by (3.21),
20 $\iff 1 \wedge (1 + ((1 \wedge (b+a)) + -(1 \wedge (1 + (1 + (-b + -a)))))) = 1 \wedge (b+a)$, by (3.21),

39 $\iff 1 \wedge (1 + (((b+a) + (0 \wedge (-(b+a)+1))) + -(1 \wedge (1 + (1 + (-b + -a)))))) = 1 \wedge (b+a)$, by (3.28),

$40 \Longleftrightarrow 1\wedge(1+(((b+a)+(0\wedge((-a+-b)+1)))+-(1\wedge(1+(1+(-b+-a)))))) = 1\wedge(b+a)$, by (NegS),
$41 \Longleftrightarrow 1\wedge(1+(((b+a)+(0\wedge((-b+-a)+1)))+-(1\wedge(1+(1+(-b+-a)))))) = 1\wedge(b+a)$, by (Scomm),
$42 \Longleftrightarrow 1\wedge(1+(((b+a)+(0\wedge(1+(-b+-a))))+-(1\wedge(1+(1+(-b+-a)))))) = 1\wedge(b+a)$, by (Scomm),
$43 \Longleftrightarrow 1\wedge(1+((b+(a+(0\wedge(1+(-b+-a)))))+-(1\wedge(1+(1+(-b+-a)))))) = 1\wedge(b+a)$, by (Sass),
$44 \Longleftrightarrow 1\wedge(1+(-(1\wedge(1+(1+(-b+-a))))+(b+(a+(0\wedge(1+(-b+-a))))))) = 1\wedge(b+a)$, by (Scomm),
$45 \Longleftrightarrow 1\wedge(1+(b+(-(1\wedge(1+(1+(-b+-a))))+(a+(0\wedge(1+(-b+-a))))))) = 1\wedge(b+a)$, by (3.21),
$46 \Longleftrightarrow 1\wedge(1+(b+(a+(-(1\wedge(1+(1+(-b+-a))))+(0\wedge(1+(-b+-a))))))) = 1\wedge(b+a)$, by (3.21),
$47 \Longleftrightarrow 1\wedge(1+(b+(a+-1))) = 1\wedge(b+a)$, by (3.27),
$48 \Longleftrightarrow 1\wedge(b+a) = 1\wedge(b+a)$, by (3.24), that is always true.
Thus, the identity 11 is true, i.e. (Z_u^R) holds. □

Note that the above proof is a 'hybrid' proof.

Proposition 3.4.10 *(See Proposition 2.2.62)*
Let $\mathcal{G}_u = (G, \vee, \wedge, +, -, 0_G, u = u_G)$ be a commutative l-group with strong unit $u = u_G$, with $x \leq y \Longleftrightarrow x \wedge y = x \Longleftrightarrow x \vee y = y$. Then, the property (YY_u^R) holds (see Remark 3.4.3):
(YY_u^R) for all $0_G \leq a, b, c \leq u_G$, if $\inf(u_G, a+b) = a$, then
$u_G + -\inf(u_G, (u_G + -a) + (u_G + -\inf(u_G, b+c))) =$
$\inf(u_G, b + (u_G + -\inf(u_G, (u_G + -a) + (u_G + -c))))$, *i.e.*

for all $0_G \leq a, b, c \leq u_G$, if $u_G \wedge (a+b) = a$, then
$u_G + -(u_G \wedge ((u_G + -a) + (u_G + -(u_G \wedge (b+c))))) =$
$u_G \wedge (b + (u_G + -(u_G \wedge ((u_G + -a) + (u_G + -c)))))$.

Proof. (Michael Kinyon, by Prover9 on Linux, Length of proof was 121, in 2.76 seconds; the expanded proof had the Length 284.) See the proof in **Appendix C.** □

• **Our goal now is to analyse the properties $(s5^R)$ in a commutative l-group with strong unit.**

Proposition 3.4.11 *Let $\mathcal{G}_u = (G, \vee, \wedge, +, -, 0_G, u = u_G)$ be a commutative l-group with strong unit $u = u_G$, with $x \leq y \Longleftrightarrow x \wedge y = x \Longleftrightarrow x \vee y = y$. Then, the following property holds (see Remark 3.4.3):*
$(s5^R)$ (V^{00}) For $0_G \leq a_0$, $b_0 \leq u_G$,

$$a_0 + b_0 = (u_G \wedge (a_0 + b_0)) + (u_G + -(u_G \wedge ((u_G + -a_0) + (u_G + -b_0)))).$$

Proof. (By Prover9, Length of proof was 26, in 2.81 seconds.)

In order to simplify the writting (less space on a line), we shall replace a_0 by a, b_0 by c.

Hence, we have to prove that, for $0_G \leq a, c \leq u_G$,
$a + c = (u_G \wedge (a + c)) + (u_G + -(u_G \wedge ((u_G + -a) + (u_G + -c))))$.

We shall denote the right side by T:
$T \overset{notation}{=} (u_G \wedge (a + c)) + (u_G + -(u_G \wedge ((u_G + -a) + (u_G + -c))))$. So, we must prove that $T = a + c$.

First, in (g3) $((x + y) \wedge (x + z) = x + (y \wedge z))$, take $Z := -x$ to obtain:
$(x + y) \wedge (x + -x) = x + (y \wedge -x)$,
which, by (m_0-Re), becomes:
$(x + y) \wedge 0_G = x + (y \wedge -x)$,
hence,
$0_G \wedge (x + y) = x + (y \wedge -x)$,
which, for $X := -x$, becomes, by (DN):
$0_G \wedge (-x + y) = -x + (y \wedge x)$,
hence, by adding x:
$x + (0_G \wedge (-x + y)) = x + (-x + (y \wedge x))$,
which, by (Sass), (m_0-Re), (SU), becomes:
$x + (0_G \wedge (-x + y)) = y \wedge x$,
hence, by interchanging x with y:

(3.29) $$x \wedge y = y + (0_G \wedge (-y + x)).$$

Then, in (g3) $((x + y) \wedge (x + z) = x + (y \wedge z))$ again, take $Y := 0_G$ to obtain, by (SU):
(a) $x \wedge (x + z) = x + (0_G \wedge z)$;
then, note that, by (Scomm), (Sass), (m_0-Re), (SU), we have:
(b) $-x + (y + x) = y$;
now, in (b), take $X := 0_G \wedge x$, to obtain:
$-(0_G \wedge x) + (y + (0_G \wedge x)) = y$,
which, by (a), becomes:
$-(0_G \wedge x) + (y \wedge (y + x)) = y$,
which implies:
$-(-(0_G \wedge x) + (y \wedge (y + x))) = -y$,
which, by (NegS), becomes, by (DN):

(3.30) $$-(y \wedge (y + x)) + (0_G \wedge x) = -y.$$

Now, note that $u_G \wedge (a + c)$ (from the goal)

$$\begin{aligned}
&\overset{(3.29)}{=} (a + c) + (0_G \wedge (-(a + c) + u_G)) \\
&\overset{(NegS)}{=} (a + c) + (0_G \wedge ((-c + -a) + u_G)) \\
&\overset{(Scomm)}{=} (a + c) + (0_G \wedge ((-a + -c) + u_G)) \\
&\overset{(Scomm)}{=} (a + c) + (0_G \wedge (u_G + (-a + -c))) \\
&\overset{(Sass)}{=} a + (c + (0_G \wedge (u_G + (-a + -c)))),
\end{aligned}$$

hence, we have:

$$u_G \wedge (a + c) = a + (c + (0_G \wedge (u_G + (-a + -c)))). \tag{3.31}$$

Note also that, by (Sass), (Scomm), (Sass), we have:

$$x + (y + z) = y + (x + z). \tag{3.32}$$

And note that, by (Sass), (Scomm), ($\mathrm{m_0}$-Re), we have:

$$x + (y + (z + -x)) = y + z. \tag{3.33}$$

Then, $T = (u_G \wedge (a + c)) + (u_G + -(u_G \wedge ((u_G + -a) + (u_G + -c))))$
$\stackrel{(Sass)}{=} (u_G \wedge (a + c)) + (u_G + -(u_G \wedge (u_G + (-a + (u_G + -c)))))$
$\stackrel{(3.32)}{=} u_G + ((u_G \wedge (a + c)) + -(u_G \wedge (u_G + (u_G + (-a + -c)))))$
$\stackrel{(3.31)}{=} u_G + ((a + (c + (0_G \wedge (u_G + (-a + -c))))) + -(u_G \wedge (u_G + (u_G + (-a + -c)))))$
$\stackrel{(Scomm)}{=} u_G + (-(u_G \wedge (u_G + (u_G + (-a + -c)))) + (a + (c + (0_G \wedge (u_G + (-a + -c))))))$
$\stackrel{(3.32)}{=} u_G + (a + (-(u_G \wedge (u_G + (u_G + (-a + -c)))) + (c + (0_G \wedge (u_G + (-a + -c))))))$
$\stackrel{(3.32)}{=} u_G + (a + (c + (-(u_G \wedge (u_G + (u_G + (-a + -c)))) + (0_G \wedge (u_G + (-a + -c))))))$
$\stackrel{(3.30)}{=} u_G + (a + (c + (-u_G)))$, for $Y := u_G$, $X := u_G + (-a + -c)$,
$\stackrel{(3.33)}{=} a + c.$ □

Proposition 3.4.12 *Let $\mathcal{G}_u = (G, \vee, \wedge, +, -, 0_G, u = u_G)$ be a commutative l-group with strong unit $u = u_G$, with $x \leq y \Longleftrightarrow x \wedge y = x \Longleftrightarrow x \vee y = y$. Then, for $0_G \leq a, b \leq u_G$,*

$$(V^{00}) \implies (Z_u^R),$$

where:
(V^{00}) $a + b = (u_G \wedge (a + b)) + (u_G + -(u_G \wedge ((u_G + -a) + (u_G + -b))))$,
(Z_u^R) $u_G \wedge (a + b) = u_G \wedge [(u_G \wedge (a + b)) + (u_G + -(u_G \wedge ((u_G + -a) + (u_G + -b))))]$.

Proof. Obviously. □

Open problem 3.4.13 *Prove that (Z_u^R) implies (V^{00}) or find an counter-example.*

Proposition 3.4.14 *Let $\mathcal{G}_u = (G, \vee, \wedge, +, -, 0_G, u = u_G)$ be a commutative l-group with strong unit $u = u_G$, with $x \leq y \Longleftrightarrow x \wedge y = x \Longleftrightarrow x \vee y = y$. Then, the following property holds (see Remark 3.4.3):*
$(s5^R)$ (V^{10}) For all $0_G \leq a_0, a_1, b_0 \leq u_G$, if $u_G \wedge (a_0 + a_1) = a_0$, then:
$(a_0 + a_1) + b_0 = ((u_G \wedge (a_0 + b_0)) + (u_G + -(u_G \wedge ((u_G + -a_0) + (u_G + -(u_G \wedge (a_1 + b_0))))))) + (u_G + -(u_G \wedge ((u_G + -a_1) + (u_G + -b_0))))$.

Proof. (By PROVER9, Length of proof was 54, in 1.83 seconds; the renumbered expanded proof had the Length 261.) The 'hybrid' proof is presented in **Appendix C.** □

Proposition 3.4.15 *Let $\mathcal{G}_u = (G, \vee, \wedge, +, -, 0_G, u = u_G)$ be a commutative l-group with strong unit $u = u_G$, with $x \leq y \Longleftrightarrow x \wedge y = x \Longleftrightarrow x \vee y = y$.*
Then, for all $0_G \leq a_0, a_1, b_0 \leq u_G$, we have (see $(s5^R)$) (see Remark 3.4.3):

$$(V^{10}) \Longrightarrow (V^{00}).$$

Proof. Suppose (V^{10}) holds, i.e. if $u_G \wedge (a_0 + a_1) = a_0$, then:
(a) $(a_0+a_1)+b_0 = ((u_G \wedge (a_0+b_0)) + (u_G + -(u_G \wedge ((u_G + -a_0) + (u_G + -(u_G \wedge (a_1+b_0))))))) + (u_G + -(u_G \wedge ((u_G + -a_1) + (u_G + -b_0))))$.
We must prove that (V^{00}) holds, i.e.:
$a_0 + b_0 = (u_G \wedge (a_0+b_0)) + (u_G + -(u_G \wedge ((u_G + -a_0) + (u_G + -b_0))))$.

Indeed, take $a_1 = 0_G$ in (V^{10}), to obtain:
if $u_G \wedge (a_0 + 0_G) = a_0$, then:
(a') $(a_0+0_G)+b_0 = ((u_G \wedge (a_0+b_0)) + (u_G + -(u_G \wedge ((u_G + -a_0) + (u_G + -(u_G \wedge (0_G+b_0))))))) + (u_G + -(u_G \wedge ((u_G + -0_G) + (u_G + -b_0))))$,
which, by (SU), (N0), becomes:
if $u_G \wedge a_0 = a_0$ (that is always true, since $a_0 \leq u_G$), then:
(a") $a_0+b_0 = ((u_G \wedge (a_0+b_0)) + (u_G + -(u_G \wedge ((u_G + -a_0) + (u_G + -(u_G \wedge b_0)))))) + (u_G + -(u_G \wedge ((u_G + 0_G) + (u_G + -b_0))))$;
then, since $b_0 \leq u_G$ too, we have $u_G \wedge b_0 = b_0$, hence (a") becomes, by (SU):
(a"') $a_0+b_0 = ((u_G \wedge (a_0+b_0)) + (u_G + -(u_G \wedge ((u_G + -a_0) + (u_G + -b_0))))) + (u_G + -(u_G \wedge (u_G + (u_G + -b_0))))$.

Now, since (Cp) ($x \leq y \Longleftrightarrow x+z \leq y+z$) is equivalent to:
(b) $x \wedge y = x \Longleftrightarrow (x+z) \wedge (y+z) = x+z$,
take in (b) $X := b_0$ and $Y := u_G$, to obtain:
(b') $b_0 \wedge u_G = b_0 \Longleftrightarrow (b_0+z) \wedge (u_G+z) = b_0+z$;
then, since $b_0 \leq u_G$, we have that $b_0 \wedge u_G = b_0$ is always true; hence, (b') becomes:
(c) $(b_0+z) \wedge (u_G+z) = b_0+z$;
now, in (c), take $Z := u_G + -b_0$, to obtain:
(c') $(b_0 + (u_G + -b_0)) \wedge (u_G + (u_G + -b_0)) = b_0 + (u_G + -b_0)$,
which, by (Sass), (Scomm), (m_0-Re), (SU), becomes:
(c") $u_G \wedge (u_G + (u_G + -b_0)) = u_G$.

Finally, in (a"'), the part $u_G \wedge (u_G + (u_G + -b_0))$ equals u_G, by (c"); hence, (a"') becomes, by (m_0-Re) and (SU):
$a_0 + b_0 = (u_G \wedge (a_0+b_0)) + (u_G + -(u_G \wedge ((u_G + -a_0) + (u_G + -b_0))))$, i.e. ($V^{00}$) holds. □

Proposition 3.4.16 *Let $\mathcal{G}_u = (G, \vee, \wedge, +, -, 0_G, u = u_G)$ be a commutative l-group with strong unit $u = u_G$.*
Then, for all $0_G \leq a_0, a_1, b_0 \leq u_G$, we have (see $(s5^R)$) (see Remark 3.4.3):

$$(V^{00}) \Longrightarrow (V^{10}).$$

Proof. (By PROVER9, Length of proof was 73, in 2.27 seconds; the renumbered expanded proof had the Length 327.) See the 'hybrid' proof in **Appendix C.** □

Corollary 3.4.17 *Let $\mathcal{G}_u = (G, \vee, \wedge, +, -, 0_G, u = u_G)$ be a commutative l-group with strong unit $u = u_G$.*
Then, for all $0_G \leq a_0, a_1, b_0 \leq u_G$, we have (see $(s5^R)$):

$$(V^{00}) \iff (V^{10}).$$

Proof. By Propositions 3.4.15 and 3.4.16. □

Open problem 3.4.18 *Let $\mathcal{G}_u = (G, \vee, \wedge, +, -, 0_G, u = u_G)$ be a commutative l-group with strong unit $u = u_G$.*
Then, for all $0_G \leq a_0, a_1, \ldots, a_n, b_0 \leq u_G$, does the property (V^{n0}) hold?

$$(V^{00}) \implies (V^{n0})\ ?$$

where (see Remark 3.4.3):
(V^{n0}) if $u_G \wedge (a_i + a_{i+1}) = a_i$, for $i = 0, 1, \ldots, n-1$, then
$\sum_{i=0}^{n} a_i + b_0 = \sum_{i=0}^{n+1} c_i$, where
$c_i = u_G + -(u_G \wedge ((u_G + -a_{i-1}) + (u_G + -(u_G \wedge (a_i + b_0)))))$, with $a_{-1} = u_G$, $a_{n+1} = 0_G$.

Lemma 3.4.19 *Let $\mathcal{G}_u = (G, \vee, \wedge, +, -, 0_G, u = u_G)$ be a commutative l-group with strong unit $u = u_G$.*
Then, for all $0 = 0_G \leq a, c \leq 1 = u_G$,

$$1 + -(1 \wedge (1 + -(1 \wedge (a + c)))) = 1 \wedge (a + c).$$

Proof. (By PROVER9, Length of proof was 31, in 0.09 seconds)

The particular hypotheses used by PROVER9 (among the given ones - see (α)) are:

(3.34) $$0 \wedge 1 = 0 \quad (i.e.\ 0 \leq 1),$$

(3.35) $$0 \wedge a = 0 \quad (i.e.\ 0 \leq a),$$

(3.36) $$0 \wedge c = 0 \quad (i.e.\ 0 \leq c).$$

The following properties are obvious, by (SU), (Scomm), (Sass), (m_0-Re), (NegS) and (Wcomm) $(x \wedge y = y \wedge x)$, (Wass) $((x \wedge y) \wedge z = x \wedge (y \wedge z))$:

(3.37) $$x + (-x + y) = y,$$

(3.38) $$x \wedge (y \wedge z) = y \wedge (x \wedge z).$$

Now, in (Wass) $((x \wedge y) \wedge z = x \wedge (y \wedge z))$, take $X := 0$, $Y := a$, to obtain:
$(0 \wedge a) \wedge z = 0 \wedge (a \wedge z)$,
which, by (3.35), becomes:

(3.39) $$0 \wedge (a \wedge z) = 0 \wedge z.$$

Then, in (Cp) ($x \le y \Longleftrightarrow x + z \le y + z$), i.e. in:
$x \wedge y = x \Longleftrightarrow (x + z) \wedge (y + z) = x + z$,
take $X := 0$, $Y := c$, to obtain:
$0 \wedge c = 0 \Longleftrightarrow (0 + z) \wedge (c + z) = 0 + z$,
which, by (3.36) and (SU), becomes:

$$z \wedge (c + z) = z. \tag{3.40}$$

Then, in (3.39), take $Z := c + a$, to obtain:
$0 \wedge (a \wedge (c + a)) = 0 \wedge (c + a)$,
which, by (3.40), becomes:
$0 \wedge a = 0 \wedge (c + a)$,
which, by (3.35) and (Scomm), becomes:

$$0 \wedge (a + c) = 0. \tag{3.41}$$

Then, in (3.38), take $Y := 0$, $Z := a + c$, to obtain:
$x \wedge (0 \wedge (a + c)) = 0 \wedge (x \wedge (a + c))$,
which, by (3.41), becomes:

$$0 \wedge (x \wedge (a + c)) = x \wedge 0. \tag{3.42}$$

Now, in (g3) ($(x + y) \wedge (x + z) = x + (y \wedge z)$), take $Y := 0$, to obtain, by (SU):

$$x \wedge (x + z) = x + (0 \wedge z). \tag{3.43}$$

Then, in (g3) again, take $Z := -x$, to obtain, by (m_0-Re):
$(x + y) \wedge 0 = x + (y \wedge -x)$,
which, by (Wcomm), becomes:

$$0 \wedge (x + y) = x + (y \wedge -x). \tag{3.44}$$

Then, in (3.44), take $Y := 0$, to obtain, by (SU):
$0 \wedge x = x + (0 \wedge -x)$,
which, for $X := -x$, becomes, by (DN):

$$0 \wedge -x = -x + (0 \wedge x). \tag{3.45}$$

Finally, $1 + -(1 \wedge (1 + -(1 \wedge (a + c))))$
$= 1 + -(1 + (0 \wedge -(1 \wedge (a + c))))$, by (3.43),
$= 1 + (-(0 \wedge -(1 \wedge (a + c))) + -1)$, by (NegS),
$= 1 + (-1 + -(0 \wedge -(1 \wedge (a + c))))$, by (Scomm),
$= -(0 \wedge -(1 \wedge (a + c)))$, by (3.37),
$= -(-(1 \wedge (a + c)) + (0 \wedge (1 \wedge (a + c))))$, by (3.45),
$= -(-(1 \wedge (a + c)) + (1 \wedge 0))$, by (3.42),
$= -(-(1 \wedge (a + c)) + 0)$, by (3.34),
$= -(-(1 \wedge (a + c)))$, by (SU),
$= 1 \wedge (a + c)$, by (DN). □

Proposition 3.4.20 *Let $\mathcal{G}_u = (G, \vee, \wedge, +, -, 0_G, u = u_G)$ be a commutative l-group with strong unit $u = u_G$, with $x \leq y \Longleftrightarrow x \wedge y = x \Longleftrightarrow x \vee y = y$.*
Then, for all $0_G \leq a_0, a_1, \ldots, a_n$, $b_0 \leq u_G$, we have (see (s5^R)) (see Remark 3.4.3):

$$(V^{n0}) \implies (V^{00}).$$

Proof. Suppose (V^{n0}) holds, i.e. if $u_G \wedge (a_i + a_{i+1}) = a_i$, for $i = 0, 1, \ldots, n-1$, then:
$\sum_{i=0}^{n} a_i + b_0 = \sum_{i=0}^{n+1} c_i$, where
$c_i = u_G + -(u_G \wedge ((u_G + -a_{i-1}) + (u_G + -(u_G \wedge (a_i + b_0)))))$, with $a_{-1} = u_G$, $a_{n+1} = 0_G$.

We must prove that (V^{00}) holds, i.e.:
$a_0 + b_0 = (u_G \wedge (a_0 + b_0)) + (u_G + -(u_G \wedge ((u_G + -a_0) + (u_G + -b_0))))$.

Indeed, take $a_1 = a_2 = \ldots = a_n = 0_G$ in (V^{n0}), to obtain, by (SU):
if $u_G \wedge (a_0 + 0_G) = a_0$, i.e. $u_G \wedge a_0 = a_0$ (that is always true, since $a_0 \leq u_G$), then:
$a_0 + b_0 = c_0 + c_1$, where:
$c_0 = u_G + -(u_G \wedge ((u_G + -u_G) + (u_G + -(u_G \wedge (a_0 + b_0)))))$
$= u_G + -(u_G \wedge (0_G + (u_G + -(u_G \wedge (a_0 + b_0)))))$, by ($m_0$-Re),
$= u_G + -(u_G \wedge (u_G + -(u_G \wedge (a_0 + b_0))))$, by (SU),
$= u_G \wedge (a_0 + b_0)$, by Lemma 3.4.19;
$c_1 = u_G + -(u_G \wedge ((u_G + -a_0) + (u_G + -(u_G \wedge b_0))))$
$= u_G + -(u_G \wedge ((u_G + -a_0) + (u_G + -b_0)))$;
$c_2 = u_G + -(u_G \wedge ((u_G + -0_G) + (u_G + -(u_G \wedge b_0))))$
$= u_G + -(u_G \wedge (u_G + (u_G + -b_0)))$, by (SU) and since $b_0 \leq u_G$,
$= u_G + -u_G = 0_G$, by (**) and (m_0-Re),
where in (Cp) ($x \leq y \Longleftrightarrow x + z \leq y + z$), i.e. in:
(*) $x \wedge y = x \Longleftrightarrow (x + z) \wedge (y + z) = x + z$,
we take $X := b_0$ and $Y := u_G$, to obtain:
$b_0 \wedge u_G = b_0 \Longleftrightarrow (b_0 + z) \wedge (u_G + z) = b_0 + z$,
which, since $b_0 \leq u_G$, becomes:
$(b_0 + z) \wedge (u_G + z) = b_0 + z$,
which, for $Z := u_G + -b_0$, becomes:
$(b_0 + (u_G + -b_0)) \wedge (u_G + (u_G + -b_0)) = b_0 + (u_G + -b_0)$,
which, by (Scomm), (Sass), (m_0-Re) and (SU), becomes:
(**) $u_G \wedge (u_G + (u_G + -b_0)) = u_G$.
Resuming,
$a_0 + b_0 = c_0 + c_1 = (u_G \wedge (a_0 + b_0)) + (u_G + -(u_G \wedge ((u_G + -a_0) + (u_G + -b_0))))$;
the proof is complete. □

• Our goal now is to analyse the properties (C_u^{1R}), (C_{pu}^{1R}), (m-Tr_u^R) in a commutative l-group with strong unit

Note that these properties are verified in any commutative l-group.

• **Our final goal is to analyse the last three properties**

Proposition 3.4.21 *Let $\mathcal{G}_u = (G, \vee, \wedge, +, -, 0_G, u = u_G)$ be a commutative l-group with strong unit $u = u_G$.*
Then, the property (NNu^{1R}) is verified, where (see Remark 3.4.3):
(NNu^{1R}) $u_G + -(u_G + -u_G) = u_G$.

Proof. $u_G + -(u_G + -u_G) \overset{(NegS)}{=} u_G + (- - u_G + -u_G) \overset{(DN)}{=} u_G + (u_G + -u_G) \overset{(m_0-Re)}{=}$ $u_G + 0_G \overset{(SU)}{=} u_G$. □

Remarks 3.4.22 *Let $\mathcal{G}_u = (G, \vee, \wedge, +, -, 0_G, u = u_G)$ be a commutative l-group with strong unit $u = u_G$.*

(1) The property (S^{1R}) is not verified, where (see Remark 3.4.3): for all $x \in G$, (S^{1R}) $u_G + -(x + u_G) = u_G + -x$.
Indeed, $u_G + -(x + u_G) \overset{(NegS)}{=} u_G + (-u_G + -x) \overset{(Sass)}{=} (u_G + -u_G) + -x \overset{(m_0-Re)}{=}$ $0_G + -x \overset{(SU)}{=} -x \neq u_G + -x$.

(2) Consequently, the property (Nu^{1R}) is not verified, where (see Remark 3.4.3): (Nu^{1R}) $u_G + -u_G = u_G$.
Indeed, $u_G + -u_G = 0_G \neq u_G$.

• **Conclusions.** A unital commutative l-group verifies all the properties of a strong m$_0$-ME structure, with the exception of (S^{1R}) (and, consequently, of (Nu^{1R})).

**

Let $\mathcal{G}_u^+ = (G^+, \vee, \wedge, +, \smile, 0_{G^+}, u = u_{G^+})$, with $x \leq y \Longleftrightarrow x \vee y = y \Longleftrightarrow x \wedge y = x$, be a positive-unital commutative l-group (Definition 1.4.1). Then, it verifies all the properties of a strong m$_0$-ME positive cone (Definition 3.3.1), with the exception of (DN) ($\smile (\smile x) = x$) on G^+.

3.5 Homomorphisms of strong m$_0$-ME structures and positive cones

We introduce now the following definitions, the analogous of those from the l-group theory.

Definitions 3.5.1 (The dual ones are omitted)

(1) Let $\mathcal{M} = (M, +, \smile, 0_M)$ and $\mathcal{N} = (N, +, \smile, 0_N)$ be two involutive m$_0$-ME algebras. A function $f : M \longrightarrow N$ is said to be an *involutive m$_0$-ME algebras homomorphism*, if for all $x, y \in M$,
(m1) $f(x + y) = f(x) + f(y)$,
(m2) $f(\smile x) = \smile f(x)$,
(m3) $f(0_M) = 0_N$.

(2) Let $\mathcal{M}_u = (M, \leq_m, +, \smile, 0_M, u = u_M)$ and $\mathcal{N}_v = (N, \leq_m, +, \smile, 0_N, v = v_N)$ be two strong m_0-ME structures. A function $f : M \longrightarrow N$ is said to be a *strong m_0-ME structures homomorphism*, if f is an involutive m_0-ME algebras homomorphism and for all $x, y \in M$,
(m4) $x \leq_m y \Longrightarrow f(x) \leq_m f(y)$,
(m5) $f(u) = v$.

(3) Let $\mathcal{M}_u^+ = (M^+, \leq_m, +, \smile, 0_{M^+}, u = u_{M^+})$ and $\mathcal{N}_v^+ = (N^+, \leq_m, +, \smile, 0_{N^+}, v = v_{N^+})$ be two strong m_0-ME positive cones. A function $f^+ : M^+ \longrightarrow N^+$ is said to be a *strong m_0-ME positive cones homomorphism*, if, for all $x, y \in M^+$,
(p1) $f^+(x + y) = f^+(x) + f^+(y)$,
(p2) $f^+(\smile x) = \smile f^+(x)$,
(p3) $f^+(0_{M^+}) = 0_{N^+}$,
(p4) $x \leq_m y \Longrightarrow f^+(x) \leq_m f^+(y)$,
(p5) $f^+(u) = v$.

Denote by $\mathcal{C}$(**s-m$_0$-ME**) the category whose objects are the strong m_0-ME structures and morphisms are the strong m_0-ME structures homomorphisms.

Denote by $\mathcal{C}$(**s-m$_0$-ME$^+$**) the category whose objects are the strong m_0-ME positive cones and morphisms are the strong m_0-ME positive cones homomorphisms.

Part II

The first step: the functors c_m^+ and T_m

Chapter 4

Construction of a strong m_0-ME positive cone $\mathbf{c}_{\mathbf{m}}^{+}(\mathcal{M}_u)$. The functor $\mathbf{c}_{\mathbf{m}}^{+}$

In this chapter, we introduce and study the functor $\mathbf{c}_{\mathbf{m}}^{+}$ which constructs, from a strong m_0-ME structure $\mathcal{M}_u$, a strong m_0-ME positive cone $\mathbf{c}_{\mathbf{m}}^{+}(\mathcal{M}_u) = \mathcal{M}_u^{+}$. The name '$\mathbf{c}_{\mathbf{m}}^{+}$' comes from 'positive cone'.

4.1 Construction of a strong m_0-ME positive cone $\mathbf{c}_{\mathbf{m}}^{+}(\mathcal{M}_u) = \mathcal{M}_u^{+}$ from a strong m_0-ME structure $\mathcal{M}_u$

Let $\mathcal{M}_u = (M, \leq_m, +, \smile, 0_M, u_M)$ be a strong m_0-ME structure and let define the positive cone as follows (see Remarks 3.3.13 (ii)):

$$M^{+} \overset{def.}{=} \{x \in M \mid x \geq_m 0_M,\ \textit{such that}\ (Nu^{+1R}),\ (S^{+1R})\ \textit{hold}\}.$$

Note that the property $(\mathrm{s}2^{R})$ restricted to M^{+} coincides with the property $(\mathrm{s}2^{+R})$.

If we denote with the same symbols $+$ and $\smile$ the restrictions on M^{+} of the operations $+$ and $\smile$ on M and if we denote with $\leq_m$ the restriction on M^{+} of the binary relation $\leq_m$ on M, then we obtain the structure

$$\mathcal{M}_u^{+} = (M^{+}, \leq_m, +, \smile, 0_M, u_M)$$

which, obviously, is a *strong m_0-ME positive cone*, where $0_{M^{+}} = 0_M$ and $u_{M^{+}} = u_M$.

Dually, let $\mathcal{M}_u = (M, \leq_m, \cdot, ^{-1}, 1_M, u_M)$ be a strong m$_1$-ME structure and let define the positive cone as follows:

$$M^+ \stackrel{def.}{=} \{x \in M \mid x \geq_m 1_M,\ such\ that\ (Nu^{+1}),\ (S^{+1})\ holds\}.$$

Note that the property (s2) restricted to M^+ coincides with the property (s2$^+$).

If we denote with the same symbols $\cdot$ and $^{-1}$ the restrictions on M^+ of the operations $\cdot$ and $^{-1}$ on M and if we denote with $\leq_m$ the restriction on M^+ of the binary relation $\leq_m$ on M, then we obtain the structure

$$\mathcal{M}_u^+ = (M^+, \leq_m, \cdot, ^{-1}, 1_M, u_M)$$

which, obviously, is a *strong* m_1*-ME positive cone*, where $1_{M^+} = 1_M$ and $u_{M^+} = u_M$.

Remark 4.1.1 *By Definitions 3.2.5, we have:*

$$0_M \in M^+ \quad and,\ dually,\ 1_M \in M^+.$$

Hence, we have proved the following result (the dual one is omitted).

Theorem 4.1.2 *For every strong* m_0*-ME structure* $\mathcal{M}_u = (M, \leq_m, +, \smile, 0_M, u_M)$, *the structure* $\mathcal{M}_u^+ = (M^+, \leq_m, +, \smile, 0_{M^+} = 0_M, u_{M^+} = u_M)$ *is a strong* m_0*-ME positive cone.*

- **The functor $\mathbf{c}_{\mathbf{m}}^{+}$**

We put:

$$\mathbf{c}_{\mathbf{m}}^{+}(\mathcal{M}_u) = \mathcal{M}_u^+,$$

i.e. we have:

$$\begin{array}{ccc} \mathcal{C}(\mathbf{s\text{-}m_0\text{-}ME}) & \xrightarrow{\mathbf{c}_{\mathbf{m}}^{+}} & \mathcal{C}(\mathbf{s\text{-}m_0\text{-}ME^+}) \\ \mathcal{M}_u & \xrightarrow{\mathbf{c}_{\mathbf{m}}^{+}} & \mathcal{M}_u^+ \end{array}$$

4.2 Morphisms of strong m$_0$-ME positive cones

Given a morphism $f : M \longrightarrow N$ of strong m$_0$-ME structures
$\mathcal{M}_u = (M, \leq_m, +, \smile, 0_M, u = u_M)$ and $\mathcal{N}_v = (N, \leq_m, +, \smile, 0_N, v = u_N)$,
f restricts to a function f^+ from M^+ to N^+ and, obviously, f^+ is a morphism of strong m$_0$-ME positive cones
$\mathcal{M}_u^+ = (M^+, \leq_m, +, \smile, 0_{M^+}, u = u_{M^+})$ and $\mathcal{N}_v^+ = (N^+, \leq_m, +, \smile, 0_{N^+}, v = u_{N^+})$.

This establishes that $\mathbf{c}_{\mathbf{m}}^{+}$ is a functor

$$\mathbf{c}_{\mathbf{m}}^{+} : \mathcal{C}(\mathbf{s} - \mathbf{m_0} - \mathbf{ME}) \longrightarrow \mathcal{C}(\mathbf{s} - \mathbf{m_0} - \mathbf{ME}^+)$$

that maps $\mathcal{M}_u$ to $\mathcal{M}_u^+$ and maps a morphism $f : M \longrightarrow N$ to its restriction

$$f^+ = \mathbf{c}_{\mathbf{m}}^{+}(f) : \mathbf{c}_{\mathbf{m}}^{+}(M) = M^+ \longrightarrow \mathbf{c}_{\mathbf{m}}^{+}(N) = N^+,$$

i.e. we have the following commutative diagram:

$$\begin{array}{ccc} M & \xrightarrow{\mathbf{c}_{\mathbf{m}}^{+}} & M^{+} = \mathbf{c}_{\mathbf{m}}^{+}(M) \\ {\scriptstyle f}\downarrow & & \downarrow{\scriptstyle f^{+} = \mathbf{c}_{\mathbf{m}}^{+}(f)} \\ N & \xrightarrow{\mathbf{c}_{\mathbf{m}}^{+}} & N^{+} = \mathbf{c}_{\mathbf{m}}^{+}(N) \end{array}$$

Chapter 5

Construction of a strong m_0-ME structure $\mathbf{T_m}(\mathcal{M}_u^+)$. The functor $\mathbf{T_m}$

In this chapter, following ideas from ([1], Section 4), we introduce and study the functor $\mathbf{T_m}$ which constructs, from a strong m_0-ME positive cone $\mathcal{M}_u^+$, a strong m_0-ME structure $\mathbf{T_m}(\mathcal{M}_u^+) = \frac{\mathcal{M}_u^+ \times \mathbf{N}}{\sim}$. The name '$\mathbf{T_m}$' comes here (versus [1]) from 'turning/tipping', because the entire construction made by this functor is a turning/tipping point (or an inflexion point): it differs essentially from the construction from [10], because a strong m_0-ME positive cone does not verify the cancellation property (C); it is analogous to that from [1].

5.1 Construction of a strong m_0-ME structure $\mathbf{T_m}(\mathcal{M}_u^+)$ from a strong m_0-ME positive cone $\mathcal{M}_u^+$

Let $\mathcal{M}_u^+ = (M^+, \leq_m, +, \smile, 0_{M^+}, u = u_{M^+})$ be a strong m_0-ME positive cone throughout this section.

The goal of this section, reached in Theorem 5.1.27 below, is to build a strong m_0-ME structure, denoted $\mathbf{T_m}(\mathcal{M}_u^+)$.

Denote $M^+ \times \mathbf{N} = \{(x, n) \mid x \in M^+, n \in \mathbf{N}\}$.

Note that $(0_{M^+}, 0)$, $(u_{M^+}, 0) \in M^+ \times \mathbf{N}$.

Consider the binary relation $\sim_u$ $(u = u_{M^+})$, or simply $\sim$, if there is no possible confusion, defined on $M^+ \times \mathbf{N}$ as follows: for all (x, n), $(y, p) \in M^+ \times \mathbf{N}$,

$$(5.1) \qquad (x, n) \sim (y, p) \overset{def.}{\Longleftrightarrow} x + pu = y + nu.$$

Lemma 5.1.1 *$\sim$ is an equivalence relation on $M^+ \times \mathbf{N}$.*

Proof. - *reflexivity*: $(x,n) \sim (x,n) \overset{def.}{\Longleftrightarrow} x + nu = x + nu$, that is true.

- *symmetry*: if $(x,n) \sim (y,p)$, i.e. $x + pu = y + nu$, then $y + nu = x + pu$, i.e. $(y,p) \sim (x,n)$.

- *transitivity*: if $(x,n) \sim (y,p)$ and $(y,p) \sim (z,r)$, i.e.
$x + pu = y + nu$ and $y + ru = z + pu$, then
$(x + ru) + pu = (x + pu) + ru = (y + nu) + ru$
$= (y + ru) + nu = (z + pu) + nu = (z + nu) + pu$,
hence, $(x + ru) + pu = (z + nu) + pu$;
then, by (C_u^R) in $\mathcal{M}_u^+$, $x + ru = z + nu$, i.e. $(x,n) \sim (z,r)$. □

Lemma 5.1.2 *We have:*
(i) $(x,n) \sim (x + mu, n + m)$,
(ii) $(x,0) \sim (x + mu, m)$,
(iii) $(mu, m) \sim (0_{M^+}, 0)$.

Proof. (i): $(x,n) \sim (x + mu, n + m)$
$\overset{def.}{\Longleftrightarrow} x + (n+m)u = (x + mu) + nu$
$\Longleftrightarrow x + (nu + mu) = (x + mu) + nu$, that is true, by (Scomm), (Sass) in $\mathcal{M}_u^+$.
(ii): By above (i), for $n = 0$.
(iii): By above (i) and (SU), for $(x,n) = (0_{M^+}, 0)$. □

Lemma 5.1.3 *If $(x,n) \sim (y,p)$ and $n < p$, say $p = n + m$, then $y = x + mu$.*

Proof.

$$\begin{array}{lcl} (x,n) \sim (y,p) & \overset{def.}{\Longleftrightarrow} & x + pu = y + nu \\ & \Longleftrightarrow & x + (n+m)u = y + nu \\ & \Longleftrightarrow & x + (nu + mu) = y + nu \\ & \Longleftrightarrow & (x + mu) + nu = y + nu, \text{ by (Scomm), (Sass)}, \\ & \overset{(C_u^R)}{\Longrightarrow} & x + mu = y, \end{array}$$

hence $y = x + mu$. □

Remarks 5.1.4
(1) for all $(x, n_x), (y, n_y) \in M^+ \times \mathbf{N}$,

$$(x, n_x) \sim (y, n_y) \overset{def.}{\Longleftrightarrow} x + n_y u = y + n_x u.$$

Note that:

- *if $n_x = n_y$, then, $x + n_y u = y + n_x u$ is equivalent with $x = y$, by (C_u^R);*

- *if $n_x < n_y$, then let $d(x,y) = n_y - n_x \in \mathbf{N}$, hence $n_y = n_x + d(x,y)$; then, $x + n_y u = y + n_x u$ becomes $x + (n_x + d(x,y))u = y + n_x u$, i.e. $(x + d(x,y)u) + n_x u = y + n_x u$, i.e. $y = x + d(x,y)u$, by (C_u^R); hence, $(x, n_x) \sim (x + (n_y - n_x)u, n_x + (n_y - n_x))$;*

- *if $n_y < n_x$, then let $d'(x,y) = n_x - n_y \in \mathbf{N}$, hence $n_x = n_y + d'(x,y)$; then, $x + n_y u = y + n_x u$ becomes $x + n_y u = y + (n_y + d'(x,y))u$, i.e. $x + n_y u =$*

$(y + d'(x,y)u) + n_y u$, i.e. $x = y + d'(x,y)u$, by (C_u^R); hence, $(x, n_y) \sim (x + (n_x - n_y)u, n_y + (n_x - n_y))$.

$$\text{Thus, } (x, n_x) \sim (y, n_y) \overset{def.}{\Longleftrightarrow} \begin{cases} x = y, & \text{if } n_x = n_y, \\ y = x + (n_y - n_x)u, & \text{if } n_x < n_y, \\ x = y + (n_x - n_y)u, & \text{if } n_y < n_x. \end{cases}$$

(2) for all $(x, 0)$, $(y, 0) \in M^+ \times \mathbf{N}$,

$$(x,0) \sim (y,0) \overset{def.}{\Longleftrightarrow} x + 0u = y + 0u \Longleftrightarrow x = y.$$

We shall denote the equivalence class of $(x, n) \in M^+ \times \mathbf{N}$ by $\overbrace{(x,n)}_u$, or simply by $\overbrace{(x,n)}$, if there is no possible confusion, and the quotient set by $\frac{M^+ \times \mathbf{N}}{\sim}$. Hence,

$$\frac{M^+ \times \mathbf{N}}{\sim} \overset{def.}{=} \{\overbrace{(X = x + mu, n)} \mid X, x \in M^+, \ m, n \in \mathbf{N}\}.$$

Remark 5.1.5 *Concerning the element $\overbrace{(x + mu, n)}$, note that:*

- if $m = n$, then $\overbrace{(x + mu, n)} = \overbrace{(x + nu, n)} = \overbrace{(x, 0)}$, by Lemma 5.1.2 (ii);

- if $m > n$, say $m = n + r$, then $\overbrace{(x + mu, n)} = \overbrace{(x + (n + r)u, n)} = \overbrace{((x + ru) + nu, n)} = \overbrace{(x + ru, 0)}$, by Lemma 5.1.2 (ii);

- if $m < n$, say $n = m + s$, then $\overbrace{(x + mu, n)} = \overbrace{(x + mu, m + s)} = \overbrace{(x, s)}$, by Lemma 5.1.2 (i).

Lemma 5.1.6 *We have:*
$\overbrace{(x, n_x)} = \{(x, n_x), (x + u, n_x + 1), (x + 2u, n_x + 2), \ldots\}$ and
$\overbrace{(x, 0)} = \{(x, 0), (x + u, 1), (x + 2u, 2), \ldots\}$.

Proof. By Lemmas 5.1.2, 5.1.3. □

We define the following operations $\mathbf{0}$, $\mathbf{u}$, $+$, $\smile$ on $\frac{M^+ \times \mathbf{N}}{\sim}$ by: for all $X, Y, x \in M^+$, for all $m, n, p \in \mathbf{N}$,

$$\mathbf{0} \overset{def.}{=} \overbrace{(0_{M^+}, 0)}, \tag{5.2}$$

$$\mathbf{u} \overset{def.}{=} \overbrace{(u, 0)}, \tag{5.3}$$

$$\overbrace{(X, n)} + \overbrace{(Y, p)} \overset{def.}{=} \overbrace{(X + Y, n + p)}, \tag{5.4}$$

$$\smile \overbrace{(x + mu, n)} \overset{def.}{=} \overbrace{(\smile x + mu, n)}. \tag{5.5}$$

We define also the binary relation $\preceq_m$ on $\frac{M^+\times\mathbf{N}}{\sim}$ by: for all $\overbrace{(X,n)}, \overbrace{(Y,p)} \in \frac{M^+\times\mathbf{N}}{\sim}$,

$$(5.6) \qquad \overbrace{(X,n)} \preceq_m \overbrace{(Y,p)} \overset{def.}{\Longleftrightarrow} X+pu \leq_m Y+nu$$

and we define the infimum related to $\preceq_m$, $\inf_m$, by (see [1], Section 4):

$$(5.7) \qquad \inf_m(\overbrace{(X,n)}, \overbrace{(Y,p)}) \overset{def.}{=} \overbrace{(\inf_m(X+pu, Y+nu), n+p)}.$$

Corollary 5.1.7 *We have:*

(0) $\overbrace{(x+mu,n)} = \overbrace{(x,n)} +m\mathbf{u}$,

(i) $\smile \overbrace{(x,n)} = \overbrace{(\smile x,n)}$,

(ii) $\smile \overbrace{(x,0)} = \overbrace{(\smile x,0)}$,

(iii) $\smile \overbrace{(x+nu,n)} = \overbrace{(\smile x+nu,n)} = \overbrace{(\smile x,0)}$,

(iv) $\smile \overbrace{(mu,n)} = \overbrace{(mu,n)}$.

Proof.

(0): $\overbrace{(x,n)} +m\mathbf{u}$
$= \overbrace{(x,n)} +m\overbrace{(u,0)}$
$= \overbrace{(x,n)} + \overbrace{(mu,m0)}$
$= \overbrace{(x,n)} + \overbrace{(mu,0)}$
$= \overbrace{(x+mu,n+0)} = \overbrace{(x+mu,n)}$.

(i): By (5.5), for $m=0$.
(ii): By (5.5), for $m=0$, $n=0$.
(iii): By (5.5), for $m=n$, and by Lemma 5.1.2 (ii).
(iv): $\smile \overbrace{(mu,n)} = \smile \overbrace{(0_{M^+}+mu,n)} \overset{def.}{=} \overbrace{(\smile 0_{M^+}+mu,n)}$
$= \overbrace{(0_{M^+}+mu,n)} = \overbrace{(mu,n)}$. □

• **Our goal in this section, reached in Theorem 5.1.27 below, is to prove that $\frac{M^+\times\mathbf{N}}{\sim}$ can be organized as a strong $\mathbf{m_0}$-ME structure $\frac{M_u^+\times\mathbf{N}}{\sim}$.**

Proposition 5.1.8 *The algebra* $(\frac{M^+\times\mathbf{N}}{\sim}, +, \smile, \mathbf{0})$ *is an involutive* m_0*-ME algebra.*

Proof. The operations are well defined. Indeed:

• If $(X',n') \in \overbrace{(X,n)}$ and $(Y',p') \in \overbrace{(Y,p)}$, i.e. $X'+nu = X+n'u$ and $Y'+pu = Y+p'u$, then
$\overbrace{(X,n)} + \overbrace{(Y,p)} \overset{def.}{=} \overbrace{(X+Y,n+p)}$ and
$\overbrace{(X',n')} + \overbrace{(Y',p')} \overset{def.}{=} \overbrace{(X'+Y',n'+p')}$.

We must prove that $\widehat{(X+Y, n+p)} = \widehat{(X'+Y', n'+p')}$, i.e.
$(X+Y)+(n'+p')u = (X'+Y')+(n+p)u$.
Indeed, $(X+Y)+(n'+p')u = (X+n'u)+(Y+p'u) = (X'+nu)+(Y'+pu) = (X'+Y')+(n+p)u$.

Thus, the addition $+$ is well defined.

• Suppose $(y,p) \in \widehat{(x,n)}$, with $n < p$, say $p = n+m$; then, $y = x+mu$, by Lemma 5.1.3. Hence, $(y,p) = (x+mu, n+m)$.

We must prove that $\smile \widehat{(y,p)} = \smile \widehat{(x,n)}$.

Indeed, $\smile \widehat{(y,p)} = \smile \widehat{(x+mu, n+m)} \overset{def.}{=} \widehat{(\smile x+mu, n+m)} = \widehat{(\smile x, n)} = \smile \widehat{(x,n)}$, by Lemma 5.1.2 (i).

Thus, the negation $\smile$ is well defined.

• (SU), (Scomm), (Sass), (N0), (DN) hold. Indeed,

(SU): $\widehat{(X,n)} + \widehat{(0_{M^+},0)} = \widehat{(X+0_{M^+}, n+0)} = \widehat{(X,n)}$, by (SU) in $\mathcal{M}_u^+$ and $\mathbf{N}$.

(Scomm): $\widehat{(X,n)} + \widehat{(Y,p)} = \widehat{(X+Y, n+p)} = \widehat{(Y+X, p+n)} = \widehat{(Y,p)} + \widehat{(X,n)}$, by commutativity of $+$ in $\mathcal{M}_u^+$ and in $\mathbf{N}$.

(Sass): similarly.

(N0): $\smile \mathbf{0} = \smile \widehat{(0_{M^+},0)} = \widehat{(\smile 0_{M^+},0)} = \widehat{(0_{M^+},0)} = \mathbf{0}$, by (N0) in $\mathcal{M}_u^+$.

(DN): If $x \neq 0_{M^+}$, then $\smile(\smile \widehat{(x+mu,n)}) = \smile \widehat{(\smile x+mu,n)}$
$= \widehat{(\smile(\smile x)+mu,n)} = \widehat{(x+mu,n)}$, by (DN) in $\mathcal{M}_u^+$.

If $x = 0_{M^+}$, then $\smile(\smile \widehat{(mu,n)}) = \smile(\smile \widehat{(0_{M^+}+mu,n)})$
$= \widehat{(\smile(\smile 0_{M^+})+mu,n)} = \widehat{(0_{M^+}+mu,n)} = \widehat{(mu,n)}$. □

Proposition 5.1.9 *The structure* $(\frac{M^+\times \mathbf{N}}{\sim}, \preceq_m, +, \smile, \mathbf{0})$ *is a* m_0*-ME structure.*

Proof. By above Proposition 5.1.8, it remains to prove that $\mathbf{0} \preceq_m \mathbf{0}$. Indeed, $\mathbf{0} \preceq_m \mathbf{0}$, i.e. $\widehat{(0_{M^+},0)} \preceq_m \widehat{(0_{M^+},0)}$ means $0_{M^+}+0u \leq_m 0_{M^+}+0u$, i.e. $0_{M^+} \leq_m 0_{M^+}$, that is true, by Definitions 3.2.5 in $\mathcal{M}_u^+$. □

The structure $(\frac{M^+\times \mathbf{N}}{\sim}, \preceq_m, +, \smile, \mathbf{0}, \mathbf{u})$ verifies the following properties.

Proposition 5.1.10 *We have:*

$$\widehat{(x+mu,n)} \succeq_m \mathbf{0} \iff n = 0.$$

Proof.
$\widehat{(x+mu,n)} \succeq_m \mathbf{0} \iff \mathbf{0} \preceq_m \widehat{(x+mu,n)} \iff \widehat{(0_{M^+},0)} \preceq_m \widehat{(x+mu,n)}$.

By Remark 5.1.5,

- if $m = n$, then $\widehat{(x+mu,n)} = \widehat{(x,0)}$,

- if $m > n$, say $m = n + r$, then $\overbrace{(x + mu, n)} = \overbrace{(x + ru, 0)}$,
- if $m < n$, say $n = m + s$, then $\overbrace{(x + mu, n)} = \overbrace{(x, s)}$.

Hence,

- if $m = n$, then $\overbrace{(0_{M^+}, 0)} \preceq_m \overbrace{(x, 0)} \iff 0_{M^+} \leq_m x$, that is true, by $(\mathrm{s}0^R)$ in $\mathcal{M}_u^+$;
- if $m > n$, say $m = n + r$, then $\overbrace{(0_{M^+}, 0)} \preceq_m \overbrace{(x + ru, 0)} \iff 0_{M^+} \leq_m x + ru$, that is true, by $(\mathrm{s}0^R)$;
- if $m < n$, say $n = m + s$, then $\overbrace{(0_{M^+}, 0)} \preceq_m \overbrace{(x, s)} \iff su \leq_m x$, that is not true.

Thus, $\overbrace{(x + mu, n)} \succeq_m \mathbf{0} \iff m \geq n$, i.e. $\overbrace{(x + mu, n)}$ has the form $\overbrace{(x, 0)}$ or $\overbrace{(x + ru, 0)}$, i.e. $n = 0$. □

Moreover, we have the following result.

Proposition 5.1.11 *If $x \leq_m u$ in $\mathcal{M}_u^+$, then*

$$\mathbf{0} \preceq_m \overbrace{(x + mu, n)} \preceq_m \mathbf{u} \iff m = n = 0.$$

Proof. We have seen, in Proposition 5.1.10, that $\mathbf{0} \preceq_m \overbrace{(x + mu, n)} \iff n = 0$.

It remains to prove that $\overbrace{(x + mu, 0)} \preceq_m \mathbf{u} \iff m = 0$. Indeed,

$\overbrace{(x + mu, 0)} \preceq_m \mathbf{u}$
$\iff \overbrace{(x + mu, 0)} \preceq_m \overbrace{(u, 0)}$
$\iff x + mu + 0u \leq_m u + 0u$
$\iff x + mu \leq_m u$
$\iff x + (m - 1)u \leq_m 0_{M^+}$, by (C_{pu}^{1R}) in $\mathcal{M}_u^+$,
that is not true.

But, $\overbrace{(x, 0)} \preceq_m \mathbf{u}$
$\iff \overbrace{(x, 0)} \preceq_m \overbrace{(u, 0)}$
$\iff x \leq_m u$, that is true. □

Note that Proposition 5.1.11 says that

$$0_{M^+} \leq_m x \leq_m u,$$

in $\mathcal{M}_u^+ = (M^+, \leq_m, +, \smile, 0_{M^+}, u)$, is equivalent to

$$\mathbf{0} \preceq_m \overbrace{(x, 0)} \preceq_m \mathbf{u},$$

in $(\frac{M^+ \times \mathbf{N}}{\sim}, \preceq_m, +, \smile, \mathbf{0}, \mathbf{u})$.

Corollary 5.1.12 *We have:*

$$(1) \quad \overbrace{(x + mu, 0)} \succeq_m \mathbf{0}, \qquad (2) \quad \overbrace{(x, n_x)} \preceq_m \mathbf{0}.$$

Proof. (1): By Proposition 5.1.10.

(2): $\widehat{(x, n_x)} \preceq_m \mathbf{0} \Longleftrightarrow \widehat{(x, n_x)} \preceq_m \widehat{(0_{M^+}, 0)} \Longleftrightarrow$
$x + 0u \leq_m 0_{M^+} + n_x u \Longleftrightarrow x \leq_m n_x u$ that is true, by ($s1^R$) in $\mathcal{M}_u^+$. □

Proposition 5.1.13 *We have:*

($s0^R$) $\mathbf{u} = \widehat{(u, 0)} \succ_m \widehat{(0_{M^+}, 0)} = \mathbf{0}$.

Proof. By Corollary 5.1.12 (1), $\mathbf{u} \succeq_m \mathbf{0}$ and $\mathbf{u} \neq \mathbf{0}$, because $u \neq 0_{M^+}$, since $u >_m 0_{M^+}$. □

Proposition 5.1.14 *We have:*

($s1^R$) for each $\widehat{(x + mu, n)} \in \frac{M^+ \times \mathbf{N}}{\sim}$, *there exists* $m' \in \mathbf{N}$ *such that*

$$\widehat{(x + mu, n)}, \smile \widehat{(x + mu, n)} \preceq_m m'\mathbf{u}.$$

Proof. Let $\widehat{(x + mu, n)} \in \frac{M^+ \times \mathbf{N}}{\sim}$; hence, $x \in M^+$; then, by ($s1^R$), there exists $n_x \in \mathbf{N}$ such that

(a) $x, \smile x \leq_m n_x u$ is true.

We have:

$$\begin{aligned}
\text{(b)} \quad \widehat{(x + mu, n)} \preceq_m m'\mathbf{u} \quad &\Longleftrightarrow \quad \widehat{(x + mu, n)} \preceq_m m' \widehat{(u, 0)} \\
&\Longleftrightarrow \quad \widehat{(x + mu, n)} \preceq_m \widehat{(m'u, m'0)} \\
&\Longleftrightarrow \quad \widehat{(x + mu, n)} \preceq_m \widehat{(m'u, 0)} \\
&\Longleftrightarrow \quad x + mu + 0u \leq_m m'u + nu \\
&\Longleftrightarrow \text{(b')} \quad x + mu \leq_m m'u + nu.
\end{aligned}$$

- If $m = n$, then (b') becomes:

$x + nu \leq_m m'u + nu \overset{(C_{pu}^R)}{\Longleftrightarrow} x \leq_m m'u$;
it follows that, if we take $m' = n_x$, then (b) holds, by (a).

- If $m > n$, say $m = n + r$, then (b') becomes:

$x + (n + r)u \leq_m m'u + nu$
$\Longleftrightarrow x + nu + ru \leq_m m'u + nu$
$\overset{(C_{pu}^R)}{\Longleftrightarrow} x + ru \leq_m m'u$; we put:
(c) $x + ru \leq_m m'u$;
it follows that, if we take $m' = n_x + r$, then (c) becomes:
$x + ru \leq_m (n_x + r)u$
$\Longleftrightarrow x + ru \leq_m n_x u + ru$
$\overset{(C_{pu}^R)}{\Longleftrightarrow} x \leq_m n_x u$, that is true, by (a); hence, (b) holds.

- If $m < n$, say $n = m + s$, then (b') becomes:

$x + mu \leq_m m'u + (m + s)u$
$\Longleftrightarrow x + mu \leq_m m'u + mu + su$
$\overset{(C_{pu}^R)}{\Longleftrightarrow} x \leq_m m'u + su$; we put:
(d) $x \leq_m m'u + su$;

it follows that:
- if $s < n_x$, then we take $m' = n_x - s$ and (d) becomes:
$x \leq_m (n_x - s)u + su = n_x u$, that is true, by (a);
- if $s \geq n_x$, then we take $m' = 0$ and (d) becomes:
$x \leq_m 0u + su = su$, that is true.
Indeed, we have: $x \leq_m n_x u$, by (a), and $n_x \leq s$ implies $n_x u \leq_m su$, by Corollary 3.2.17 (iii); it follows, by (m-Tr_u^R), that $x \leq_m su$.
It follows that (b) holds.

Let now $\smile \overbrace{(x+mu, n)} \in \frac{M^+ \times \mathbf{N}}{\sim}$; but $\smile \overbrace{(x+mu, n)} = \overbrace{(\smile x + mu, n)}$, with $\smile x \in M^+$; then, $x \in M^+$ and, by ($\mathrm{s1}^R$), there exists $n_x \in \mathbf{N}$ such that
(a) $x, \smile x \leq_m n_x u$ is true.
Like before, (B) $\overbrace{(\smile x + mu, n)} \preceq_m m'\mathbf{u} \Longleftrightarrow$ (B') $\smile x + mu \leq_m m'u + nu$ and the rest follows similarly. □

Remark 5.1.15 *For* $\mathbf{0} = \overbrace{(0_{M^+}, 0)}$, *there exists* $m' = 0 \in \mathbf{N}$ *such that:* $\mathbf{0} \preceq_m m'\mathbf{u}$.
Indeed, $\overbrace{(0_{M^+}, 0)} \preceq_m 0\mathbf{u} = \mathbf{0} = \overbrace{(0_{M^+}, 0)}$.

Proposition 5.1.16 *We have (see Proposition 5.1.10):*
($s2^R$) *(i) For each element* $\mathbf{0} \preceq_m \overbrace{(x+mu, 0)} \in \frac{M^+ \times \mathbf{N}}{\sim}$, *there exists* $m' \in \mathbf{N}$ *and there exist the unique elements of* $\frac{M^+ \times \mathbf{N}}{\sim}$: $\mathbf{0} \preceq_m \overbrace{(x_1, 0)}, \ldots, \overbrace{(x_{m'}, 0)} \preceq_m \mathbf{u}$ *s.t.:*
$\overbrace{(x+mu, 0)} = \overbrace{(x_1, 0)} + \ldots + \overbrace{(x_{m'}, 0)}$ $(\preceq_m m'\mathbf{u})$,
$\smile \overbrace{(x+mu, 0)} = \smile \overbrace{(x_1, 0)} \smile \ldots \smile \overbrace{(x_{m'}, 0)}$ $(\preceq_m m'\mathbf{u})$.

(ii) If $\overbrace{(x+ru, 0)} = \overbrace{(x_1, 0)} + \ldots + \overbrace{(x_{m'}, 0)}$
and $\overbrace{(y+pu, 0)} = \overbrace{(y_1, 0)} + \ldots + \overbrace{(y_{m'}, 0)}$,
then $\overbrace{(x+ru, 0)} = \overbrace{(y+pu, 0)} \Longleftrightarrow \overbrace{(x_i, 0)} = \overbrace{(y_i, 0)}$, $i = 1, \ldots, m'$.

(iii) If $\overbrace{(x+ru, 0)} = \overbrace{(x_1, 0)} + \ldots + \overbrace{(x_{m'}, 0)}$
and $\overbrace{(y+pu, 0)} = \overbrace{(y_1, 0)} + \ldots + \overbrace{(y_{m'}, 0)}$,
then $\overbrace{(x+ru, 0)} \leq_m \overbrace{(y+pu, 0)} \Longleftrightarrow \overbrace{(x_i, 0)} \leq_m \overbrace{(y_i, 0)}$, $i = 1, \ldots, m'$.

Proof. (i): • If $m = 0$, let $\overbrace{(x, 0)} \succeq_m \mathbf{0}$, where $x \in M^+$; then, by ($\mathrm{s2}^{+R}$) (i) in $\mathcal{M}_u^+$, there exists $n_x \in \mathbf{N}$ and there exist the unique elements of M^+:
$0_{M^+} \leq_m x_1, x_2, \ldots, x_{n_x} \leq_m u$ such that
$x = x_1 + x_2 + \ldots + x_{n_x}$ $(\leq_m n_x u)$ and $\smile x = \smile x_1 \smile x_2 \smile \ldots \smile x_{n_x}$ $(\leq_m n_x u)$.
Then, $\overbrace{(x, 0)} = \overbrace{(x_1 + x_2 + \ldots + x_{n_x}, 0)} = \overbrace{(x_1, 0)} + \ldots + \overbrace{(x_{n_x}, 0)}$ $(\preceq_m n_x \mathbf{u} = n_x \overbrace{(u, 0)})$ and
$\smile \overbrace{(x, 0)} = \overbrace{(\smile x, 0)} = \overbrace{(\smile x_1 \smile x_2 \smile \ldots \smile x_{n_x}, 0)} = \overbrace{(\smile x_1, 0)} + \ldots + \overbrace{(\smile x_{n_x}, 0)}$
$= \smile \overbrace{(x_1, 0)} \smile \ldots \smile \overbrace{(x_{n_x}, 0)}$ $(\preceq_m n_x \mathbf{u})$.

Thus, there exists $m' = n_x \in \mathbf{N}$ and there exist the unique elements
$\mathbf{0} = \overbrace{(0_{M^+},0)} \preceq_m \overbrace{(x_1,0)}, \ldots, \overbrace{(x_{m'},0)} \preceq_m \overbrace{(u,0)} = \mathbf{u}$ such that
$\overbrace{(x,0)} = \overbrace{(x_1,0)} + \ldots + \overbrace{(x_{m'},0)}$ $(\preceq_m m'\mathbf{u})$ and
$\smile \overbrace{(x,0)} = \smile \overbrace{(x_1,0)} \smile \ldots \smile \overbrace{(x_{m'},0)}$ $(\preceq_m m'\mathbf{u})$.

• If $m \neq 0$, let $\overbrace{(x+mu,0)} \succeq_m \mathbf{0}$, where $x \in M^+$; then, by (s2^{+R}) (i) in $\mathcal{M}_u^+$, there exists $n_x \in \mathbf{N}$ and there exist the unique elements of M^+:
$0_{M^+} \leq_m x_1, x_2, \ldots, x_{n_x} \leq_m u$ such that
$x = x_1 + x_2 + \ldots + x_{n_x}$ $(\leq_m n_x u)$ and $\smile x = \smile x_1 \smile x_2 \smile \ldots \smile x_{n_x}$ $(\leq_m n_x u)$.
Then, $\overbrace{(x+mu,0)}$
$= \overbrace{(x,0)} + m\mathbf{u}$, by Corollary 5.1.7 (0)
$= \overbrace{(x_1 + x_2 + \ldots + x_{n_x},0)} + m\mathbf{u}$
$= \overbrace{(x_1,0)} + \ldots + \overbrace{(x_{n_x},0)} + m\mathbf{u}$.
Then, there exists $m' = n_x + m \in \mathbf{N}$ and there exists the unique m' elements of $\frac{M^+ \times \mathbf{N}}{\sim}$:

$$\mathbf{0} \preceq_m \overbrace{(x_1,0)}, \ldots, \overbrace{(x_{n_x},0)}, \underbrace{\mathbf{u}, \ldots, \mathbf{u}}_{m \text{ times}} \preceq_m \mathbf{u} \text{ such that}$$

$$\overbrace{(x+mu,0)} = \overbrace{(x_1,0)} + \ldots + \overbrace{(x_{n_x},0)} + \underbrace{\mathbf{u} + \ldots + \mathbf{u}}_{m \text{ times}} = \overbrace{(x_1,0)} + \ldots + \overbrace{(x_{n_x},0)} + m\mathbf{u}.$$

Also, $\smile \overbrace{(x+mu,0)} = \overbrace{(\smile (x+mu),0)} = \overbrace{(\smile x + mu,0)}$

$$= \overbrace{(\smile x_1 \smile x_2 \smile \ldots \smile x_{n_x} + mu,0)} = \overbrace{(\smile x_1,0)} + \ldots + \overbrace{(\smile x_{n_x},0)} + \underbrace{\mathbf{u} + \ldots + \mathbf{u}}_{m \text{ times}}$$

$$= \smile \overbrace{(x_1,0)} \smile \ldots \smile \overbrace{(x_{n_x},0)} \underbrace{\smile (\smile \mathbf{u}) \smile \ldots \smile (\smile \mathbf{u})}_{m \text{ times}}.$$

(ii): Let $\overbrace{(x+ru,0)} = \overbrace{(x_1,0)} + \ldots + \overbrace{(x_{m'},0)}$
and $\overbrace{(y+pu,0)} = \overbrace{(y_1,0)} + \ldots + \overbrace{(y_{m'},0)}$,
i.e. $x + ru = x_1 + \ldots + x_{m'}$ and $y + pu = y_1 + \ldots + y_{m'}$ in $\mathcal{M}_u^+$.
Then, $\overbrace{(x+ru,0)} = \overbrace{(y+pu,0)}$
$\Longleftrightarrow x + ru = y + pu$
$\Longleftrightarrow x_1 + \ldots + x_{m'} = y_1 + \ldots + y_{m'}$
$\Longleftrightarrow x_i = y_i$, $i = 1, \ldots, m'$, by (s2^{+R}) (ii) in $\mathcal{M}_u^+$,
$\Longleftrightarrow \overbrace{(x_i,0)} = \overbrace{(y_i,0)}$, $i = 1, \ldots, m'$.

(iii): Let $\overbrace{(x+ru,0)} = \overbrace{(x_1,0)} + \ldots + \overbrace{(x_{m'},0)}$
and $\overbrace{(y+pu,0)} = \overbrace{(y_1,0)} + \ldots + \overbrace{(y_{m'},0)}$,
i.e. $x + ru = x_1 + \ldots + x_{m'}$ and $y + pu = y_1 + \ldots + y_{m'}$ in $\mathcal{M}_u^+$.

Then, $\overbrace{(x+ru,0)} \preceq_m \overbrace{(y+pu,0)}$
$\Longleftrightarrow x+ru \leq_m y+pu$
$\Longleftrightarrow x_1+\ldots+x_{m'} \leq_m y_1+\ldots+y_{m'}$
$\Longleftrightarrow x_i \leq_m y_i$, $i=1,\ldots,m'$, by (s2^{+R}) (iii) in $\mathcal{M}_u^+$,
$\Longleftrightarrow \overbrace{(x_i,0)} \leq_m \overbrace{(y_i,0)}$, $i=1,\ldots,m'$. □

Proposition 5.1.17 *For all* $\mathbf{0} \preceq_m \overbrace{(x,0)}, \overbrace{(y,0)} \preceq_m \mathbf{u}$,

(s3^R) *(i) there exist* $\inf_m(\mathbf{u}, \overbrace{(x,0)}+\overbrace{(y,0)}) = \overbrace{(\inf_m(u,x+y),0)}$ *and*

$$\smile \inf_m(\mathbf{u}, \smile \overbrace{(x,0)} + \smile \overbrace{(y,0)}) = \overbrace{(\smile \inf_m(u, \smile x + \smile y),0)}$$

and belong to $[\mathbf{0},\mathbf{u}]$;

(ii) $\inf_m(\mathbf{u}, \overbrace{(x,0)}+\mathbf{u}) = \mathbf{u}$,

$\smile \inf_m(\mathbf{u}, \smile \overbrace{(x,0)} \smile \mathbf{0}) = \mathbf{0}$, $\smile \inf_m(\mathbf{u}, \smile \overbrace{(x,0)} \smile \mathbf{u}) = \overbrace{(x,0)}$;

(iii) $\overbrace{(x,0)} \preceq_m \overbrace{(y,0)}$ $(\overbrace{(x,0)}, \overbrace{(y,0)} \neq \mathbf{0}) \Longleftrightarrow \mathbf{u} \preceq_m \overbrace{(y,0)} \smile \overbrace{(x,0)}$.

Proof. By Proposition 5.1.11, $0_{M^+} \leq_m x, y \leq_m u$.

(i): $\inf_m(\mathbf{u}, \overbrace{(x,0)}+\overbrace{(y,0)}) = \inf_m(\overbrace{(u,0)}, \overbrace{(x+y,0)})$
$\stackrel{(5.7)}{=} \overbrace{(\inf_m(u+0u,(x+y)+0u),0+0)} = \overbrace{(\inf_m(u,x+y),0)}$
and $\inf_m(u,x+y)$ exists and belong to $[0_{M^+},u]$ in $\mathcal{M}_u^+$, by (s3^R) (i);

it follows that $\inf_m(\mathbf{u}, \overbrace{(x,0)}+\overbrace{(y,0)})$ exists and belong to $[\mathbf{0},\mathbf{u}]$, by Proposition 5.1.11.

Also, $\smile \inf_m(\mathbf{u}, \smile \overbrace{(x,0)} \smile \overbrace{(y,0)}) = \smile \inf_m(\overbrace{(u,0)}, \overbrace{(\smile x \smile y,0)})$
$\stackrel{(5.7)}{=} \smile \overbrace{(\inf_m(u+0u,(\smile x \smile y)+0u),0+0)}$
$= \smile \overbrace{(\inf_m(u,\smile x \smile y),0)} = \overbrace{(\smile \inf_m(u,\smile x \smile y),0)}$
and $\smile \inf_m(u, \smile x \smile y)$ exists and belong to $[0_{M^+},u]$ in $\mathcal{M}_u^+$, by (s3^R) (i);

it follows that $\smile \inf_m(\mathbf{u}, \smile \overbrace{(x,0)} \smile \overbrace{(y,0)})$ exists and belong to $[\mathbf{0},\mathbf{u}]$, by Proposition 5.1.11.

(ii): Indeed,

• $\inf_m(\mathbf{u}, \overbrace{(x,0)}+\mathbf{u})$
$= \inf_m(\overbrace{(u,0)}, \overbrace{(x,0)}+\overbrace{(u,0)})$
$= \inf_m(\overbrace{(u,0)}, \overbrace{(x+u,0)})$
$\stackrel{(5.7)}{=} \overbrace{(\inf_m(u+0u,(x+u)+0u),0+0)}$
$= \overbrace{(\inf_m(u,x+u),0)}$
$\stackrel{(s3^R)(ii)}{=} \overbrace{(u,0)}$

$= \mathbf{u}.$

• $\smile \inf_m(\mathbf{u}, \smile \overbrace{(x,0)} \smile \mathbf{0})$

$= \smile \inf_m(\overbrace{(u,0)}, \smile \overbrace{(x,0)} \smile \overbrace{(0_{M^+},0)})$

$= \smile \inf_m(\overbrace{(u,0)}, \overbrace{(\smile x,0)} + \overbrace{(\smile 0_{M^+},0)})$

$= \smile \inf_m(\overbrace{(u,0)}, \overbrace{(\smile x \smile 0_{M^+},0)})$

$\stackrel{(5.7)}{=} \smile \overbrace{(\inf_m(u+0u, (\smile x \smile 0_{M^+}) + 0u), 0+0)}$

$= \smile \overbrace{(\inf_m(u, \smile x \smile 0_{M^+}), 0)}$

$= \overbrace{(\smile \inf_m(u, \smile x \smile 0_{M^+}), 0)}$

$\stackrel{(s3^R)(ii)}{=} \overbrace{(0_{M^+},0)}$

$= \mathbf{0}.$

• $\smile \inf_m(\mathbf{u}, \smile \overbrace{(x,0)} \smile \mathbf{u})$

$= \smile \inf_m(\overbrace{(u,0)}, \smile \overbrace{(x,0)} \smile \overbrace{(u,0)})$

$= \smile \inf_m(\overbrace{(u,0)}, \overbrace{(\smile x,0)} + \overbrace{(\smile u,0)})$

$= \smile \inf_m(\overbrace{(u,0)}, \overbrace{(\smile x \smile u,0)})$

$\stackrel{(5.7)}{=} \smile \overbrace{(\inf_m(u+0u, (\smile x \smile u) + 0u), 0+0)}$

$= \smile \overbrace{(\inf_m(u, \smile x \smile u), 0)}$

$= \overbrace{(\smile \inf_m(u, \smile x \smile u), 0)}$

$\stackrel{(s3^R)(ii)}{=} \overbrace{(x,0)}.$

(iii): $\overbrace{(x,0)} \preceq_m \overbrace{(y,0)}$ $(\overbrace{(x,0)}, \overbrace{(y,0)} \neq \mathbf{0}) \iff x \leq_m y$ $(x, y \neq 0_{M^+}) \iff u \leq_m y \smile x$, by (s3^R) (iii) in $\mathcal{M}_u^+$, i.e. $\mathbf{u} \preceq_m \overbrace{(y,0)} \smile \overbrace{(x,0)}$. □

Proposition 5.1.18 *For all* $\mathbf{0} \preceq_m \overbrace{(x,0)}, \overbrace{(y,0)}, \overbrace{(z,0)} \preceq_m \mathbf{u}$,

(s4^R) *(i) there exist and are equal* $\inf_m(\mathbf{u}, \overbrace{(x,0)} + \inf_m(\mathbf{u}, \overbrace{(y,0)} + \overbrace{(z,0)}))$

$$= \inf_m(\mathbf{u}, \overbrace{(x,0)} + (\overbrace{(y,0)} + \overbrace{(z,0)})) = \overbrace{(\inf_m(u, x+(y+z)), 0)},$$

$$\text{(ii) } \inf_m(\mathbf{u}, \smile \inf_m(\mathbf{u}, \smile \overbrace{(x,0)} \smile \overbrace{(y,0)})) = \smile \inf_m(\mathbf{u}, \smile \overbrace{(x,0)} \smile \overbrace{(y,0)}).$$

Proof. By Proposition 5.1.11, $0_{M^+} \leq_m x, y, z \leq_m u$.

(i): $\inf_m(\mathbf{u}, \overbrace{(x,0)} + \inf_m(\mathbf{u}, \overbrace{(y,0)} + \overbrace{(z,0)}))$

$= \inf_m(\mathbf{u}, \overbrace{(x,0)} + \inf_m(\overbrace{(u,0)}, \overbrace{(y+z,0)}))$

$\stackrel{(5.7)}{=} \inf_m(\mathbf{u}, \overbrace{(x,0)} + \overbrace{(\inf_m(u+0u, y+z+0u), 0+0)})$

$= \inf_m(\mathbf{u}, \overbrace{(x,0)} + \overbrace{(\inf_m(u, y+z), 0)})$

$$= \inf_m(\mathbf{u}, \ \widehat{(x + \inf_m(u, y+z), 0+0)}\ )$$

$$= \inf_m(\widehat{(u,0)}, \widehat{(x + \inf_m(u, y+z), 0)}\ )$$

$$\overset{(5.7)}{=} \widehat{(\inf_m(u + 0u, (x + \inf_m(u, y+z)) + 0u), 0+0)}$$

$$= \widehat{(\inf_m(u, x + \inf_m(u, y+z)), 0)}$$

$$\overset{(s4^R)}{=} \widehat{(\inf_m(u, x + (y+z)), 0)}$$

$$\overset{(5.7)}{=} \inf_m(\widehat{(u,0)}, \widehat{(x + (y+z), 0)})$$

$$= \inf_m(\mathbf{u}, \widehat{(x,0)} + (\widehat{(y,0)} + \widehat{(z,0)})),$$

and $\inf_m(u, x + (y+z))$ exists in $\mathcal{M}_u^+$, by ($s3^R$) (i).

(ii): $\inf_m(\mathbf{u}, \ \smile \inf_m(\mathbf{u}, \ \smile \widehat{(x,0)} \smile \widehat{(y,0)})\ )$

$$= \inf_m(\mathbf{u}, \ \smile \inf_m(\widehat{(u,0)}, \ \widehat{(\smile x \smile y, 0)})\ )$$

$$\overset{(5.7)}{=} \inf_m(\mathbf{u}, \ \smile \widehat{(\inf_m(u + 0u, \ (\smile x \smile y) + 0u), 0+0)}\ )$$

$$= \inf_m(\mathbf{u}, \ \smile \widehat{(\inf_m(u, \ \smile x \smile y), 0)}\ )$$

$$= \inf_m(\widehat{(u,0)}, \ \widehat{(\smile \inf_m(u, \ \smile x \smile y), 0)}\ )$$

$$\overset{(5.7)}{=} \widehat{(\inf_m(u + 0u, \smile \inf_m(u, \smile x \smile y) + 0u), 0+0)}$$

$$= \widehat{(\inf_m(u, \smile \inf_m(u, \smile x \smile y)), 0)}$$

$$\overset{(s4^R)(ii)}{=} \widehat{(\smile \inf_m(u, \smile x \smile y), 0)}$$

$$= \smile \widehat{(\inf_m(u, \smile x \smile y), 0)}$$

and

$$\smile \inf_m(\mathbf{u}, \smile \widehat{(x,0)} \smile \widehat{(y,0)})$$

$$= \smile \inf_m(\widehat{(u,0)}, \ \widehat{(\smile x \smile y, 0)})$$

$$\overset{(5.7)}{=} \smile \widehat{(\inf_m(u, \smile x \smile y), 0)}.$$

Thus, $\inf_m(\mathbf{u}, \ \smile \inf_m(\mathbf{u}, \ \smile \widehat{(x,0)} \smile \widehat{(y,0)})\ ) = \smile \inf_m(\mathbf{u}, \smile \widehat{(x,0)} \smile \widehat{(y,0)})$. □

Proposition 5.1.19 *For all* $\mathbf{0} \preceq_m \widehat{(x,0)}, \widehat{(y,0)}, \widehat{(z,0)} \preceq_m \mathbf{u}$,
(XX_u^R) there exist and are equal

$$\smile \inf_m(\mathbf{u}, \smile \inf_m(\mathbf{u}, \widehat{(x,0)} + \widehat{(y,0)}) \smile \inf_m(\mathbf{u}, \widehat{(z,0)} \smile \inf_m(\mathbf{u}, \smile \widehat{(x,0)} \smile \widehat{(y,0)}))) =$$

$$\smile \inf_m(\mathbf{u}, \smile \inf_m(\mathbf{u}, \widehat{(y,0)} + \widehat{(z,0)}) \smile \inf_m(\mathbf{u}, \widehat{(x,0)} \smile \inf_m(\mathbf{u}, \smile \widehat{(y,0)} \smile \widehat{(z,0)}))).$$

Proof. By Proposition 5.1.11, $0_{M^+} \leq_m x, y, z \leq_m u$. We have:

$$A \overset{not.}{=} \smile \inf_m(\mathbf{u}, \smile \inf_m(\mathbf{u}, \overbrace{(x,0)} + \overbrace{(y,0)}) \smile \inf_m(\mathbf{u}, \overbrace{(z,0)} \smile \inf_m(\mathbf{u}, \smile \overbrace{(x,0)} \smile \overbrace{(y,0)}))) \overset{(s3^R)(i)}{=}$$

$$\smile \inf_m(\mathbf{u}, \smile \overbrace{(\inf_m(u, x+y), 0)} \smile \inf_m(\mathbf{u}, \overbrace{(z,0)} \smile \inf_m(\mathbf{u}, \overbrace{(\smile x, 0)} + \overbrace{(\smile y, 0)})))$$

$$\overset{(s3^R)}{=} \smile \inf_m(\mathbf{u}, \smile \overbrace{(\inf_m(u, x+y), 0)} \smile \inf_m(\mathbf{u}, \overbrace{(z,0)} \smile \overbrace{(\inf_m(u, \smile x \smile y), 0)}))$$

$$= \smile \inf_m(\mathbf{u}, \smile \overbrace{(\inf_m(u, x+y), 0)} \smile \inf_m(\mathbf{u}, \overbrace{(z,0)} + \overbrace{(\smile \inf_m(u, \smile x \smile y), 0)}))$$

$$\overset{(s3^R)}{=} \smile \inf_m(\mathbf{u}, \smile \overbrace{(\inf_m(u, x+y), 0)} \smile \overbrace{(\inf_m(u, z \smile \inf_m(u, \smile x \smile y)), 0)} =$$

$$\smile \inf_m(\overbrace{(u,0)}, \overbrace{(\smile \inf_m(u, x+y), 0)} + \overbrace{(\smile \inf_m(u, z \smile \inf_m(u, \smile x \smile y)), 0)}$$

$$\overset{(5.7)}{=} \smile \overbrace{(\inf_m(u, \smile \inf_m(u, x+y) \smile \inf_m(u, z \smile \inf_m(u, \smile x \smile y))), 0)}$$

$$= \overbrace{(\smile \inf_m(u, \smile \inf_m(u, x+y) \smile \inf_m(u, z \smile \inf_m(u, \smile x \smile y))), 0)}$$

and, similarly,

$$B \overset{not.}{=} \smile \inf_m(\mathbf{u}, \smile \inf_m(\mathbf{u}, \overbrace{(y,0)} + \overbrace{(z,0)}) \smile \inf_m(\mathbf{u}, \overbrace{(x,0)} \smile \inf_m(\mathbf{u}, \smile \overbrace{(y,0)} \smile \overbrace{(z,0)})))$$

$$= \overbrace{(\smile \inf_m(u, \smile \inf_m(u, y+z) \smile \inf_m(u, x \smile \inf_m(u, \smile y \smile z))), 0)}.$$

But, by (XX_u^R) in $\mathcal{M}_u^+$, there exist and are equal:
$\smile \inf_m(u, \smile \inf_m(u, x+y) \smile \inf_m(u, z \smile \inf_m(u, \smile x \smile y))) =$
$\smile \inf_m(u, \smile \inf_m(u, y+z) \smile \inf_m(u, x \smile \inf_m(u, \smile y \smile z)))$,
hence, there exist and are equal, in $\frac{M^+ \times \mathbf{N}}{\sim}$:

$$\overbrace{(\smile \inf_m(u, \smile \inf_m(u, x+y) \smile \inf_m(u, z \smile \inf_m(u, \smile x \smile y))), 0)} =$$

$$\overbrace{(\smile \inf_m(u, \smile \inf_m(u, y+z) \smile \inf_m(u, x \smile \inf_m(u, \smile y \smile z))), 0)};$$

thus, $A = B$. □

Proposition 5.1.20 *For all* $\mathbf{0} \preceq_m \overbrace{(x,0)}, \overbrace{(y,0)}, \overbrace{(z,0)} \preceq_m \mathbf{u}$,

(YY_u^R) *if* $\inf_m(\mathbf{u}, \overbrace{(x,0)} + \overbrace{(y,0)}) = \overbrace{(x,0)}$, *then there exist and are equal*

$$\smile \inf_m(\mathbf{u}, \smile \overbrace{(x,0)} \smile \inf_m(\mathbf{u}, \overbrace{(y,0)} + \overbrace{(z,0)}))$$

$$= \inf_m(\mathbf{u}, \overbrace{(y,0)} \smile \inf_m(\mathbf{u}, \smile \overbrace{(x,0)} \smile \overbrace{(z,0)})).$$

Proof. By Proposition 5.1.11, $0_{M^+} \leq_m x, y, z \leq_m u$. We have:

$$\begin{aligned}
A &\overset{not.}{=} \smile \inf_m(\mathbf{u},\ \smile \overbrace{(x,0)} \smile \inf_m(\mathbf{u},\ \overbrace{(y,0)} + \overbrace{(z,0)})\ )\\
&= \smile \inf_m(\mathbf{u},\ \smile \overbrace{(x,0)} \smile \overbrace{(\inf_m(u, y+z), 0)}\ ), && \text{by } (s3^R),\\
&= \smile \inf_m(\mathbf{u},\ \overbrace{(\smile x, 0)} + \overbrace{(\smile \inf_m(u, y+z), 0)}\ )\\
&= \smile \inf_m(\mathbf{u},\ \overbrace{(\smile x \smile \inf_m(u, y+z), 0+0)}\ )\\
&= \smile \inf_m(\overbrace{(u,0)},\ \overbrace{(\smile x \smile \inf_m(u, y+z), 0)}\ )\\
&\overset{(5.7)}{=} \smile \overbrace{(\inf_m(u+0u,\ (\smile x \smile \inf_m(u, y+z)) + 0u), 0+0)}\\
&= \smile \overbrace{(\inf_m(u,\ \smile x \smile \inf_m(u, y+z)\ ), 0)}\\
&= \overbrace{(\smile \inf_m(u,\ \smile x \smile \inf_m(u, y+z)\ ), 0)}
\end{aligned}$$

and

$$\begin{aligned}
B &\overset{not.}{=} \inf_m(\mathbf{u},\ \overbrace{(y,0)} \smile \inf_m(\mathbf{u},\ \smile \overbrace{(x,0)} \smile \overbrace{(z,0)})\ )\\
&= \inf_m(\mathbf{u},\ \overbrace{(y,0)} \smile \inf_m(\mathbf{u},\ \overbrace{(\smile x,0)} + \overbrace{(\smile z,0)})\ )\\
&= \inf_m(\mathbf{u},\ \overbrace{(y,0)} \smile \overbrace{(\inf_m(u,\ \smile x \smile z), 0)}\ ), && \text{by } (s3^R),\\
&= \inf_m(\mathbf{u},\ \overbrace{(y,0)} + \overbrace{(\smile \inf_m(u,\ \smile x \smile z), 0)}\ )\\
&= \inf_m(\mathbf{u},\ \overbrace{(y \smile \inf_m(u,\ \smile x \smile z), 0+0)}\ )\\
&= \inf_m(\overbrace{(u,0)},\ \overbrace{(y \smile \inf_m(u,\ \smile x \smile z), 0)}\ )\\
&\overset{(5.7)}{=} \overbrace{(\inf_m(u+0u,\ (y \smile \inf_m(u,\ \smile x \smile z)) + 0u), 0+0)}\\
&= \overbrace{(\inf_m(u,\ y \smile \inf_m(u, \smile x \smile z)), 0)}.
\end{aligned}$$

If $\inf_m(\mathbf{u},\ \overbrace{(x,0)} + \overbrace{(y,0)}) = \overbrace{(x,0)}$

$\Longleftrightarrow \inf_m(\overbrace{(u,0)},\ \overbrace{(x+y,0)}) = \overbrace{(x,0)}$

$\overset{(5.7)}{\Longleftrightarrow} \overbrace{(\inf_m(u,\ x+y), 0+0)} = \overbrace{(x,0)}$

$\Longleftrightarrow \overbrace{(\inf_m(u,\ x+y), 0)} = \overbrace{(x,0)}$

$\Longleftrightarrow \inf_m(u,\ x+y) = x$, in M^+;

then, by (YY_u^R) in $\mathcal{M}_u^+$, there exist and are equal:

$\smile \inf_m(u,\ \smile x \smile \inf_m(u, y+z)\ ) =$

$\inf_m(u,\ y \smile \inf_m(u, \smile x \smile z)\ )$,

hence

$\overbrace{(\smile \inf_m(u,\ \smile x \smile \inf_m(u, y+z)\ ), 0)} = \overbrace{(\inf_m(u,\ y \smile \inf_m(u, \smile x \smile z)), 0)}$,

thus $A = B$. □

Proposition 5.1.21 *For all* $\mathbf{0} \preceq_m \overbrace{(x_0,0)}, \overbrace{(x_1,0)}, \ldots, \overbrace{(x_n,0)},\ \overbrace{(y_0,0)} \preceq_m \mathbf{u}$, *(s5^R)* *($V^{00}$)* $\overbrace{(x_0,0)} + \overbrace{(y_0,0)} =$

$$\inf_m(\mathbf{u}, \overbrace{(x_0,0)} + \overbrace{(y_0,0)}) \smile \inf_m(\mathbf{u}, \smile \overbrace{(x_0,0)} \smile \overbrace{(y_0,0)});$$

(V^{10}) if $\overbrace{(\inf_m(\mathbf{u}, x_0 + x_1), 0)} = \overbrace{(x_0,0)}$, *then*

$$\overbrace{(x_0,0)} + \overbrace{(x_1,0)} + \overbrace{(y_0,0)} = \inf_m(\mathbf{u}, \overbrace{(x_0,0)} + \overbrace{(y_0,0)})$$
$$\smile \inf_m(\mathbf{u}, \smile \overbrace{(x_0,0)} \smile \inf_m(\mathbf{u}, \overbrace{(x_1,0)} + \overbrace{(y_0,0)}))$$
$$\smile \inf_m(\mathbf{u}, \smile \overbrace{(x_1,0)} \smile \overbrace{(y_0,0)});$$

(V^{n0}) if $\overbrace{(\inf_m(\mathbf{u}, x_i + x_{i+1}), 0)} = \overbrace{(x_i,0)}$, *for* $i = 0, 1, \ldots, n \smile 1$, *then*

$$\sum_{i=0}^{n} \overbrace{(x_i,0)} + \overbrace{(y_0,0)} = \sum_{i=0}^{n+1} \overbrace{(c_i,0)}, \text{ where:}$$

$$\overbrace{(c_i,0)} = \smile \inf_m(\,\mathbf{u}, \smile \overbrace{(x_{i\smile 1},0)} \smile \inf_m(\mathbf{u}, \overbrace{(x_i,0)} + \overbrace{(y_0,0)})\,),\ i = 0, 1, \ldots, n+1,$$

with $\overbrace{(x_{\smile 1},0)} = \mathbf{u}$, $\overbrace{(x_{n+1},0)} = \mathbf{0}$.

Proof. By Proposition 5.1.11, $0_{M^+} \le_m x_0, x_1, \ldots, x_n,\ y_0 \le_m u$.

(V^{00}): $\inf_m(\mathbf{u}, \overbrace{(x_0,0)} + \overbrace{(y_0,0)}) \smile \inf_m(\mathbf{u}, \smile \overbrace{(x_0,0)} \smile \overbrace{(y_0,0)})$

$= inf_m(\overbrace{(u,0)}, \overbrace{(x_0 + y_0, 0+0)}) \smile \inf_m(\overbrace{(u,0)}, \overbrace{(\smile x_0, 0)} + \overbrace{(\smile y_0, 0)})$

$= inf_m(\overbrace{(u,0)}, \overbrace{(x_0 + y_0, 0)}) \smile \inf_m(\overbrace{(u,0)}, \overbrace{(\smile x_0 \smile y_0, 0)})$

$\overset{(5.7)}{=} \overbrace{(\inf_m(u, x_0 + y_0), 0+0)} \smile \overbrace{(\inf_m(u, \smile x_0 \smile y_0), 0+0)}$

$= \overbrace{(\inf_m(u, x_0 + y_0), 0)} + \overbrace{(\smile \inf_m(u, \smile x_0 \smile y_0), 0)}$

$= \overbrace{(\inf_m(u, x_0 + y_0) \smile \inf_m(u, \smile x_0 \smile y_0), 0+0)}$

$= \overbrace{(\inf_m(u, x_0 + y_0) \smile \inf_m(u, \smile x_0 \smile y_0), 0)}$

$\overset{(s5^R)(i)}{=} \overbrace{(x_0 + y_0, 0)} = \overbrace{(x_0,0)} + \overbrace{(y_0,0)}$.

(V^{10}): $\inf_m(\mathbf{u}, \overbrace{(x_0,0)} + \overbrace{(y_0,0)})$

$\smile \inf_m(\mathbf{u}, \smile \overbrace{(x_0,0)} \smile \inf_m(\mathbf{u}, \overbrace{(x_1,0)} + \overbrace{(y_0,0)}))$

$\smile \inf_m(\mathbf{u}, \smile \overbrace{(x_1,0)} \smile \overbrace{(y_0,0)})$

$= \inf_m(\overbrace{(u,0)}, \overbrace{(x_0 + y_0, 0+0)})$

$\smile \inf_m(\overbrace{(u,0)}, \overbrace{(\smile x_0, 0)} \smile \inf_m(\overbrace{(u,0)}, \overbrace{(x_1 + y_0, 0+0)}))$

$\smile \inf_m(\overbrace{(u,0)}, \overbrace{(\smile x_1, 0)} + \overbrace{(\smile y_0, 0)})$

$= \inf_m(\overbrace{(u,0)}, \overbrace{(x_0 + y_0, 0)})$

$\smile \inf_m(\overbrace{(u,0)}, \overbrace{(\smile x_0, 0)} \smile \inf_m(\overbrace{(u,0)}, \overbrace{(x_1 + y_0, 0)}))$

$\smile \inf_m(\overbrace{(u,0)}, \overbrace{(\smile x_1 \smile y_0, 0)})$

$$\overset{(5.7)}{=} \overbrace{(\inf_m(u, x_0 + y_0), 0 + 0)}$$
$$\smile \inf_m(\overbrace{(u, 0)}, \overbrace{(\smile x_0, 0)} \smile \overbrace{(\inf_m(u, x_1 + y_0), 0 + 0)}\,)$$
$$\smile \overbrace{(\inf_m(u, \smile x_1 \smile y_0), 0 + 0)}$$
$$\overset{(5.7)}{=} \overbrace{(\inf_m(u, x_0 + y_0), 0)}$$
$$\smile \inf_m(\overbrace{(u, 0)}, \overbrace{(\smile x_0 \smile \inf_m(u, x_1 + y_0), 0)})$$
$$\smile \overbrace{(\inf_m(u, \smile x_1 \smile y_0), 0)}$$
$$\overset{(5.7)}{=} \overbrace{(\inf_m(u, x_0 + y_0), 0)}$$
$$\smile \overbrace{(\inf_m(u, \smile x_0 \smile \inf_m(u, x_1 + y_0)\,), 0 + 0)}$$
$$\smile \overbrace{(\inf_m(u, \smile x_1 \smile y_0), 0)}$$
$$= \overbrace{(\inf_m(u, x_0 + y_0), 0)}$$
$$+ \overbrace{(\smile \inf_m(u, \smile x_0 \smile \inf_m(u, x_1 + y_0)\,), 0)}$$
$$+ \overbrace{(\smile \inf_m(u, \smile x_1 \smile y_0), 0)}$$
$$= \overbrace{(\inf_m(u, x_0 + y_0) \smile \inf_m(u, \smile x_0 \smile \inf_m(u, x_1 + y_0)) \smile \inf_m(u, \smile x_1 \smile y_0), 0)}$$
$$\overset{(s5^R)(V^{10})}{=} \overbrace{(x_0 + x_1 + y_0, 0)} = \overbrace{(x_0, 0)} + \overbrace{(x_1, 0)} + \overbrace{(y_0, 0)},$$

since, if $\overbrace{(\inf_m(\mathbf{u}, x_0 + x_1), 0)} = \overbrace{(x_0, 0)}$, then $\inf_m(u, x_0 + x_1) = x_0$.

$$(\mathrm{V}^{n0}): \smile \inf_m(\,\mathbf{u}, \smile \overbrace{(x_{i-1}, 0)} \smile \inf_m(\mathbf{u}, \overbrace{(x_i, 0)} + \overbrace{(y_0, 0)})\,)$$
$$= \smile \inf_m(\,\overbrace{(u, 0)}, \overbrace{(\smile x_{i-1}, 0)} \smile \inf_m(\overbrace{(u, 0)}, \overbrace{(x_i + y_0, 0)})\,)$$
$$\overset{(5.7)}{=} \smile \inf_m(\,\overbrace{(u, 0)}, \overbrace{(\smile x_{i-1}, 0)} \smile \overbrace{(\inf_m(u, x_i + y_0), 0 + 0)}\,)$$
$$= \smile \inf_m(\,\overbrace{(u, 0)}, \overbrace{(\smile x_{i-1}, 0)} \smile \overbrace{(\inf_m(u, x_i + y_0), 0)}\,)$$
$$= \smile \inf_m(\,\overbrace{(u, 0)}, \overbrace{(\smile x_{i-1} \smile \inf_m(u, x_i + y_0), 0 + 0)}\,)$$
$$= \smile \inf_m(\,\overbrace{(u, 0)}, \overbrace{(\smile x_{i-1} \smile \inf_m(u, x_i + y_0), 0)}\,)$$
$$\overset{(5.7)}{=} \smile \overbrace{(\inf_m(u, \smile x_{i-1} \smile \inf_m(u, x_i + y_0)), 0 + 0)}$$
$$= \overbrace{(\smile \inf_m(u, \smile x_{i-1} \smile \inf_m(u, x_i + y_0)), 0)}$$
$$\overset{(s5^R)(V^{n0})}{=} \overbrace{(c_i, 0)}, \text{ hence,}$$

$$\sum_{i=0}^{n+1}\overbrace{(c_i,0)} = \overbrace{(\sum_{i=0}^{n+1} c_i, \sum_{i=0}^{n+1} 0)} = \overbrace{(\sum_{i=0}^{n+1} c_i, 0)} \overset{(s5^R)(V^{n0})}{=} \overbrace{(\sum_{i=0}^{n} x_i + y_0, 0)}$$

$$= \overbrace{(\sum_{i=0}^{n} x_i, 0)} + \overbrace{(y_0,0)} = \sum_{i=0}^{n}\overbrace{(x_i,0)} + \overbrace{(y_0,0)}, \text{ where}$$

$\overbrace{(x_{-1},0)} = \overbrace{(u,0)} = \mathbf{u}$, $\overbrace{(x_{n+1},0)} = \overbrace{(0_{M^+},0)} = \mathbf{0}$,
since, if $\overbrace{(\inf_m(\mathbf{u}, x_i + x_{i+1}),0)} = \overbrace{(x_i,0)}$, for $i = 0,1,\ldots,n-1$, then
$\inf_m(u, x_i + x_{i+1}) = x_i$, for $i = 0,1,\ldots,n-1$. □

Proposition 5.1.22 *For every $m' \geq 1$,*

(C_u^R) $\quad \overbrace{(x,n)} + m'\mathbf{u} = \overbrace{(y,p)} + m'\mathbf{u} \implies \overbrace{(x,n)} = \overbrace{(y,p)}$.

Proof. Indeed,

$$\begin{array}{llll} & \overbrace{(x,n)} + m'\mathbf{u} & = & \overbrace{(y,p)} + m'\mathbf{u} \\ \iff & \overbrace{(x+m'u,n)} & = & \overbrace{(y+m'u,p)}, \quad \text{by Corollary 5.1.7 (0),} \\ \iff & (x+m'u)+pu & = & (y+m'u)+nu \\ \iff & (x+pu)+m'u & = & (y+nu)+m'u \\ \iff & x+pu & = & y+nu, \quad \text{by } (\mathrm{C}_u^R) \text{ in } \mathcal{M}_u^+, \\ \iff & \overbrace{(x,n)} & = & \overbrace{(y,p)}. \end{array}$$

□

Proposition 5.1.23 *For every $m' \geq 1$,*

(C_{pu}^R) $\quad \overbrace{(x,n)} \preceq_m \overbrace{(y,p)} \iff \overbrace{(x,n)} + m'\mathbf{u} \preceq_m \overbrace{(y,p)} + m'\mathbf{u}$.

Proof. Indeed,

$$\begin{array}{llll} & \overbrace{(x,n)} + m'\mathbf{u} & \preceq_m & \overbrace{(y,p)} + m'\mathbf{u} \\ \iff & \overbrace{(x+m'u,n)} & \preceq_m & \overbrace{(y+m'u,p)}, \quad \text{by Corollary 5.1.7 (0),} \\ \iff & (x+m'u)+pu & \leq_m & (y+m'u)+nu \\ \iff & (x+pu)+m'u & \leq_m & (y+nu)+m'u \\ \iff & x+pu & \leq_m & y+nu, \quad \text{by } (\mathrm{C}_{pu}^R) \text{ in } \mathcal{M}_u^+, \\ \iff & \overbrace{(x,n)} & \preceq_m & \overbrace{(y,p)}. \end{array}$$

□

Proposition 5.1.24 *We have:*

$(m\text{-}Tr_u^R)$ *$\overbrace{(x+mu,n)} \preceq_m p\mathbf{u}$ and $p\mathbf{u} \preceq_m r\mathbf{u}$ imply $\overbrace{(x+mu,n)} \preceq_m r\mathbf{u}$.*

Proof. • $\overbrace{(x+mu,n)} \preceq_m p\mathbf{u} \iff$
$\overbrace{(x+mu,n)} \preceq_m p\overbrace{(u,0)} \iff$
$\overbrace{(x+mu,n)} \preceq_m \overbrace{(pu,p0)} \iff$

$\overbrace{(x+mu,n)} \preceq_m \overbrace{(pu,0)} \Longleftrightarrow$
$x+mu+0u \leq_m pu+nu \Longleftrightarrow$
(a) $x+mu \leq_m pu+nu$.

• $p\mathbf{u} \preceq_m r\mathbf{u} \Longleftrightarrow$
$p\overbrace{(u,0)} \preceq_m r\overbrace{(u,0)} \Longleftrightarrow$
$\overbrace{(pu,p0)} \preceq_m \overbrace{(ru,r0)} \Longleftrightarrow$
$\overbrace{(pu,0)} \preceq_m \overbrace{(ru,0)} \Longleftrightarrow$
$pu \leq_m ru \Longleftrightarrow$
(b) $pu+nu \leq_m ru+nu$, by (C^R_{pu}).

• From (a) and (b) and (m-Tr^R_u) in $\mathcal{M}^+_u$, we obtain:
$x+mu \leq_m ru+nu \Longleftrightarrow$
$\overbrace{(x+mu,n)} \preceq_m \overbrace{(ru,0)} \Longleftrightarrow$
$\overbrace{(x+mu,n)} \preceq_m r\overbrace{(u,0)} \Longleftrightarrow$
$\overbrace{(x+mu,n)} \preceq_m r\mathbf{u}$. □

Proposition 5.1.25 *For every* $n \geq 1$,
(Nu^R) $\smile n\mathbf{u} = n\mathbf{u}$,
(NNu^R) $\smile(\smile n\mathbf{u}) = n\mathbf{u}$.

Proof. Indeed,

$$\begin{array}{rcll}
\smile n\mathbf{u} & = & \smile n\overbrace{(u,0)} & \\
& = & \smile \overbrace{(nu,n0)} & \\
& = & \smile \overbrace{(nu,0)} & \\
& = & \smile \overbrace{(0_{M^+}+nu,0)}, & \text{by (SU)},\\
& = & \overbrace{(\smile 0_{M^+}+nu,0)}, & \text{by (5.5)},\\
& = & \overbrace{(0_{M^+}+nu,0)}, & \text{by (N0) in } \mathcal{M}^+_u,\\
& = & \overbrace{(nu,0)}, & \text{by (SU)},\\
& = & n\overbrace{(u,0)}, & \\
& = & n\mathbf{u} &
\end{array}$$

and
$\smile(\smile n\mathbf{u}) = \smile n\mathbf{u} = n\mathbf{u}$. □

Proposition 5.1.26 *For every* $m \geq 1$,

(S^R) $\quad \smile(\overbrace{(x,n)} + m\mathbf{u}) = \smile \overbrace{(x,n)} + m\mathbf{u}$.

Proof. Indeed,

$$\begin{aligned}
& \smile (\widehat{(x,n)} + m\mathbf{u}) \\
= & \smile \widehat{(x+mu,n)}, && \text{by Corollary 5.1.7 (0)}, \\
= & \widehat{(\smile x+mu,n)}, && \text{by (5.5)}, \\
= & \widehat{(\smile x,n)} + m\mathbf{u}, && \text{by Corollary 5.1.7 (0)}, \\
= & \smile \widehat{(x,n)} + m\mathbf{u}.
\end{aligned}$$

□

Theorem 5.1.27 *The structure*

$$\frac{\mathcal{M}_u^+ \times \mathbf{N}}{\sim} = (\frac{M^+ \times \mathbf{N}}{\sim}, \preceq_m, +, \smile, \mathbf{0}, \mathbf{u})$$

is a strong m_0*-ME structure.*

Proof. By Propositions 5.1.8, 5.1.9, 5.1.13, 5.1.14, 5.1.16, 5.1.17, 5.1.18, 5.1.19, 5.1.20, 5.1.21, 5.1.22, 5.1.23, 5.1.24, 5.1.25, 5.1.26. □

- **The functor $\mathbf{T_m}$**

We put

$$\mathbf{T_m}(\mathcal{M}_u^+) = \frac{\mathcal{M}_u^+ \times \mathbf{N}}{\sim},$$

i.e. we have:

$$\begin{array}{lcl}
\mathcal{C}(\mathbf{s\text{-}m_0\text{-}ME}) & \overset{\mathbf{T_m}}{\longleftarrow} & \mathcal{C}(\mathbf{s\text{-}m_0\text{-}ME^+}) \\
\mathbf{T_m}(\mathcal{M}_u^+) = \frac{\mathcal{M}_u^+ \times \mathbf{N}}{\sim} & \overset{\mathbf{T_m}}{\longleftarrow} & \mathcal{M}_u^+
\end{array}$$

5.1.1 The strong m_0-ME positive cone $\mathbf{c}_{\mathbf{m}}^+(\frac{\mathcal{M}_u^+ \times \mathbf{N}}{\sim}) = (\frac{\mathcal{M}_u^+ \times \mathbf{N}}{\sim})^+$

By above Corollary 5.1.12, we can see that the positive cone of the strong m_0-ME structure $\mathbf{T_m}(\mathcal{M}_u^+) = \frac{\mathcal{M}_u^+ \times \mathbf{N}}{\sim}$ is $\mathbf{c}_{\mathbf{m}}^+(\mathbf{T_m}(\mathcal{M}_u^+)) = (\frac{\mathcal{M}_u^+ \times \mathbf{N}}{\sim})^+$, where:

$$(\frac{M^+ \times \mathbf{N}}{\sim})^+ = \{\widehat{(x+mu,n)} \mid \widehat{(x+mu,n)} \succeq_m \mathbf{0},\ s.t.\ (Nu^{+1R}),\ (S^{+1R})\ hold\}$$

$$= \{\widehat{(x+mu,0)} \mid x \in M^+,\ s.t.\ (Nu^{+1R}),\ (S^{+1R})\ hold\} = \{\widehat{(x,0)} \mid x \in M^+\}.$$

Indeed,

$$\begin{aligned}
\smile \mathbf{u} &= \smile \widehat{(u,0)} \\
&= \widehat{(\smile u,0)}, \\
&= \widehat{(0_{M^+},0)}, && \text{by } (\mathrm{Nu}^{+1R}) \text{ in } \mathcal{M}_u^+, \\
&= \mathbf{0};
\end{aligned}$$

thus, (Nu^{+1R}) holds in $(\frac{M^+ \times \mathbf{N}}{\sim})^+$
and

$$
\begin{aligned}
\smile(\overbrace{(x,0)}+\mathbf{u}) &= \smile(\overbrace{(x,0)}+\overbrace{(u,0)}) \\
&= \smile(\overbrace{(x+u,0)}), \\
&= \overbrace{(\smile(x+u),0)}, \\
&= \overbrace{(\smile x \smile u,0)}, \\
&= \overbrace{(\smile x,0)}, \qquad\qquad \text{by } (\mathrm{S}^{+1R}) \text{ in } \mathcal{M}_u^+, \\
&= \smile \overbrace{(x,0)};
\end{aligned}
$$

thus, (S^{+1R}) holds in $(\frac{M^+\times\mathbf{N}}{\sim})^+$.

Note that $\mathbf{c}^+_{\mathbf{m}}$ maps the strong m$_0$-ME structure $\mathbf{T_m}(\mathcal{M}_u^+)$ to the strong m$_0$-ME positive cone $\mathbf{c}^+_{\mathbf{m}}(\mathbf{T_m}(\mathcal{M}_u^+)) = (\mathbf{T_m}(\mathcal{M}_u^+))^+$, where,

$$(\mathbf{T_m}(M^+))^+ = (\frac{M^+\times\mathbf{N}}{\sim})^+ = \{\overbrace{(x,0)} \mid x \in M^+\}.$$

5.2 Morphisms of strong m$_0$-ME structures

For a morphism $f^+ : M^+ \longrightarrow N^+$ of strong m$_0$-ME positive cones
$\mathcal{M}_u^+ = (M^+, \leq_m, +, \smile, 0_{M^+}, u = u_{M^+})$ and $\mathcal{N}_v^+ = (N^+, \leq_m, +, \smile, 0_{N^+}, v = u_{N^+})$, we set

$$\mathbf{T_m}(f^+) : \mathbf{T_m}(M^+) \longrightarrow \mathbf{T_m}(N^+),$$

defined as follows: for $X = x + mu \in M^+$,

$$\overbrace{(X,n)_u} \longmapsto \overbrace{(f^+(X),n)_v}, \tag{5.8}$$

i.e. $\mathbf{T_m}(f^+)(\overbrace{(X,n)_u}) \overset{def.}{=} \overbrace{(f^+(X),n)_v}$,
i.e. we have the following commutative diagram:

$$
\begin{array}{ccc}
\mathbf{T_m}(M^+) & \xleftarrow{\quad\mathbf{T_m}\quad} & M^+ \\
\mathbf{T_m}(f^+)\big\downarrow & & \big\downarrow f^+ \\
\mathbf{T_m}(N^+) & \xleftarrow[\quad\mathbf{T_m}\quad]{} & N^+
\end{array}
$$

Proposition 5.2.1 *The function $\mathbf{T_m}(f^+)$ is a morphism of strong m$_0$-ME structures.*

Proof. Since f^+ is a morphism of strong m$_0$-ME positive cones (i.e. (p1) - (p5) hold, see Definitions 3.5.1 (3)), then we have:

- The function $\mathbf{T_m}(f^+)$ is well defined.

Indeed, if $\overbrace{(X,n)_u} = \overbrace{(Y,p)_u}$, i.e. $(X,n) \sim_u (Y,p)$, i.e. $X + pu = Y + nu$, in $\mathcal{M}_u^+$,

then we must prove that
$\overbrace{(f^+(X), n)_v} = \overbrace{(f^+(Y), p)_v}$ i.e. $(f^+(X), n) \sim_v (f^+(Y), p)$,
i.e. $f^+(X) + pv = f^+(Y) + nv$.
Indeed, we have:

$$\begin{aligned} f^+(X) + pv &= f^+(X) + pf^+(u) \\ &= f^+(X) + f^+(pu) \\ &= f^+(X + pu) \\ &= f^+(Y + nu) \\ &= f^+(Y) + f^+(nu) \\ &= f^+(Y) + nf^+(u) \\ &= f^+(Y) + nv. \end{aligned}$$

It follows that $\overbrace{(f^+(X), n)_v} = \overbrace{(f^+(Y), p)_v}$.

• The function $\mathbf{T}_{\mathbf{m}}(f^+)$ is a morphism of strong m$_0$-ME structures, i.e. is a strong m$_0$-ME structures homomorphism.
Indeed,

(m1) $\mathbf{T}_{\mathbf{m}}(f^+)(\overbrace{(X, n)_u} + \overbrace{(Y, p)_u}) = \mathbf{T}_{\mathbf{m}}(f^+)(\overbrace{(X, n)_u}) + \mathbf{T}_{\mathbf{m}}(f^+)(\overbrace{(Y, p)_u})$.
Indeed,

$$\begin{aligned} \mathbf{T}_{\mathbf{m}}(f^+)(\overbrace{(X, n)_u} + \overbrace{(Y, p)_u}) &= \mathbf{T}_{\mathbf{m}}(f^+)(\overbrace{(X + Y, n + p)_u}) \\ &= \overbrace{(f^+(X + Y), n + p)_v} \\ &= \overbrace{(f^+(X) + f^+(Y), n + p)_v} \\ &= \overbrace{(f^+(X), n)_v} + \overbrace{(f^+(Y), p)_v} \\ &= \mathbf{T}_{\mathbf{m}}(f^+)(\overbrace{(X, n)_u}) + \mathbf{T}_{\mathbf{m}}(f^+)(\overbrace{(Y, p)_u}). \end{aligned}$$

(m2) $\mathbf{T}_{\mathbf{m}}(f^+)(\smile \overbrace{(X, n)_u}) = \smile \mathbf{T}_{\mathbf{m}}(f^+)(\overbrace{(X, n)_u})$.
Indeed,

$$\begin{aligned} \mathbf{T}_{\mathbf{m}}(f^+)(\smile \overbrace{(X, n)_u}) &= \mathbf{T}_{\mathbf{m}}(f^+)(\smile \overbrace{(x + mu, n)_u}) \\ &= \mathbf{T}_{\mathbf{m}}(f^+)(\overbrace{(\smile x + mu, n)_u}) \\ &= \overbrace{(f^+(\smile x + mu), n)_v} \\ &= \overbrace{(f^+(\smile x) + f^+(mu), n)_v} \\ &= \overbrace{(f^+(\smile x) + mf^+(u), n)_v} \\ &= \overbrace{(f^+(\smile x) + mv, n)_v} \end{aligned}$$

and

$$\begin{aligned}
\smile \mathbf{T_m}(f^+)(\overbrace{(X,n)_u}) &= \smile \mathbf{T_m}(f^+)(\overbrace{(x+mu,n)_u}) \\
&= \smile \overbrace{(f^+(x+mu),n)_v} \\
&= \smile \overbrace{(f^+(x)+f^+(mu),n)_v} \\
&= \smile \overbrace{(f^+(x)+mf^+(u),n)_v} \\
&= \smile \overbrace{(f^+(x)+mv,n)_v} \\
&= \overbrace{(\smile f^+(x)+mv,n)_v} \\
&= \overbrace{(f^+(\smile x)+mv,n)_v}, \qquad \text{by (p2).}
\end{aligned}$$

(m3) $\mathbf{T_m}(f^+)(\mathbf{0}_u) = \mathbf{0}_v$.

Indeed, $\mathbf{T_m}(f^+)(\mathbf{0}_u) = \mathbf{T_m}(f^+)(\overbrace{(0_{M^+},0)_u}) = \overbrace{(f^+(0_{M^+}),0)_v} = \overbrace{(0_{N^+},0)_v} = \mathbf{0}_v$.

(m4) If $\overbrace{(X,n)_u} \preceq_m \overbrace{(Y,p)_u}$, then $\mathbf{T_m}(f^+)(\overbrace{(X,n)_u}) \preceq_m \mathbf{T_m}(f^+)(\overbrace{(Y,p)_u})$, i.e. $\overbrace{(f^+(X),n)_v} \preceq_m \overbrace{(f^+(Y),p)_v}$.

Indeed, suppose that $\overbrace{(X,n)_u} \preceq_m \overbrace{(Y,p)_u}$, i.e. $X+pu \leq_m Y+nu$.

$$\begin{aligned}
\overbrace{(f^+(X),n)_v} \preceq_m \overbrace{(f^+(Y),p)_v} &\iff f^+(X)+pv \leq_m f^+(Y)+nv \\
&\iff f^+(X)+pf^+(u) \leq_m f^+(Y)+nf^+(u) \\
&\iff f^+(X)+f^+(pu) \leq_m f^+(Y)+f^+(nu) \\
&\iff f^+(X+pu) \leq_m f^+(Y+nu),
\end{aligned}$$

that is true, by (p4).

(m5) $\mathbf{T_m}(f^+)(\mathbf{u}) = \mathbf{v}$.

Indeed, $\mathbf{T_m}(f^+)(\mathbf{u}) = \mathbf{T_m}(f^+)(\overbrace{(u,0)_u}) = \overbrace{(f^+(u),0)_v} = \overbrace{(v,0)_v} = \mathbf{v}$. □

This establishes that $\mathbf{T_m}$ is a functor

$$\mathbf{T_m} : \mathcal{C}(\mathbf{s}-\mathbf{m_0}-\mathbf{ME}^+) \longrightarrow \mathcal{C}(\mathbf{s}-\mathbf{m_0}-\mathbf{ME})$$

that maps $\mathcal{M}_u^+$ to $\mathbf{T_m}(\mathcal{M}_u^+)$ and maps a morphism $f^+ : M^+ \longrightarrow N^+$ to the morphism

$$\mathbf{T_m}(f^+) : \mathbf{T_m}(M^+) \longrightarrow \mathbf{T_m}(N^+).$$

Chapter 6

Connections between the functors $\mathbf{c}^+_{\mathbf{m}}$ and $\mathbf{T}_{\mathbf{m}}$

In this chapter, we prove (Corollary 6.3.3) that the functors $\mathbf{c}^+_{\mathbf{m}}$ and $\mathbf{T}_{\mathbf{m}}$ are not quasi-inverses, hence that the categories of strong m_0-ME structures and of strong m_0-ME positive cones are not equivalent.

Note that $\mathbf{c}^+_{\mathbf{m}}$ maps the strong m_0-ME structure $\mathcal{M}_u$ to the strong m_0-ME positive cone $\mathbf{c}^+_{\mathbf{m}}(\mathcal{M}_u)$ and f to $\mathbf{c}^+_{\mathbf{m}}(f)$.

Note that $\mathbf{c}^+_{\mathbf{m}}$ maps also the strong m_0-ME structure $\mathbf{T}_{\mathbf{m}}(\mathcal{M}^+_u)$ to the strong m_0-ME positive cone $\mathbf{c}^+_{\mathbf{m}}(\mathbf{T}_{\mathbf{m}}(\mathcal{M}^+_u))$ and $\mathbf{T}_{\mathbf{m}}(f^+)$ to $\mathbf{c}^+_{\mathbf{m}}(\mathbf{T}_{\mathbf{m}}(f^+))$.

Hence, we have the following situation:

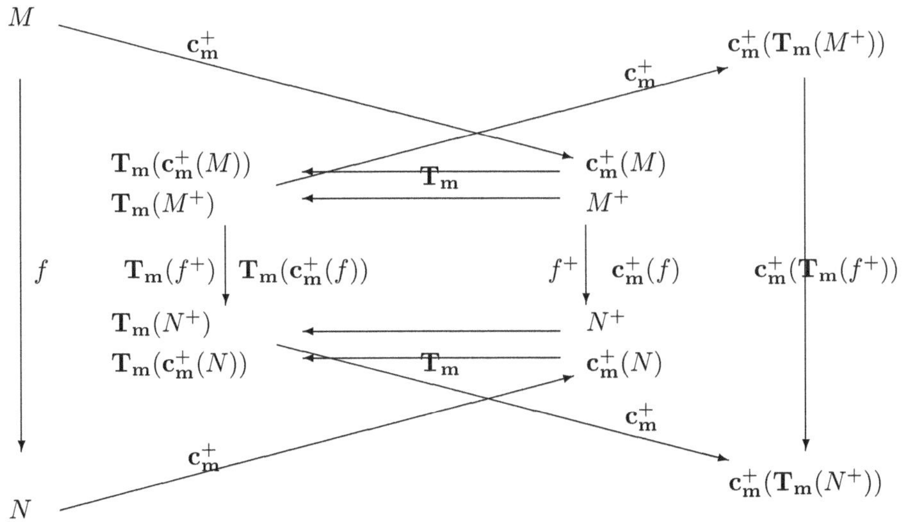

We shall analyse now if the functors $\mathbf{c}^+_{\mathbf{m}}$ and $\mathbf{T}_{\mathbf{m}}$ are quasi-inverses.

6.1 The isomorphism η_u^+ between $\mathcal{M}_u^+$ and $\mathbf{c}_{\mathbf{m}}^{+}(\mathbf{T}_{\mathbf{m}}(\mathcal{M}_u^+))$

For each strong m_0-ME positive cone

$$\mathcal{M}_u^+ = (M^+, \leq_m, +, \smile, 0_{M^+}, u = u_{M^+}),$$

consider the strong m_0-ME structure $\mathbf{T}_{\mathbf{m}}(\mathcal{M}_u^+)$ obtained by $\mathbf{T}_{\mathbf{m}}$

$$\mathbf{T}_{\mathbf{m}}(\mathcal{M}_u^+) = (\mathbf{T}_{\mathbf{m}}(M^+) = \frac{M^+ \times \mathbf{N}}{\sim}, \preceq_m, +, \smile, \mathbf{0} = \overbrace{(0_{M^+}, 0)}, \mathbf{u} = \overbrace{(u_{M^+}, 0)})$$

and then the strong m_0-ME positive cone $\mathbf{c}_{\mathbf{m}}^{+}(\mathbf{T}_{\mathbf{m}}(\mathcal{M}_u^+))$ obtained by $\mathbf{c}_{\mathbf{m}}^{+}$

$$\mathbf{c}_{\mathbf{m}}^{+}(\mathbf{T}_{\mathbf{m}}(\mathcal{M}_u^+)) = (\mathbf{c}_{\mathbf{m}}^{+}(\mathbf{T}_{\mathbf{m}}(M^+)) = (\frac{M^+ \times \mathbf{N}}{\sim})^+, \preceq_m, +, \smile, \mathbf{0}, \mathbf{u}).$$

Consider now the function

$$\eta_u^+ : M^+ \longrightarrow \mathbf{c}_{\mathbf{m}}^{+}(\mathbf{T}_{\mathbf{m}}(M^+)) = (\frac{M^+ \times \mathbf{N}}{\sim})^+$$

defined as follows: for each $x \in M^+$,

$$x \longmapsto \overbrace{(x, 0)}, \tag{6.1}$$

i.e. $\eta_u^+(x) \overset{def.}{=} \overbrace{(x, 0)}$.

Proposition 6.1.1 *(See ([1], Proposition 4.11))*
The function η_u^+ is a morphism of strong m_0-ME positive cones, i.e. a strong m_0-ME positive cones homomorphism.

Proof. • The function η_u^+ is well defined, obviously.
• The function η_u^+ is a strong m_0-ME positive cones homomorphism. Indeed,

(p1): $\eta_u^+(x + y) = \overbrace{(x + y, 0)} = \overbrace{(x, 0)} + \overbrace{(y, 0)} = \eta_u^+(x) + \eta_u^+(y)$.

(p2): $\eta_u^+(\smile x) = \overbrace{(\smile x, 0)} = \smile \overbrace{(x, 0)} = \smile \eta_u^+(x)$.

(p3): $\eta_u^+(0_{M^+}) = \overbrace{(0_{M^+}, 0)} = \mathbf{0}$.

(p4): Suppose $x \leq_m y$, in $\mathcal{M}_u^+$; we must prove that $\eta_u^+(x) \preceq_m \eta_u^+(y)$, in $(\mathbf{T}_{\mathbf{m}}(\mathcal{M}_u^+))^+$.

Indeed, $\eta_u^+(x) \preceq_m \eta_u^+(y)$

$\Longleftrightarrow \overbrace{(x, 0)} \preceq_m \overbrace{(y, 0)}$,

$\Longleftrightarrow x + 0u_{M^+} \leq_m y + 0u_{M^+}$,

$\Longleftrightarrow x + 0_{M^+} \leq_m y + 0_{M^+}$,

$\Longleftrightarrow x \leq_m y$.

(p5): $\eta_u^+(u_{M^+}) = \overbrace{(u_{M^+}, 0)} = \mathbf{u}$. □

Proposition 6.1.2 *(See ([1], Proposition 4.15))*
The function η_u^+ is bijective.

Proof. • To prove *injectivity*, suppose we have $\eta_u^+(x) = \eta_u^+(y)$, i.e. $\widetilde{(x,0)_u} = \widetilde{(y,0)_u}$; then, $x = y$.

• To prove *surjectivity*, let $\widetilde{(x,0)} \in \mathbf{c_m^+}((\mathbf{T_m}(M^+)) = (\frac{M^+ \times \mathbf{N}}{\sim_u})^+$; then, there is $x \in M^+$ such that $\eta_u^+(x) = \widetilde{(x,0)}$. $\qquad\square$

Proposition 6.1.3 *(See ([1], Proposition 4.12))*
   The function $\eta^+ : 1_{\mathcal{C}(\mathbf{s-m_o-ME^+})} \longrightarrow \mathbf{c_m^+} \circ T$ is a natural transformation, i.e. for every morphism $f^+ : M^+ \longrightarrow N^+$ of strong positive cones $\mathcal{M}_u^+ = (M^+, \leq_m, +, \smile, 0_{M^+}, u = u_{M^+})$ and $\mathcal{N}_v^+ = (N^+, \leq_m, +, \smile, 0_{N^+}, v = u_{N^+})$, the following diagram commutes:

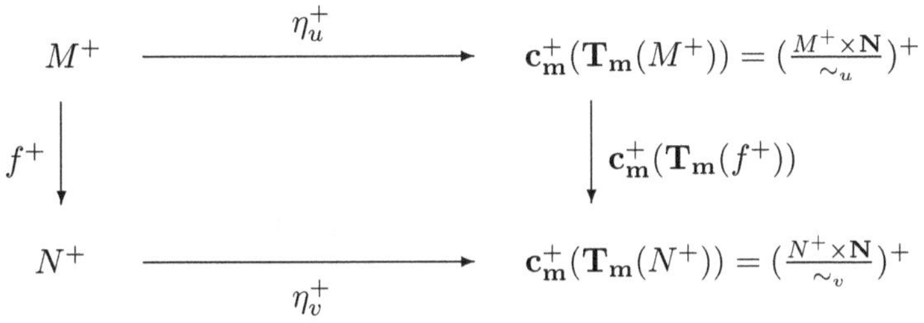

$$
\begin{array}{ccc}
M^+ & \xrightarrow{\ \ \eta_u^+\ \ } & \mathbf{c_m^+}(\mathbf{T_m}(M^+)) = (\frac{M^+ \times \mathbf{N}}{\sim_u})^+ \\[1em]
{\scriptstyle f^+}\downarrow & & \downarrow {\scriptstyle \mathbf{c_m^+}(\mathbf{T_m}(f^+))} \\[1em]
N^+ & \xrightarrow[\ \ \eta_v^+\ \ ]{} & \mathbf{c_m^+}(\mathbf{T_m}(N^+)) = (\frac{N^+ \times \mathbf{N}}{\sim_v})^+
\end{array}
$$

Proof. $\eta_v^+(f^+(x)) = \widetilde{(f^+(x),0)_v}$ and
$\mathbf{c_m^+}(\mathbf{T_m}(f^+))(\eta_u^+(x)) = \mathbf{c_m^+}(\mathbf{T_m}(f^+))(\widetilde{(x,0)_u}) = \mathbf{T_m}(f^+)(\widetilde{(x,0)_u}) \overset{(5.8)}{=} \widetilde{(f^+(x),0)_v}$.
$\square$

Hence we have:

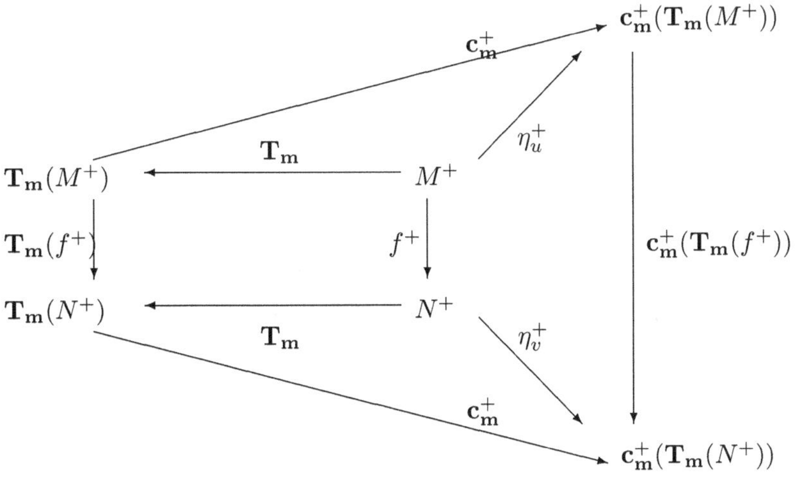

6.2 There exists also a function ϵ_u?

For each strong m_0-ME structure

$$\mathcal{M}_u = (M, \leq_m, +, \smile, 0_M, u = u_M),$$

consider its associated strong m_0-ME positive cone $\mathbf{c}^+_\mathbf{m}(\mathcal{M}_u)$ obtained by $\mathbf{c}^+_\mathbf{m}$

$$\mathbf{c}^+_\mathbf{m}(\mathcal{M}_u) = (\mathbf{c}^+_\mathbf{m}(M) = M^+, \leq_m, +, \smile, 0_M, u_M)$$

and then the strong m_0-ME structure $\mathbf{T}_\mathbf{m}(\mathbf{c}^+_\mathbf{m}(\mathcal{M}_u))$ obtained by $\mathbf{T}_\mathbf{m}$

$$\mathbf{T}_\mathbf{m}(\mathbf{c}^+_\mathbf{m}(\mathcal{M}_u)) =$$

$$(\mathbf{T}_\mathbf{m}(\mathbf{c}^+_\mathbf{m}(M)) = \mathbf{T}_\mathbf{m}(M^+) = \frac{M^+ \times \mathbf{N}}{\sim}, \preceq_m, +, \smile, \mathbf{0} = \overbrace{(0_M, 0)}, \mathbf{u} = \overbrace{(u_M, 0)}).$$

Consider naturally the function

$$\epsilon_u : \mathbf{T}_\mathbf{m}(\mathbf{c}^+_\mathbf{m}(M)) = \mathbf{T}_\mathbf{m}(M^+) \longrightarrow M$$

defined as follows: for each $\overbrace{(x + mu_M, n)} \in \mathbf{T}_\mathbf{m}(\mathbf{c}^+_\mathbf{m}(M)) = \mathbf{T}_\mathbf{m}(M^+)$,

$$\overbrace{(x + mu_M, n)} \longmapsto x + mu_M \smile nu_M. \tag{6.2}$$

i.e. $\epsilon_u(\overbrace{(x + mu_M, n)}) \overset{def.}{=} x + mu_M \smile nu_M \overset{(Nu^R)}{=} x + mu_M + nu_M$.

Is the function ϵ_u well defined?
For the case $m < n$, say $n = m + s \in \mathbf{N}$, we have:
$\overbrace{(x + mu_M, n)} = \overbrace{(x + mu_M, m + s)} = \overbrace{(x, s)}$;
but, $\epsilon_u(\overbrace{(x + mu_M, n)}) \overset{def.}{=} x + mu_M + nu_M = x + mu_M + (m + s)u_M$,
while $\epsilon_u(\overbrace{(x, s)}) \overset{def.}{=} x + su_M$, and $x + mu_M + (m + s)u_M \neq x + su_M$.
Hence, the function ϵ_u is not well defined.

It follows that there exists not a function ϵ_u which be a strong m_0-ME structures isomorphism such that we have:

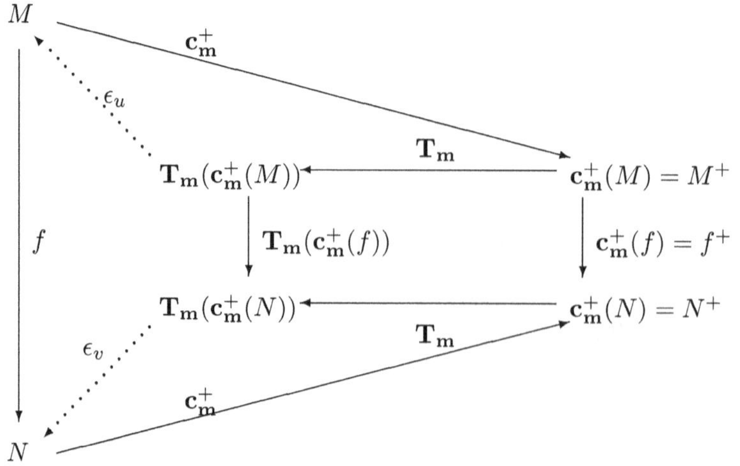

6.3 Conclusions

It follows that we have the following result.

Theorem 6.3.1 *The functors*
$1_{\mathcal{C}(\mathbf{s}-\mathbf{m_0}-\mathbf{ME}^+)} : \mathcal{C}(\mathbf{s}-\mathbf{m_0}-\mathbf{ME}^+) \longrightarrow \mathcal{C}(\mathbf{s}-\mathbf{m_0}-\mathbf{ME}^+)$
and
$\mathbf{c}^+_{\mathbf{m}} \circ \mathbf{T_m} : \mathcal{C}(\mathbf{s}-\mathbf{m_0}-\mathbf{ME}^+) \longrightarrow \mathcal{C}(\mathbf{s}-\mathbf{m_0}-\mathbf{ME}^+)$
are naturally isomorphic.

Proof. By Propositions 6.1.1, 6.1.2, 6.1.3. □

Remark 6.3.2 *The functors*
$1_{\mathcal{C}(\mathbf{s}-\mathbf{m_0}-\mathbf{ME})} : \mathcal{C}(\mathbf{s}-\mathbf{m_0}-\mathbf{ME}) \longrightarrow \mathcal{C}(\mathbf{s}-\mathbf{m_0}-\mathbf{ME})$
and
$\mathbf{T} \circ \mathbf{c}^+_{\mathbf{m}} : \mathcal{C}(\mathbf{s}-\mathbf{m_0}-\mathbf{ME}) \longrightarrow \mathcal{C}(\mathbf{s}-\mathbf{m_0}-\mathbf{ME})$
are not naturally isomorphic.

Corollary 6.3.3 *(See ([1], Theorem 4.16))*
The functors
$\mathcal{C}(\mathbf{s}\text{-}\mathbf{m_0}\text{-}\mathbf{ME}) \xrightarrow{\mathbf{c}^+_{\mathbf{m}}} \mathcal{C}(\mathbf{s}\text{-}\mathbf{m_0}\text{-}\mathbf{ME}^+)$ *and* $\mathcal{C}(\mathbf{s}\text{-}\mathbf{m_0}\text{-}\mathbf{ME}^+) \xrightarrow{\mathbf{T_m}} \mathcal{C}(\mathbf{s}\text{-}\mathbf{m_0}\text{-}\mathbf{ME})$
are not quasi-inverses, thus the categories $\mathcal{C}(\mathbf{s}\text{-}\mathbf{m_0}\text{-}\mathbf{ME})$ *and* $\mathcal{C}(\mathbf{s}\text{-}\mathbf{m_0}\text{-}\mathbf{ME}^+)$ *are not equivalent (the functor* $\mathbf{c}^+_{\mathbf{m}}$ *is not an equivalence).*

Part III

The second step: the functors G_m^+ and U_m

Chapter 7

Construction of a strong m_0-ME positive cone $\mathbf{G}^+_{\mathbf{m}}(\mathcal{A})$. The functor $\mathbf{G}^+_{\mathbf{m}}$

In this chapter, we introduce and study the functor $\mathbf{G}^+_{\mathbf{m}}$ which constructs, from a right-XY algebra $\mathcal{A}$, a strong m_0-ME positive cone $\mathbf{G}^+_{\mathbf{m}}(\mathcal{A})$. The name '$\mathbf{G}^+_{\mathbf{m}}$' comes from 'good' - the good pairs and the good sequences it builds.

7.1 Construction of a strong m_0-ME positive cone $\mathbf{G}^+_{\mathbf{m}}(\mathcal{A})$ from an XY algebra $\mathcal{A}$

Let $\mathcal{A} = (A, \oplus, ^-, 0 = 0_A)$ be a right-XY algebra (Definition 2.2.13 (1)) throughout this section, with $1 = 1_A \overset{def.}{=} 0^-$.

The goal of this section, reached in Theorem 7.1.77 below, is to build from $\mathcal{A}$ a strong m_0-ME positive cone (Definition 3.3.1 (1)) $\mathbf{G}^+_{\mathbf{m}}(\mathcal{A})$.

Many of Abbadini's results from [1] are used in this section, because they are valid for right-XY algebras.

7.1.1 Good pairs, good sequences of $\mathcal{A}$. The elements $0_{G^+(A)}$ and $u_{G^+(A)}$ of $G^+(A)$

While the notion of good sequence is old [10], the notion of good pair was introduced by M. Abbadini in [1].

Definitions 7.1.1 (See ([1], Definition 5.1), ([10], section 2.2))

(1) A *good pair* of A is a pair (x, y) of elements of A such that $x \oplus y = x$.

(2) A *good sequence* of A is a sequence $\mathbf{a} = (a_0, a_1, \ldots)$ of elements of A such that:

(i) for each $i = 0, 1, \ldots$, the pair (a_i, a_{i+1}) is good and
(ii) there is $n \in \mathbf{N}$, such that $a_k = 0$ for all $k > n$, i.e. $\mathbf{a} = (a_0, a_1, \ldots, a_n, 0, 0, \ldots)$.
We shall simply write:

$$\mathbf{a} = (a_0, a_1, \ldots, a_n)$$

and we shall say that *the length of* $\mathbf{a}$ is $n+1$ and write: $l(\mathbf{a}) = n+1$.

We shall denote by $G^+(A)$ the set of good sequences of $\mathcal{A}$ ('G' comes from 'good').

Note that, like in MV algebras, (x, y) is a good pair iff $x \odot y = y$, by Proposition 2.2.54.

Note that, due to the special role of the paranthesis '(,)', in the sequel, if necessary, we shall use the brackets '[,]', to avoid any confusion.

Note that, if 0^m denotes an m-tuple of zeros (which is a good sequence), then we have the **identic** good sequences:

$$(a_0, a_1, \ldots, a_n) = (a_0, a_1, \ldots, a_n, 0^m).$$

But, if 1^m denotes an m-tuple of 1 (which is a good sequence), then we have the **different** good sequences:

$$(a_0, a_1, \ldots, a_n) \neq (1^m, a_0, a_1, \ldots, a_n).$$

We shall denote the good sequence $(x, 0, 0, \ldots)$ by (x), for each $x \in A$.

We shall denote the good sequence $(0, 0, 0, \ldots)$ by (0).
We shall denote the good sequence $(1, 0, 0, \ldots)$ by (1).
Define:

$$(7.1) \qquad 0_{G^+(A)} \stackrel{def.}{=} (0) = (0_A), \qquad u_{G^+(A)} \stackrel{def.}{=} (1) = (1_A).$$

Remark 7.1.2 *The sequence $(1, 1, 1, \ldots)$ is not good, because (ii) of above Definition does not hold. But, the sequence $\underbrace{(1, 1, \ldots, 1)}_{n \text{ times}}$ is good.*

7.1.2 Results on good pairs, good sequences of $\mathcal{A}$

Lemma 7.1.3 *(See ([1], Lemma 6.10))*
For all $x, y \in A$, the pair $(x \oplus y, x \odot y)$ is good.

Proof. By property (Z), $(x \oplus y) \oplus (x \odot y) = x \oplus y$, i.e. $(x \oplus y, x \odot y)$ is a good pair. □

Lemma 7.1.4 *(See ([1], Lemma 7.6))*
For every good pair (a, b) of $\mathcal{A}$ and every element $x \in A$, the pairs

$$(a \oplus x, b) \quad \text{and} \quad (a, b \odot x)$$

are good.

Proof. We have that $a \oplus b = a$ and, consequently, by Proposition 2.2.54, we also have that $a \odot b = b$.

(1) $(a \oplus x) \oplus b = (a \oplus b) \oplus x = a \oplus x$; thus, the pair $(a \oplus x, b)$ is good.

(2) $a \oplus (b \odot x) = a$. Indeed (based on a proof by PROVER9, Length of proof was 26, in 58.50 seconds):
since $a \odot b = b$, i.e. $(a^- \oplus b^-)^- = b$, by (2.1), then, by (DN), we have:
(*) $a^- \oplus b^- = b^-$;
now, in (Sass) $((x \oplus y) \oplus z = x \oplus (y \oplus z))$, take $Y := b^-$, $Z := a^-$, to obtain:
$(x \oplus b^-) \oplus a^- = x \oplus (b^- \oplus a^-)$,
which, by (*), becomes:
(**) $(x \oplus b^-) \oplus a^- = x \oplus b^-$;
now, in (YY) $(x \oplus y = x \Longrightarrow x \odot (y \oplus z) = y \oplus (x \odot z))$, take $X := x \oplus b^-$ and $Y := a^-$, to obtain:
(***) $(x \oplus b^-) \oplus a^- = x \oplus b^- \Longrightarrow (x \oplus b^-) \odot (a^- \oplus z) = a^- \oplus ((x \oplus b^-) \odot z)$;
but, in (***), the part $(x \oplus b^-) \oplus a^- = x \oplus b^-$ is always true, by (**); hence, (***) becomes:
$(x \oplus b^-) \odot (a^- \oplus z) = a^- \oplus ((x \oplus b^-) \odot z)$,
which, for $Z := 0$, becomes:
$(x \oplus b^-) \odot (a^- \oplus 0) = a^- \oplus ((x \oplus b^-) \odot 0)$,
which, by (SU), (m-La), becomes:
$(x \oplus b^-) \odot a^- = a^-$,
which, by (DN), becomes:
$((x \oplus b^-) \odot a^-)^- = a$,
i.e. $(x \oplus b^-)^- \oplus a = a$, i.e. $(x^- \odot b) \oplus a = a$, hence $a \oplus (b \odot x) = a$, for $X := x^-$, by (DN). □

Lemma 7.1.5 *If (x_0, y_0), (x_1, y_0), (x_0, y_1) are good pairs, then*

$$(x_0 \odot x_1, y_0) \quad \textit{and} \quad (x_0, y_0 \oplus y_1)$$

are good pairs.

Proof. We have that $x_0 \oplus y_0 = x_0$, $x_1 \oplus y_0 = x_1$ and $x_0 \oplus y_1 = x_0$.

In (YY) $(x \oplus y = x \Longrightarrow x \odot (y \oplus z) = y \oplus (x \odot z))$, take $X := x_0$, $Y := y_0$, $Z := x_1$, to obtain:
$x_0 \oplus y_0 = x_0 \Longrightarrow x_0 \odot (y_0 \oplus x_1) = y_0 \oplus (x_0 \odot x_1)$,
which becomes, since $x_0 \oplus y_0 = x_0$:
$x_0 \odot (y_0 \oplus x_1) = y_0 \oplus (x_0 \odot x_1)$,
which becomes, since $x_1 \oplus y_0 = x_1$:
$x_0 \odot x_1 = y_0 \oplus (x_0 \odot x_1)$, hence the pair $(x_0 \odot x_1, y_0)$ is good.

$x_0 \oplus (y_0 \oplus y_1) = (x_0 \oplus y_0) \oplus y_1 = x_0 \oplus y_1 = x_0$, hence the pair $(x_0, y_0 \oplus y_1)$ is good. □

Lemma 7.1.6 *(See ([1], Lemma 7.5))*

Let $x_0, \ldots, x_n$, $y_0, \ldots, y_m \in A$ and suppose that, for every $i \in \{0, \ldots, n\}$ and for every $j \in \{0, \ldots, m\}$, the pair (x_i, y_j) is good. Then, the pair

$$(x_0 \odot \ldots \odot x_n,\ y_0 \oplus \ldots \oplus y_m)$$

is good.

Proof. By hypothesis, $x_i \oplus y_j = x_i$, for every $i \in \{0, \dots, n\}$, $j \in \{0, \dots, m\}$.

• The statement is true obviously for $(n, m) = (0, 0)$.

• The statement is true for $(n, m) = (1, 0)$, i.e. the pair $(x_0 \odot x_1, y_0)$ is good, by Lemma 7.1.5.

• The statement is true for $(n, m) = (0, 1)$, i.e. the pair $(x_0, y_0 \oplus y_1)$ is good, by Lemma 7.1.5 again.

• Let $(n, m) \in \mathbf{N} \times \mathbf{N} \setminus \{(0,0), (1,0), (0,1)\}$ and suppose that the statement is true for each $(h, k) \in \mathbf{N} \times \mathbf{N}$ such that $(h, k) \neq (n, m)$, $h \leq n$, $k \leq m$. We shall prove that the statement is true for (n, m). Indeed, at least one of the two conditions $n \neq 0$ and $m \neq 0$ holds.

- Suppose $n \neq 0$. Then, by induction hypothesis, the pairs:

$$(x_0 \odot \dots \odot x_{n-1},\ y_0 \oplus \dots \oplus y_m) \quad and \quad (x_n,\ y_0 \oplus \dots \oplus y_m)$$

are good. Then, by Lemma 7.1.5, the pair:

$$((x_0 \odot \dots \odot x_{n-1}) \odot x_n,\ y_0 \oplus \dots \oplus y_m)$$

is good.

- Suppose $m \neq 0$. Then, by induction hypothesis, the pairs:

$$(x_0 \odot \dots \odot x_n,\ y_0 \oplus \dots \oplus y_{m-1}) \quad and \quad (x_0 \odot \dots \odot x_n,\ y_m)$$

are good. Then, by Lemma 7.1.5, the pair:

$$(x_0 \odot \dots \odot x_n,\ (y_0 \oplus \dots \oplus y_{m-1}) \oplus y_m)$$

is good. □

Lemma 7.1.7 *(Transitivity) (See ([1], Lemma 7.7))*

If (a, b) and (b, c) are good pairs, then (a, c) is a good pair.

Proof. We have that $a \oplus b = a$ and $b \oplus c = b$; we must prove that $a \oplus c = a$.

Indeed, in (Sass) ($(x \oplus y) \oplus z = x \oplus (y \oplus z)$), take $X := b$, $Y := c$, $Z := a$, to obtain:
$(b \oplus c) \oplus a = b \oplus (c \oplus a)$,
which becomes, by (Scomm):
$a = b \oplus (a \oplus c)$,
which becomes, by (Sass):
$a = (b \oplus a) \oplus c$, i.e. $a = a \oplus c$. □

Lemma 7.1.8 *If $(a_0, a_1, \dots, a_n)$ is a good sequence, then $(1, a_0, a_1, \dots, a_n)$ is a good sequence too.*

Proof. Obviously, by (m-La^R). □

Lemma 7.1.9 *Let* $\mathbf{a} = (a_0, a_1, \ldots, a_n)$ *and* $\mathbf{b} = (b_0, b_1, \ldots, b_n)$ *be two good sequences of* $\mathcal{A}$. *Then,*

$$\mathbf{a} = \mathbf{b} \iff a_i = b_i,\ i = 0, 1, \ldots, n.$$

Proof. Obviously. □

We shall endow $G^+(A)$ with the structure of a strong m_0-ME positive cone in below Theorem 7.1.77.

7.1.3 The sum + on the set $G^+(A)$ of good sequences

Like in [1], we define the sum of two good sequences in two natural ways.

Given the good sequences $\mathbf{a}$ and $\mathbf{b}$, $\mathbf{a} = (a_0, a_1, \ldots)$ and $\mathbf{b} = (b_0, b_1, \ldots)$, we can suppose that their "length" is the same (by adding elements 0), $n+1$:
$\mathbf{a} = (a_0, a_1, \ldots, a_n)$ and $\mathbf{b} = (b_0, b_1, \ldots, b_n)$.

Denote $\mathbf{c} = \mathbf{a} + \mathbf{b}$, with $\mathbf{c} = (c_0, c_1, \ldots)$ defined either by [1]:

$$(7.2) \qquad c_n \overset{def.1}{=} (a_0 \oplus b_n) \odot (a_1 \oplus b_{n-1}) \odot \ldots \odot (a_{n-1} \oplus b_1) \odot (a_n \oplus b_0)$$

or by [10], [1]:

$$(7.3)\ c_n \overset{def.2}{=} b_n \oplus (a_0 \odot b_{n-1}) \oplus (a_1 \odot b_{n-2}) \oplus \ldots \oplus (a_{n-2} \odot b_1) \oplus (a_{n-1} \odot b_0) \oplus a_n.$$

• **Our first goal, reached in Proposition 7.1.11 below, is to prove that these two definitions coincide.**

Proposition 7.1.10 *(See ([1], Lemma 7.8))*

Let $\mathbf{a} = (a_0, a_1, \ldots, a_n)$ *and* $\mathbf{b} = (b_0, b_1, \ldots, b_n)$ *be two good sequences of* $\mathcal{A}$. *Then,*

$$(P) \qquad (a_0 \odot b_n) \oplus \ldots \oplus (a_{n-2} \odot b_2) \oplus (a_{n-1} \odot b_1) \oplus (a_n \odot b_0)$$
$$= a_0 \odot (a_1 \oplus b_n) \odot \ldots \odot (a_{n-1} \oplus b_2) \odot (a_n \oplus b_1) \odot b_0$$

Proof. (By the proof of ([1], Lemma 7.8))

• The property (P) is obviously true for $n = 0$.

• Let $n \in \mathbf{N} \setminus \{0\}$ and suppose that (P) holds for $n-1$.

• We shall prove that (P) holds for n. Indeed, by inductive hypothesis, we obtain:
$T \overset{notation}{=} (a_0 \odot b_n) \oplus \ldots \oplus (a_{n-2} \odot b_2) \oplus (a_{n-1} \odot b_1) \oplus (a_n \odot b_0)$
$\overset{(Sass)}{=} [(a_0 \odot b_n) \oplus \ldots \oplus (a_{n-2} \odot b_2) \oplus (a_{n-1} \odot b_1)] \oplus (a_n \odot b_0)$
$= [a_0 \odot (a_1 \oplus b_n) \odot \ldots \odot (a_{n-1} \oplus b_2) \odot b_1] \oplus (a_n \odot b_0)$.

Let us denote:
$B \overset{notation}{=} a_0 \odot (a_1 \oplus b_n) \odot \ldots \odot (a_{n-1} \oplus b_2)$ and
$A \overset{notation}{=} a_0 \odot (a_1 \oplus b_n) \odot \ldots \odot (a_{n-1} \oplus b_2) \odot b_1$,

so we have:
$A = B \odot b_1$ and $T = A \oplus (a_n \odot b_0)$.

Since **b** is a good sequence, it follows that (b_0, b_1) is a good pair; then, by Lemma 7.1.4, the pair $(b_0, B \odot b_1)$, i.e. (b_0, A) is a good pair, hence we have:
(i) $b_0 \oplus A = b_0$.
Then, in (YY) $(x \oplus y = x \Longrightarrow x \odot (y \oplus z) = y \oplus (x \odot z))$, take $x := b_0$, $y := A$, $z := a_n$, to obtain:
$b_0 \oplus A = b_0 \Longrightarrow b_0 \odot (A \oplus a_n) = A \oplus (b_0 \odot a_n)$,
which, by (i), becomes:
$b_0 \odot (A \oplus a_n) = A \oplus (b_0 \odot a_n)$, hence, by (Pcomm):
(ii) $T = b_0 \odot (A \oplus a_n)$.

Since **a** is a good sequence, it follows that the pairs (a_0, a_1), (a_1, a_2), $\ldots$ (a_{n-1}, a_n) are good; then, by Lemma 7.1.7, the pairs (a_0, a_n), (a_1, a_n), $\ldots$ (a_{n-2}, a_n), (a_{n-1}, a_n) are good; then, by Lemma 7.1.4, the pairs

$$(a_0, a_n),\ (a_1 \oplus b_n, a_n),\ \ldots\ (a_{n-2} \oplus b_3, a_n),\ (a_{n-1} \oplus b_2, a_n)$$

are good. Then, by Lemma 7.1.5, the pair

$$(a_0 \odot (a_1 \oplus b_n) \odot \ldots \odot (a_{n-2} \oplus b_3) \odot (a_{n-1} \oplus b_2),\ a_n),$$

i.e. (B, a_n) is a good, hence we have:
(iii) $B \oplus a_n = B$.
Then, in (YY) $(x \oplus y = x \Longrightarrow x \odot (y \oplus z) = y \oplus (x \odot z))$, take $x := B$, $y := a_n$, $z := b_1$, to obtain:
$B \oplus a_n = B \Longrightarrow B \odot (a_n \oplus b_1) = a_n \oplus (B \odot b_1)$,
which, by (iii), becomes:
$B \odot (a_n \oplus b_1) = a_n \oplus (B \odot b_1)$, hence:
(iv) $B \odot (a_n \oplus b_1) = a_n \oplus A$.

Now, from (ii) and (iv), we obtain:
$T = b_0 \odot (A \oplus a_n) = b_0 \odot (B \odot (a_n \oplus b_1))$
$\overset{(Pcomm)}{=} (B \odot (a_n \oplus b_1)) \odot b_0$
$= ([a_0 \odot (a_1 \oplus b_n) \odot \ldots \odot (a_{n-1} \oplus b_2)] \odot (a_n \oplus b_1)) \odot b_0$.
Thus, (P) holds for n and the proof is complete. □

Proposition 7.1.11 *(See ([1], Lemma 7.8))*

Let $\mathbf{a} = (a_0, a_1, \ldots, a_n)$ *and* $\mathbf{b} = (b_0, b_1, \ldots, b_n)$ *be two good sequences of* $\mathcal{A}$. *Then,*

$$(a_0 \oplus b_n) \odot \ldots \odot (a_n \oplus b_0) = b_n \oplus (a_0 \odot b_{n-1}) \oplus \ldots \oplus (a_{n-1} \odot b_0) \oplus a_n.$$

Proof. (By the proof of ([1], Lemma 7.8))

$b_n \oplus (a_0 \odot b_{n-1}) \oplus (a_1 \odot b_{n-2}) \oplus \ldots \oplus (a_{n-1} \odot b_0) \oplus a_n$
$\overset{(PU)}{=} (1 \odot b_n) \oplus (a_0 \odot b_{n-1}) \oplus (a_1 \odot b_{n-2}) \oplus \ldots \oplus (a_{n-1} \odot b_0) \oplus (a_n \odot 1)$
$= (a_{-1} \odot b_n) \oplus (a_0 \odot b_{n-1}) \oplus (a_1 \odot b_{n-2}) \oplus \ldots \oplus (a_{n-1} \odot b_0) \oplus (a_n \odot b_{-1})$, for $a_{-1} = 1$ and $b_{-1} = 1$,

$\overset{(P)}{=} a_{-1} \odot (a_0 \oplus b_n) \odot \ldots \odot (a_n \oplus b_0) \odot b_{-1}$
$= 1 \odot (a_0 \oplus b_n) \odot \ldots \odot (a_n \oplus b_0) \odot 1$
$\overset{(PU)}{=} (a_0 \oplus b_n) \odot \ldots \odot (a_n \oplus b_0)$,
because, by Lemma 7.1.8, the sequences $(1, a_0, a_1, \ldots, a_n)$ and $(1, b_0, b_1, \ldots, b_n)$ are good. □

- **Examples of sums**

Lemma 7.1.12 *(See Corollary 7.1.82) (m=0)*

$$(Sum^{00}) \qquad (a_0) + (b_0) = (a_0 \oplus b_0, a_0 \odot b_0).$$

Proof. By Definition 1 of the sum, (7.2),
$(a_0) + (b_0) = (c_0, c_1, \ldots)$, where:
$c_0 = a_0 \oplus b_0$,
$c_1 = (a_0 \oplus b_1) \odot (a_1 \oplus b_0) = (a_0 \oplus 0) \odot (0 \oplus b_0) \overset{(SU)}{=} a_0 \odot b_0$,
$c_2 = (a_0 \oplus b_2) \odot (a_1 \oplus b_1) \odot (a_2 \oplus b_0) = (a_0 \oplus 0) \odot (0 \oplus 0) \odot (0 \oplus b_0) \overset{(SU),(m-La)}{=} 0.$
Hence, $(a_0) + (b_0) = (a_0 \oplus b_0, a_0 \odot b_0)$ of length 2. □

Corollary 7.1.13

$$(a_0) + (0) = (a_0) \quad \textit{and} \quad (a_0) + (1) = (1, a_0).$$

Proof. By Lemma 7.1.12 and by (SU), (m-La) and (m-La^R), (PU). □

Lemma 7.1.14 *(See Corollary 7.1.85) (m=1)*
If (a_0, a_1) and (b_0) are good sequences, then

$$(Sum^{10}) \qquad (a_0, a_1) + (b_0) = (c_0, c_1, c_2),$$

where:
$c_0 = a_0 \oplus b_0$,
$c_1 = a_0 \odot (a_1 \oplus b_0)$,
$c_2 = a_1 \odot b_0$.

Proof. We have $a_0 \oplus a_1 = a_0$ and also $a_0 \odot a_1 = a_1$. By Definition 1 of the sum,
$c_0 = a_0 \oplus b_0$.
$c_1 = (a_0 \oplus b_1) \odot (a_1 \oplus b_0) = (a_0 \oplus 0) \odot (a_1 \oplus b_0) \overset{(SU)}{=} a_0 \odot (a_1 \oplus b_0)$.
$c_2 = (a_0 \oplus b_2) \odot (a_1 \oplus b_1) \odot (a_2 \oplus b_0)$
$= (a_0 \oplus 0) \odot (a_1 \oplus 0) \odot (0 \oplus b_0)$
$\overset{(SU)}{=} a_0 \odot a_1 \odot b_0 = a_1 \odot b_0$.
$c_3 = (a_0 \oplus b_3) \odot (a_1 \oplus b_2) \odot (a_2 \oplus b_1) \odot (a_3 \oplus b_0)$
$= (a_0 \oplus 0) \odot (a_1 \oplus 0) \odot (0 \oplus 0) \odot (0 \oplus b_0) \overset{(SU),(m-La)}{=} 0.$ □

Lemma 7.1.15 *(See Corollary 7.1.90) (m=2)*

If (a_0, a_1, a_2) *and* (b_0) *are good sequences, then*

$$(Sum^{20}) \qquad (a_0, a_1, a_2) + (b_0) = (c_0, c_1, c_2, c_3),$$

where:
$c_0 = a_0 \oplus b_0,$
$c_1 = a_0 \odot (a_1 \oplus b_0),$
$c_2 = a_1 \odot (a_2 \oplus b_0),$
$c_3 = a_2 \odot b_0.$

Proof. We have $a_0 \oplus a_1 = a_0$, $a_1 \oplus a_2 = a_1$ and also $a_0 \odot a_1 = a_1$, $a_1 \odot a_2 = a_2$. By Definition 1 of the sum,
$c_0 = a_0 \oplus b_0.$
$c_1 = (a_0 \oplus b_1) \odot (a_1 \oplus b_0) = (a_0 \oplus 0) \odot (a_1 \oplus b_0) \overset{(SU)}{=} a_0 \odot (a_1 \oplus b_0).$
$c_2 = (a_0 \oplus b_2) \odot (a_1 \oplus b_1) \odot (a_2 \oplus b_0)$
$= (a_0 \oplus 0) \odot (a_1 \oplus 0) \odot (a_2 \oplus b_0)$
$\overset{(SU)}{=} a_0 \odot a_1 \odot (a_2 \oplus b_0) = a_1 \odot (a_2 \oplus b_0).$
$c_3 = (a_0 \oplus b_3) \odot (a_1 \oplus b_2) \odot (a_2 \oplus b_1) \odot (a_3 \oplus b_0)$
$= (a_0 \oplus 0) \odot (a_1 \oplus 0) \odot (a_2 \oplus 0) \odot (0 \oplus b_0)$
$\overset{(SU)}{=} a_0 \odot a_1 \odot a_2 \odot b_0 = a_2 \odot b_0.$
$c_4 = (a_0 \oplus b_4) \odot (a_1 \oplus b_3) \odot (a_2 \oplus b_2) \odot (a_3 \oplus b_1) \odot (a_4 \oplus b_0)$
$= (a_0 \oplus 0) \odot (a_1 \oplus 0) \odot (a_2 \oplus 0) \odot (0 \oplus 0) \odot (0 \oplus b_0) \overset{(SU),(m-La)}{=} 0.$ □

Lemma 7.1.16 *(See Corollary 7.1.92) (m=n)*
If $(a_0, a_1, \ldots, a_n)$ *and* (b_0) *are good sequences, then*

$$(Sum^{n0}) \qquad (a_0, a_1, \ldots, a_n) + (b_0) = (c_0, c_1 \ldots, c_n, c_{n+1}),$$

where:
$c_i = a_{i-1} \odot (a_i \oplus b_0)$, *for* $i = 0, 1, \ldots, n+1$, *with* $a_{-1} = 1$.

Proof. We have $a_i \oplus a_{i+1} = a_i$ and $a_i \odot a_{i+1} = a_{i+1}$, for $i = 0, 1, \ldots, n-1$. Then, for $i = 0, 1, \ldots, n+1$ and $a_{-1} = 1$, we have:
$c_i \overset{def.1}{=} (a_0 \oplus b_i) \odot (a_1 \oplus b_{i-1}) \odot \ldots \odot (a_{i-1} \oplus b_1) \odot (a_i \oplus b_0)$
$= (a_0 \oplus 0) \odot (a_1 \oplus 0) \odot \ldots \odot (a_{i-1} \oplus 0) \odot (a_i \oplus b_0)$
$= a_0 \odot a_1 \odot \ldots \odot a_{i-1} \odot (a_i \oplus b_0)$
$= a_{i-1} \odot (a_i \oplus b_0)$ and
$c_{n+2} \overset{def.1}{=} (a_0 \oplus b_{n+2}) \odot (a_1 \oplus b_{n+1}) \odot \ldots \odot (a_{n+1} \oplus b_1) \odot (a_{n+2} \oplus b_0)$
$= (a_0 \oplus 0) \odot (a_1 \oplus 0) \odot \ldots \odot (0 \oplus 0) \odot (0 \oplus b_0)$
$= a_0 \odot a_1 \odot \ldots \odot 0 \odot b_0 = 0.$ □

Lemma 7.1.17 *(See Corollary 7.1.94) If* (a_0, a_1) *and* (b_0, b_1) *are good sequences, then*

$$(Sum^{11}) \qquad (a_0, a_1) + (b_0, b_1) = (c_0, c_1, c_2, c_3),$$

where:
$c_0 = a_0 \oplus b_0,$

$c_1 = (a_0 \oplus b_1) \odot (a_1 \oplus b_0)$,
$c_2 = a_0 \odot b_0 \odot (a_1 \oplus b_1)$,
$c_3 = a_1 \odot b_1$.

Proof. We have $a_0 \oplus a_1 = a_0$, $b_0 \oplus b_1 = b_0$ and also $a_0 \odot a_1 = a_1$, $b_0 \odot b_1 = b_1$. By Definition 1 of the sum,
$c_0 = a_0 \oplus b_0$.
$c_1 = (a_0 \oplus b_1) \odot (a_1 \oplus b_0)$.
$c_2 = (a_0 \oplus b_2) \odot (a_1 \oplus b_1) \odot (a_2 \oplus b_0)$
$= (a_0 \oplus 0) \odot (a_1 \oplus b_1) \odot (0 \oplus b_0)$
$\overset{(SU)}{=} a_0 \odot (a_1 \oplus b_1) \odot b_0 = a_0 \odot b_0 \odot (a_1 \oplus b_1)$.
$c_3 = (a_0 \oplus b_3) \odot (a_1 \oplus b_2) \odot (a_2 \oplus b_1) \odot (a_3 \oplus b_0)$
$= (a_0 \oplus 0) \odot (a_1 \oplus 0) \odot (0 \oplus b_1) \odot (0 \oplus b_0) \overset{(SU),(Pcomm)}{=} a_1 \odot b_1$.
$c_4 = (a_0 \oplus b_4) \odot (a_1 \oplus b_3) \odot (a_2 \oplus b_2) \odot (a_3 \oplus b_1) \odot (a_4 \oplus b_0)$
$= (a_0 \oplus 0) \odot (a_1 \oplus 0) \odot (0 \oplus 0) \odot (0 \oplus b_1) \odot (0 \oplus b_0) \overset{(SU),(m-La)}{=} 0$. □

Remark 7.1.18 *Let $\mathbf{a} \neq 0_{G^+(A)}$ and $\mathbf{b} \neq 0_{G^+(A)}$ be two good sequences of $\mathcal{A}$ and let $\mathbf{c} = \mathbf{a} + \mathbf{b}$. Then,*

$$l(\mathbf{c}) = l(\mathbf{a}) + l(\mathbf{b}).$$

• **Our second goal, reached in Proposition 7.1.20 below, is to prove that a + b is a good sequence.**

Lemma 7.1.19 *(See ([1], Lemma 7.9))*
Let $\mathbf{a} = (a_0, a_1, \ldots, a_n)$ and $\mathbf{b} = (b_0, b_1, \ldots, b_n)$ be two good sequences of $\mathcal{A}$. Then, the pair

$$((a_0 \oplus b_n) \odot \ldots \odot (a_n \oplus b_0),\ (a_0 \odot b_n) \oplus \ldots \oplus (a_n \odot b_0))$$

is good.

Proof. (By the proof of ([1], Lemma 7.9))
By Lemma 7.1.6, it is enough to prove that, for all $i, j \in \{0, 1, \ldots, n\}$, the pair

$$(a_i \oplus b_{n-i},\ a_j \odot b_{n-j})$$

is good. Indeed,
- if $i = j$, it follows by Lemma 7.1.3;
- if $i < j$, then, since $\mathbf{a}$ is a good sequence, the pair (a_i, a_j) is good, by Lemma 7.1.7; then, by Lemma 7.1.4, the pair $(a_i \oplus b_{n-i},\ a_j \odot b_{n-j})$ is good;
- if $i > j$, hence $n-i < n-j$, then, since $\mathbf{b}$ is a good sequence, the pair (b_{n-i}, b_{n-j}) is good, by Lemma 7.1.7; then, by Lemma 7.1.4, the pair $(a_i \oplus b_{n-i},\ a_j \odot b_{n-j})$ is good. □

Proposition 7.1.20 *(See ([1], Proposition 7.10))*
For all good sequences $\mathbf{a}$ and $\mathbf{b}$ of $\mathcal{A}$, their sum $\mathbf{a} + \mathbf{b}$ is a good sequence.

Proof. (By the proof of ([1], Proposition 7.10))

Let $\mathbf{a} = (a_0, a_1, \ldots, a_n)$ and $\mathbf{b} = (b_0, b_1, \ldots, b_n)$ be two good sequences of $\mathcal{A}$. Let $\mathbf{c} = \mathbf{a} + \mathbf{b}$ be their sum, $\mathbf{c} = (c_0, c_1, \ldots, c_{2n+1})$.

We must prove that $\mathbf{c}$ is a good sequence, i.e. that the pair (c_m, c_{m+1}) is a good pair, for all $m = 0, 1, \ldots, 2n$.

Indeed, following the first definition of sum, (7.2), we have:

$c_m = (a_0 \oplus b_m) \odot (a_1 \oplus b_{m-1}) \odot \ldots \odot (a_{m-1} \oplus b_1) \odot (a_m \oplus b_0)$
$\overset{(PU)}{=} 1 \odot (a_0 \oplus b_m) \odot (a_1 \oplus b_{m-1}) \odot \ldots \odot (a_{m-1} \oplus b_1) \odot (a_m \oplus b_0) \odot 1$
$\overset{(m-La^R)}{=} (1\oplus b_{m+1})\odot(a_0\oplus b_m)\odot(a_1\oplus b_{m-1})\odot\ldots\odot(a_{m-1}\oplus b_1)\odot(a_m\oplus b_0)\odot(a_{m+1}\oplus 1)$,
for $a_{-1} = 1$ and $b_{-1} = 1$.

Following the second definition of sum, (7.3), we have:

$c_{m+1} = b_{m+1} \oplus (a_0 \odot b_m) \oplus (a_1 \odot b_{m-1}) \oplus \ldots \oplus (a_{m-1} \odot b_1) \oplus (a_m \odot b_0) \oplus a_{m+1}$
$\overset{(PU)}{=} (1\odot b_{m+1})\oplus(a_0\odot b_m)\oplus(a_1\odot b_{m-1})\oplus\ldots\oplus(a_{m-1}\odot b_1)\oplus(a_m\odot b_0)\oplus(a_{m+1}\odot 1)$,
for $a_{-1} = 1$ and $b_{-1} = 1$.

By Lemma 7.1.8, the sequences $(1, a_0, a_1, \ldots, a_n)$ and $(1, b_0, b_1, \ldots, b_n)$ are good.

Then, by Lemma 7.1.19, the pair (c_m, c_{m+1}) is good. □

• **Our third goal, reached in next Proposition 7.1.21 below, is to prove that the sum + of good sequences is commutative.**

Proposition 7.1.21 *(See ([1], Proposition 7.11))*
For all good sequences $\mathbf{a}$ *and* $\mathbf{b}$ *of* $\mathcal{A}$,

$$(Scomm) \qquad \mathbf{a} + \mathbf{b} = \mathbf{b} + \mathbf{a}.$$

Proof. Obviously, by (Scomm), (Pcomm) in A. □

Remark 7.1.22 *(See ([1], Remark 7.12))*
Let $\mathbf{a}$ *be a good sequence of* $\mathcal{A}$*. Then, we obviously have (see Corollary 7.1.13):*

$$(SU) \qquad \mathbf{a} + 0_{G^+(A)} = \mathbf{a}.$$

Hence, $(G^+(A), +, 0_{G^+(A)})$ *is a unital commutative magma.*

• **Our fourth goal is to prove that the sum + of good sequences is associative,** i.e. given any $\mathbf{a}$, $\mathbf{b}$, $\mathbf{c}$ good sequences of a right-XY algebra $\mathcal{A}$, we have:

$$(Sass) \qquad (\mathbf{a} + \mathbf{b}) + \mathbf{c} = \mathbf{a} + (\mathbf{b} + \mathbf{c}).$$

We shall present two proofs: Abbadini's proof from [1], which remains valid for right-XY algebras too, and our proof.

– **Abbadini's proof from [1]**

Lemma 7.1.23 *Let* $\mathbf{a} = (a_0, a_1, \ldots, a_{n-1}, a_n)$ *be a good sequence of* $\mathcal{A}$*. Then, for* $k = 0, 1, \ldots, n-1$,

$$a_k \oplus a_n = a_k.$$

Proof. $a_k \oplus a_n = (a_k \oplus a_{k+1}) \oplus a_n = (a_k \oplus (a_{k+1} \oplus a_{k+2})) \oplus a_n$
$\overset{(Sass)}{=} (a_k \oplus a_{k+1} \oplus a_{k+2}) \oplus a_n = \ldots = (a_k \oplus a_{k+1} \oplus \ldots \oplus a_{n-1}) \oplus a_n$
$\overset{(Sass)}{=} a_k \oplus a_{k+1} \oplus \ldots \oplus (a_{n-1} \oplus a_n) = a_k \oplus a_{k+1} \oplus \ldots \oplus a_{n-1} = \ldots = a_k.$ □

Lemma 7.1.24 *(See ([1], Lemma 7.13))*
Let $\mathbf{a} = (a_0, a_1, \ldots, a_{n-1}, a_n)$ *be a good sequence of* $\mathcal{A}$. *Then,*

$$(a_0, a_1, \ldots, a_{n-1}, a_n) = (a_0, a_1, \ldots, a_{n-1}) + (a_n).$$

Proof. Set $(c_0, c_1, c_2, \ldots, c_n) = (a_0, a_1, \ldots, a_{n-1}) + (a_n)$. Then, following Definition 1 of the sum, we have: for $(b_0, b_1, \ldots, b_{n-1}) := (a_n, 0, \ldots, 0)$,
$c_0 = a_0 \oplus b_0 = a_0 \oplus a_n = a_0$, by Lemma 7.1.23.
$c_1 = (a_0 \oplus b_1) \odot (a_1 \oplus b_0) = a_0 \odot (a_1 \oplus a_n) = a_0 \odot a_1 = a_1$, by Lemma 7.1.23.
$c_2 = (a_0 \oplus b_2) \odot (a_1 \oplus b_1) \odot (a_2 \oplus b_0) = a_0 \odot a_1 \odot (a_2 \oplus a_n) = a_0 \odot a_1 \odot a_2 = a_2$, by Lemma 7.1.23.
...
$c_n = (a_0 \oplus b_n) \odot (a_1 \oplus b_{n-1}) \odot \ldots \odot (a_{n-1} \oplus b_1) \odot (a_n \oplus b_0) = a_0 \odot a_1 \odot \ldots \odot a_{n-1} \odot (0 \oplus a_n) = a_{n-1} \odot a_n = a_n$.
Thus, $(c_0, c_1, c_2, \ldots, c_n) = (a_0, a_1, a_2, \ldots, a_n)$. □

Notation 7.1.25 *([1], Notation 7.14)*
A magma $(X, +)$ *consists of a set* X *and a binary operation* $+$ *on* X. *Given a subset* T *of* X, *we define, inductively on* $n > 0$, $n \in \mathbf{N}$, *the subset* T_n; *we set* $T_1 := T$ *and, for* $n > 0$,

$$T_n := \{t + z \mid t \in T,\ z \in T_{n-1}\} \cup \{z + t \mid t \in T,\ z \in T_{n-1}\}.$$

Roughly speaking, T_n *is the set of elements of* X *that can be obtained with at most* n *occurrences of elements of* T *via application of the operation* $+$.
We say that T *generates* X, *if* $\cup_{n>0} T_n = X$.

Lemma 7.1.26 *(See ([1], Lemma 7.15))*
For every right-XY algebra $\mathcal{A}$, *the set* $\{(x) \in G^+(A) \mid x \in A\}$ *generates the commutative magma* $(G^+(A), +)$.

Proof. By induction, using Lemma 7.1.24. □

Lemma 7.1.27 *(Light's associativity test) ([1], Lemma 7.16)*
Let $(X, +)$ *be a magma and let* $T \subset X$ *be a subset of* X *that generates* X. *Suppose that, for all* $x, z \in X$, $t \in T$, *we have:*

$$(x + t) + z \;=\; x + (t + z).$$

Then, the operation $+$ *is associative.*

Proof. Since T generates X, we have: $\cup_{n>0} T_n = X$, $n \in \mathbf{N}$. We prove by induction on $n > 0$ that, for every $y \in T_n$ and every $x, z \in X$, we have: $(x+y)+z = x+(y+z)$.

• The case $n = 1$ (for every $y \in T_1$ and every $x, z \in X$, we have: $(x + y) + z = x + (y + z)$) is true by hypothesis.

• Suppose that the statement is true for up to $n - 1$ ($n \geq 2$).

• We shall prove that the statement is true for n.

Indeed, we have either $y = t + y'$ or $y = y' + t$, for some $t \in T$ and $y' \in Tn$.

Suppose $y = t + y'$; then, we have:
$(x + y) + z = (x + (t + y')) + z = ((x + t) + y') + z = (x + t) + (y' + z) = x + (t + (y' + z)) = x + ((t + y') + z) = x + (y + z)$.

Similarly, suppose $y = y' + t$; then, we have:
$(x + y) + z = (x + (y' + t)) + z = ((x + y') + t) + z = (x + y') + (t + z) = x + (y' + (t + z)) = x + ((y' + t) + z) = x + (y + z)$. □

Lemma 7.1.28 *(See ([1], Lemma 7.17))*

Let $\mathcal{A} = (A, \oplus, ^-, 0)$ be a left-XY algebra, let (a_0, a_1) and (b_0, b_1) be good pairs of $\mathcal{A}$ and let $x \in A$. Then,

$$(a_0 \oplus (x \odot b_0) \oplus b_1) \odot (a_1 \oplus x \oplus b_0) = (a_0 \oplus x \oplus b_1) \odot (a_1 \oplus (a_0 \odot x) \oplus b_0)$$

and both are equal to $a_1 \oplus \sigma(a_0, x, b_0) \oplus b_1$.

Proof. Since the pair (a_0, a_1) is good, it follows from Lemma 7.1.4 that the pair

$$(a_0 \oplus (x \odot b_0) \oplus b_1,\ a_1) \tag{7.4}$$

is good.

Since the pair (b_0, b_1) is good, it follows from Lemma 7.1.4 that the pair

$$(x \oplus b_0,\ b_1) \tag{7.5}$$

is good.

Recall that property (YY) of $\mathcal{A}$ is: $x \oplus y = x \implies x \odot (y \oplus z) = y \oplus (x \odot z)$.

Consequently, we have:

$$\begin{array}{rcll} T1 & \overset{not.}{=} & (a_0 \oplus (x \odot b_0) \oplus b_1) \odot (a_1 \oplus x \oplus b_0) & \\ & = & a_1 \oplus ([a_0 \oplus (x \odot b_0) \oplus b_1] \odot (x \oplus b_0)), & \text{by (YY)}, \\ & & \text{for } X := a_0 \oplus (x \odot b_0) \oplus b_1,\ Y := a_1, & \\ & & Z := x \oplus b_0, \text{ since (7.4) is good}, & \\ & = & a_1 \oplus ((x \oplus b_0) \odot [b_1 \oplus (a_0 \oplus (x \odot b_0))]), & \text{by (Pcomm), (Sass), (Scomm)}, \\ & = & a_1 \oplus (b_1 \oplus [(x \oplus b_0) \odot (a_0 \oplus (x \odot b_0))]), & \text{by (YY)}, \\ & & \text{for } X := x \odot b_0,\ Y := b_1, & \\ & & Z := a_0 \oplus (x \odot b_0), \text{ since (7.5) is good}, & \\ & = & a_1 \oplus ([(x \oplus b_0) \odot ((x \odot b_0) \oplus a_0)] \oplus b_1), & \text{by (Scomm)}, \\ & = & a_1 \oplus \sigma(x, b_0, a_0) \oplus b_1, & \text{by (Sass)}, \end{array}$$

where $\sigma(x, y, z) = \sigma_1(x, y, z) = (x \oplus y) \odot ((x \odot y) \oplus z)$, by Notation 2.2.47.

Similarly, since the pair (a_0, a_1) is good, it follows from Lemma 7.1.4 that the pair

$$(a_0 \oplus x \oplus b_1,\ a_1) \tag{7.6}$$

is good.

Since the pair (b_0, b_1) is good, it follows from Lemma 7.1.4 that the pair

$$((a_0 \odot x) \oplus b_0,\ b_1) \tag{7.7}$$

is good.

Consequently, we have:

$$\begin{array}{lll} T2 & \overset{not.}{=} & (a_0 \oplus x \oplus b_1) \odot (a_1 \oplus (a_0 \odot x) \oplus b_0) \\ & = & a_1 \oplus ([a_0 \oplus x \oplus b_1] \odot ((a_0 \odot x) \oplus b_0)), \quad \text{by (YY)}, \\ & & \text{for } X := a_0 \oplus x \oplus b_1,\ Y := a_1, \\ & & Z := (a_0 \odot x) \oplus b_0, \text{ since (7.6) is good}, \\ & = & a_1 \oplus (((a_0 \odot x) \oplus b_0) \odot [b_1 \oplus (a_0 \oplus x)]), \quad \text{by (Pcomm), (Sass), (Scomm)}, \\ & = & a_1 \oplus (b_1 \oplus [((a_0 \odot x) \oplus b_0) \odot (a_0 \oplus x)]), \quad \text{by (YY)}, \\ & & \text{for } X := (a_0 \odot x) \odot b_0,\ Y := b_1, \\ & & Z := a_0 \oplus x, \text{ since (7.7) is good}, \\ & = & a_1 \oplus ([(a_0 \oplus x) \odot ((a_0 \odot x) \oplus b_0)] \oplus b_1), \quad \text{by (Scomm), (Pcomm)}, \\ & = & a_1 \oplus \sigma(a_0, x, b_0) \oplus b_1, \quad \text{by (Sass)}. \end{array}$$

Then, $T1 = T2$, because $\sigma(x, b_0, a_0) = \sigma(a_0, x, b_0)$, by Remark 2.2.50. □

Lemma 7.1.29 *(See ([1], Lemma 7.18))*

Let $\mathcal{A}$ be a right-XY algebra and let $n \in \mathbf{N}$, $n > 0$. Let $(a_0, \ldots, a_n)$ and $(b_0, \ldots, b_n)$ be good sequences of $\mathcal{A}$ and let $x \in A$. Then

$$\bigodot_{i=0}^{n}(a_i \oplus (x \odot b_{n-i-1}) \oplus b_{n-i}) = \bigodot_{i=0}^{n}(a_i \oplus (a_{i-1} \odot x) \oplus b_{n-i}),$$

with $a_{-1} = 1$ and $b_{-1} = 1$.

Proof. By induction on $n > 0$.

• For $n = 1$, we have:
$(a_0 \oplus (x \odot b_0) \oplus b_1) \odot (a_1 \oplus (x \odot b_{-1}) \oplus b_0) = (a_0 \oplus (a_{-1} \odot x) \oplus b_1) \odot (a_1 \oplus (a_0 \odot x) \oplus b_0)$,
that is true, by (PU) and Lemma 7.1.28.

• Suppose that the statement holds for $n-1$ ($n \geq 2$), for the good sequences $(a_0, \ldots, a_{n-1})$ and $(b_1, \ldots, b_n)$, i.e. we have:

$$\bigodot_{i=0}^{n-1}(a_i \oplus (x \odot b_{n-i-1}) \oplus b_{n-i}) = \bigodot_{i=0}^{n-1}(a_i \oplus (a_{i-1} \odot x) \oplus b_{n-i}), \tag{7.8}$$

with $a_{-1} = 1$ and $b_0 = 1$.

It follows that:
$\bigodot_{i=0}^{n-1}(a_i \oplus (x \odot b_{n-i-1}) \oplus b_{n-i})$
$= (\bigodot_{i=0}^{n-2}(a_i \oplus (x \odot b_{n-i-1}) \oplus b_{n-i})) \odot (a_{n-1} \oplus (x \odot b_0) \oplus b_1)$,
hence we have, by (PU):

$$\bigodot_{i=0}^{n-1}(a_i \oplus (x \odot b_{n-i-1}) \oplus b_{n-i}) = (\bigodot_{i=0}^{n-2}(a_i \oplus (x \odot b_{n-i-1}) \oplus b_{n-i})) \odot (a_{n-1} \oplus x \oplus b_1). \tag{7.9}$$

• We prove that the statement holds for n.
Indeed,

$$\bigodot_{i=0}^{n}(a_i \oplus (x \odot b_{n-i-1}) \oplus b_{n-i}),$$

$= (\bigodot_{i=0}^{n-2}(a_i \oplus (x \odot b_{n-i-1}) \oplus b_{n-i})) \odot (a_{n-1} \oplus (x \odot b_0) \oplus b_1) \odot (a_n \oplus x \oplus b_0),$
by (PU), since $b_{-1} = 1$,

$= (\bigodot_{i=0}^{n-2}(a_i \oplus (x \odot b_{n-i-1}) \oplus b_{n-i})) \odot (a_{n-1} \oplus x \oplus b_1) \odot (a_n \oplus (a_{n-1} \odot x) \oplus b_0),$
by Lemma 7.1.28 for the good pairs (a_{n-1}, a_n) and (b_0, b_1),

$= (\bigodot_{i=0}^{n-1}(a_i \oplus (x \odot b_{n-i-1}) \oplus b_{n-i})) \odot (a_n \oplus (a_{n-1} \odot x) \oplus b_0),$
by (Sass) and (7.9),

$= (\bigodot_{i=0}^{n-1}(a_i \oplus (a_{i-1} \odot x) \oplus b_{n-i})) \odot (a_n \oplus (a_{n-1} \odot x) \oplus b_0),$
by (7.8),

$= \bigodot_{i=0}^{n}(a_i \oplus (a_{i-1} \odot x) \oplus b_{n-i}).$
The proof is complete. □

Lemma 7.1.30 *(See ([1], Lemma 7.19))*
Let $\mathcal{A}$ be a right-XY algebra, let $\mathbf{a} = (a_0, \ldots, a_n)$ and $\mathbf{b} = (b_0, \ldots, b_n)$ be good sequences of $\mathcal{A}$ and let $x \in A$. Then,

$$(\mathbf{a} + (x)) + \mathbf{b} \ = \ \mathbf{a} + ((x) + \mathbf{b}).$$

Proof. We set:

$$\mathbf{d} = \mathbf{a} + (x), \qquad \mathbf{e} = (x) + \mathbf{b},$$

namely:

$$(d_0, \ldots, d_n, \ldots) = (a_0, \ldots, a_n) + (x, 0, \ldots, 0),$$
$$(e_0, \ldots, e_n, \ldots) = (x, 0, \ldots, 0) + (b_0, \ldots, b_n).$$

Hence, for every $n \in \mathbf{N}$, by denoting $(b'_0, b'_1, \ldots, b'_n) := (x, 0, \ldots, 0)$, we have:
$d_n \overset{def.2}{=} b'_n \oplus (a_0 \odot b'_{n-1}) \oplus (a_1 \odot b'_{n-2}) \oplus \ldots \oplus (a_{n-2} \odot b'_1) \oplus (a_{n-1} \odot b'_0) \oplus a_n$
$= 0 \oplus (a_0 \odot 0) \oplus (a_1 \odot 0) \oplus \ldots \oplus (a_{n-2} \odot 0) \oplus (a_{n-1} \odot x) \oplus a_n$
$= (a_{n-1} \odot x) \oplus a_n$, by (m-La), (SU)
$= a_n \oplus (a_{n-1} \odot x)$, by (Scomm).
Similarly, for every $n \in \mathbf{N}$, by denoting $(a'_0, a'_1, \ldots, a'_n) := (x, 0, \ldots, 0)$, we have:
$e_n \overset{def.2}{=} b_n \oplus (a'_0 \odot b_{n-1}) \oplus (a'_1 \odot b_{n-2}) \oplus \ldots \oplus (a'_{n-2} \odot b_1) \oplus (a'_{n-1} \odot b_0) \oplus a'_n$
$= b_n \oplus (x \odot b_{n-1}) \oplus (0 \odot b_{n-2}) \oplus \ldots \oplus (0 \odot b_1) \oplus (0 \odot b_0) \oplus 0$
$= b_n \oplus (x \odot b_{n-1})$, by (m-La), (SU),
$= (x \odot b_{n-1}) \oplus b_n$, by (Scomm).
Now, we set:

$$\mathbf{f} = (\mathbf{a} + (x)) + \mathbf{b} = \mathbf{d} + \mathbf{b}, \qquad \mathbf{g} = \mathbf{a} + ((x) + \mathbf{b}) = \mathbf{a} + \mathbf{e}.$$

We must prove that $\mathbf{f} = \mathbf{g}$.

Indeed, for every $n \in \mathbf{N}$, we have:

$$\begin{aligned}
f_n &\overset{def.1}{=} \bigodot_{i=0}^n (d_i \oplus b_{n-i}) \\
&= \bigodot_{i=0}^n ([a_i \oplus (a_{i-1} \odot x)] \oplus b_{n-i}) \\
&= \bigodot_{i=0}^n (a_i \oplus (a_{i-1} \odot x) \oplus b_{n-i}), \text{ by (Sass)}, \\
&= \bigodot_{i=0}^n (a_i \oplus (x \odot b_{n-i-1}) \oplus b_{n-i}), \text{ by Lemma 7.1.29}, \\
&= \bigodot_{i=0}^n (a_i \oplus [(x \odot b_{n-i-1}) \oplus b_{n-i}]), \text{ by (Sass)}, \\
&= \bigodot_{i=0}^n (a_i + e_{n-i}) \\
&= g_n.
\end{aligned}$$

Hence, $\mathbf{f} = \mathbf{g}$. □

The proof by Abbadini that the addition of good sequences is associative: ([1], Proposition 7.20) By Lemmas 7.1.26, 7.1.27, 7.1.30. □

– **Our proof.**

Proposition 7.1.31 *For any (a), (b), (c) good sequences of $\mathcal{A}$, we have:*

$$(Sass) \qquad [(a) + (b)] + (c) = (a) + [(b) + (c)].$$

Proof. $[(a)+(b)]+(c) = (a\oplus b, a\odot b)+(c) = (c_0, c_1, c_2)$, where, following Definition 1 of sum (see Lemma 7.1.14):
$c_0 = (a \oplus b) \oplus c$,
$c_1 = (a \oplus b) \odot ((a \odot b) \oplus c)$,
$c_2 = (a \odot b) \odot c$
and
$(a) + [(b) + (c)] = (a) + (b \oplus c, b \odot c) = (b \oplus c, b \odot c) + (a) = (d_0, d_1, d_2)$, by (Scomm) in $G^+(A)$, where (see Lemma 7.1.14):
$d_0 = (b \oplus c) \oplus a$,
$d_1 = (b \oplus c) \odot ((b \odot c) \oplus a)$,
$d_2 = (b \odot c) \odot a$.
Note that $c_0 = d_0$ by (Scomm) and (Sass) in A; $c_1 = d_1$, by (XX); $c_2 = d_2$, by (Pcomm) and (Pass) in A. □

Proposition 7.1.32 *For the good sequences (x_0), (x_1), ..., (x_n) of $\mathcal{A}$,*

$$(x_0) + (x_1) + \ldots + (x_n) = (c_0^{n+1}, c_1^{n+1}, \ldots, c_n^{n+1}),$$

where $c_0^{n+1} = x_0 \oplus x_1 \oplus \ldots \oplus x_n$.

Proof. By induction on $n \geq 1$.

• The statement is true for $n = 1$, by Lemma 7.1.12:
$(x_0) + (x_1) = (x_0 \oplus x_1, x_0 \odot x_1)$, where $c_0^2 = x_0 + x_1$.

• Suppose that the statement is true for $n-1$, i.e.
$(x_0) + (x_1) + \ldots + (x_{n-1}) = (c_0^n, c_1^n, \ldots, c_{n-1}^n)$,
where $c_0^n = x_0 \oplus x_1 \oplus \ldots \oplus x_{n-1}$.

• We shall prove that the statement is true for n. Indeed,
$(x_0) + (x_1) + \ldots + (x_{n-1}) + (x_n)$
$= [(x_0) + (x_1) + \ldots + (x_{n-1})] + (x_n)$, by Proposition 7.1.31,
$= (c_0^n, c_1^n, \ldots, c_{n-1}^n) + (x_n)$
$= (c_0^{n+1}, c_1^{n+1}, \ldots, c_n^{n+1})$, where, by inductive hypothesis,
$c_0^n = x_0 \oplus x_1 \oplus \ldots \oplus x_{n-1}$.
Then, by (7.2),
$c_0^{n+1} \stackrel{def.1}{=} c_0^n \oplus x_n = [x_0 \oplus x_1 \oplus \ldots \oplus x_{n-1}] \oplus x_n = x_0 \oplus x_1 \oplus \ldots \oplus x_{n-1} \oplus x_n$,
by (Sass) in A. □

Corollary 7.1.33 *Let* $\mathbf{a} = (a_0, a_1, \ldots, a_{n-1}, a_n)$ *be a good sequence of* $\mathcal{A}$. *Then,*

$$(a_0, a_1, \ldots, a_{n-1}, a_n) = [[\ldots[(a_0)+(a_1)]+\ldots]+(a_{n-1})]+(a_n) = (a_0)+(a_1)+\ldots+(a_n).$$

Proof. Obviously, by Lemma 7.1.24 and by Proposition 7.1.31. □

Example 7.1.34 Let (a_0, a_1, a_2) be a good sequence, i.e. $a_0 \oplus a_1 = a_0$ and $a_1 \oplus a_2 = a_1$, hence $a_0 \odot a_1 = a_1$ and $a_1 \odot a_2 = a_2$. Then,

$$(a_0, a_1, a_2) = [(a_0) + (a_1)] + (a_2) = (a_1) + [(a_0) + (a_2)].$$

Indeed, $[(a_0) + (a_1)] + (a_2) = (a_0 \oplus a_1, a_0 \odot a_1) + (a_2) = (a_0, a_1) + (a_2) = (c_0, c_1, c_2)$, where, following Definition 1 of sum (see Lemma 7.1.14): for $(b_0, b_1) := (a_2, 0)$,
$c_0 = a_0 \oplus b_0 = a_0 \oplus a_2 = a_0$, by Lemma 7.1.23,
$c_1 = (a_0 \oplus b_1) \odot (a_1 \oplus b_0) = a_0 \odot (a_1 \oplus a_2) = a_0 \odot a_1 = a_1$,
$c_2 = a_1 \odot a_2 = a_2$;
thus, $(c_0, c_1, c_2) = (a_0, a_1, a_2)$.

Also, $(a_1) + [(a_0) + (a_2)] = (a_1) + (a_0 \oplus a_2, a_0 \odot a_2) = (a_1) + (a_0, a_0 \odot (a_1 \odot a_2)) = (a_1) + (a_0, a_2) = (a_0, a_2) + (a_1) = (d_0, d_1, d_2)$, where (see Lemma 7.1.14): for $(b_0, b_1) := (a_1, 0)$,
$d_0 = a_0 \oplus b_0 = a_0 \oplus a_1 = a_0$,
$d_1 = (a_0 \oplus b_1) \odot (a_2 \oplus b_0) = (a_0 \oplus 0) \odot (a_2 \oplus a_1) = a_0 \odot a_1 = a_1$,
$d_2 = a_2 \odot b_0 = a_2 \odot a_1 = a_2$;
thus, $(d_0, d_1, d_2) = (a_0, a_1, a_2)$.

Our proof that the addition of good sequences is associative:

Let $\mathbf{a} = (a_0, a_1, \ldots, a_n)$, $\mathbf{b} = (b_0, b_1, \ldots, b_n)$, $\mathbf{c} = (c_0, c_1, \ldots, c_n)$ be good sequences of $\mathcal{A}$, hence
$\mathbf{a} = (a_0) + (a_1) + \ldots + (a_n)$, $\mathbf{b} = (b_0) + (b_1) + \ldots + (b_n)$, $\mathbf{c} = (c_0) + (c_1) + \ldots + (c_n)$,
by Corollary 7.1.33. Then,

$(\mathbf{a}+\mathbf{b})+\mathbf{c}$
$= [(a_0)+(a_1)+\ldots+(a_n) \ + \ (b_0)+(b_1)+\ldots+(b_n)] \ + \ (c_0)+(c_1)+\ldots+(c_n)$
$= (a_0)+(a_1)+\ldots+(a_n) \ + \ [(b_0)+(b_1)+\ldots+(b_n) \ + \ (c_0)+(c_1)+\ldots+(c_n)]$
$= \mathbf{a}+(\mathbf{b}+\mathbf{c})$, by Proposition 7.1.31. □

Since (SU), (Scomm), (Sass) hold in $G^+(A)$, by Proposition 7.1.21, Remark 7.1.22, we obtain:

Corollary 7.1.35 *The algebra* $(G^+(A), +, 0_{G^+(A)})$ *is a commutative monoid.*

7.1.4 The cancellation problem on $G^+(A)$

Recall that any MV algebra $\mathcal{A}$ is cancellative and that its commutative monoid $G^+(A) = M_A$ is cancellative too ([10], Proposition 2.3.1 (i)), i.e. for any **a**, **b**, **c** in $G^+(A) = M_A$,

$$(C) \qquad \mathbf{a} + \mathbf{b} = \mathbf{a} + \mathbf{c} \Longrightarrow \mathbf{b} = \mathbf{c}.$$

The proof of ([10], Proposition 2.3.1 (i)) uses Chang's subdirect representation theorem (hence the axiom of choice) and is made in some lines. The cancellation of $G^+(A)$ makes the special binary relation $\sim$ (1.10) be transitive, hence an equivalence relation (see Section 1.3).

The proper XY algebras (i.e. XY algebras that are not MV algebras) are non-cancellative or cancellative. We shall study the two cases.

- **The particular case of non-cancellative XY algebras**

There exist non-cancellative right-XY algebras $\mathcal{A} = (A, \oplus, ^-, 0)$ whose $G^+(A)$ is not cancellative, as the following example provided by MACE4 shows.

Example 7.1.36
Consider the right-XY algebra $\mathcal{A} = (A = \{0, 2, 3, 4, 5, 6, 1\}, \oplus, ^-, 0)$ with the following tables:

$\oplus$	0	2	3	4	5	6	1
0	0	2	3	4	5	6	1
2	2	4	4	5	4	5	1
3	3	4	4	5	4	5	1
4	4	5	5	4	5	4	1
5	5	4	4	5	4	5	1
6	6	5	5	4	5	4	1
1	1	1	1	1	1	1	1

,

x	x^-
0	1
2	2
3	3
4	4
5	6
6	5
1	0

, with

$\odot$	0	2	3	4	5	6	1
0	0	0	0	0	0	0	0
2	0	4	4	6	6	4	2
3	0	4	4	6	6	4	3
4	0	6	6	4	4	6	4
5	0	6	6	4	4	6	5
6	0	4	4	6	6	4	6
1	0	2	3	4	5	6	1

.

$\mathcal{A}$ is not cancellative: $2 \oplus 2 = 2 \oplus 3$ and $2 \odot 2 = 2 \odot 3$ and $2 \neq 3$. Hence, it is not an MV algebra (because any MV algebra is cancellative).

$G^+(A)$ is not cancellative, because there exist the good sequences $\mathbf{a} = (a0, a1) = (1, 2)$, $\mathbf{b} = (b0, b1) = (1, 2)$, $\mathbf{c} = (c0, c1) = (1, 3)$ such that $\mathbf{a} + \mathbf{b} = \mathbf{a} + \mathbf{c}$ and $\mathbf{b} \neq \mathbf{c}$.

Indeed, the following conditions were imposed to MACE4:
$(f0, f1, f2, f3) = (a0, a1) + (b0, b1) = (a0, a1) + (c0, c1) = (g0, g1, g2, g3)$,

$a0 \oplus a1 = a0$, $b0 \oplus b1 = b0$, $c0 \oplus c1 = c0$, $a0 \odot a1 = a1$, $b0 \odot b1 = b1$, $c0 \odot c1 = c1$;

$$\begin{array}{ll} f0 = a0 \oplus b0, & g0 = a0 \oplus c0, \\ f1 = (a0 \oplus b1) \odot (a1 \oplus b0), & g1 = (a0 \oplus c1) \odot (a1 \oplus c0), \\ f2 = (a0 \odot b0) \odot (a1 \oplus b1), & g2 = (a0 \odot c0) \odot (a1 \oplus c1), \\ f3 = a1 \odot b1, & g3 = a1 \odot c1, \end{array}$$

and

$f0 = g0$, $f1 = g1$, $f2 = g2$, $f3 = g3$.

MACE4 provided:

$(a0, a1) = (1, 2)$, $(b0, b1) = (1, 2)$, $(c0, c1) = (1, 3)$, hence $(b0, b1) \neq (c0, c1)$;

$f0 = 1$, $f1 = 1$, $f2 = 4$, $f3 = 4$;

$g0 = 1$, $g1 = 1$, $g2 = 4$, $g3 = 4$.

But, all non-cancellative XY algebras verify a particular cancellation property, $(\mathbf{C}^R_u)$, as Proposition 7.1.40 below proves.

Lemma 7.1.37

$$0u_{G^+(A)} = 0(1) \overset{def.}{=} (0) \quad and \quad nu_{G^+(A)} = n(1) = (\underbrace{1, 1, \ldots, 1}_{n\ times}), \ for\ n = 1, 2 \ldots.$$

Proof. By Corollary 7.1.33, since $n(1)$ means $\underbrace{(1) + (1) + \ldots + (1)}_{n \text{ times}}$. □

Lemma 7.1.38 *Let* $\mathbf{a} = (a_0, a_1, \ldots, a_{n-1}, a_n)$ *be a good sequence of* $\mathcal{A}$. *Then,*

$$(a_0, a_1, \ldots, a_{n-1}, a_n) + (1) = (1, a_0, a_1, \ldots, a_{n-1}, a_n).$$

Proof. Put $(c_0, c_1, \ldots, c_{n-1}, c_n, c_{n+1}) = (a_0, a_1, \ldots, a_{n-1}, a_n) + (1)$. Then, by Definition 1 of the sum, (7.2), we obtain:

$c_0 = a_0 \oplus b_0 = a_0 \oplus 1 = 1$.

$c_1 = (a_0 \oplus b_1) \odot (a_1 \oplus b_0) = (a_0 \oplus 0) \odot (a_1 \oplus 1) = a_0 \odot 1 = a_0$.

$c_2 = (a_0 \oplus b_2) \odot (a_1 \oplus b_1) \odot (a_2 \oplus b_0) = (a_0 \oplus 0) \odot (a_1 \oplus 0) \odot (a_2 \oplus 1) = a_0 \odot a_1 \odot 1 = a_1$.

...

$c_n = (a_0 \oplus b_n) \odot (a_1 \oplus b_{n-1}) \odot \ldots \odot (a_{n-1} \oplus b_1) \odot (a_n \oplus b_0) = (a_0 \oplus 0) \odot (a_1 \oplus 0) \odot \ldots \odot (a_{n-1} \oplus 0) \odot (a_n \oplus 1) = a_0 \odot a_1 \odot \ldots \odot a_{n-1} \odot 1 = a_{n-1}$.

$c_{n+1} = a_n$, similarly.

$c_{n+2} = 0$. □

Corollary 7.1.39 *Let* $\mathbf{a} = (a_0, a_1, \ldots, a_{n-1}, a_n)$ *be a good sequence of* $\mathcal{A}$. *Then, for any* $m \in \mathbf{N}$,

$$(a_0, a_1, \ldots, a_{n-1}, a_n) + m(1) = (\underbrace{1, 1, \ldots, 1}_{m\ times}, a_0, a_1, \ldots, a_{n-1}, a_n).$$

Proof. By Lemma 7.1.38 and by (Sass) in $G^+(A)$, since $m(1)$ means $\underbrace{(1) + (1) + \ldots + (1)}_{m \text{ times}}$. □

Proposition 7.1.40 *Let* $\mathbf{a} = (a_0, a_1, \ldots, a_{n-1}, a_n)$ *and* $\mathbf{b} = (b_0, b_1, \ldots, b_{n-1}, b_n)$ *be two good sequences of* $\mathcal{A}$. *Then, the following particular cancellation property holds: for any* $m \geq 1$,

$$(C_u^R) \qquad \mathbf{a} + mu_{G^+(A)} = \mathbf{b} + mu_{G^+(A)} \Longrightarrow \mathbf{a} = \mathbf{b}.$$

Proof. $\mathbf{a} + mu_{G^+(A)} = \mathbf{b} + mu_{G^+(A)}$
$\Longleftrightarrow (a_0, a_1, \ldots, a_{n-1}, a_n) + m(1) = (b_0, b_1, \ldots, b_{n-1}, b_n) + m(1)$
$\Longleftrightarrow (\underbrace{1, 1, \ldots, 1}_{m \text{ times}}, a_0, a_1, \ldots, a_{n-1}, a_n) = (\underbrace{1, 1, \ldots, 1}_{m \text{ times}}, b_0, b_1, \ldots, b_{n-1}, b_n)$,
$\Longleftrightarrow a_i = b_i$, for $i = 0, 1, \ldots, n$
$\Longleftrightarrow \mathbf{a} = \mathbf{b}$,
by Corollary 7.1.39. □

• The particular case of cancellative proper XY algebras

Let $\mathcal{A} = (A, \oplus, ^-, 0)$ be a cancellative proper right-XY algebra (Definition 2.2.16), i.e. verifying the cancellation property: for all $x, y, z \in A$,
(C) $z \oplus x = z \oplus y$ and $z \odot x = z \odot y$ imply $x = y$.

Proposition 7.1.41 *For any* (a_0), (b_0), (c_0) *in* $G^+(A)$, *we have:*

$$(C^0) \qquad (a_0) + (b_0) = (a_0) + (c_0) \Longrightarrow (b_0) = (c_0).$$

Proof. If $(a_0) + (b_0) = (a_0) + (c_0)$, hence $(a_0 \oplus b_0, a_0 \odot b_0) = (a_0 \oplus c_0, a_0 \odot c_0)$, by Lemma 7.1.12, i.e. $a_0 \oplus b_0 = a_0 \oplus c_0$ and $a_0 \odot b_0 = a_0 \odot c_0$, then, by property (C) of $\mathcal{A}$, we obtain $b_0 = c_0$, i.e. $(b_0) = (c_0)$. □

Proposition 7.1.42 *For any* $\mathbf{a} = (a_0, a_1)$ *and* (b_0), (c_0) *in* $G^+(A)$, *we have:*

$$(C^1) \qquad \mathbf{a} + (b_0) = \mathbf{a} + (c_0) \Longrightarrow (b_0) = (c_0).$$

Proof. We have $a_0 \oplus a_1 = a_0$, hence $a_0 \odot a_1 = a_1$. Then,
$\mathbf{a} + (b_0) = (a_0, a_1) + (b_0) = (f_0, f_1, f_2)$,
$\mathbf{a} + (c_0) = (a_0, a_1) + (c_0) = (g_0, g_1, g_2)$,
where, by Lemma 7.1.14,
$f_0 = a_0 \oplus b_0, \qquad g_0 = a_0 \oplus c_0,$
$f_1 = a_0 \odot (a_1 \oplus b_0), \qquad g_1 = a_0 \odot (a_1 \oplus c_0),$
$f_2 = a_1 \odot b_0, \qquad g_2 = a_1 \odot c_0.$
Since $\mathbf{a} + (b_0) = \mathbf{a} + (c_0)$, it follows that $f_0 = g_0$, $f_1 = g_1$, $f_2 = g_2$. Then,

$$a_0 \oplus (a_1 \oplus b_0) \overset{(Sass)}{=} (a_0 \oplus a_1) \oplus b_0 = a_0 \oplus b_0$$
$$\overset{f_0 = g_0}{=} a_0 \oplus c_0 = (a_0 \oplus a_1) \oplus c_0 \overset{(Sass)}{=} a_0 \oplus (a_1 \oplus c_0)$$

and, since we also have:
$f_1 = a_0 \odot (a_1 \oplus b_0) = a_0 \odot (a_1 \oplus c_0) = g_1$,
it follows, by (C) in $\mathcal{A}$, that:
$a_1 \oplus b_0 = a_1 \oplus c_0$
and, since we also have:
$f_2 = a_1 \odot b_0 = a_1 \odot c_0 = g_2$,
it follows, by (C) in $\mathcal{A}$, that: $b_0 = c_0$, i.e. $(b_0) = (c_0)$. □

Proposition 7.1.43 *For any* $\mathbf{a} = (a_0, a_1, a_2)$ *and* (b_0), (c_0) *in* $G^+(A)$, *we have:*

$$(C^2) \qquad \mathbf{a} + (b_0) = \mathbf{a} + (c_0) \implies (b_0) = (c_0).$$

Proof. We have $a_0 \oplus a_1 = a_0$, $a_1 \oplus a_2 = a_1$, hence $a_0 \odot a_1 = a_1$, $a_1 \odot a_2 = a_2$. Then,

$\mathbf{a} + (b_0) = (a_0, a_1, a_2) + (b_0) = (f_0, f_1, f_2, f_3)$,
$\mathbf{a} + (c_0) = (a_0, a_1, a_2) + (c_0) = (g_0, g_1, g_2, g_3)$,

where, by Lemma 7.1.15,

$f_0 = a_0 \oplus b_0, \qquad g_0 = a_0 \oplus c_0,$
$f_1 = a_0 \odot (a_1 \oplus b_0), \quad g_1 = a_0 \odot (a_1 \oplus c_0),$
$f_2 = a_1 \odot (a_2 \oplus b_0), \quad g_2 = a_1 \odot (a_2 \oplus c_0),$
$f_3 = a_2 \odot b_0, \qquad g_3 = a_2 \odot c_0.$

Since $\mathbf{a} + (b_0) = \mathbf{a} + (c_0)$, it follows that $f_0 = g_0$, $f_1 = g_1$, $f_2 = g_2$, $f_3 = g_3$. Then,

$a_0 \oplus (a_1 \oplus b_0) \overset{(Sass)}{=} (a_0 \oplus a_1) \oplus b_0 = a_0 \oplus b_0$
$\overset{f_0=g_0}{=} a_0 \oplus c_0 = (a_0 \oplus a_1) \oplus c_0 \overset{(Sass)}{=} a_0 \oplus (a_1 \oplus c_0)$

and, since we also have:

$f_1 = a_0 \odot (a_1 \oplus b_0) = a_0 \odot (a_1 \oplus c_0) = g_1$,

it follows, by (C) in $\mathcal{A}$, that:

(*) $a_1 \oplus b_0 = a_1 \oplus c_0$.

$a_1 \oplus (a_2 \oplus b_0) \overset{(Sass)}{=} (a_1 \oplus a_2) \oplus b_0 = a_1 \oplus b_0$
$\overset{(*)}{=} a_1 \oplus c_0 = (a_1 \oplus a_2) \oplus c_0 \overset{(Sass)}{=} a_1 \oplus (a_2 \oplus c_0)$

and, since we also have:

$f_2 = a_1 \odot (a_2 \oplus b_0) = a_1 \odot (a_2 \oplus c_0) = g_2$,

it follows, by (C) in $\mathcal{A}$, that:

(**) $a_2 \oplus b_0 = a_2 \oplus c_0$

and, since we also have:

$f_3 = a_2 \odot b_0 = a_2 \odot c_0 = g_3$,

it follows, by (C) in $\mathcal{A}$, that: $b_0 = c_0$, i.e. $(b_0) = (c_0)$. □

Remarks 7.1.44 *Let* $\mathcal{A} = (A, \oplus, ^-, 0)$ *be a cancellative proper right-XY algebra.*

(1) For any $a0$ *and* $\mathbf{b} = (b0, b1)$, $\mathbf{c} = (c0, c1)$ *in* $G^+(A)$,

$$a0 + \mathbf{b} = a0 + \mathbf{c} \implies \mathbf{b} = \mathbf{c}?$$

i.e. does $(f0, f1, f2) = a0 + (b0, b1) = a0 + (c0, c1) = (g0, g1, g2)$ *imply* $(b0, b1) = (c0, c1)$*?*

Hence, we have the following conditions in a cancellative proper XY algebra:

$b0 \oplus b1 = b0$, $c0 \oplus c1 = c0$, $b0 \odot b1 = b1$, $c0 \odot c1 = c1$;

$f0 = a0 \oplus b0, \qquad g0 = a0 \oplus c0,$
$f1 = b0 \odot (b1 \oplus a0), \quad g1 = c0 \odot (c1 \oplus a0),$
$f2 = a0 \odot b1, \qquad g2 = a0 \odot c1$

and

$f0 = g0$, $f1 = g1$, $f2 = g2$.

PROVER9 **has run for days and did not provided a proof and** MACE4 **also has run for days and did not provided a counter-example.**

If we put the condition that the XY algebra be an MV algebra (i.e. $x \vee^M_m y = y \vee^M_m x$), then PROVER9 *provided a proof of Length 134 (the length of the expanded renumbered proof was 283)!*

(2) For any $\mathbf{a} = (a0, a1)$ *and* $\mathbf{b} = (b0, b1)$, $\mathbf{c} = (c0, c1)$ *in* $G^+(A)$,

$$\mathbf{a} + \mathbf{b} = \mathbf{a} + \mathbf{c} \implies \mathbf{b} = \mathbf{c}?$$

i.e. does $(f0, f1, f2, f3) = (a0, a1) + (b0, b1) = (a0, a1) + (c0, c1) = (g0, g1, g2, g3)$ *imply* $(b0, b1) = (c0, c1)$*?*
Hence, we have the following conditions in a cancellative proper XY algebra:
$a0 \oplus a1 = a0,\ b0 \oplus b1 = b0,\ c0 \oplus c1 = c0,\ a0 \odot a1 = a1,\ b0 \odot b1 = b1,\ c0 \odot c1 = c1;$

$$\begin{array}{ll} f0 = a0 \oplus b0, & g0 = a0 \oplus c0, \\ f1 = (a0 \oplus b1) \odot (a1 \oplus b0), & g1 = (a0 \oplus c1) \odot (a1 \oplus c0), \\ f2 = (a0 \odot b0) \odot (a1 \oplus b1), & g2 = (a0 \odot c0) \odot (a1 \oplus c1), \\ f3 = a1 \odot b1, & g3 = a1 \odot c1, \end{array}$$

and
$f0 = g0,\ f1 = g1,\ f2 = g2,\ f3 = g3.$

PROVER9 **has run for days and did not provided a proof and** MACE4 **also has run for days and did not provided a counter-example.**

If we put the condition that the XY algebra be an MV algebra (i.e. $x \vee^M_m y = y \vee^M_m x$), then PROVER9 *provided a proof of Length 491 (the length of the expanded renumbered proof was 1546) in 8357.05 seconds!*

Open problems 7.1.45

1. Find an example of cancellative proper XY algebra $\mathcal{A}$ whose $G^+(A)$ verifies the cancellative property (C).

2. Find an example of cancellative XY algebra $\mathcal{A}$ whose $G^+(A)$ does not verify the above property (C) for some $\mathbf{a}$, $\mathbf{b}$, $\mathbf{c}$ *of $G^+(A)$ or prove that the commutative monoid $G^+(A)$ of any cancellative XY algebra $\mathcal{A}$ verifies the cancellative property (C).*

Note that, if there is at least one example of cancellative proper XY algebra $\mathcal{A}$ whose $G^+(A)$ verifies the cancellative property (C), then we can consider the subcategory of such cancellative XY algebras denoted $\mathcal{C}(CXY^R)$.

7.1.5 The negation $\smile$ on the set $G^+(A)$

Given the good sequence $\mathbf{a} = (a_0, a_1, \ldots, a_{n-1}, a_n) \neq (0)$ of $\mathcal{A}$, we define the negation $\smile$ of $\mathbf{a}$ by (see (1.11):

$$(7.10) \qquad \smile \mathbf{a} = \smile (a_0, a_1, \ldots, a_{n-1}, a_n) \overset{def.}{=} (a_n^-, a_{n-1}^-, \ldots, a_1^-, a_0^-)$$

and given the good sequence $0_{G^+(A)} = (0)$ of $\mathcal{A}$, we define (see Remark 7.1.2):

$$(7.11) \qquad (N0) \qquad \smile 0_{G^+(A)} \overset{def.}{=} 0_{G^+(A)}.$$

Hence, given the good sequence (x), for each $x \in A \setminus \{0\}$, we have:

$$(7.12) \qquad \smile (x) \overset{def.}{=} (x^-).$$

Lemma 7.1.46

$$\smile(\smile 0_{G^+(A)}) = 0_{G^+(A)}.$$

Proof. By (N0). □

Proposition 7.1.47 *We have:*

$(a_0, a_1, \ldots, a_{n-1}, a_n)$ *is a good sequence* $\Longleftrightarrow$ $(a_n^-, a_{n-1}^-, \ldots, a_1^-, a_0^-)$ *is a good sequence.*

Proof. By Proposition 2.2.55. □

Proposition 7.1.48 *Let* $\mathbf{a} = (a_0, a_1, \ldots, a_{n-1}, a_n)$ *be a good sequence of* $\mathcal{A}$. *Then, for* $\mathbf{a} \neq (n+1)u_{G^+(A)}$, $n \in \mathbf{N}$*:*

$$(DN) \qquad \smile(\smile \mathbf{a}) = \mathbf{a}.$$

Proof. Obviously, since $a_i^{=} = a_i$, for all $i = 0, 1, \ldots, n$, by (DN) in $\mathcal{A}$. □

Proposition 7.1.49 *For* $u_{G^+(A)} = (1) \in G^+(A)$ *and for all* $n \geq 1$, *we have:*
(Nu^{+R}*)* $\smile nu_{G^+(A)} = 0_{G^+(A)}$,
(NNu^R*)* $\smile(\smile nu_{G^+(A)}) = nu_{G^+(A)}$.

Proof. (Nu^{+R}): $\smile nu_{G^+(A)}$
$=\smile n(1)$
$=\smile \underbrace{(1, 1, \ldots, 1)}_{n \text{ times}}$, by Lemma 7.1.37,
$= \underbrace{(1^-, 1^-, \ldots, 1^-)}_{n \text{ times}}$, by definition of $-$,
$= \underbrace{(0, 0, \ldots, 0)}_{n \text{ times}}$, by (Neg1-0) in A,
$= (0) = 0_{G^+(A)}$.
(NNu^R): $\smile(\smile nu_{G^+(A)})$
$=\smile(\smile n(1))$
$=\smile(\smile \underbrace{(1, 1, \ldots, 1)}_{n \text{ times}})$, by Lemma 7.1.37,
$=\smile \underbrace{(1^-, 1^-, \ldots, 1^-)}_{n \text{ times}}$
$= \underbrace{(1^=, 1^=, \ldots, 1^=)}_{n \text{ times}}$
$= \underbrace{(1, 1, \ldots, 1)}_{n \text{ times}}$, by (DN) in A,
$= n(1) = nu_{G^+(A)}$. □

Corollary 7.1.50 *For any* $\mathbf{a} \in G^+(A)$, *we have:*
(DN) $\smile(\smile \mathbf{a}) = \mathbf{a}$.

Proof. By Lemma 7.1.46, Proposition 7.1.48 and (NNu^R). □

Corollary 7.1.51 *The algebra $(G^+(A), +, \smile, 0_{G^+(A)})$ is an involutive m_0-ME algebra.*

Proof. By Corollary 7.1.35 and since (N0) holds, by (7.11), and (DN) holds, by Corollary 7.1.50. □

Proposition 7.1.52 *For all $\mathbf{a} \in G^+(A)$, $n \geq 1$, we have:*

$$(S^{+R}) \qquad \smile(\mathbf{a} + nu_{G^+(A)}) = \smile \mathbf{a} \smile nu_{G^+(A)} = \smile \mathbf{a}.$$

Proof. Let $\mathbf{a} = (a_0, a_1, \ldots, a_{n-1}, a_n)$. Then,

$\smile(\mathbf{a} + nu_{G^+(A)}) = \smile(\mathbf{a} + n(1))$

$= \smile((a_0, a_1, \ldots, a_{n-1}, a_n) + (\underbrace{1, 1, \ldots, 1}_{n \text{ times}}))$, by Lemma 7.1.37,

$= \smile(\underbrace{1, 1, \ldots, 1}_{n \text{ times}}, a_0, a_1, \ldots, a_{n-1}, a_n)$, by Corollary 7.1.39,

$= (a_n^-, a_{n-1}^-, \ldots, a_1^-, a_0^-, \underbrace{0, 0, \ldots, 0}_{n \text{ times}})$

$= (a_n^-, a_{n-1}^-, \ldots, a_1^-, a_0^-) = \smile \mathbf{a}$

and

$\smile \mathbf{a} \smile nu_{G^+(A)} \overset{(Nu^{+R})}{=} \smile \mathbf{a} + 0_{G^+(A)} = \smile \mathbf{a}$, by (SU). □

7.1.6 Two binary relations $\leq_m$ and $\leq_m^M$ on $G^+(A)$

Let $\mathbf{a} = (a_0, a_1, \ldots, a_{n-1}, a_n)$ and $\mathbf{b} = (b_0, b_1, \ldots, b_{n-1}, b_n)$ be two good sequences of $\mathcal{A}$. We can define two external (i.e. not defined by the operations on $G^+(A)$) binary relations, $\leq_m$ and $\leq_m^M$, on $G^+(A)$ by:

$$\mathbf{a} \leq_m \mathbf{b} \overset{def.}{\Longleftrightarrow} a_i \leq_m b_i, \ for \ i = 0, 1, \ldots, n, \quad in \ \mathcal{A}$$

$$\mathbf{a} \leq_m^M \mathbf{b} \overset{def.}{\Longleftrightarrow} a_i \leq_m^M b_i, \ for \ i = 0, 1, \ldots, n, \quad in \ \mathcal{A}.$$

Remark 7.1.53 *By Proposition 2.2.56, in a right-XY algebra $\mathcal{A} = (A, \oplus, ^-, 0)$, for all $x \in A$, we have $0 \leq_m x \leq_m 1$ and $0 \leq_m^M x$, therefore only the binary relation $\leq_m$ on $G^+(A)$ is interesting in the sequel.*

Lemma 7.1.54

$$(i)\ 0_{G^+(A)} \leq_m 0_{G^+(A)}, \quad (ii)\ u_{G^+(A)} \leq_m u_{G^+(A)}, \quad (iii)\ 0_{G^+(A)} \leq_m u_{G^+(A)}.$$

Proof. (i): $0_{G^+(A)} \leq_m 0_{G^+(A)}$, i.e. $(0) \leq_m (0) \overset{def.}{\Longleftrightarrow} 0 \leq_m 0$, in $\mathcal{A}$, that is true, by Lemma 2.1.6.

(ii): $u_{G^+(A)} \leq_m u_{G^+(A)}$, i.e. $(1) \leq_m (1) \overset{def.}{\Longleftrightarrow} 1 \leq_m 1$, in $\mathcal{A}$, that is true, by Lemma 2.1.6.

(iii): $0_{G^+(A)} \leq_m u_{G^+(A)}$ means $(0) \leq_m (1)$ in $G^+(A)$, i.e. $0 \leq_m 1$ in $\mathcal{A}$, that is true, by Proposition 2.2.56. □

Proposition 7.1.55 *The structure $(G^+(A), \leq_m, +, \smile, 0_{G^+(A)})$ is a m_0-ME structure.*

Proof. By Corollary 7.1.51 and Lemma 7.1.54 (i). □

Proposition 7.1.56 *For every $\mathbf{a} \in G^+(A)$,*

$$0_{G^+(A)} \leq_m \mathbf{a} \leq_m u_{G^+(A)} \iff \mathbf{a} = (a).$$

Proof. Obviously, since $0_{G^+(A)} \leq_m \mathbf{a} \leq_m u_{G^+(A)}$ means $(0) \leq_m \mathbf{a} \leq_m (1)$. □

Note that Proposition 7.1.56 says that

$$0_{G^+(A)} \leq_m \mathbf{a} \leq_m u_{G^+(A)}$$

is equivalent to

$$0_{G^+(A)} \leq_m (a) \leq_m u_{G^+(A)}.$$

Proposition 7.1.57 *Let $\mathbf{a} = (a_0, a_1, \ldots, a_{n-1}, a_n) \neq 0_{G^+(A)}$ be a good sequence of $\mathcal{A}$. We have:*

$$(s0^R) \qquad 0_{G^+(A)} \leq_m \mathbf{a}.$$

Proof. $0_{G^+(A)} \leq_m \mathbf{a}$ means $(0) \leq_m \mathbf{a}$, i.e.
$\underbrace{(0, 0, \ldots, 0)}_{n+1 \text{ times}} \leq_m (a_0, a_1, \ldots, a_{n-1}, a_n)$ in $G^+(A)$, i.e.
$0 \leq_m a_i$, for $i = 0, 1, \ldots, n$, in $\mathcal{A}$, that is true, by Proposition 2.2.56. □

Proposition 7.1.58 *For every $\mathbf{a} \in G^+(A)$, there exists $n_\mathbf{a} \in \mathbf{N}$, such that:*

$$(s1^R) \qquad \mathbf{a}, \smile \mathbf{a} \leq_m n_\mathbf{a} u_{G^+(A)}.$$

Proof. Indeed, let $\mathbf{a} = (a_0, a_1, \ldots, a_n)$ be a good sequence of length $n+1$; since $a_i \leq_m 1$, for $i = 0, 1, \ldots, n$, in $\mathcal{A}$, by Proposition 2.2.56, it follows that
$\mathbf{a} = (a_0, a_1, \ldots, a_n) \leq_m \underbrace{(1, 1, \ldots, 1)}_{n+1 \text{ times}} = (n+1)(1) = (n+1)u_{G^+(A)}$, by Lemma 7.1.37;
hence, there exists $n_\mathbf{a} = n+1 \in \mathbf{N}$ (i.e. $n_\mathbf{a}$ is the length of $\mathbf{a}$), such that $\mathbf{a} \leq_m n_\mathbf{a} u_{G^+(A)}$.

Also, $\smile \mathbf{a} = \smile (a_0, a_1, \ldots, a_n) = (a_n^-, \ldots, a_0^-) \leq_m \underbrace{(1, 1, \ldots, 1)}_{n+1 \text{ times}} = (n+1)(1) = (n+1)u_{G^+(A)}$, hence, there exists $n_\mathbf{a} = n+1 \in \mathbf{N}$, such that $\smile \mathbf{a} \leq_m n_\mathbf{a} u_{G^+(A)}$.
□

Proposition 7.1.59 *We have:*
$(s2^{+R})$ (i) For every $\mathbf{a} = (a_0, a_1, \ldots, a_n) \in G^+(A)$, there exists $n_\mathbf{a} = n+1 \in \mathbf{N}$ and there exist the unique $n+1$ elements of $G^+(A)$:
$(a_0), (a_1), \ldots, (a_n) \leq_m u_{G^+(A)} = (1)$, such that

$\mathbf{a} = (a_0) + (a_1) + \ldots + (a_n)$ $(\leq_m (n+1)u_{G^+(A)})$ *and*
$\smile \mathbf{a} = \smile (a_0) \smile (a_1) \smile \ldots \smile (a_n)$ $(\leq_m (n+1)u_{G^+(A)})$.

(ii) If $\mathbf{a} = (a_0) + (a_1) + \ldots + (a_n) \in G^+(A)$ *and* $\mathbf{b} = (b_0) + (b_1) + \ldots + (b_n) \in G^+(A)$, *then* $\mathbf{a} = \mathbf{b} \Longleftrightarrow (a_i) = (b_i)$, $i = 0, \ldots, n$.

(iii) If $\mathbf{a} = (a_0) + (a_1) + \ldots + (a_n) \in G^+(A)$ *and* $\mathbf{b} = (b_0) + (b_1) + \ldots + (b_n) \in G^+(A)$, *then* $\mathbf{a} \leq_m \mathbf{b} \Longleftrightarrow (a_i) \leq_m (b_i)$, $i = 0, \ldots, n$.

Proof. (i): Let $\mathbf{a} = (a_0, a_1, \ldots, a_n)$ be a good sequence of $\mathcal{A}$ of length $n+1$; by (s1^R), there exists $n_{\mathbf{a}} = n+1 \in \mathbf{N}$ such that $\mathbf{a}, -\mathbf{a} \leq_m n_{\mathbf{a}} u_{G^+(A)} = (n+1)(1)$; then, $\mathbf{a} = (a_0) + (a_1) + \ldots + (a_n)$, by Corollary 7.1.33.
Then, $\smile \mathbf{a} = \smile (a_0, a_1, \ldots, a_n) = (a_n^-, \ldots, a_1^-, a_0^-) = (a_n^-) + \ldots + (a_1^-) + (a_0^-)$
$= \smile (a_n) \smile \ldots \smile (a_1) \smile (a_0) = \smile (a_0) \smile (a_1) \smile \ldots \smile (a_n)$, by Corollary 7.1.33 and by definition of negation in $G^+(A)$.

(ii): $\mathbf{a} = \mathbf{b} \Longleftrightarrow (a_0, a_1, \ldots, a_n) = (b_0, b_1, \ldots, b_n)$
$\Longleftrightarrow a_i = b_i$, $i = 0, 1, \ldots, n$, by Lemma 7.1.9,
$\Longleftrightarrow (a_i) = (b_i)$, $i = 0, 1, \ldots, n$.

(iii): $\mathbf{a} \leq_m \mathbf{b} \Longleftrightarrow (a_0, a_1, \ldots, a_n) \leq_m (b_0, b_1, \ldots, b_n)$
$\Longleftrightarrow a_i \leq_m b_i$, $i = 0, 1, \ldots, n$, by definition,
$\Longleftrightarrow (a_i) \leq_m (b_i)$, $i = 0, 1, \ldots, n$, by definition. □

Remark 7.1.60 *The binary relation* $\leq_m$ *on* $G^+(A)$ *does not verify the compatibility property (Cp) for any* $\mathbf{a}$, $\mathbf{b}$, $\mathbf{c}$ *in* $G^+(A)$:

$$(Cp) \qquad \mathbf{a} \leq_m \mathbf{b} \Longrightarrow \mathbf{a} + \mathbf{c} \leq_m \mathbf{b} + \mathbf{c}$$

because the binary relation $\leq_m$ *on* $\mathcal{A}$ *does not verify this property, as Example 2.2.64 has showed.*

But, $\leq_m$ on $G^+(A)$ verifies a particular compatibility property:

Proposition 7.1.61 *Let* $\mathbf{a} = (a_0, a_1, \ldots, a_{n-1}, a_n)$ *and* $\mathbf{b} = (b_0, b_1, \ldots, b_{n-1}, b_n)$ *be two good sequences of* $\mathcal{A}$. *Then, the following particular compatibility property holds: for any* $m' \geq 1$,

$$(C_{pu}^R) \quad \mathbf{a} \leq_m \mathbf{b} \Longleftrightarrow \mathbf{a} + m' u_{G^+(A)} \leq_m \mathbf{b} + m' u_{G^+(A)}.$$

Proof. $\mathbf{a} + m' u_{G^+(A)} \leq_m \mathbf{b} + m' u_{G^+(A)}$
$\Longleftrightarrow (a_0, a_1, \ldots, a_{n-1}, a_n) + m'(1) \leq_m (b_0, b_1, \ldots, b_{n-1}, b_n) + m'(1)$
$\Longleftrightarrow (\underbrace{1, 1, \ldots, 1}_{m' \text{ times}}, a_0, a_1, \ldots, a_{n-1}, a_n) \leq_m (\underbrace{1, 1, \ldots, 1}_{m' \text{ times}}, b_0, b_1, \ldots, b_{n-1}, b_n)$, by Corollary 7.1.39,
$\Longleftrightarrow a_i \leq_m b_i$, for $i = 0, 1, \ldots, n$, by Lemma 2.1.6,
$\Longleftrightarrow \mathbf{a} \leq_m \mathbf{b}$. □

Proposition 7.1.62 *Let* $\mathbf{a} = (a_0, a_1, \ldots, a_n) \in G^+(A)$ *of length* $n+1$ *and* $r \in \mathbf{N}, n+1 < r$. *Then, the following particular transitivity property holds:*

$$(m - Tr_u^R) \quad \mathbf{a} \leq_m (n+1)u_{G^+(A)},\ (n+1)u_{G^+(A)} \leq_m r u_{G^+(A)} \Longrightarrow \mathbf{a} \leq_m r u_{G^+(A)}.$$

Proof. • $\mathbf{a} \leq_m (n+1)u_{G^+(A)}$
$\Longleftrightarrow (a_0, a_1, \ldots, a_n) \leq_m (n+1)(1)$
$\Longleftrightarrow (a_0, a_1, \ldots, a_n) \leq_m (\underbrace{1, 1, \ldots, 1}_{n+1 \text{ times}})$
$\Longleftrightarrow a_i \leq_m 1, i = 0, 1, \ldots, n$, in $\mathcal{A}$.

• Let $p = r - (n+1) \in \mathbf{N}$, hence $r = (n+1) + p$.
$(n+1)u_{G^+(A)} \leq_m ru_{G^+(A)}$
$\Longleftrightarrow (n+1)(1) \leq_m r(1)$
$\Longleftrightarrow (\underbrace{1, 1, \ldots, 1}_{n+1 \text{ times}}) \leq_m (\underbrace{1, 1, \ldots, 1}_{r \text{ times}})$
$\Longleftrightarrow (\underbrace{1, 1, \ldots, 1}_{n+1 \text{ times}}, \underbrace{0, 0, \ldots, 0}_{p \text{ times}}) \leq_m (\underbrace{1, 1, \ldots, 1}_{r \text{ times}})$,
since $(\underbrace{1, 1, \ldots, 1}_{n+1 \text{ times}}) = (\underbrace{1, 1, \ldots, 1}_{n+1 \text{ times}}, \underbrace{0, 0, \ldots, 0}_{p \text{ times}})$.

• Hence, $\mathbf{a} = (a_0, a_1, \ldots, a_n)$
$= (\underbrace{a_0, a_1, \ldots, a_n}_{n+1}, \underbrace{0, 0, \ldots, 0}_{p \text{ times}})$
$\leq_m (\underbrace{1, 1, \ldots, 1}_{n+1 \text{ times}}, \underbrace{0, 0, \ldots, 0}_{p \text{ times}})$
$\leq_m (\underbrace{1, 1, \ldots, 1}_{r \text{ times}}) = ru_{G^+(A)}$. Thus, $\mathbf{a} \leq_m ru_{G^+(A)}$. □

7.1.7 The $\inf_m$ and $\sup_m$ on the set $G^+(A)$ of good sequences

For the good sequences $\mathbf{a} = (a_0, a_1, \ldots, a_{n-1}, a_n)$ and $\mathbf{b} = (b_0, b_1, \ldots, b_{n-1}, b_n)$ and the binary relation $\leq_m$ on $G^+(A)$, we set

$$\inf_m(\mathbf{a}, \mathbf{b}) \stackrel{def.}{=} (\inf_m(a_0, b_0), \inf_m(a_1, b_1), \ldots, \inf_m(a_n, b_n)), \tag{7.13}$$

$$\sup_m(\mathbf{a}, \mathbf{b}) \stackrel{def.}{=} (\sup_m(a_0, b_0), \sup_m(a_1, b_1), \ldots, \sup_m(a_n, b_n)), \tag{7.14}$$

and they exist, if $\inf_m(a_0, b_0), \inf_m(a_1, b_1), \ldots, \inf_m(a_n, b_n)$ and $\sup_m(a_0, b_0), \sup_m(a_1, b_1), \ldots, \sup_m(a_n, b_n)$, respectively, do exist.

When they exist, the problem is if they are good sequences.

Lemma 7.1.63 *For the good sequences*

$$\mathbf{a} = u_{G^+(A)} = (1) = (1, 0) \quad \textit{and} \quad \mathbf{b} = (x) + (y) = (x \oplus y, x \odot y),$$

$\inf_m(\mathbf{a}, \mathbf{b})$ *and* $\sup_m(\mathbf{a}, \mathbf{b})$ *exist and are good sequences.*

Proof. Indeed,

$$\begin{array}{lcl}
\inf_m(\mathbf{a},\mathbf{b}) & = & \inf_m((1,0),(x\oplus y, x\odot y)) \\
& = & \inf_m((a_0,a_1),(b_0,b_1)) \\
& \stackrel{def.}{=} & (\inf_m(a_0,b_0), \inf_m(a_1,b_1)) \\
& = & (\inf_m(1,x\oplus y), \inf_m(0,x\odot y)) \\
& = & (\min_m(1,x\oplus y), \min_m(0,x\odot y)), \text{ that exist,} \\
& = & (x\oplus y, 0) \\
& = & (x\oplus y), \\
\sup_m(\mathbf{a},\mathbf{b}) & = & \inf_m((a_0,a_1),(b_0,b_1)) \\
& \stackrel{def.}{=} & (\sup_m(a_0,b_0), \sup_m(a_1,b_1)) \\
& = & (\sup_m(1,x\oplus y), \sup_m(0,x\odot y)) \\
& = & (\max_m(1,x\oplus y), \max_m(0,x\odot y)), \text{ that exist,} \\
& = & (1, x\odot y),
\end{array}$$

since $0 \leq_m x\oplus y$, $x\odot y \leq_m 1$ in $\mathcal{A}$; hence, $\inf_m(\mathbf{a},\mathbf{b})$ and $\sup_m(\mathbf{a},\mathbf{b})$ exist and are good sequences. □

Note that, in the sequel, we shall analyse only $\inf_m$.

Lemma 7.1.64 *For any* $0_{G^+(A)} \leq_m (x),(y) \leq_m u_{G^+(A)}$ *in* $G^+(A)$, *there exists*

$$\inf_m(u_{G^+(A)}, (x)+(y)) = (x\oplus y)$$

and is a good sequence.

Proof. $\inf_m(u_{G^+(A)}, (x)+(y)) = \inf_m((1),(x)+(y))$
$= \inf_m((1,0),\ (x\oplus y, x\odot y)) \stackrel{def.}{=} (\inf_m(1,x\oplus y),\ \inf_m(0,x\odot y)) = (x\oplus y)$,
by Lemma 7.1.63. □

Corollary 7.1.65 *For each* $0_{G^+(A)} \leq_m (x) \leq_m u_{G^+(A)}$ *in* $G^+(A)$,

$$\inf_m(u_{G^+(A)}, (x)+u_{G^+(A)}) = u_{G^+(A)}.$$

Proof. By above Lemma 7.1.64,
$\inf_m(u_{G^+(A)}, (x)+u_{G^+(A)}) = \inf_m(u_{G^+(A)}, (x)+(1_A) = (x\oplus 1_A) = (1_A) = u_{G^+(A)}$,
by (m-La^R) in $\mathcal{A}$. □

Corollary 7.1.66 *For any* $0_{G^+(A)} \leq_m (x),(y) \leq_m u_{G^+(A)}$ *in* $G^+(A)$, *there exists*

$$\smile \inf_m(u_{G^+(A)}, \smile(x) \smile (y)) = (x\odot y)$$

and is a good sequence.

Proof. $0_{G^+(A)} \leq_m (x),(y) \leq_m u_{G^+(A)}$, in $G^+(A)$
$\Longleftrightarrow (0) \leq_m (x),(y) \leq_m (1)$, in $G^+(A)$,
$\Longleftrightarrow 0 \leq_m x, y \leq_m 1$, in A,
$\Longleftrightarrow 0 \leq_m x^-, y^- \leq_m 1$, in A,
$\Longleftrightarrow (0) \leq_m (x^-),(y^-) \leq_m (1)$, in $G^+(A)$,

$\Longleftrightarrow 0_{G^+(A)} \leq_m (x^-), (y^-) \leq_m u_{G^+(A)}$, in $G^+(A)$.
Then, by Lemma 7.1.64, there exists $\inf_m(u_{G^+(A)}, (x^-) + (y^-)) = (x^- \oplus y^-)$ and is a good sequence; then, there exists
$\smile \inf_m(u_{G^+(A)}, \smile (x) \smile (y)) = \smile (x^- \oplus y^-) = ([x^- \oplus y^-]^-) = (x \odot y)$
and is a good sequence. □

Corollary 7.1.67 *For each* $0_{G^+(A)} \leq_m (x) \leq_m u_{G^+(A)}$ *in* $G^+(A)$,

$$\smile \inf_m(u_{G^+(A)}, \smile (x) \smile 0_{G^+(A)}) = 0_{G^+(A)}, \quad \smile \inf_m(u_{G^+(A)}, \smile (x) \smile u_{G^+(A)}) = (x).$$

Proof. By above Corollary 7.1.66,
$\smile \inf_m(u_{G^+(A)}, \smile (x) \smile 0_{G^+(A)}) = \smile \inf_m(u_{G^+(A)}, \smile (x) \smile (0_A)) = (x \odot 0_A) = (0_A) = 0_{G^+(A)}$, by (m-La) in $\mathcal{A}$, and
$\smile \inf_m(u_{G^+(A)}, \smile (x) \smile u_{G^+(A)}) = \smile \inf_m(u_{G^+(A)}, \smile (x) \smile (1_A)) = (x \odot 1_A) = (x)$, by (PU) in $\mathcal{A}$. □

Proposition 7.1.68 *For any* $0_{G^+(A)} \leq_m (x), (y) \leq_m u_{G^+(A)}$ *in* $G^+(A)$,

(7.15) $(s3^R)$ *(i) there exists* $\inf_m(u_{G^+(A)}, (x) + (y)) = (x \oplus y)$,

there exists $\smile \inf_m(u_{G^+(A)}, \smile (x) \smile (y)) = (x \odot y)$

and are good sequences belonging to $[0_{G^+(A)}, u_{G^+(A)}]$;

$$(ii)\ \inf_m(u_{G^+(A)}, (x) + u_{G^+(A)}) = u_{G^+(A)},$$

$$\smile \inf_m(u_{G^+(A)}, \smile (x) \smile 0_{G^+(A)}) = 0_{G^+(A)}, \ \smile \inf_m(u_{G^+(A)}, \smile (x) \smile u_{G^+(A)}) = (x);$$

$$(iii)\ (x) \leq_m (y)\ ((x), (y) \neq 0_{G^+(A)}) \iff u_{G^+(A)} \leq_m (y) \smile (x).$$

Proof. (i): By Lemma 7.1.64 and Corollary 7.1.66.
(ii): By Corollaries 7.1.65 and 7.1.67.
(iii): $(x) \leq_m (y) \iff x \leq_m y \overset{in\ \mathcal{A}}{\iff} y \oplus x^- = 1$; since we also have $y \odot x^- \geq_m 0$, by Proposition 2.1.5, it follows that we have: $y \oplus x^- = 1$ and $y \odot x^- \geq_m 0$.
Note now that $(y) \smile (x) = (y) + (x^-) = (y \oplus x^-, y \odot x^-)$; hence, $(y) \smile (x) \geq_m u_{G^+(A)}$, i.e. $(y) \smile (x) \geq_m (1)$ means $(y \oplus x^-, y \odot x^-) \geq_m (1, 0)$, i.e. $y \oplus x^- \geq_m 1$ and $y \odot x^- \geq_m 0$ in A; then, by Lemma 2.1.6 (iii), it follows that we also have: $y \oplus x^- = 1$ and $y \odot x^- \geq_m 0$. □

Proposition 7.1.69 *For any* $0_{G^+(A)} \leq_m (x), (y) \leq_m u_{G^+(A)}$ *in* $G^+(A)$,

$$(x) \leq_m (y) \text{ and } (y) \leq_m (x) \iff (y) \smile (x) = u_{G^+(A)} = (x) \smile (y).$$

Proof. $(x) \leq_m (y)$ *and* $(y) \leq_m (x)$, in $G^+(A)$
$\iff x \leq_m y$ *and* $y \leq_m x$, in A
$\iff y \oplus x^- = 1$ *and* $x \oplus y^- = 1$
$\iff y \oplus x^- = 1$ *and* $x^- \odot y = 0$
$\iff (y \oplus x^-, y \odot x^-) = (1, 0) = (1) = u_{G^+(A)}$

and

$$\begin{aligned}
&(y) \smile (x) = u_{G^+(A)} = (x) \smile (y)\\
&\Longleftrightarrow (y) + (x^-) = (1) = (x) + (y^-)\\
&\Longleftrightarrow (y \oplus x^-, y \odot x^-) = (1,0) = (x \oplus y^-, x \odot y^-)\\
&\Longleftrightarrow y \oplus x^- = 1,\ y \odot x^- = 0,\ x \oplus y^- = 1,\ x \odot y^- = 0\\
&\Longleftrightarrow y \oplus x^- = 1,\ y \odot x^- = 0,\ x^- \odot y = 0,\ x^- \oplus y = 1\\
&\Longleftrightarrow y \oplus x^- = 1,\ y \odot x^- = 0, \text{ by (Pcomm), (Scomm)}\\
&\Longleftrightarrow (y \oplus x^-, y \odot x^-) = (1,0) = (1) = u_{G^+(A)}.
\end{aligned}$$

□

Consider the good sequences (x), (y), (z) $\in G^+(A)$ and let

$$\mathbf{b} = (x) + ((y) + (z)) = (x) + (y \oplus z, y \odot z) \overset{(Scomm)}{=} (y \oplus z, y \odot z) + (x) = (b_0, b_1, b_2),$$

where: $b_0 = (y \oplus z) \oplus x$, $b_1 = (y \oplus z) \odot ((y \odot z) \oplus x)$, $b_2 = (y \odot z) \odot x$, by Lemma 7.1.14.

Then, we have the following result.

Lemma 7.1.70 *For the good sequences*

$$\mathbf{a} = u_{G^+(A)} = (1) = (1,0,0) \quad \textit{and} \quad \mathbf{b} = (x) + ((y) + (z)) = (b_0, b_1, b_2),$$

there exists $\inf_m(\mathbf{a}, \mathbf{b})$ *and is a good sequence:*

$$\inf_m(\mathbf{a}, \mathbf{b}) = \inf_m(u_{G^+(A)}, (x) + ((y) + (z))) = (x \oplus (y \oplus z)).$$

Proof. Indeed,

$$\begin{aligned}
\inf_m(\mathbf{a}, \mathbf{b}) &= \inf_m((1), (x) + ((y) + (z)))\\
&= \inf_m((1,0,0), (b_0, b_1, b_2))\\
&= \inf_m((a_0, a_1, a_2), (b_0, b_1, b_2))\\
&\overset{def.}{=} (\inf_m(a_0, b_0), \inf_m(a_1, b_1), \inf_m(a_2, b_2))\\
&= (\inf_m(1, b_0), \inf_m(0, b_1), \inf_m(0, b_2))\\
&= (b_0, 0, 0)\\
&= (b_0)\\
&= ((y \oplus z) \oplus x)\\
&\overset{(Scomm)}{=} (x \oplus (y \oplus z)),
\end{aligned}$$

since $0 = 0_A \leq_m b_0$, b_1, $b_2 \leq_m 1 = 1_A$ in $\mathcal{A}$. □

Corollary 7.1.71 *Consider the good sequences (a), (b), (c)* $\in G^+(A)$. *Then, we have:*

$$(Sass) \qquad (a \oplus [b \oplus c]) = ([a \oplus b] \oplus c).$$

Proof. By Lemma 7.1.70 and Proposition 7.1.31, we obtain:

$$\begin{aligned}
(a \oplus (b \oplus c)) &= \inf_m(u_{G^+(A)}, (a) + [(b) + (c)])\\
&= \inf_m(u_{G^+(A)}, [(a) + (b)] + (c))\\
&\overset{(Scomm)}{=} \inf_m(u_{G^+(A)}, (c) + [(a) + (b)])\\
&= (c \oplus [a \oplus b]).
\end{aligned}$$

□

Lemma 7.1.72 *For the good sequences (x), (y), (z) and*

$$\mathbf{a} = u_{G^+(A)} = (1) = (1,0) \quad \textit{and} \quad \mathbf{b} = (y) + (z) = (y \oplus z, y \odot z),$$

there exists $\inf_m(\mathbf{a}, (x) + \inf_m(\mathbf{a}, \mathbf{b}))$ *and is a good sequence:*

$$\inf_m(\mathbf{a}, (x) + \inf_m(\mathbf{a}, \mathbf{b}))$$

$$= \inf_m(u_{G^+(A)}, (x) + \inf_m(u_{G^+(A)}, (y) + (z))\ ) = (x \oplus (y \oplus z)).$$

Proof. Indeed,

$$\begin{array}{lcl} \inf_m(\mathbf{a}, (x) + \inf_m(\mathbf{a}, \mathbf{b})) & = & \inf_m(u_{G^+(A)}, (x) + \inf_m(u_{G^+(A)}, (y) + (z))) \\ & = & \inf_m((1), (x) + (y \oplus z)), \text{ by Lemma 7.1.63}, \\ & = & \inf_m((1,0), (x \oplus (y \oplus z), x \odot (y \oplus z))\ ) \\ & \stackrel{def.}{=} & (\inf_m(1, x \oplus (y \oplus z)), \inf_m(0, x \odot (y \oplus z))\ ) \\ & \stackrel{def.}{=} & (x \oplus (y \oplus z), 0) \\ & \stackrel{def.}{=} & (x \oplus (y \oplus z)). \end{array}$$

□

Corollary 7.1.73 *For any* $0_{G^+(A)} \leq_m (x), (y), (z) \leq_m u_{G^+(A)}$ *in* $G^+(A)$,

(7.16) $\quad (s4^R)$ *(i) there exist* $\inf_m(u_{G^+(A)}, (x) + \inf_m(u_{G^+(A)}, (y) + (z))\ )$

$$= \inf_m(u_{G^+(A)}, (x) + ((y) + (z))\ ) = (x \oplus (y \oplus z))$$

are equal and is a good sequence.

(ii) $\inf_m(u_{G^+(A)}, \smile \inf_m(u_{G^+(A)}, \smile (x) \smile (y))\ ) = \smile \inf_m(u_{G^+(A)}, \smile (x) \smile (y))$.

Proof. Indeed,

(i)
$$\begin{array}{lll} & \inf_m(u_{G^+(A)}, (x) + \inf_m(u_{G^+(A)}, (y) + (z))\ ) & \\ = & (x \oplus (y \oplus z)), & \text{by Lemma 7.1.72}, \\ = & \inf_m(u_{G^+(A)}, (x) + ((y) + (z))\ ), & \text{by Lemma 7.1.70}. \end{array}$$

(ii)
$$\begin{array}{lll} & \inf_m(u_{G^+(A)}, \smile \inf_m(u_{G^+(A)}, \smile (x) \smile (y))\ ) & \\ = & \inf_m(u_{G^+(A)}, (x \odot y)\ ), & \text{by Corollary 7.1.66}, \\ = & (x \odot y) & \\ = & \smile \inf_m(u_{G^+(A)}, \smile (x) \smile (y)\ ), & \text{by Corollary 7.1.66}. \end{array}$$

□

Corollary 7.1.74 *For any* $0_{G^+(A)} \leq_m (x), (y), (z) \leq_m u = u_{G^+(A)}$ *in* $G^+(A)$,

(7.17) $\quad (XX_u^R)$ *there exist and are equal good sequences*

$$\smile \inf_m(u, \smile \inf_m(u, (x) + (y)) \smile \inf_m(u, (z) \smile \inf_m(u, \smile (x) \smile (y))))$$

$$= \smile \inf_m(u, \smile \inf_m(u, (y) + (z)) \smile \inf_m(u, (x) \smile \inf_m(u, \smile (y) \smile (z)))).$$

Proof. Indeed,
$A \overset{not.}{=} \smile \inf_m(u, \smile \inf_m(u, (x) + (y)) \smile \inf_m(u, (z) \smile \inf_m(u, \smile (x) \smile (y))))$
$= \smile \inf_m(u, \ \smile (x \oplus y) \smile \inf_m(u, \ (z) \smile \inf_m(u, (x^-) + (y^-)))\ )$, by (s3^R) (i)
$= \smile \inf_m(u, \ \smile (x \oplus y) \smile \inf_m(u, \ (z) \smile (x^- \oplus y^-))\ )$, by (s3^R) (i)
$= \smile \inf_m(u, \ \smile (x \oplus y) \smile \inf_m(u, \ (z) + ((x^- \oplus y^-)^-))\ )$
$= \smile \inf_m(u, \ \smile (x \oplus y) \smile \inf_m(u, \ (z) + (x \odot y))\ )$
$= \smile \inf_m(u, \ \smile (x \oplus y) \smile (z \oplus (x \odot y))\ )$, by (s3^R) (i)
$= \smile \inf_m(u, \ ((x \oplus y)^-) + ((z \oplus (x \odot y))^-)\ )$
$= \smile (\ (x \oplus y)^- \oplus (z \oplus (x \odot y))^-\ )$, by (s3^R) (i)
$= (\ ((x \oplus y)^- \oplus (z \oplus (x \odot y))^-)^-\ )$
$= (\ (x \oplus y) \odot (z \oplus (x \odot y))\ )$
and, similarly,
$B \overset{not.}{=} \smile \inf_m(u, \smile \inf_m(u, (y) + (z)) \smile \inf_m(u, (x) \smile \inf_m(u, \smile (y) \smile (z))))$
$= (\ (y \oplus z) \odot (x \oplus (y \odot z))\ )$.
But, in $\mathcal{A}$, by (XX),
$(x \oplus y) \odot ((x \odot y) \oplus z) = (y \oplus z) \odot ((y \odot z) \oplus x)$, hence, in $G^+(A)$, we have:
$((x \oplus y) \odot ((x \odot y) \oplus z)) = ((y \oplus z) \odot ((y \odot z) \oplus x))$; thus, $A = B$. □

Corollary 7.1.75 *For any* $0_{G^+(A)} \leq_m (x), (y), (z) \leq_m u = u_{G^+(A)}$ *in* $G^+(A)$,

$$(7.18) \qquad (YY^R_u) \quad if \ \inf_m(u, (x) + (y)) = (x \oplus y) = (x), \quad then$$

there exist and are equal

$$\smile \inf_m(u, \ \smile (x) \smile \inf_m(u, \ (y) + (z))\ ) = (x \odot (y \oplus z))$$

$$= \inf(u, \ (y) \smile \inf_m(u, \ \smile (x) \smile (z))\ ) = (y \oplus (x \odot z)).$$

Proof. Indeed,

$$\begin{array}{llr} A \overset{notation}{=} & \smile \inf_m(u, \ \smile (x) \smile \inf_m(u, \ (y) + (z))\ ) & \\ = & \smile \inf_m(u, \ \smile (x) \smile (y \oplus z)\ ), & \text{by (s3}^R\text{) (i),} \\ = & \smile \inf_m(u, \ (x^-) + ((y \oplus z)^-)\ ) & \\ = & \smile (x^- \oplus (y \oplus z)^-), & \text{by (s3}^R\text{) (i),} \\ = & ((x^- \oplus (y \oplus z)^-)^-) & \\ = & (x \odot (y \oplus z)) & \end{array}$$

and

$$\begin{array}{llr} B \overset{notation}{=} & \inf(u, \ (y) \smile \inf_m(u, \ \smile (x) \smile (z))\ ) & \\ = & \inf(u, \ (y) \smile \inf_m(u, \ (x^-) + (z^-))\ ) & \\ = & \inf(u, \ (y) \smile (x^- \oplus z^-)\ ), & \text{by (s3}^R\text{) (i),} \\ = & \inf(u, \ (y) + ((x^- \oplus z^-)^-)\ ) & \\ = & \inf(u, \ (y) + (x \odot z)\ ) & \\ = & (y \oplus (x \odot z)), & \text{by (s3}^R\text{) (i).} \end{array}$$

Then, if $(x \oplus y) = (x)$ in $G^+(A)$, it follows that $x \oplus y = x$ in A.
Then, by (YY) ($x \oplus y = x \implies x \odot (y \oplus z) = y \oplus (x \odot z)$) in $\mathcal{A}$, we obtain: $x \odot (y \oplus z) = y \oplus (x \odot z)$ in A, hence we have $(x \odot (y \oplus z)) = (y \oplus (x \odot z))$ in $G^+(A)$, and, thus, $A = B$. □

Corollary 7.1.76
For any $0_{G^+(A)} \leq_m (x_0), (x_1), \ldots, (x_n), (y_0) \leq_m u = u_{G^+(A)}$ *in* $G^+(A)$,
(s5^R) (V^{00}) $(x_0) + (y_0) = \inf_m(u, (x_0) + (y_0)) \smile \inf_m(u, \smile (x_0) \smile (y_0))$,
(V^{10}) $(x_0, x_1) + (y_0) = \inf_m(u, (x_0) + (y_0))$
$\smile \inf_m(u, \smile (x_0) \smile \inf_m(u, (x_1) + (y_0)))$
$\smile \inf_m(u, \smile (x_1) \smile (y_0))$,
(V^{n0}) $(x_0, x_1, \ldots, x_n) + (y_0) =$
$= \sum_{i=0}^{n+1} \smile \inf_m(u, \smile (x_{i\smile 1}) \smile \inf_m(u_{G^+(A)}, (x_i) + (y_0)))$,
with $(x_{-1}) = u = (1_A)$, $(x_{n+1}) = 0_{G^+(A)} = (0_A)$.

Proof. (V^{00}): $(x_0) + (y_0) = (x_0 \oplus y_0, x_0 \odot y_0)$, by Lemma 7.1.12,
$= (x_0 \oplus y_0) + (x_0 \odot y_0)$, by Lemma 7.1.33,
$= \inf_m(u_{G^+(A)}, (x_0) + (y_0)) + (x_0 \odot y_0)$, by (s3^R) (i),
$= \inf_m(u_{G^+(A)}, (x_0) + (y_0)) \smile \inf_m(u_{G^+(A)}, \smile (x_0) \smile (y_0))$, by Corollary 7.1.66.
(V^{10}): $(x_0, x_1) + (y_0) = (c_0, c_1, c_2)$, by Lemma 7.1.14,
$= (c_0) + (c_1) + (c_2)$, by Corollary 7.1.33,
$= (x_0 \oplus y_0) + (x_0 \odot (x_1 \oplus y_0)) + (x_1 \odot y_0)$
$= \inf_m(u, (x_0) + (y_0))$
$\smile \inf_m(u, \smile (x_0) \smile \inf_m(u_{G^+(A)}, (x_1) + (y_0)))$
$\smile \inf_m(u, \smile (x_1) \smile (y_0))$, by (s3^R) (i) and Corollary 7.1.66.
(V^{n0}): $(x_0, x_1, \ldots, x_n) + (y_0) = (c_0, c_1, \ldots, c_{n+1})$, by Lemma 7.1.16,
$= (c_0) + (c_1) + \ldots + (c_{n+1})$, by Lemma 7.1.33,
$= \sum_{i=0}^{n+1} (c_i)$,
where:
$(c_i) = (x_{i-1} \odot (x_i \oplus y_0))$
$= \smile \inf_m(u, \smile (x_{i-1}) \smile (x_i \oplus y_0))$, by Corollary 7.1.66,
$= \smile \inf_m(u, \smile (x_{i-1}) \smile \inf_m(u, (x_i) + (y_0)))$,
with $(x_{-1}) = u = (1_A)$, $(x_{n+1}) = 0_{G^+(A)} = (0_A)$. □

Theorem 7.1.77 *Let* $\mathcal{A} = (A, \oplus, ^-, 0)$ *be a right-XY algebra. Then, the structure*

$$(G^+(A), \leq_m, +, \smile, 0_{G^+(A)}, u_{G^+(A)})$$

is a strong m_0*-ME positive cone.*

Proof. By Propositions 7.1.55, 7.1.57, 7.1.58, 7.1.59, 7.1.68, by Corollaries 7.1.73, 7.1.74, 7.1.75, 7.1.76, by Propositions 7.1.40, 7.1.61, 7.1.62, 7.1.49, 7.1.52. □

Remark 7.1.78 *(See Open problems 7.1.45) If there is at least one example of cancellative proper XY algebra* $\mathcal{A}$ *whose* $G^+(A)$ *verifies the cancellative property (C), then its strong* m_0*-ME positive cone* $(G^+(A), \leq_m, +, \smile, 0_{G^+(A)}, u_{G^+(A)})$ *verifies (C), and hence another construction of the functor* $\mathbf{T_m}$ *can be made, similar to that of* $\mathbf{T}$ *in MV algebras case (see Subsection 1.3).*

- **The functor $\mathbf{G}^+_{\mathbf{m}}$**

We put

$$\mathbf{G}^+_{\mathbf{m}}(\mathcal{A}) = (G^+(A), \leq_m, +, \smile, 0_{G^+(A)}, u_{G^+(A)}),$$

i.e. we have:

$$\begin{array}{ccc} \mathcal{C}(\textbf{s-m}_0\textbf{-ME}^+) & \stackrel{\mathbf{G}^+_{\mathbf{m}}}{\longleftarrow} & \mathcal{C}(\mathbf{XY}^R) \\ \mathbf{G}^+_{\mathbf{m}}(\mathcal{A}) & \stackrel{\mathbf{G}^+_{\mathbf{m}}}{\longleftarrow} & \mathcal{A} \end{array}$$

7.1.8 Additional results: Theorem 7.1.81

Lemma 7.1.79 *For the good sequences*

$$\mathbf{a} = u_{G^+(A)} = (1) = (\underbrace{1, 0, \ldots, 0}_{n+1 \text{ times}}) \text{ and } \mathbf{b} = (x_0) + \ldots + (x_n) = (c_0^{n+1}, c_1^{n+1}, \ldots, c_n^{n+1}),$$

$\inf_m(\mathbf{a}, \mathbf{b})$ *exists and is a good sequence.*

Proof. Indeed,

$$\begin{array}{rcl} \inf_m(\mathbf{a}, \mathbf{b}) & = & \inf_m(u_{G^+(A)}, (x_0) + \ldots + (x_n)) \\ & = & \inf_m((\underbrace{1, 0, \ldots, 0}_{n+1 \text{ times}}), (c_0^{n+1}, c_1^{n+1}, \ldots, c_n^{n+1})) \\ & \stackrel{def.}{=} & (\inf_m(1, c_0^{n+1}), \inf_m(0, c_1^{n+1}), \ldots, \inf_m(0, c_n^{n+1})) \\ & = & (\min(1, c_0^{n+1}), \min(0, c_1^{n+1}), \ldots, \min(0, c_n^{n+1})), \text{ that exist,} \\ & = & (\underbrace{c_0^{n+1}, 0, \ldots, 0}_{n+1 \text{ times}}) \\ & = & (c_0^{n+1}) = (x_0 \oplus x_1 \oplus \ldots \oplus x_n), \text{ by Proposition 7.1.32.} \end{array}$$

□

Corollary 7.1.80 *For any* $0_{G^+(A)} \leq_m (x_0), (x_1), \ldots, (x_n) \leq_m u_{G^+(A)}$ *in* $G^+(A)$,
(i) there exists $\inf_m(u_{G^+(A)}, (x_0) + (x_1) + \ldots + (x_n)) = (x_0 \oplus x_1 \oplus \ldots \oplus x_n)$ *and is a good sequence;*
(ii) there exists $\smile \inf_m(u_{G^+(A)}, \smile (x_0) \smile (x_1) \smile \ldots \smile (x_n)) = (x_0 \odot x_1 \odot \ldots \odot x_n)$ *and is a good sequence.*

Proof. (i): By Lemma 7.1.79, there exists $\inf_m(u_{G^+(A)}, (x_0) + \ldots + (x_n)) = (x_0 \oplus x_1 \oplus \ldots \oplus x_n)$.

(ii): Since, $0_{G^+(A)} \leq_m (x_0^-), (x_1^-), \ldots, (x_n^-) \leq_m u_{G^+(A)}$ in $G^+(A)$, by above (i), we have:
there exists $\inf_m(u_{G^+(A)}, (x_0^-) + (x_1^-) + \ldots + (x_n^-)) = (x_0^- \oplus x_1^- \oplus \ldots \oplus x_n^-)$,
hence there exists $\smile \inf_m(u_{G^+(A)}, (x_0^-) + (x_1^-) + \ldots + (x_n^-)) = \smile (x_0^- \oplus x_1^- \oplus \ldots \oplus x_n^-)$,
i.e. $\smile \inf_m(u_{G^+(A)}, \smile (x_0) \smile (x_1) \smile \ldots \smile (x_n)) = ([x_0^- \oplus x_1^- \oplus \ldots \oplus x_n^-]^-)$,
hence $\smile \inf_m(u_{G^+(A)}, \smile (x_0) \smile (x_1) \smile \ldots \smile (x_n)) = (x_0 \odot x_1 \odot \ldots \odot x_n)$. □

Theorem 7.1.81 *Let* $0_{G^+(A)} \leq_m (x_i), (y_i) \leq_m u_{G^+(A)}$, $i = 0, 1, \ldots, n$, *in* $G^+(A)$.

If $(x_0) + (x_1) + \ldots + (x_n) = (y_0) + (y_1) + \ldots + (y_n)$, *in* $G^+(A)$, *then*
(i) $x_0 \oplus x_1 \oplus \ldots \oplus x_n = y_0 \oplus y_1 \oplus \ldots \oplus y_n$, *in* A,
(ii) $x_0 \odot x_1 \odot \ldots \odot x_n = y_0 \odot y_1 \odot \ldots \odot y_n$, *in* A.

Proof. (i): By Corollary 7.1.80 (i),
there exist $\inf_m(u_{G^+(A)}, (x_0) + (x_1) + \ldots + (x_n)) = (x_0 \oplus x_1 \oplus \ldots \oplus x_n)$ and
there exist $\inf_m(u_{G^+(A)}, (y_0) + (y_1) + \ldots + (y_n)) = (y_0 \oplus y_1 \oplus \ldots \oplus y_n)$.
If $(x_0) + (x_1) + \ldots + (x_n) = (y_0) + (y_1) + \ldots + (y_n)$, then
$\inf_m(u_{G^+(A)}, (x_0)+(x_1)+\ldots+(x_n)) = \inf_m(u_{G^+(A)}, (y_0)+(y_2)+\ldots+(y_n))$, hence
$(x_0 \oplus x_1 \oplus \ldots \oplus x_n) = (y_0 \oplus y_1 \oplus \ldots \oplus y_n)$, in $G^+(A)$, hence
$x_0 \oplus x_2 \oplus \ldots \oplus x_n = y_0 \oplus y_1 \oplus \ldots \oplus y_n$, in A.

(ii): By Corollary 7.1.80 (ii),
there exist $\smile \inf_m(u_{G^+(A)}, \smile (x_0) \smile (x_1) \smile \ldots \smile (x_n)) = (x_0 \odot x_1 \odot \ldots \odot x_n)$ and
there exist $\smile \inf_m(u_{G^+(A)}, \smile (y_0) \smile (y_1) \smile \ldots \smile (y_n)) = (y_0 \odot y_1 \odot \ldots \odot y_n)$.
If $(x_0) + (x_1) + \ldots + (x_n) = (y_0) + (y_1) + \ldots + (y_n)$, then
$\smile (x_0) \smile (x_1) \smile \ldots \smile (x_n) = \smile (y_0) \smile (y_1) \smile \ldots \smile (y_n)$, by $(s2^{+R})$ (i), hence
$\smile \inf_m(u_{G^+(A)}, \smile (x_0) \smile (x_1) \smile \ldots \smile (x_n)) =$
$\smile \inf_m(u_{G^+(A)}, \smile (y_0) \smile (y_1) \smile \ldots \smile (y_n))$, i.e.
$(x_0 \odot x_1 \odot \ldots \odot x_n) = (y_0 \odot y_1 \odot \ldots \odot y_n)$, in $G^+(A)$, hence
$x_0 \odot x_1 \odot \ldots \odot x_n = y_0 \odot y_1 \odot \ldots \odot y_n$, in A. □

- **Corollaries of Theorem 7.1.81**

Let $\mathcal{A} = (A, \oplus, ^-, 0)$ be a right-XY algebra and
$\mathbf{G}_{\mathbf{m}}^{+}(\mathcal{A}) = (G^+(A), \leq_m, +, \smile, 0_{G^+(A)}, u_{G^+(A)})$ be the strong m_0-ME positive cone associated to $\mathcal{A}$ by the functor $\mathbf{G}_{\mathbf{m}}^{+}$.

Corollary 7.1.82 *(See Lemma 7.1.12) (m=0)*
If (a_0) *and* (b_0) *are good sequences, then*
(i) $(a_0) + (b_0) = (a_0 \oplus b_0, a_0 \odot b_0)$.
(ii) $a_0 \oplus b_0 = [a_0 \oplus b_0] \oplus [a_0 \odot b_0]$.
(iii) $a_0 \odot b_0 = [a_0 \oplus b_0] \odot [a_0 \odot b_0]$.

Proof. (i): By Lemma 7.1.12.

(ii): By property (Z). Alternatively, by Corollary 7.1.33, $(a_0 \oplus b_0, a_0 \odot b_0) = (a_0 \oplus b_0) + (a_0 \odot b_0)$; hence, by above (i), $(a_0) + (b_0) = (a_0 \oplus b_0) + (a_0 \odot b_0)$; then, apply Theorem 7.1.81 (i).

(iii): By property (Zd). Alternatively, by Corollary 7.1.33, $(a_0 \oplus b_0, a_0 \odot b_0) = (a_0 \oplus b_0) + (a_0 \odot b_0)$; then, apply Theorem 7.1.81 (ii). □

Remark 7.1.83 *(See Remark 2.2.38)*

The property from Corollary 7.1.82 (ii) is just the property (Z) studied in subsection 2.2.3. We have here an alternative proof in $G^+(A)$ *of the fact that (Z) holds in* $\mathcal{A}$.

Remark 7.1.84 *The property from Corollary 7.1.82 (iii) is obviously just (Zd), the dual of the property (Z) from Corollary 7.1.82 (ii).*

Corollary 7.1.85 *(See Lemma 7.1.14) (m=1)*

If (a_0, a_1) and (b_0) are good sequences (i.e. $a_0 \oplus a_1 = a_0$), then

(i) $(a_0, a_1) + (b_0) = (c_0, c_1, c_2)$,

where: $c_0 = a_0 \oplus b_0$, $c_1 = a_0 \odot (a_1 \oplus b_0)$, $c_2 = a_1 \odot b_0$.

(ii) $a_0 \oplus a_1 \oplus b_0 = c_0 \oplus c_1 \oplus c_2$.

(iii) $a_0 \odot a_1 \odot b_0 = c_0 \odot c_1 \odot c_2$.

Proof. (i): By Lemma 7.1.14, since $a_0 \oplus a_1 = a_0$.

(ii): By Corollary 7.1.33, $(a_0, a_1) = (a_0)+(a_1)$ and $(c_0, c_1, c_2) = (c_0)+(c_1)+(c_2)$; hence, $(a_0) + (a_1) + (b_0) = (c_0) + (c_1) + (c_2)$; then, apply Theorem 7.1.81 (i).

(iii): Similarly, using Theorem 7.1.81 (ii). □

Remark 7.1.86 *(See Remark 2.2.40)*

The property from Corollary 7.1.85 (ii) is the property denoted (Z^{10}) in subsection 2.2.3 and proved, by a complicated proof, to be verified in any XY algebra $\mathcal{A}$. We have here a simple alternative proof, in $G^+(A)$, of the fact that (Z^{10}) holds in $\mathcal{A}$. More, we have seen that (Z^{10}) $\iff$ (Z) (Corollary 2.2.22).

Remarks 7.1.87

(1) The property from Corollary 7.1.85 (ii) is:

(Z^{10}) $a_0 \oplus a_1 = a_0 \Longrightarrow a_0 \oplus a_1 \oplus b_0 = [a_0 \oplus b_0] \oplus [a_0 \odot [a_1 \oplus b_0]] \oplus [a_1 \odot b_0]$.

(2) The dual property of the property from Corollary 7.1.85 (ii) is:

($Z^{10}d$) $a_0 \odot a_1 = a_0 \Longrightarrow a_0 \odot a_1 \odot b_0 = [a_0 \odot b_0] \odot [a_0 \oplus [a_1 \odot b_0]] \odot [a_1 \oplus b_0]$ (D).

(3) The property from Corollary 7.1.85 (iii) is (since $a_0 \oplus a_1 = a_0 \iff a_0 \odot a_1 = a_1$):

$a_0 \odot a_1 = a_1 \Longrightarrow a_0 \odot a_1 \odot b_0 = [a_0 \oplus b_0] \odot [a_0 \odot [a_1 \oplus b_0]] \odot [a_1 \odot b_0]$ (H).

(4) Denote $D \overset{notation}{=} [a_0 \odot b_0] \odot [a_0 \oplus [a_1 \odot b_0]] \odot [a_1 \oplus b_0]$, the extreme right side of (2) and

denote $H \overset{notation}{=} [a_0 \oplus b_0] \odot [a_0 \odot [a_1 \oplus b_0]] \odot [a_1 \odot b_0]$, the extreme right side of (3).

We shall prove, in the next Proposition 7.1.89, that, in any XY algebra, the property (DH) holds, where:

(DH) $a_0 \odot a_1 = a_0 \Longrightarrow a_0 \odot a_1 \odot b_0 = H$.

It follows that the following property holds:

$a_0 \odot a_1 = a_0 \Longrightarrow D = H$

i.e. we can say that, roughly speaking, Corollary 7.1.85 (iii) is the dual of Corollary 7.1.85 (ii), abstraction made of the hypotheses, which are different.

Recall the properties (Z) and (Zd) verified by any right-XY algebra $\mathcal{A} = (A, \oplus, ^-, 0)$ (Corollary 2.2.37):

(Z) $(x \oplus y) \oplus (x \odot y) = x \oplus y$ and, dually,

(Zd) $(x \odot y) \odot (x \oplus y) = x \odot y$.

Lemma 7.1.88 *We have:*

$(Z) + (Zd) + (Scomm) + (Sass) + (Pcomm) + (Pass) + (DN) \Longrightarrow$

(DH) $x \odot y = x \Longrightarrow x \odot y \odot z = (x \oplus z) \odot (x \odot (y \oplus z)) \odot (y \odot z)$.

Proof. (Based on a proof by PROVER9 of Length 31)

First, note that, if $x \odot y = x$, then $x \odot y \odot z = x \odot z$.

Then, note that, from (Pcomm) and (Pass), we obtain:

$$(7.19) \qquad x \odot (y \odot z) = y \odot (x \odot z).$$

Suppose we have:

$$(7.20) \qquad c1 \odot c2 = c1.$$

By (7.20) and by (Pass), we obtain:

$$(7.21) \qquad c1 \odot x = c1 \odot (c2 \odot x).$$

We have to prove:
$c1 \odot c3 = (c1 \oplus c3) \odot (c1 \odot (c2 \oplus c3)) \odot (c2 \odot c3)$,
where: $(c1 \oplus c3) \odot [(c1 \odot (c2 \oplus c3)) \odot (c2 \odot c3)]$
$\overset{(7.19)}{=} (c1 \odot (c2 \oplus c3)) \odot [(c1 \oplus c3) \odot (c2 \odot c3)]$
$\overset{(Pass)}{=} c1 \odot ((c2 \oplus c3) \odot [(c1 \oplus c3) \odot (c2 \odot c3)])$
$\overset{(Pcomm)}{=} c1 \odot ((c2 \oplus c3) \odot [(c2 \odot c3) \odot (c1 \oplus c3)])$
$\overset{(7.19)}{=} c1 \odot ((c2 \odot c3) \odot [(c2 \oplus c3) \odot (c1 \oplus c3)])$
$\overset{(Pcomm)}{=} c1 \odot ((c2 \odot c3) \odot [(c1 \oplus c3) \odot (c2 \oplus c3)])$
$\overset{(Pass)}{=} c1 \odot (c2 \odot (c3 \odot [(c1 \oplus c3) \odot (c2 \oplus c3)]))$
$\overset{(7.21)}{=} c1 \odot (c3 \odot [(c1 \oplus c3) \odot (c2 \oplus c3)])$.
hence, we have to prove that:

$$(7.22) \qquad c1 \odot (c3 \odot [(c1 \oplus c3) \odot (c2 \oplus c3)]) = c1 \odot c3.$$

Indeed, first, from (Zd) ($(x \odot y) \odot (x \oplus y) = x \odot y$), by (Pass), we obtain:

$$(7.23) \qquad x \odot (y \odot (x \oplus y)) = x \odot y.$$

Then, in (Pass) ($(x \odot y) \odot z = x \odot (y \odot z)$), take $Y := y \odot (x \oplus y)$, to obtain:
$(x \odot [y \odot (x \oplus y)]) \odot z = x \odot ([y \odot (x \oplus y)] \odot z)$,
which, by (7.23), becomes:
$(x \odot y) \odot z = x \odot ([y \odot (x \oplus y)] \odot z)$,
which, by (Pass) twice, becomes:

$$(7.24) \qquad x \odot (y \odot z) = x \odot (y \odot [(x \oplus y) \odot z]).$$

Now, in (7.21), take $X := x \odot (c2 \oplus x)$, to obtain:
$c1 \odot [x \odot (c2 \oplus x)] = c1 \odot (c2 \odot [x \odot (c2 \oplus x)])$,
which, by (7.23) on right side, becomes:
$c1 \odot [x \odot (c2 \oplus x)] = c1 \odot (c2 \odot x)$,
which, by (7.21), becomes:

$$(7.25) \qquad c1 \odot [x \odot (c2 \oplus x)] = c1 \odot x.$$

Finally, we can prove that (7.22) holds.

Indeed, $c1 \odot (c3 \odot [(c1 \oplus c3) \odot (c2 \oplus c3)]) \overset{(7.24)}{=} c1 \odot (c3 \odot (c2 \oplus c3)) \overset{(7.25)}{=} c1 \odot c3$. □

Proposition 7.1.89 *(See Remarks 7.1.87)*
Let $\mathcal{A} = (A, \oplus, ^-, 0)$ be a right-XY algebra. Then, (DH) holds.

Proof. By Lemma 7.1.88. □

Corollary 7.1.90 *(See Lemma 7.1.15) (m=2)*
If (a_0, a_1, a_2) and (b_0) are good sequences (i.e. $a_0 \oplus a_1 = a_0$, $a_1 \oplus a_2 = a_1$), then
(i) $(a_0, a_1, a_2) + (b_0) = (c_0, c_1, c_2, c_3)$,
where: $c_0 = a_0 \oplus b_0$, $c_1 = a_0 \odot (a_1 \oplus b_0)$, $c_2 = a_1 \odot (a_2 \oplus b_0)$, $c_3 = a_2 \odot b_0$.
(ii) $a_0 \oplus a_1 \oplus a_2 \oplus b_0 = c_0 \oplus c_1 \oplus c_2 \oplus c_3$.
(iii) $a_0 \odot a_1 \odot a_2 \odot b_0 = c_0 \odot c_1 \odot c_2 \odot c_3$.

Proof. (i): By Lemma 7.1.15, since $a_0 \oplus a_1 = a_0$ and $a_1 \oplus a_2 = a_1$.
(ii): By Corollary 7.1.33, $(a_0, a_1, a_2) = (a_0) + (a_1) + (a_2)$ and $(c_0, c_1, c_2, c_3) = (c_0) + (c_1) + (c_2) + (c_3)$; hence, $(a_0) + (a_1) + (a_2) + (b_0) = (c_0) + (c_1) + (c_2) + (c_3)$; then, apply Theorem 7.1.81 (i).
(iii): Similarly. □

Remark 7.1.91 *(See Remark 2.2.42)*
The property from Corollary 7.1.90 (ii) is the property denoted (Z^{20}) in subsection 2.2.3 and proved, by a complicated proof, to be verified in any XY algebra $\mathcal{A}$. We have here a simple alternative proof in $G^+(A)$ of the fact that (Z^{20}) holds in $\mathcal{A}$. More, we have seen that $(Z^{20}) \Longleftrightarrow (Z)$ (Corollary 2.2.25).

Corollary 7.1.92 *(See Lemma 7.1.16) (m=n)*
If $(a_0, a_1, \ldots, a_n)$ and (b_0) are good sequences (i.e. $a_i \oplus a_{i+1} = a_i$, $i = 0, 1, \ldots, n-1$), then
(i) $(a_0, a_1, \ldots, a_n) + (b_0) = (c_0, c_1, \ldots, c_n, c_{n+1})$,
where: $c_i = a_{i-1} \odot (a_i \oplus b_0)$, for $i = 0, 1, \ldots, n+1$, with $a_{-1} = 1$.
(ii) $a_0 \oplus a_1 \oplus \ldots \oplus a_n \oplus b_0 = c_0 \oplus c_1 \oplus \ldots \oplus c_n \oplus c_{n+1}$.
(iii) $a_0 \odot a_1 \odot \ldots \odot a_n \odot b_0 = c_0 \odot c_1 \odot \ldots \odot c_n \odot c_{n+1}$.

Proof. (i): By Lemma 7.1.16, since $a_i \oplus a_{i+1} = a_i$, for $i = 0, 1, \ldots, n-1$.
(ii): By Corollary 7.1.33, $(a_0, a_1, \ldots, a_n) = (a_0) + (a_1) + \ldots + (a_n)$ and $(c_0, c_1, \ldots, c_n, c_{n+1}) = (c_0) + (c_1) + \ldots + (c_n) + (c_{n+1})$;
hence, $(a_0) + (a_1) + \ldots + (a_n) + (b_0) = (c_0) + (c_1) + \ldots + (c_n) + (c_{n+1})$;
then, apply Theorem 7.1.81 (i).
(iii): Similarly. □

Remark 7.1.93 *(See Remark 2.2.43)*
The property from Corollary 7.1.92 (ii) is the property denoted (Z^{n0}) in subsection 2.2.3 and not proved there to be verified in any XY algebra $\mathcal{A}$. We have here a simple indirect proof in $G^+(A)$ of the fact that (Z^{n0}) holds in $\mathcal{A}$. More, we have seen that $(Z^{n0}) \Longleftrightarrow (Z)$ (Corollary 2.2.28).

Corollary 7.1.94 *(See Lemma 7.1.17)*
If (a_0, a_1) *and* (b_0, b_1) *are good sequences (i.e.* $a_0 \oplus a_1 = a_0$, $b_0 \oplus b_1 = b_0$*), then*
(i) $(a_0, a_1) + (b_0, b_1) = (c_0, c_1, c_2, c_3)$,
where: $c_0 = a_0 \oplus b_0$, $c_1 = (a_0 \oplus b_1) \odot (a_1 \oplus b_0)$, $c_2 = a_0 \odot b_0 \odot (a_1 \oplus b_1)$, $c_3 = a_1 \odot b_1$.
(ii) $a_0 \oplus a_1 \oplus b_0 \oplus b_1 = c_0 \oplus c_1 \oplus c_2 \oplus c_3$.
(iii) $a_0 \odot a_1 \odot b_0 \odot b_1 = c_0 \odot c_1 \odot c_2 \odot c_3$.

Proof. (i): By Lemma 7.1.17, since $a_0 \oplus a_1 = a_0$ and $b_0 \oplus b_1 = b_0$.
(ii): By Corollary 7.1.33, $(a_0, a_1) = (a_0) + (a_1)$, $(b_0, b_1) = (b_0) + (b_1)$ and $(c_0, c_1, c_2, c_3) = (c_0) + (c_1) + (c_2) + (c_3)$;
hence, $(a_0) + (a_1) + (b_0) + (b_1) = (c_0) + (c_1) + (c_2) + (c_3)$;
then, apply Theorem 7.1.81 (i).
(iii): Similarly. □

Remark 7.1.95 *(See Remark 2.2.45)*
The property from Corollary 7.1.94 (ii) is the property denoted (Z^{11}) *in subsection 2.2.3 and proved, by a complicated proof, to be verified in any XY algebra* $\mathcal{A}$. *You have here a simple alternative proof in* $G^+(A)$ *of the fact that* (Z^{11}) *holds in* $\mathcal{A}$. *More, we have seen that* $(Z^{11}) \iff (Z)$ *(Corollary 2.2.32).*

7.2 Morphisms of strong m_0-ME positive cones

For a morphism h of right-XY algebras $\mathcal{A} = (A, \oplus, ^-, 0_A)$ and $\mathcal{B} = (B, \oplus, ^-, 0_B)$

$$h : A \longrightarrow B$$

we set

$$\mathbf{G}^+_\mathbf{m}(h) : G^+(A) \longrightarrow G^+(B)$$

defined as follows: for a good sequence $\mathbf{x} = (x_0, x_1, \ldots, x_n) \in G^+(A)$,

$$\mathbf{x} = (x_0, x_1, \ldots, x_n) \longmapsto (h(x_0), h(x_1), \ldots, h(x_n)), \tag{7.26}$$

i.e. $\mathbf{G}^+_\mathbf{m}(h)(\mathbf{x}) \stackrel{def.}{=} (h(x_0), h(x_1), \ldots, h(x_n))$;
in particular, for all $x \in A$,

$$\mathbf{G}^+_\mathbf{m}(h)(\mathbf{G}^+_\mathbf{m}(x)) \stackrel{def.}{=} \mathbf{G}^+_\mathbf{m}(h(x)), \tag{7.27}$$

i.e. we have the following commutative diagram:

$$\begin{array}{ccc} G^+(A) & \xleftarrow{\quad \mathbf{G}^+_\mathbf{m} \quad} & A \\ \mathbf{G}^+_\mathbf{m}(h) \downarrow & & \downarrow h \\ G^+(B) & \xleftarrow[\quad \mathbf{G}^+_\mathbf{m} \quad]{} & B \end{array}$$

Proposition 7.2.1 *The function $\mathbf{G}^+_{\mathbf{m}}(h)$ is a morphism of strong m_0-ME positive cones.*

Proof. • The function $\mathbf{G}^+_{\mathbf{m}}(h)$ is well defined.
Indeed, if $\mathbf{x}$=$\mathbf{y}$, i.e. $(x_0, x_1, \ldots, x_n) = (y_0, y_1, \ldots, y_n)$ in $G^+(A)$,
hence $x_i = y_i$, $i = 0, 1, \ldots, n$, in A,
then $h(x_i) = h(y_i)$, in B,
hence $(h(x_0), h(x_1), \ldots, h(x_n)) = (h(y_0), h(y_1), \ldots, h(y_n))$ in $G^+(B)$,
i.e. $\mathbf{G}^+_{\mathbf{m}}(h)(\mathbf{x}) = G^+(h)(\mathbf{y})$.

• The function $\mathbf{G}^+_{\mathbf{m}}(h)$ is a homomorphism of strong m_0-ME positive cones, i.e. (p1) - (p5) hold.
Indeed, first note that (H1) - (H5) hold. Then, we have:

(p1): $\mathbf{G}^+_{\mathbf{m}}(h)(\mathbf{x}+\mathbf{y}) = \mathbf{G}^+_{\mathbf{m}}(h)(\mathbf{x}) + \mathbf{G}^+(h)(\mathbf{y})$.
Indeed, if $\mathbf{x} = (x_0, x_1, \ldots, x_n)$ and $\mathbf{y} = (y_0, y_1, \ldots, y_n)$,
let $\mathbf{z} = \mathbf{x} + \mathbf{y} = (z_0, z_1, \ldots z_{2n+1})$, where:

$$z_n \overset{(7.2)}{=} (x_0 \oplus y_n) \odot (x_1 \oplus y_{n-1}) \odot \ldots \odot (x_n \oplus y_0).$$

Let $\mathbf{u} = \mathbf{G}^+(h)(\mathbf{x}+\mathbf{y})$
$= \mathbf{G}^+_{\mathbf{m}}(h)(\mathbf{z}) \overset{def.}{=} (h(z_0), h(z_1), \ldots, h(z_{2n+1}))$
$= (u_0, u_1, \ldots, u_{2n+1})$.

Let $\mathbf{w} = \mathbf{G}^+_{\mathbf{m}}(h)(\mathbf{x}) + \mathbf{G}^+_{\mathbf{m}}(h)(\mathbf{y})$
$= (h(x_0), h(x_1), \ldots, h(x_n)) + (h(y_0), h(y_1), \ldots, h(y_n))$
$= (w_0, w_1, \ldots, w_{2n+1})$, where:

$$w_n \overset{(7.2)}{=} (h(x_0) \oplus h(y_n)) \odot (h(x_1) \oplus h(y_{n-1})) \odot \ldots \odot (h(x_n) \oplus h(y_0)).$$

We must prove that $\mathbf{u}$=$\mathbf{w}$, i.e. that $u_i = w_i$, $i = 0, 1, \ldots 2n+1$.
Indeed, $u_n = h(z_n) = h((x_0 \oplus y_n) \odot (x_1 \oplus y_{n-1}) \odot \ldots \odot (x_n \oplus y_0))$
$\overset{(H1')}{=} h(x_0 \oplus y_n) \odot h(x_1 \oplus y_{n-1}) \odot \ldots \odot h(x_n \oplus y_0)$
$\overset{(H1)}{=} (h(x_0) \oplus h(y_n)) \odot (h(x_1) \oplus h(y_{n-1})) \odot \ldots \odot (h(x_n) \oplus h(y_0)) = w_n$.

(p2): $\mathbf{G}^+_{\mathbf{m}}(h)(\smile \mathbf{x}) = \smile \mathbf{G}^+_{\mathbf{m}}(h)(\mathbf{x})$.
Indeed, $\mathbf{G}^+_{\mathbf{m}}(h)(\smile \mathbf{x}) = \mathbf{G}^+_{\mathbf{m}}(h)(\smile (x_0, x_1, \ldots, x_n))$
$= \mathbf{G}^+_{\mathbf{m}}(h)((x_n^-, \ldots, x_1^-, x_0^-))$, by definition of $\smile$ in $G^+(A)$
$= (h(x_n^-), \ldots, h(x_1^-), h(x_0^-))$, by definition of $\mathbf{G}^+_{\mathbf{m}}(h)$,
$= (h(x_n)^-, \ldots, h(x_1)^-, h(x_0)^-)$, by (H2),
$= \smile (h(x_0), h(x_1), \ldots, h(x_n))$, by definition of $\smile$ in $G^+(B)$,
$= \smile \mathbf{G}^+_{\mathbf{m}}(h)((x_0, x_1, \ldots, x_n))$, by definition of $\mathbf{G}^+_{\mathbf{m}}(h)$,
$= \smile \mathbf{G}^+_{\mathbf{m}}(h)(\mathbf{x})$.

(p3): $\mathbf{G}^+_{\mathbf{m}}(h)(0_{G^+(A)}) = 0_{G^+(B)}$.
Indeed, $\mathbf{G}^+_{\mathbf{m}}(h)(0_{G^+(A)}) = \mathbf{G}^+_{\mathbf{m}}(h)((0_A))$, by definition of $0_{G^+(A)}$,
$= (h(0_A))$, by definition of $\mathbf{G}^+_{\mathbf{m}}(h)$,
$= (0_B)$, by (H3),
$= 0_{G^+(B)}$, by definition of $0_{G^+(B)}$.

(p4): If $\mathbf{x} \leq_m \mathbf{y}$, then $\mathbf{G}^+_{\mathbf{m}}(h)(\mathbf{x}) \leq_m \mathbf{G}^+_{\mathbf{m}}(h)(\mathbf{y})$.
Indeed, if $\mathbf{x} \leq_m \mathbf{y}$, i.e. $(x_0, x_1, \ldots, x_n) \leq_m (y_0, y_1, \ldots, y_n)$ in $\mathbf{G}^+_{\mathbf{m}}(\mathcal{A})$,

i.e. $x_i \leq_m y_i$, $i = 0, 1, \ldots, n$, in $\mathcal{A}$,
then $h(x_i) \leq_m h(y_i)$, $i = 0, 1, \ldots, n$, in $\mathcal{B}$, by (H4),
i.e. $(h(x_0), h(x_1), \ldots, h(x_n)) \leq_m (h(y_0), h(y_1), \ldots, h(y_n))$ in $\mathbf{G}^+_{\mathbf{m}}(\mathcal{B})$,
i.e. $\mathbf{G}^+_{\mathbf{m}}(h)(\mathbf{x}) \leq_m \mathbf{G}^+_{\mathbf{m}}(h)(\mathbf{y})$.

(p5): $\mathbf{G}^+_{\mathbf{m}}(h)(u_{G^+(A)}) = u_{G^+(B)}$.
Indeed, $\mathbf{G}^+_{\mathbf{m}}(h)(u_{G^+(A)}) = \mathbf{G}^+_{\mathbf{m}}(h)((1_A))$, by definition of $u_{G^+(A)}$,
$= (h(1_A))$, by definition of $\mathbf{G}^+_{\mathbf{m}}(h)$,
$= (1_B)$, by (H5),
$= u_{G^+(B)}$, by definition of $u_{G^+(B)}$. □

This establishes that $\mathbf{G}^+_{\mathbf{m}}$ is a functor

$$\mathbf{G}^+_{\mathbf{m}} : \mathcal{C}(\mathbf{XY^R}) \longrightarrow \mathcal{C}(\mathbf{s} - \mathbf{m_0} - \mathbf{ME}^+)$$

that maps $\mathcal{A}$ to $\mathbf{G}^+_{\mathbf{m}}(\mathcal{A})$ and maps a morphism $h : A \longrightarrow B$ to the morphism

$$\mathbf{G}^+_{\mathbf{m}}(h) : G^+(A) \longrightarrow G^+(B).$$

Chapter 8

Construction of the XY algebra $\mathbf{U_m}(\mathcal{M}_u^+)$. The unit interval functor $\mathbf{U_m}$

In this chapter, we introduce and study the unit interval functor $\mathbf{U_m}$ which constructs, from a strong m_0-ME positive cone $\mathcal{M}_u^+$, a right-XY algebra $\mathbf{U_m}(\mathcal{M}_u^+)$.

8.1 Construction of the XY algebra $\mathrm{U}(\mathcal{M}_u^+)$ from the strong m_0-ME positive cone $\mathcal{M}_u^+$

Let $\mathcal{M}_u^+ = (M^+, \leq_m, +, \smile, 0_{M^+}, u = u_{M^+})$ be a strong m_0-ME positive cone (Definition 3.3.1 (1)) throughout this section.

The goal of this section, reached in Theorem 8.1.2 below, is to build from $\mathcal{M}_u^+$ a right-XY algebra (Definition 2.2.13 (1)).

Let

$$[0_{M^+}, u_{M^+}] \overset{def.}{=} \{x \in M^+ \mid 0_{M^+} \leq_m x \leq_m u_{M^+}\} \subset M^+$$

be the unit interval in $\mathcal{M}_u^+$. We shall endow $[0_{M^+}, u_{M^+}]$ with a structure of right-XY algebra.

Clearly, 0_{M^+}, $u_{M^+} \in [0_{M^+}, u_{M^+}]$ by Lemma 7.1.54. Moreover, we define $\leq_m$ on $[0_M, u_M]$ by restriction of $\leq_m$ from $\mathcal{M}_u^+$.

Definitions 8.1.1 (See the analogous ([10], Definition 2.1.1))

For all $x, y \in [0_{M^+}, u_{M^+}]$, define, by ($\mathrm{s}3^R$) (i),

$$x \oplus y \overset{def.}{=} \inf_m(u_{M^+}, x + y), \tag{8.1}$$

$$\neg x \overset{def.}{=} \smile x, \quad x \neq 0_{M^+}, \tag{8.2}$$

(8.3)
$(Neg0\text{–}1)$ $$\neg 0_{M^+} \stackrel{def.}{=} u_{M^+}.$$

We must check that $x \oplus y,\ \neg x \in [0_{M^+}, u_{M^+}]$. Indeed, $x \oplus y \leq_m u_{M^+}$, by definition, and $x \oplus y \geq_m 0_{M^+}$, as element of M^+. Also, $\neg x = \smile x$ and $\smile x \leq_m u_{M^+}$, by (s1^R), since $x \leq_m u_{M^+}$, and $\smile x \geq_m 0_{M^+}$, as element of M^+.

Theorem 8.1.2 *(See the analogous ([10], Proposition 2.1.2))*
The algebra $([0_{M^+}, u_{M^+}], \oplus, \neg, 0_{M^+})$ *is a right-XY algebra.*

Proof. For all $x, y, z \in [0_{M^+}, u = u_{M^+}]$, we have:
(SU): $x \oplus 0_{M^+} = x$. Indeed,
$x \oplus 0_{M^+} = \inf_m(u, x + 0_{M^+}) = \inf_m(u, x) = x$, by (SU) in $\mathcal{M}_u^+$.
(Scomm): $x \oplus y = y \oplus x$. Indeed,
$x \oplus y = \inf_m(u, x + y) = \inf_m(u, y + x) = y \oplus x$, by (Scomm) in $\mathcal{M}_u^+$.
(Sass): $(x \oplus y) \oplus z = x \oplus (y \oplus z)$. Indeed,

$$\begin{array}{rcll}
x \oplus (y \oplus z) & \stackrel{def.}{=} & x \oplus \inf_m(u_{M^+}, y + z) & \\
& \stackrel{def.}{=} & \inf_m(u, x + \inf_m(u, y + z)) & \\
& = & \inf_m(u, x + (y + z)), & \text{by } (s4^R) \text{ in } \mathcal{M}_u^+, \\
& = & \inf_m(u, (x + y) + z), & \text{by (Sass) in } \mathcal{M}_u^+, \\
& = & \inf_m(u, z + (x + y)), & \text{by (Scomm) in } \mathcal{M}_u^+, \\
& = & \inf_m(u, z + \inf_m(u, x + y)), & \text{by } (s4^R) \text{ in } \mathcal{M}_u^+, \\
& \stackrel{def.}{=} & z \oplus \inf_m(u, x + y) & \\
& \stackrel{def.}{=} & z \oplus (x \oplus y) & \\
& = & (x \oplus y) \oplus z, & \text{by above (Scomm).}
\end{array}$$

(Neg1-0): $\neg u_{M^+} = 0_{M^+}$. Indeed,
$\neg u_{M^+} = \smile u_{M^+} = 0_{M^+}$, by (Nu^{+1R}) in $\mathcal{M}_u^+$.
(m-La^R): $x \oplus u_{M^+} = u_{M^+}$. Indeed,
$x \oplus u_{M^+} = \inf_m(u, x + u_{M^+}) = u_{M^+}$, by (s3^R) (ii).
A direct proof:
$0_{M^+} \leq_m x$ implies $0_{M^+} + u \leq_m x + u$, by (C_{pu}^{1R}) in $\mathcal{M}_u^+$;
hence, $u \leq_m x + u$, by (SU) in $\mathcal{M}_u^+$;
hence, $x \oplus u = \inf_m(u, x + u) = \min_m(u, x + u) = u$.
(DN): If $x \neq u_{M^+}$, $\neg(\neg x) = \smile(\smile x) = x$, by (DN) in $\mathcal{M}_u^+$, and $\neg(\neg u_{M^+}) = \smile(\smile u_{M^+}) = u_{M^+}$, by (NNu^{1R}) in $\mathcal{M}_u^+$.
(XX): $(x \oplus y) \odot ((x \odot y) \oplus z) = (y \oplus z) \odot ((y \odot z) \oplus x)$. Indeed,

$$\begin{array}{rcl}
A & \stackrel{not.}{=} & (x \oplus y) \odot ((x \odot y) \oplus z) \\
& = & \neg(\neg(x \oplus y) \oplus \neg((x \odot y) \oplus z)) \\
& = & \neg(\neg(x \oplus y) \oplus \neg(\neg(\neg x \oplus \neg y) \oplus z)) \\
& = & \smile(\smile(x \oplus y) \oplus \smile(\smile(\smile x \oplus \smile y) \oplus z)) \\
& = & \smile(\smile \inf_m(u, x + y) \oplus \smile(\smile \inf_m(u, \smile x \smile y) \oplus z)) \\
& = & \smile(\smile \inf_m(u, x + y) \oplus \smile \inf_m(u, \smile \inf_m(u, \smile x \smile y) + z)) \\
& = & \smile \inf_m(u, \smile \inf_m(u, x + y) \smile \inf_m(u, \smile \inf_m(u, \smile x \smile y) + z))
\end{array}$$

and, similarly,

$$B \overset{not.}{=} (y \oplus z) \odot ((y \odot z) \oplus x)$$
$$= \smile \inf_m(u, \smile \inf_m(u, y+z) \smile \inf_m(u, \smile \inf_m(u, \smile y \smile z) + x)).$$

But, by (XX_u^R) in $\mathcal{M}_u^+$,

$\smile \inf_m(u, \smile \inf_m(u, x+y) \smile \inf_m(u, z \smile \inf_m(u, \smile x \smile y)))$
$= \smile \inf_m(u, \smile \inf_m(u, y+z) \smile \inf_m(u, x \smile \inf_m(u, \smile y \smile z))),$
hence, by (Scomm), $A = B$.

(YY): $x \oplus y = x \implies x \odot (y \oplus z) = y \oplus (x \odot z)$. Indeed,
If $x \oplus y = x$, i.e. if $\inf_m(u, x+y) = x$, then, since we have:

$$\begin{aligned} x \odot (y \oplus z) &= x \odot \inf_m(u, y+z) \\ &= \neg(\neg x \oplus \neg(\inf_m(u, y+z)) \\ &= \smile(\smile x \oplus \smile \inf_m(u, y+z)) \\ &= \smile \inf_m(u_{M^+}, \smile x \smile \inf_m(u, y+z)) \end{aligned}$$

and

$$\begin{aligned} y \oplus (x \odot z) &= y \oplus \neg(\neg x \oplus \neg z) \\ &= y \oplus \smile(\smile x \oplus \smile z) \\ &= y \oplus \smile \inf_m(u, \smile x \smile z) \\ &= \inf_m(u, y \smile \inf_m(u, \smile x \smile z)); \end{aligned}$$

it follows, by (YY_u^R) in $\mathcal{M}_u^+$, that:
$\smile \inf_m(u, \smile x \smile \inf_m(u, y+z)) = \inf_m(u_{M^+}, y \smile \inf_m(u, \smile x \smile z))$, hence
$x \odot (y \oplus z) = y \oplus (x \odot z)$. Thus, (YY) holds. □

Corollary 8.1.3 *For all* $x, y \in [0_{M^+}, u = u_{M^+}]$,

$$x \odot y = \smile \inf_m(u, \smile x \smile y).$$

Proof. $x \odot y = \neg(\neg x \oplus \neg y) = \smile(\smile x \oplus \smile y) = \smile \inf_m(u, \smile x \smile y)$. □

Corollary 8.1.4 *For all* $x_0, x_1, \ldots x_n \in [0_{M^+}, u = u_{M^+}]$,

$$x_0 \oplus x_1 \oplus \ldots \oplus x_n = \inf_m(u, x_0 + x_1 + \ldots + x_n).$$

Proof. By induction on $n \geq 1$.

- The statement is true for $n = 1$, by (8.1): $x_0 \oplus x_1 \overset{def.}{=} \inf_m(u, x_0 + x_1)$.
- Suppose that the statement is true for $n-1$, i.e. we have:
$x_0 \oplus x_1 \oplus \ldots \oplus x_{n-1} = \inf_m(u, x_0 + x_1 + \ldots + x_{n-1})$.
- We shall prove that the statement is true for n. Indeed,

$x_0 \oplus x_1 \oplus \ldots \oplus x_{n-1} \oplus x_n$
$= [x_0 \oplus x_1 \oplus \ldots \oplus x_{n-1}] \oplus x_n$, by (Sass) in A,
$= x_n \oplus [x_0 \oplus x_1 \oplus \ldots \oplus x_{n-1}]$, by (Scomm) in A,
$= x_n \oplus \inf_m(u, x_0 + x_1 + \ldots + x_{n-1})$, by the inductive hypothesis,
$= \inf_m(u, x_n + \inf_m(u, x_0 + x_1 + \ldots + x_{n-1}))$, by (8.1),
$= \inf_m(u, x_n + [x_0 + x_1 + \ldots + x_{n-1}])$, by Corollary 3.3.15,
$= \inf_m(u, x_0 + x_1 + \ldots + x_{n-1} + x_n)$, by (Scomm), (Sass) in $\mathcal{M}_u^+$. □

Corollary 8.1.5 *In the right-XY algebra* $([0_{M^+}, u = u_{M^+}], \oplus, \neg, 0_{M^+})$, *the property (Z) holds.*

Proof 1. By Corollary 2.2.37.
Proof 2. (Z), i.e. $(x \oplus y) \oplus (x \odot y) = x \oplus y$ means, by (8.1):
$\inf_m(u, x \oplus y + x \odot y) = \inf_m(u, x + y)$,
hence, by Corollary 8.1.3,
$\inf_m(u, \inf_m(u, x + y) + \smile \inf_m(u, \smile x \smile y)) = \inf_m(u, x + y)$,
which is (Z_u^R), that holds in $\mathcal{M}_u^+$ (see Corollary 3.3.3). □

Proposition 8.1.6 *In the right-XY algebra* $([0_{M^+}, u = u_{M^+}], \oplus, \neg, 0_{M^+})$, *the basic internal binary relation* $\leq'_m$ *coincides with the binary relation* $\leq_m$ *inherited from* $\mathcal{M}_u^+$.

Proof. For $x, y \in [0_{M^+}, u]$,

$$\begin{array}{llll} x \leq'_m y & \stackrel{(2.2)}{\Longleftrightarrow} & y \oplus \neg x = u & \\ & \Longleftrightarrow & \inf_m(u, y \smile x) = u & \\ & \Longleftrightarrow & u \leq_m y \smile x & \\ & \Longleftrightarrow & x \leq_m y, & \text{by } (s3^R)(iii) \text{ in } \mathcal{M}_u^+. \end{array}$$

□

- **The unit interval functor $\mathbf{U_m}$**

We put:

$$\mathbf{U_m}(\mathcal{M}_u^+) = ([0_{M^+}, u_{M^+}], \oplus, \neg, 0_{M^+}),$$

i.e. we have:

$$\begin{array}{ccc} \mathcal{C}(\textbf{s-m}_0\textbf{-ME}^+) & \stackrel{\mathbf{U_m}}{\longrightarrow} & \mathcal{C}(\mathbf{XY}^R) \\ \mathcal{M}_u^+ & \stackrel{\mathbf{U_m}}{\longrightarrow} & \mathbf{U_m}(\mathcal{M}_u^+) \end{array}$$

8.2 Morphisms of XY algebras

Given a morphism $f^+ : M^+ \longrightarrow N^+$ of strong m_0-ME positive cones $\mathcal{M}_u^+ = (M^+, \leq_m, +, \smile, 0_{M^+}, u = u_{M^+})$ and $\mathcal{N}_v^+ = (N^+, \leq_m, +, \smile, 0_{N^+}, v = u_{N^+})$, we denote with $\mathbf{U_m}(f^+)$ its restriction

$$\mathbf{U_m}(f^+) : \mathbf{U_m}(M^+) \longrightarrow \mathbf{U_m}(N^+),$$

defined as follows: for all $x \in M^+$,

$$\mathbf{U_m}(f^+)(\mathbf{U_m}(x)) \stackrel{def.}{=} \mathbf{U_m}(f^+(x)), \quad (8.4)$$

i.e. we have the following commutative diagram:

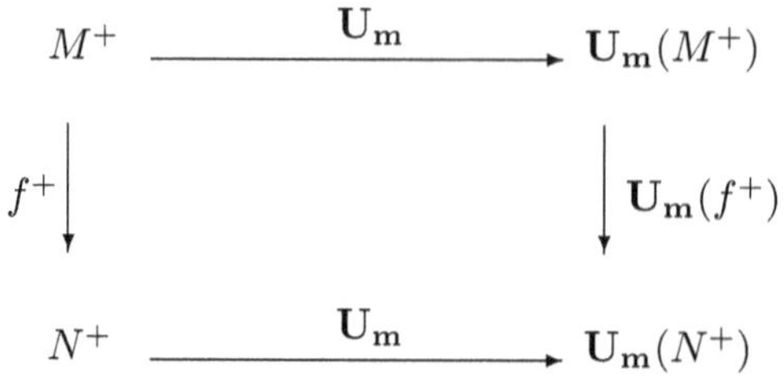

Proposition 8.2.1 *The function $\mathbf{U_m}(f^+)$ is a morphism of right-XY algebras.*

Proof. • The function $\mathbf{U_m}(f^+)$ is well defined.
Indeed, if $x = y$ in M^+, then $f^+(x) = f^+(y)$ in N^+, hence $\mathbf{U_m}(f^+(x)) = \mathbf{U_m}(f^+(y))$ in $\mathbf{U_m}(N^+)$.

• The function $\mathbf{U_m}(f^+)$ is a homomorphism of right-XY algebras, i.e. (H1) - (H3) hold.
Indeed, first note that f^+ verifies (p1) - (p5). Then, we have:

(H1): $\mathbf{U_m}(f^+)(x \oplus y) = \mathbf{U_m}(f^+)(x) \oplus \mathbf{U_m}(f^+)(y)$.

Indeed, $\mathbf{U_m}(f^+)(x \oplus y) \overset{(8.1)}{=} \mathbf{U_m}(f^+)(\mathbf{U_m}(x+y)) \overset{(8.4)}{=} \mathbf{U_m}(f^+(x+y))$
$\overset{(p1)}{=} \mathbf{U_m}(f^+(x) + f^+(y)) \overset{(8.1)}{=} f^+(x) \oplus f^+(y) = \mathbf{U_m}(f^+(x)) \oplus \mathbf{U_m}(f^+(y))$
$\overset{(8.4)}{=} \mathbf{U_m}(f^+)(\mathbf{U_m}(x)) \oplus \mathbf{U_m}(f^+)(\mathbf{U_m}(y)) = \mathbf{U_m}(f^+)(x) \oplus \mathbf{U_m}(f^+)(y)$.

(H2): $\mathbf{U_m}(f^+)(\neg x) = \neg \mathbf{U_m}(f^+)(x)$, $x \neq 0_{M^+}$
and $\mathbf{U_m}(f^+)(\neg 0_{M^+}) = \neg \mathbf{U_m}(f^+)(0_{M^+})$.

Indeed,
$\mathbf{U_m}(f^+)(\neg x) \overset{(8.2)}{=} \mathbf{U_m}(f^+)(U(\smile x)) \overset{(8.4)}{=} \mathbf{U_m}(f^+(\smile x)) \overset{(p2)}{=} \mathbf{U_m}(\smile f^+(x))$
$\overset{(8.2)}{=} \neg f^+(x) = \neg \mathbf{U_m}(f^+(x)) \overset{(8.4)}{=} \neg \mathbf{U_m}(f^+)(\mathbf{U_m}(x)) = \neg \mathbf{U_m}(f^+)(x)$
and
$\mathbf{U_m}(f^+)(\neg 0_{M^+}) \overset{(8.3)}{=} \mathbf{U_m}(f^+)(\mathbf{U_m}(u_{M^+})) \overset{(8.4)}{=} \mathbf{U_m}(f^+(u_{M^+})) \overset{(p5)}{=} \mathbf{U_m}(u_{N^+}) \overset{(8.3)}{=}$
$\neg 0_{N^+} \overset{(p3)}{=} \neg f^+(0_{M^+}) = \neg \mathbf{U_m}(f^+(0_{M^+})) \overset{(8.4)}{=} \neg \mathbf{U_m}(f^+)(\mathbf{U_m}(0_{M^+})) = \neg \mathbf{U_m}(f^+)(0_{M^+})$.

(H3): $\mathbf{U_m}(f^+)(0_{M^+}) = 0_{N^+}$.

Indeed, $\mathbf{U_m}(f^+)(0_{M^+}) = \mathbf{U_m}(f^+)(\mathbf{U_m}(0_{M^+})) \overset{(8.4)}{=} \mathbf{U_m}(f^+(0_{M^+})) \overset{(p3)}{=} \mathbf{U_m}(0_{N^+}) = 0_{N^+}$. □

This establishes that $\mathbf{U_m}$ is a functor

$$\mathbf{U_m} : \mathcal{C}(\mathbf{s} - \mathbf{m_0} - \mathbf{ME}^+) \longrightarrow \mathcal{C}(\mathbf{XY^R})$$

that maps $\mathcal{M}_u^+$ to $\mathbf{U_m}(\mathcal{M}_u^+)$ and maps a morphism $f^+ : M^+ \longrightarrow N^+$ to the morphism $\mathbf{U_m}(f^+) : \mathbf{U_m}(M^+) \longrightarrow \mathbf{U_m}(N^+)$.

Chapter 9

Connections between the functors $\mathbf{G}_{\mathbf{m}}^{+}$ and $\mathbf{U}_{\mathbf{m}}$

In this chapter, we prove (Corollary 9.3.3) that the functors $\mathbf{G}_{\mathbf{m}}^{+}$ and $\mathbf{U}_{\mathbf{m}}$ are quasi-inverses, and hence that the categories of strong m_0-ME positive cones and of right-XY algebras are equivalent.

Note that $\mathbf{U}_{\mathbf{m}}$ maps the strong m_0-ME positive cone $\mathcal{M}_u^+$ to the right-XY algebra $U(\mathcal{M}_u^+)$ and f^+ to $\mathbf{U}_{\mathbf{m}}(f^+)$.

Note also that $\mathbf{U}_{\mathbf{m}}$ maps the strong m_0-ME positive cone $\mathbf{G}_{\mathbf{m}}^{+}(\mathcal{A})$ to the right-XY algebra $\mathbf{U}_{\mathbf{m}}(\mathbf{G}_{\mathbf{m}}^{+}(\mathcal{A}))$ and $\mathbf{G}_{\mathbf{m}}^{+}(h)$ to $\mathbf{U}_{\mathbf{m}}(\mathbf{G}_{\mathbf{m}}^{+}(h))$.

Hence, we have the following situation:

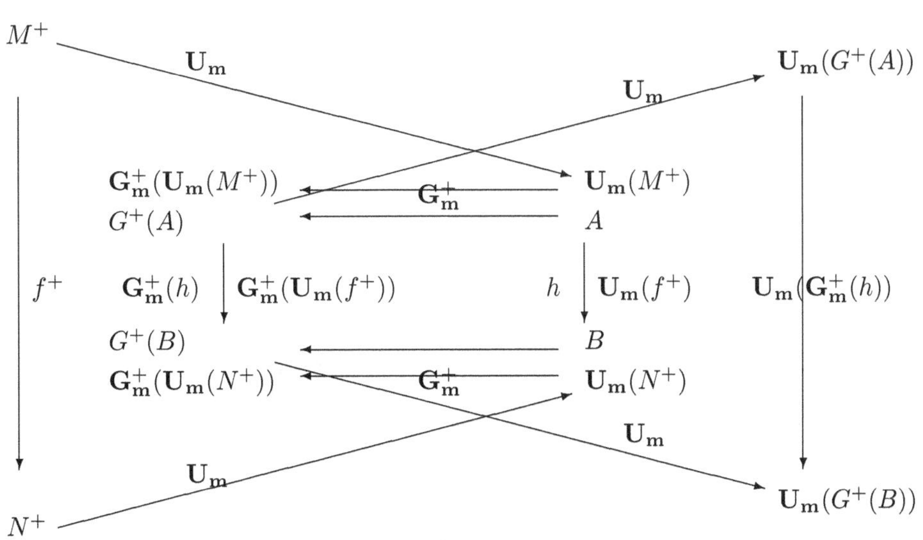

9.1 The isomorphisms α^+_u and α^+_A

9.1.1 The isomorphism α^+_u between $\mathbf{G}^+_{\mathbf{m}}(\mathbf{U}_{\mathbf{m}}(\mathcal{M}^+_u))$ and $\mathcal{M}^+_u$

For each strong m_0-ME positive cone

$$\mathcal{M}^+_u = (M^+, \leq_m, +, \smile, 0_{M^+}, u = u_{M^+}),$$

consider the right-XY algebra $\mathbf{U}_{\mathbf{m}}(\mathcal{M}^+_u)$ obtained by $\mathbf{U}_{\mathbf{m}}$

$$\mathbf{U}_{\mathbf{m}}(\mathcal{M}^+_u) = (\mathbf{U}_{\mathbf{m}}(M^+) = [0_{M^+}, u], \oplus, \neg, 0_{M^+})$$

and then the strong m_0-ME positive cone $\mathbf{G}^+_{\mathbf{m}}(\mathbf{U}_{\mathbf{m}}(\mathcal{M}^+_u))$ obtained by $\mathbf{G}^+_{\mathbf{m}}$

$$\mathbf{G}^+_{\mathbf{m}}(\mathbf{U}_{\mathbf{m}}(\mathcal{M}^+_u)) = (G^+([0_{M^+}, u]), +, \smile, (0_{M^+}), (u)).$$

Consider now the function

$$\alpha^+_u : \mathbf{G}^+_{\mathbf{m}}(\mathbf{U}_{\mathbf{m}}(M^+)) \longrightarrow M^+$$

defined by: for each $\mathbf{x} = (x_0, x_1, \ldots, x_n) \in \mathbf{G}^+_{\mathbf{m}}(\mathbf{U}_{\mathbf{m}}(M^+))$, i.e. $x_i \in \mathbf{U}_{\mathbf{m}}(M^+)$, $i = 0, 1, \ldots, n$, i.e. $0_{M^+} \leq_m x_i \leq_m u_{M^+}$, $i = 0, 1, \ldots, n$,

$$\mathbf{x} = (x_0, x_1, \ldots, x_n) \longmapsto x_0 + x_1 + \ldots + x_n, \tag{9.1}$$

i.e. $\alpha^+_u(\mathbf{x}) = \alpha^+_u((x_0, x_1, \ldots, x_n)) \overset{def.}{=} x_0 + x_1 + \ldots + x_n$;
in particular, for each $(x) \in \mathbf{G}^+_{\mathbf{m}}(\mathbf{U}_{\mathbf{m}}(M^+))$,

$$\alpha^+_u(\,(x)\,) \overset{def.}{=} x. \tag{9.2}$$

• The function α^+_u is well defined.
Indeed, if $\mathbf{x} = \mathbf{y}$, i.e. $(x_0, x_1, \ldots, x_n) = (y_0, y_1, \ldots, y_n)$ in $\mathbf{G}^+_{\mathbf{m}}(\mathbf{U}_{\mathbf{m}}(M^+))$
i.e. $x_i = y_i$, $i = 0, 1, \ldots, n$, in $\mathbf{U}_{\mathbf{m}}(M^+) = [0_{M^+}, u_{M^+}]$,
i.e. $x_i = y_i$, $i = 0, 1, \ldots, n$, in M^+,
then $x_0 + x_1 + \ldots + x_n = y_0 + y_1 + \ldots + y_n$ in M^+,
i.e. $\alpha^+_u((x_0, x_1, \ldots, x_n)) = \alpha^+_u((y_0, y_1, \ldots, y_n))$, i.e. $\alpha^+_u(\mathbf{x}) = \alpha^+_u(\mathbf{y})$.

Lemma 9.1.1 *(See Lemma 9.1.7) (m=0)*
For all (x_0) and $(y_0) \in \mathbf{G}^+_{\mathbf{m}}(\mathbf{U}_{\mathbf{m}}(M^+))$, we have:

$$\alpha^+_u((x_0) + (y_0)) = \alpha^+_u((x_0)) + \alpha^+_u((y_0)).$$

Proof. $T1 \overset{notation}{=} \alpha^+_u((x_0) + (y_0))$
$= \alpha^+_u((x_0 \oplus y_0, x_0 \odot y_0))$, by Lemma 7.1.12,
$\overset{(9.1)}{=} x_0 \oplus y_0 + x_0 \odot y_0$,
$\overset{(8.1)}{=} \inf_m(u, x_0 + y_0) + x_0 \odot y_0$
$= \inf_m(u, x_0 + y_0) \smile \inf_m(u \smile x_0 \smile y_0)$, by Corollary 8.1.3,
$\overset{(s5^R)(V^{00})}{=} x_0 + y_0$.
$T2 \overset{notation}{=} \alpha^+_u((x_0)) + \alpha^+_u((y_0)) \overset{(9.2)}{=} x_0 + y_0$.
Thus, $T1 = T2$. □

Lemma 9.1.2 *(See Lemma 9.1.8) (m=1)*
For all (x_0, x_1) and $(y_0) \in \mathbf{G}^+_\mathbf{m}(\mathbf{U_m}(M^+))$, we have:

$$\alpha_u^+((x_0, x_1) + (y_0)) = \alpha_u^+((x_0, x_1)) + \alpha_u^+((y_0)).$$

Proof. $T1 \overset{notation}{=} \alpha_u^+((x_0, x_1) + (y_0))$
$= \alpha_u^+((c_0, c_1, c_2))$
$\overset{(9.1)}{=} c_0 + c_1 + c_2$
$= x_0 \oplus y_0 + x_0 \odot (x_1 \oplus y_0) + x_1 \odot y_0$, by Lemma 7.1.14,
$= \inf_m(u, x_0 + y_0) \smile \inf_m(u, \smile x_0 \smile \inf_m(u, x_1 + y_0)) \smile \inf_m(u, \smile x_1 \smile y_0)$, by (8.1) and Corollary 8.1.3,
$\overset{(s5^R)(V^{10})}{=} [x_0 + x_1] + y_0$.
$T2 \overset{notation}{=} \alpha_u^+((x_0, x_1)) + \alpha_u^+((y_0)) \overset{(9.1),(9.2)}{=} [x_0 + x_1] + y_0$.
Thus, $T1 = T2$. □

Lemma 9.1.3 *(See Lemma 9.1.9) (m=n)*
For all $\mathbf{x} = (x_0, x_1, \ldots, x_n)$ and $(y_0) \in \mathbf{G}^+_\mathbf{m}(\mathbf{U_m}(M^+))$, we have:

$$\alpha_u^+(\mathbf{x} + (y_0)) = \alpha_u^+(\mathbf{x}) + \alpha_u^+((y_0)).$$

Proof. $T1 \overset{notation}{=} \alpha_u^+(\mathbf{x} + (y_0))$
$= \alpha_u^+(\,(x_0, x_1, \ldots, x_m) + (y_0)\,)$
$= \alpha_u^+(\,(c_0, c_1, \ldots, c_n, c_{n+1})\,)$, by Lemma 7.1.16,
$\overset{(9.1)}{=} c_0 + c_1 + \ldots + c_{n+1}$
$= \sum_{i=0}^{n+1} c_i$
$= \sum_{i=0}^{n+1} x_{i-1} \odot (x_i \oplus y_0)$, with $x_{-1} = u_{M^+}$, $x_{n+1} = 0_{M^+}$,
$= \sum_{i=0}^{n+1} x_{i-1} \odot \inf_m(u, x_i + y_i)$, by (8.1),
$= \sum_{i=0}^{n+1} \smile \inf_m(u, \smile x_{i-1} \smile \inf_m(u, x_i + y_i))$, by Corollary 8.1.3,
$= [\sum_{i=0}^{n} x_i] + y_0$, by (s5^R) (V^{n0}).
$T2 \overset{notation}{=} \alpha_u^+(\mathbf{x}) + \alpha_u^+((y_0))$
$= \alpha_u^+(\,(x_0, x_1, \ldots, x_n)\,) + \alpha_u^+((y_0))$
$\overset{(9.1),(9.2)}{=} [x_0 + x_1 + \ldots + x_n] + y_0$
$= (\sum_{i=0}^{n} x_i) + y_0$.
Thus, $T1 = T2$. □

Proposition 9.1.4 *(See ([1], Proposition 8.18))*
The function α_u^+ is a morphism of strong m_0-ME positive cones.

Proof. The function α_u^+ is a strong m$_0$-ME positive cones homomorphism. Indeed,

(p1): (See ([1], Proposition 8.18))
For each $\mathbf{x} = (x_0, x_1, \ldots, x_n)$ and $\mathbf{y} = (y_0, y_1, \ldots, y_m)$ in $\mathbf{G}^+(\mathbf{U_m}(M^+))$, we have: $\alpha_u^+(\mathbf{x} + \mathbf{y}) = \alpha_u^+(\mathbf{x}) + \alpha_u^+(\mathbf{y})$.
Indeed, we prove by induction on m.

• For $m = 0$, the statement is true, by Lemma 9.1.3.

• Suppose that the statement is true for a fixed m, i.e. we have:
(*) $\alpha_u^+(\,(x_0, x_1, \ldots, x_n) + (y_0, y_1, \ldots, y_m)\,) = \alpha_u^+((x_0, x_1, \ldots, x_n)) + \alpha_u^+((y_0, y_1, \ldots, y_m))$.

• We shall prove that the statement is true for $m + 1$. Indeed,
$\alpha_u^+(\,(x_0, x_1, \ldots, x_n) + (y_0, y_1, \ldots, y_m, y_{m+1})\,)$
$= \alpha_u^+(\,(x_0, x_1, \ldots, x_n) + [(y_0, y_1, \ldots, y_m) + (y_{m+1})]\,)$, by Lemma 7.1.24,
$= \alpha_u^+(\,[(x_0, x_1, \ldots, x_n) + (y_0, y_1, \ldots, y_m)] + (y_{m+1})\,)$, by (Sass) in $\mathbf{G}^+(\mathbf{U_m}(M^+))$,
$= \alpha_u^+(\,(x_0, x_1, \ldots, x_n) + (y_0, y_1, \ldots, y_m)\,) + \alpha_u^+(\,(y_{m+1})\,)$, by Lemma 9.1.3,
$= [\alpha_u^+(\,(x_0, x_1, \ldots, x_n)\,) + \alpha_u^+(\,(y_0, y_1, \ldots, y_m)\,)] + \alpha_u^+(\,(y_{m+1})\,)$ by inductive hypothesis, (*),
$= \alpha_u^+(\,(x_0, x_1, \ldots, x_n)\,) + [\alpha_u^+(\,(y_0, y_1, \ldots, y_m)\,) + \alpha_u^+(\,(y_{m+1})\,)]$, by (Sass) in M^+,
$= \alpha_u^+(\,(x_0, x_1, \ldots, x_n)\,) + \alpha_u^+(\,(y_0, y_1, \ldots, y_m) + (y_{m+1})\,)$, by Lemma 9.1.3,
$= \alpha_u^+(\,(x_0, x_1, \ldots, x_n)\,) + \alpha_u^+(\,(y_0, y_1, \ldots, y_m, y_{m+1})\,)$, by Lemma 7.1.24 again.
The proof by induction is complete.

(p2): For each $\mathbf{x} = (x_0, x_1, \ldots, x_n)$ in $\mathbf{G}^+_\mathbf{m}(\mathbf{U_m}(M^+))$, we have:
$\alpha_u^+(\smile \mathbf{x}) = \smile \alpha_u^+(\mathbf{x})$. Indeed,
$\alpha_u^+(\smile \mathbf{x}) = \alpha_u^+(\smile (x_0, x_1, \ldots, x_n))$
$= \alpha_u^+((\neg x_n, \ldots, \neg x_1, \neg x_0))$
$= \neg x_n + \ldots + \neg x_1 + \neg x_0$
$= \smile x_n \smile \ldots \smile x_1 \smile x_0$
$= \smile x_0 \smile x_1 \smile \ldots \smile x_n$
and
$\smile \alpha_u^+(\mathbf{x}) = \smile \alpha_u^+((x_0, x_1, \ldots, x_n))$
$= \smile (x_0 + x_1 + \ldots + x_n)$
$= \smile x_0 \smile x_1 \smile \ldots \smile x_n$, by $(s2^{+R})$ (i) in M^+.

(p3): $\alpha_u^+(0_{\mathbf{G}^+_\mathbf{m}(\mathbf{U_m}(M^+))}) = \alpha_u^+((0_{M^+})) = 0_{M^+}$.

(p4): For each $\mathbf{x} = (x_0, x_1, \ldots, x_n)$ and $\mathbf{y} = (y_0, y_1, \ldots, y_n)$ in $\mathbf{G}^+_\mathbf{m}(\mathbf{U_m}(M^+))$, we have:
$\mathbf{x} \leq_m \mathbf{y} \implies \alpha_u^+(\mathbf{x}) \leq_m \alpha_u^+(\mathbf{y})$. Indeed,
$\mathbf{x} \leq_m \mathbf{y} \iff (x_0, x_1, \ldots, x_n) \leq_m (y_0, y_1, \ldots, y_n)$ in $\mathbf{G}^+_\mathbf{m}(\mathbf{U_m}(M^+))$,
$\iff x_i \leq_m y_i$, $i = 0, 1, \ldots, n$, in $\mathbf{U_m}(M^+)$,
$\iff x_i \leq_m y_i$, $i = 0, 1, \ldots, n$, in M^+,
$\iff x_0 + x_1 + \ldots + x_n \leq_m y_0 + y_1 + \ldots + y_n$, by $(s2^{+R})$ (iii) in M^+,
$\iff \alpha_u^+(\mathbf{x}) \leq_m \alpha_u^+(\mathbf{y})$.

(p5): $\alpha_u^+(u_{\mathbf{G}^+_\mathbf{m}(\mathbf{U_m}(M^+))}) = \alpha_u^+((u_{M^+})) = u_{M^+}$. □

Proposition 9.1.5 *(See ([1], Proposition 8.13))*
The function α_u^+ is a bijection.

Proof. • α_u^+ is injective. Indeed,
if $\alpha_u^+(\mathbf{x}) = \alpha_u^+(\mathbf{y})$, i.e.
$\alpha_u^+((x_0, x_1, \ldots, x_n)) = \alpha_u^+((y_0, y_1, \ldots, y_n))$, i.e.
$x_0 + x_1 + \ldots + x_n = y_0 + y_1 + \ldots + y_n$ in M^+,
then, by $(s2^{+R})$ (ii), we obtain $x_i = y_i$, $i = 0, 1, \ldots, n$ in M^+, with $0_{M^+} \leq_m x_i, y_i \leq_m u_{M^+}$; hence, $x_i, y_i \in \mathbf{U_m}(M^+)$, $i = 0, 1, \ldots, n$, hence $(x_0, x_1, \ldots, x_n) = (y_0, y_1, \ldots, y_n)$ in $\mathbf{G}^+_\mathbf{m}(\mathbf{U_m}(M^+))$, i.e. $\mathbf{x} = \mathbf{y}$.

- α_u^+ is surjective. Indeed,

let $x \in M^+$; by (s2^{+R}) (i), there exists $n_x \in \mathbf{N}$ and there exist unique elements of M^+:

$0_{M^+} \leq_m x_1, x_2, \ldots, x_{n_x} \leq_m u_{M^+}$, such that $x = x_1 + x_2 + \ldots + x_{n_x}$ $(\leq_m n_x u_{M^+})$; hence, there exist unique $x_1, \ldots, x_{n_x} \in \mathbf{U_m}(M^+)$, such that $x = x_1 + x_2 + \ldots + x_{n_x}$, i.e. there exists unique $\mathbf{x} = (x_0, x_1, \ldots, x_{n_x}) \in \mathbf{G_m^+}(\mathbf{U_m}(M^+))$ such that $\alpha_u^+(\mathbf{x}) = \alpha_u^+((x_0, x_1, \ldots, x_{n_x})) = x_1 + x_2 + \ldots + x_{n_x} = x.$ $\square$

Proposition 9.1.6 *(See ([1], Proposition 8.19))*

The function α^+ is a natural transformation, i.e. for every morphism of strong m_0-ME positive cones $f^+ : M^+ \longrightarrow N^+$, the following diagram commutes:

$$
\begin{array}{ccc}
\mathbf{G_m^+}(\mathbf{U_m}(M^+)) & \xrightarrow{\ \alpha_u^+\ } & M^+ \\[2mm]
{\scriptstyle \mathbf{G_m^+}(\mathbf{U_m}(f^+))}\Big\downarrow & & \Big\downarrow{\scriptstyle f^+} \\[2mm]
\mathbf{G_m^+}(\mathbf{U_m}(N^+)) & \xrightarrow[\ \alpha_v^+\ ]{} & N^+
\end{array}
$$

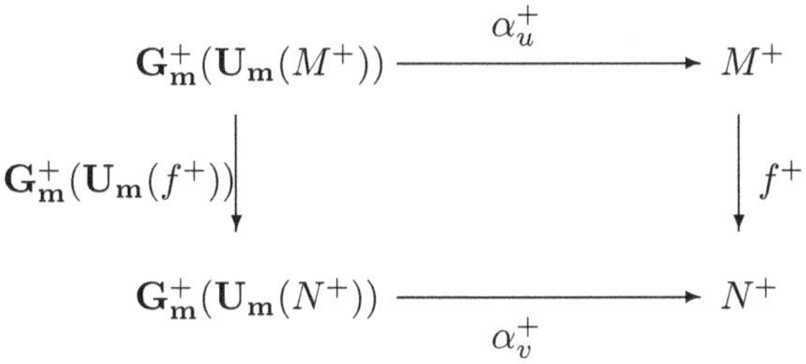

Proof. Let f^+ be the morphism of strong m_0-ME positive cones $\mathcal{M}_u^+ = (M^+, \leq_m, +, \smile, 0_{M^+}, u = u_{M^+})$ and $\mathcal{N}_v^+ = (N^+, \leq_m, +, \smile, 0_{N^+}, v = u_{N^+})$. We have:

$f^+(\alpha_u^+((x_0, x_1, \ldots, x_n))) \overset{(9.1)}{=} f^+(x_0 + x_1 + \ldots + x_n)$
$\overset{(p1)}{=} f^+(x_0) + f^+(x_1) + \ldots + f^+(x_n)$
and
$\alpha_v^+(\mathbf{G_m^+}(\mathbf{U_m}(f^+))((x_0, x_1, \ldots, x_n)))$
$\overset{(7.26)}{=} \alpha_v^+((\mathbf{U_m}(f^+)(x_0), \mathbf{U_m}(f^+)(x_1), \ldots, \mathbf{U_m}(f^+)(x_n)))$
$= \alpha_v^+((f^+(x_0), f^+(x_1), \ldots, f^+(x_n)))$
$\overset{(9.1)}{=} f^+(x_0) + f^+(x_1) + \ldots + f^+(x_n).$ $\square$

Hence, we have:

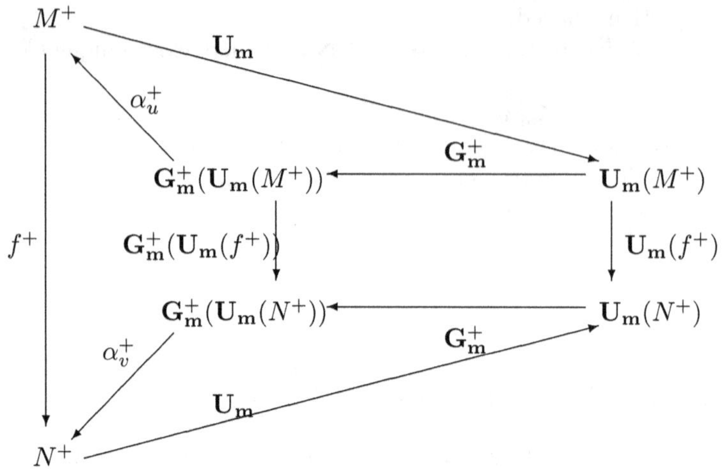

9.1.2 The isomorphism α_A^+ between $\mathbf{G}^+_{\mathbf{m}}(\mathbf{U}_{\mathbf{m}}(\mathbf{G}^+_{\mathbf{m}}(\mathcal{A})))$ and $\mathbf{G}^+_{\mathbf{m}}(\mathcal{A})$

For each right-XY algebra

$$\mathcal{A} = (A, \oplus, ^-, 0_A), \qquad with \quad 1_A \overset{def.}{=} 0_A^-,$$

let us consider its associated strong m_0-ME positive cone $\mathbf{G}^+_{\mathbf{m}}(\mathcal{A})$ obtained by $\mathbf{G}^+_{\mathbf{m}}$

$$\mathbf{G}^+_{\mathbf{m}}(\mathcal{A}) = (G^+(A), \leq_m, +, \smile, 0_{G^+(A)} = (0_A), u_{G^+(A)} = (1_A)).$$

Now, for this particular strong m_0-ME positive cone $\mathcal{M}_u^+ = \mathbf{G}^+_{\mathbf{m}}(\mathcal{A})$, let us consider the right-XY algebra $\mathbf{U}_{\mathbf{m}}(\mathcal{M}_u^+)$ obtained by $\mathbf{U}_{\mathbf{m}}$

$$\mathbf{U}_{\mathbf{m}}(\mathcal{M}_u^+) = \mathbf{U}_{\mathbf{m}}(\mathbf{G}^+_{\mathbf{m}}(\mathcal{A})) = ([(0_A), (1_A)], \oplus, \neg, (0_A))$$

and, finally, let us consider its particular associated strong m_0-ME positive cone $\mathbf{G}^+_{\mathbf{m}}(\mathbf{U}_{\mathbf{m}}(\mathcal{M}_u^+))$ obtained by $\mathbf{G}^+_{\mathbf{m}}$

$$\mathbf{G}^+_{\mathbf{m}}(\mathbf{U}_{\mathbf{m}}(\mathcal{M}_u^+)) = \mathbf{G}^+_{\mathbf{m}}(\mathbf{U}_{\mathbf{m}}(\mathbf{G}^+_{\mathbf{m}}(\mathcal{A})))$$

$$= (G^+([(0_A), (1_A)]), \leq_m, +, \smile, 0_{G^+(\mathbf{U}_{\mathbf{m}}(G^+(A)))} = ((0_A)), u_{G^+(\mathbf{U}_{\mathbf{m}}(G^+(A)))} = ((1_A))).$$

Consider now the function

$$\alpha_A^+ : G^+(\mathbf{U}_{\mathbf{m}}(G^+(A))) \longrightarrow G^+(A)$$

defined by: for each $((x_0), (x_1), \ldots, (x_n)) \in G^+(\mathbf{U}_{\mathbf{m}}(G^+(A)))$,

$$(9.3) \qquad ((x_0), (x_1), \ldots, (x_n)) \longmapsto (x_0) + (x_1) + \ldots + (x_n) \in G^+(A),$$

i.e. $\alpha_A^+(\,((x_0), (x_1), \ldots, (x_n))\,) = \alpha_A^+(\,((x_0)) + ((x_1)) + \ldots + ((x_n))\,) \overset{def.}{=} (x_0) + (x_1) + \ldots + (x_n)$, by Lemma 7.1.33;
in particular, for each $((x)) \in G^+(\mathbf{U}_{\mathbf{m}}(G^+(A)))$,

$$(9.4) \qquad \alpha_A^+(\,((x))\,) \overset{def.}{=} (x).$$

Note that the function α_A^+ is a particular case of the function α_u^+, for $\mathcal{M}_u^+ = \mathbf{G}_{\mathbf{m}}^+(\mathcal{A})$. Consequently, it is well defined and it is an isomorphism.

Hence, we have:

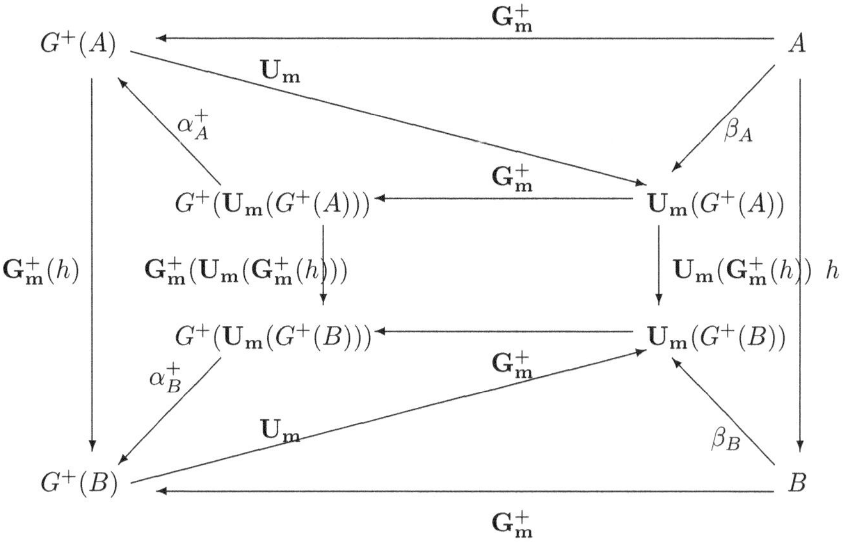

Lemma 9.1.7 *(m=0) For all* $((x_0))$, $((y_0)) \in G^+(\mathbf{U}_{\mathbf{m}}(G^+(A)))$, *we have:*

$$\alpha_A^+(\,((x_0)) + ((y_0))\,) = \alpha_A^+(\,((x_0))\,) + \alpha_A^+(\,((y_0))\,).$$

Proof. $T1 \overset{notation}{=} \alpha_A^+(\,((x_0)) + ((y_0))\,)$
$= \alpha_A^+(\,((x_0) \oplus (y_0), (x_0) \odot (y_0))\,)$, by Lemma 7.1.12,
$= \alpha_A^+(\,((x_0 \oplus y_0), (x_0 \odot y_0))\,)$, by Corollary 9.2.2,
$\overset{(9.3)}{=} (x_0 \oplus y_0) + (x_0 \odot y_0)$, in $G^+(A)$.

$T2 \overset{notation}{=} \alpha_A^+(\,((x_0))\,) + \alpha_A^+(\,((y_0))\,)$
$\overset{def.}{=} (x_0) + (y_0)$, in $G^+(A)$
$= (x_0 \oplus y_0, x_0 \odot y_0)$, in $G^+(A)$, by Lemma 7.1.12,
$= (x_0 \oplus y_0) + (x_0 \odot y_0)$, in $G^+(A)$, by Lemma 7.1.33.

Thus, $T1 = T2$. □

Lemma 9.1.8 *(m=1) For all* $((x_0), (x_1))$, $((y_0)) \in G^+(\mathbf{U}_{\mathbf{m}}(G^+(A)))$, *we have:*

$$\alpha_A^+(\,((x_0), (x_1)) + ((y_0))\,) = \alpha_A^+(\,((x_0), (x_1))\,) + \alpha_A^+(\,((y_0))\,).$$

Proof. First, note that: $(x_0) \oplus (x_1) = (x_0)$ and hence $(x_0) \odot (x_1) = (x_1)$, by the definition of the good sequence $((x_0), (x_1))$ in $G^+(\mathbf{U}_{\mathbf{m}}(G^+(A)))$;
then, we have:
$(x_0 \oplus x_1) = (x_0)$ and $(x_0 \odot x_1) = (x_1)$, in $G^+(A)$, by Corollary 9.2.2;
hence, we have:
$x_0 \oplus x_1 = x_0$ and $x_0 \odot x_1 = x_1$, in A, by the definition of equality in $G^+(A)$;
hence, we have:

$(x_0, x_1) \in G^+(A)$, i.e. it is a good sequence of A
and, consequently, we have, by Lemma 7.1.33, that:

$$(x_0, x_1) = (x_0) + (x_1). \tag{9.5}$$

$T1 \overset{notation}{=} \alpha_A^+(\,((x_0),(x_1)) \,+\, ((y_0))\,)$
$= \alpha_A^+(\,((c_0),(c_1),(c_2))\,)$, by Lemma 7.1.14,
$\overset{(9.3)}{=} (c_0) + (c_1) + (c_2)$, where:
$(c_0) = (x_0) \oplus (y_0) = (x_0 \oplus y_0)$, by Corollary 9.2.2,
$(c_1) = (x_0) \odot [(x_1) \oplus (y_0)] = (x_0 \odot [x_1 \oplus y_0])$, by Corollary 9.2.2,
$(c_2) = (x_1) \odot (y_0) = (x_1 \odot y_0)$, by Corollary 9.2.2.
$T2 \overset{notation}{=} \alpha_A^+(\,((x_0),(x_1))\,) + \alpha_A^+(\,((y_0))\,)$
$\overset{(9.3)}{=} [(x_0) + (x_1)] + (y_0)$, in $G^+(A)$,
$\overset{(9.5)}{=} (x_0, x_1) + (y_0)$, in $G^+(A)$,
$= (d_0, d_1, d_2)$, in $G^+(A)$, by Lemma 7.1.14,
$= (d_0) + (d_1) + (d_2)$, in $G^+(A)$, by Lemma 7.1.33,
where:
$d_0 = x_0 \oplus y_0$,
$d_1 = x_0 \odot [x_1 \oplus y_0]$,
$d_2 = x_1 \odot y_0$.
Thus, $T1 = T2$. □

Lemma 9.1.9 *(m=n) For all* $((x_0),(x_1),\ldots,(x_n))$, $((y_0)) \in G^+(\mathbf{U_m}(G^+(A)))$, *we have:*

$$\alpha_A^+(\,((x_0),(x_1),\ldots,(x_n)) \,+\, ((y_0))\,) = \alpha_A^+(\,((x_0),(x_1),\ldots,(x_n))\,) + \alpha_A^+(\,((y_0))\,).$$

Proof. First, note that: $(x_i) \oplus (x_{i+1}) = (x_i)$ and hence $(x_i) \odot (x_{i+1}) = (x_{i+1})$, for $i = 0, 1, \ldots, n-1$, by the definition of the good sequence $((x_0),(x_1),\ldots,(x_n))$ in $G^+(\mathbf{U_m}(G^+(A)))$;
then, we have:
$(x_i \oplus x_{i+1}) = (x_i)$ and $(x_i \odot x_{i+1}) = (x_{i+1})$, in $G^+(A)$, by Corollary 9.2.2;
hence, we have:
$x_i \oplus x_{i+1} = x_i$ and $x_i \odot x_{i+1} = x_{i+1}$, in A, by the definition of equality in $G^+(A)$;
hence, we have:
$(x_0, x_1, \ldots, x_n) \in G^+(A)$, i.e. it is a good sequence of A
and, consequently, we have, by Lemma 7.1.33, that:

$$(x_0, x_1, \ldots, x_n) = (x_0) + (x_1) + \ldots + (x_n). \tag{9.6}$$

$T1 \overset{notation}{=} \alpha_A^+(\,((x_0),(x_1),\ldots,(x_n)) \,+\, ((y_0))\,)$
$= \alpha_A^+(\,((c_0),(c_1),\ldots,(c_{n+1}))\,)$, by Lemma 7.1.16,
$\overset{(9.3)}{=} (c_0) + (c_1) + \ldots + (c_{n+1})$, in $G^+(A)$,
$= \sum_{i=0}^{n+1} (c_i)$,
where, for $i = 0, 1, \ldots, n+1$, $x_{-1} = 1_A$, $x_{n+1} = 0_A$:
$(c_i) = (x_{i-1}) \odot [(x_i) \oplus (y_0)] = (x_{i-1} \odot [x_i \oplus y_0])$, by Corollary 9.2.2.

$T2 \overset{notation}{=} \alpha_A^+(\,((x_0),(x_1),\ldots,(x_n))\,) + \alpha_A^+(\,((y_0))\,)$
$\overset{(9.3)}{=} [(x_0)+(x_1)+\ldots+(x_n)] + (y_0)$, in $G^+(A)$,
$\overset{(9.6)}{=} (x_0,x_1,\ldots,x_n) + (y_0)$, in $G^+(A)$,
$= (d_0,d_1,\ldots,d_{n+1})$, in $G^+(A)$, by Lemma 7.1.16,
$= (d_0)+(d_1)+\ldots+(d_{n+1})$, in $G^+(A)$, by Lemma 7.1.33,
$= \sum_{i=0}^{n+1}(d_i)$,
where, for $i = 0,1,\ldots,n+1$, $x_{-1} = 1_A$, $x_{n+1} = 0_A$:
$d_i = x_{i-1} \odot [x_i \oplus y_0]$.
Thus, $T1 = T2$. □

9.2 The isomorphisms β_A and β_A^+

9.2.1 The isomorphism β_A between $\mathcal{A}$ and $\mathbf{U_m}(G^+(\mathcal{A}))$

For each right-XY algebra

$$\mathcal{A} = (A,\oplus,^-,0_A), \qquad 1_A \overset{def.}{=} 0_A^-,$$

consider the strong m_0-ME positive cone $\mathbf{G_m^+}(\mathcal{A})$ obtained by $\mathbf{G_m^+}$

$$\mathbf{G_m^+}(\mathcal{A}) = (G^+(A), \le_m, +, \smile, 0_{G^+(A)} = (0_A), u_{G^+(A)} = (1_A))$$

and then the right-XY algebra $\mathbf{U_m}(\mathbf{G_m^+}(\mathcal{A}))$ obtained by $\mathbf{U_m}$

$$\mathbf{U_m}(\mathbf{G_m^+}(\mathcal{A})) = ([0_{G^+(A)} = (0_A), u_{G^+(A)} = (1_A)], \oplus, \neg, (0_A)).$$

Consider now the function

$$\beta_A : A \longrightarrow \mathbf{U_m}(G^+(A))$$

defined as follows: for each $x \in A$,

$$x \longmapsto (x), \tag{9.7}$$

i.e. $\beta_A(x) \overset{def.}{=} (x)$.

Proposition 9.2.1 *(See ([1], Proposition 8.1))*
The function β_A is a morphism of right-XY algebras.

Proof. • The function β_A is well defined, obviously.
• The function β_A is a right-XY algebras homomorphism. Indeed,
(H1): $\beta_A(x\oplus y) = \beta_A(x)\oplus\beta_A(y)$. Indeed,
$\beta_A(x\oplus y) \overset{def.}{=} (x\oplus y) \overset{(7.15)}{=} \inf_m(u_{G^+(A)}, (x)+(y)) \overset{(8.1)}{=} (x)\oplus(y) = \beta_A(x)\oplus\beta_A(y)$.
(H2): $\beta_A(x^-) = \neg(\beta_A(x))$. Indeed,
$\beta_A(x^-) \overset{def.}{=} (x^-) =\smile (x) \overset{(8.2)}{=} \neg((x)) = \neg(\beta_A(x))$.

(H3): $\beta_A(0_A) = 0_{\mathbf{U_m}(G^+(A))}$. Indeed,

$\beta_A(0_A) \overset{def.}{=} (0_A) \overset{def.}{=} 0_{G^+(A)} \overset{def.}{=} 0_{U(G^+(A))}$.

(H1'): $\beta_A(x \odot y) = \beta_A(x) \odot \beta_A(y)$. Indeed,

$(x \odot y) \overset{def.}{=} \beta_A(x \odot y) = \beta_A((x^- \oplus y^-)^-) \overset{(H2)}{=} \neg(\beta_A(x^- \oplus y^-))$

$\overset{(H1)}{=} \neg(\beta_A(x^-) \oplus \beta_A(y^-)) \overset{(H2)}{=} \neg(\neg(\beta_A(x)) \oplus \neg(\beta_A(y))) = \beta_A(x) \odot \beta_A(y) = (x) \odot (y)$.
$\square$

Corollary 9.2.2 In $\mathbf{U_m}(G^+(A))$, we have: for all $x, y \in A$,

$$(x) \oplus (y) = (x \oplus y), \qquad (x) \odot (y) = (x \odot y).$$

Proof. By the proof of Proposition 9.2.1, (H1) and (H1'). $\square$

Proposition 9.2.3 The function β_A is bijective.

Proof. Immediate. $\square$

Proposition 9.2.4 (See ([1], Proposition 8.2))

The function $\beta : 1_{\mathcal{C}(\mathbf{XYR})} \longrightarrow U \circ G^+$ is a natural transformation, i.e. for every morphism of right-XY algebras $h : A \longrightarrow B$, the following diagram commutes:

$$
\begin{array}{ccc}
A & \xrightarrow{\;\;\beta_A\;\;} & \mathbf{U_m}(G^+(A)) \\
\downarrow{\scriptstyle h} & & \downarrow{\scriptstyle \mathbf{U_m}(\mathbf{G}_{\mathbf{m}}^+(h))} \\
B & \xrightarrow[\;\;\beta_B\;\;]{} & \mathbf{U_m}(G^+(B))
\end{array}
$$

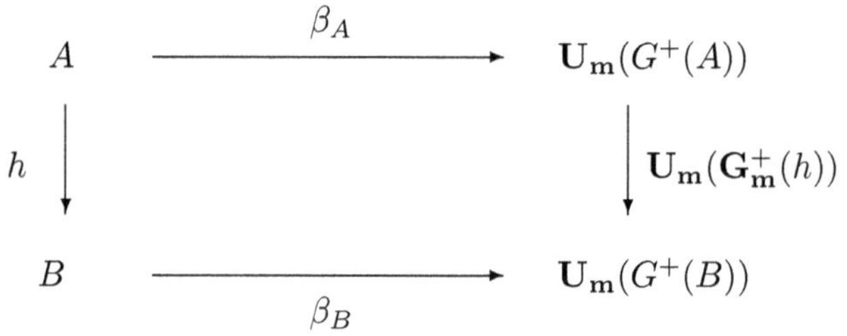

Proof. $\mathbf{U_m}(\mathbf{G}_{\mathbf{m}}^+(h))(\beta_A(x)) = \mathbf{U_m}(\mathbf{G}_{\mathbf{m}}^+(h))((x)) = \mathbf{U_m}(\mathbf{G}_{\mathbf{m}}^+(h))(\mathbf{U_m}((x)))$

$\overset{(8.4)}{=} \mathbf{U_m}(\mathbf{G}_{\mathbf{m}}^+(h)((x))) \overset{(7.27)}{=} \mathbf{U_m}((h(x))) = (h(x)) = \beta_B(h(x))$. $\square$

Hence we have:

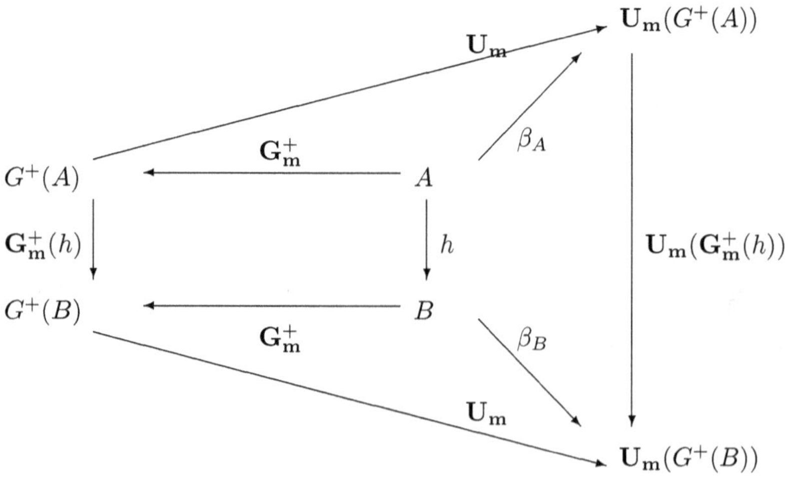

Corollary 9.2.5 *For all $x_0, x_1, \ldots, x_n \in A$, we have:*
$\beta_A(x_0 \oplus x_1 \oplus \ldots \oplus x_n) = \beta_A(x_0) \oplus \beta_A(x_1) \oplus \ldots \oplus \beta_A(x_n)$.

Proof. $\beta_A(x_0 \oplus x_1 \oplus \ldots \oplus x_n) \overset{(9.7)}{=} (x_0 \oplus x_1 \oplus \ldots \oplus x_n)$
$\overset{Cor.\ (7.1.80)(i)}{=} \inf_m(u_{G^+(A)}, (x_0) + (x_1) + \ldots + (x_n)) \overset{(8.1)}{=} (x_0) \oplus (x_1) \oplus \ldots \oplus (x_n)$
$= \beta_A(x_0) \oplus \beta_A(x_1) \oplus \ldots \oplus \beta_A(x_n)$, by Corollary 7.1.80 (i). □

Note that Corollary 9.2.5 says that β_A is a mapping:

$$x_0 \oplus x_1 \oplus \ldots \oplus x_n \longmapsto (x_0 \oplus x_1 \oplus \ldots \oplus x_n) = (x_0) \oplus (x_1) \oplus \ldots \oplus (x_n).$$

9.2.2 The isomorphism β_A^+ between $\mathbf{G_m^+}(\mathcal{A})$ and $\mathbf{G_m^+}(\mathbf{U_m}(\mathbf{G_m^+}(\mathcal{A})))$

For each right-XY algebra

$$\mathcal{A} = (A, \oplus, ^-, 0_A), \qquad with \quad 1_A \overset{def.}{=} 0_A^-,$$

let us consider its associated strong m_0-ME positive cone $\mathbf{G_m^+}(\mathcal{A})$ obtained by $\mathbf{G_m^+}$

$$\mathbf{G_m^+}(\mathcal{A}) = (G^+(A), \leq_m, +, \smile, 0_{G^+(A)} = (0_A), u_{G^+(A)} = (1_A)).$$

Now, for this particular strong m_0-ME positive cone $\mathcal{M}_u^+ = \mathbf{G_m^+}(\mathcal{A})$,
let us consider the right-XY algebra $\mathbf{U_m}(\mathcal{M}_u^+)$ obtained by $\mathbf{U_m}$

$$\mathbf{U_m}(\mathcal{M}_u^+) = \mathbf{U_m}(\mathbf{G_m^+}(\mathcal{A})) = ([(0_A), (1_A)], \oplus, \neg, (0_A))$$

and, finally, let us consider its particular associated strong m_0-ME positive cone $\mathbf{G_m^+}(\mathbf{U_m}(\mathcal{M}_u^+))$ obtained by $\mathbf{G_m^+}$

$$\mathbf{G_m^+}(\mathbf{U_m}(\mathcal{M}_u^+)) = \mathbf{G_m^+}(\mathbf{U_m}(\mathbf{G_m^+}(\mathcal{A})))$$

$$= (G^+([(0_A), (1_A)]), \leq_m, +, \smile, 0_{G^+(U(G^+(A)))} = ((0_A)), u_{G^+(U(G^+(A)))} = ((1_A))).$$

Consider now the function

$$\beta_A^+ \overset{def.}{=} \mathbf{G_m^+} \circ \beta_A : G^+(A) \longrightarrow G^+(\mathbf{U_m}(G^+(A)))$$

defined by: for each $(x_0) + (x_1) + \ldots + (x_n) \in G^+(A)$,

(9.8) $\quad (x_0) + (x_1) + \ldots + (x_n) \mapsto ((x_0), (x_1), \ldots, (x_n)) \in G^+(\mathbf{U_m}(G^+(A)))$,

i.e. $\beta_A^+((x_0) + (x_1) + \ldots + (x_n)) \overset{def.}{=} ((x_0), (x_1), \ldots, (x_n))$;
in particular, for each $(x) \in G^+(A)$,

(9.9) $$\beta_A^+((x)) \overset{def.}{=} ((x)).$$

Hence, we have:

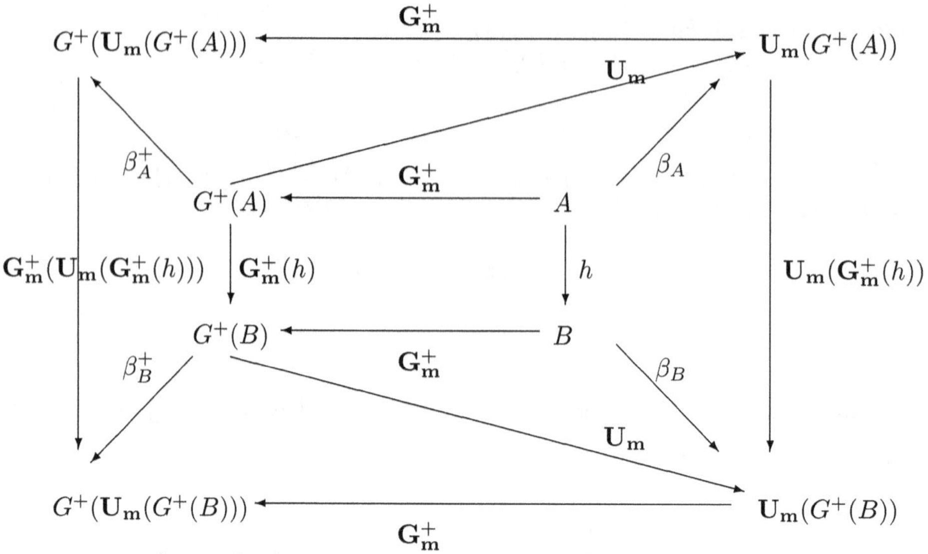

Note that $\beta^+_A = (\alpha^+_A)^{-1}$ (the inverse function of α^+_A). Consequently, β^+_A is well defined and is an isomorphism.

9.3 Conclusions

Theorem 9.3.1 *The functors*
$1_{\mathcal{C}(\mathbf{s-m_0-ME^+})} : \mathcal{C}(\mathbf{s} - \mathbf{m_0} - \mathbf{ME}^+) \longrightarrow \mathcal{C}(\mathbf{s} - \mathbf{m_0} - \mathbf{ME}^+)$
and
$\mathbf{G}^+_\mathbf{m} \circ \mathbf{U_m} : \mathcal{C}(\mathbf{s} - \mathbf{m_0} - \mathbf{ME}^+) \longrightarrow \mathcal{C}(\mathbf{s} - \mathbf{m_0} - \mathbf{ME}^+)$
are naturally isomorphic.

Proof. By Propositions 9.1.4, 9.1.5, 9.1.6. □

Theorem 9.3.2 *The functors*
$1_{\mathcal{C}(\mathbf{XY^R})} : \mathcal{C}(\mathbf{XY^R}) \longrightarrow \mathcal{C}(\mathbf{XY^R})$
and
$\mathbf{U} \circ \mathbf{G}^+_\mathbf{m} : \mathcal{C}(\mathbf{XY^R}) \longrightarrow \mathcal{C}(\mathbf{XY^R})$
are naturally isomorphic.

Proof. By Propositions 9.2.1, 9.2.3, 9.2.4. □

Corollary 9.3.3 *(See ([1], Theorem 8.20))*

The functors

$\mathcal{C}(\mathbf{XY}^R) \xrightarrow{\mathbf{G}^+_\mathbf{m}} \mathcal{C}(\mathbf{s}\text{-}\mathbf{m}_0\text{-}\mathbf{ME}^+)$
and
$\mathcal{C}(\mathbf{s}\text{-}\mathbf{m}_0\text{-}\mathbf{ME}^+) \xrightarrow{\mathbf{U_m}} \mathcal{C}(\mathbf{XY}^R)$
are quasi-inverses.

Thus, the categories $\mathcal{C}(\mathbf{XY}^R)$ and $\mathcal{C}(\mathbf{s}\text{-}\mathbf{m}_0\text{-}\mathbf{ME}^+)$ are equivalent.

Proof. By above Theorems 9.3.1 and 9.3.2. □

Remark 9.3.4 *Following this equivalence between the categories of strong m_0-ME positive cones and right-XY algebras, some proofs in one category can be made easier in the other category. See for examples the cases discussed in Remarks 2.2.38, 2.2.40, 2.2.42, 2.2.43, 2.2.45.*

Part IV

Connections between the four functors

Chapter 10

Final connections

In this chapter, we introduce the 'ancestor' functors Γ_m and Ξ_m of Γ and Ξ, respectively, and we introduce and study the isomorphisms η_A^+, of strong m_0-ME positive cones, and φ_A, of right-XY algebras.

10.1 The main result

Put $\mathbf{\Gamma_m} = \mathbf{U_m} \circ \mathbf{c_m^+}$ and $\mathbf{\Xi_m} = \mathbf{T_m} \circ \mathbf{G_m^+}$, i.e. we have:

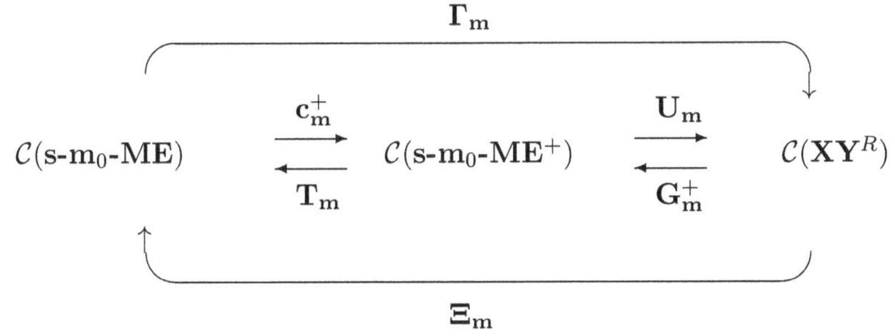

We are ready to state the main result of the work.

Theorem 10.1.1 *(See ([1], Theorem 8.21))*
The functor $\mathbf{\Gamma_m} = \mathbf{U_m} \circ \mathbf{c_m^+} : \mathcal{C}(\textbf{s-m}_0\textbf{-ME}) \longrightarrow \mathcal{C}(\mathbf{XY}^R)$ is not an equivalence of categories, but the functor $\mathbf{U} : \mathcal{C}(\textbf{s-m}_0\textbf{-ME}^+) \longrightarrow \mathcal{C}(\mathbf{XY}^R)$ is an equivalence of categories.

Proof. By Corollaries 6.3.3 and 9.3.3, Γ_m is not an equivalence. By Corollary 9.3.3, $\mathbf{U_m}$ is an equivalence. □

10.2 The isomorphism η_A^+ between $\mathbf{G}_\mathbf{m}^+(\mathcal{A})$ and $\mathbf{c}_\mathbf{m}^+(\mathbf{T}_\mathbf{m}(\mathbf{G}_\mathbf{m}^+(\mathcal{A})))$

For each right-XY algebra

$$\mathcal{A} = (A, \oplus, ^-, 0_A), \qquad \textit{with} \quad 1_A \overset{def.}{=} 0_A^-,$$

consider its associated strong m_0-ME positive cone $\mathbf{G}_\mathbf{m}^+(\mathcal{A})$ obtained by $\mathbf{G}_\mathbf{m}^+$

$$\mathbf{G}_\mathbf{m}^+(\mathcal{A}) = (G^+(A), \leq_m, +, \smile, 0_{G^+(A)} = (0_A), u = u_{G^+(A)} = (1_A)).$$

Now, for this particular strong m_0-ME positive cone $\mathcal{M}_u^+ = \mathbf{G}_\mathbf{m}^+(\mathcal{A})$, consider the strong m_0-ME structure $\mathbf{T}_\mathbf{m}(\mathcal{M}_u^+) = \mathbf{T}_\mathbf{m}(\mathbf{G}_\mathbf{m}^+(\mathcal{A}))$ obtained by $\mathbf{T}_\mathbf{m}$

$$\mathbf{T}_\mathbf{m}(\mathbf{G}_\mathbf{m}^+(\mathcal{A})) = (\mathbf{T}_\mathbf{m}(G^+(A)) = \frac{G^+(A) \times \mathbf{N}}{\sim}, \preceq_m, +, \smile, \mathbf{0} = \widehat{(0_{G^+(A)}, 0)}, \mathbf{u} = \widehat{(u, 0)})$$

and then the strong m_0-ME positive cone $\mathbf{c}_\mathbf{m}^+(\mathbf{T}_\mathbf{m}(\mathcal{M}_u^+)) = \mathbf{c}_\mathbf{m}^+(\mathbf{T}_\mathbf{m}(\mathbf{G}_\mathbf{m}^+(\mathcal{A})))$ obtained by $\mathbf{c}_\mathbf{m}^+$

$$\mathbf{c}_\mathbf{m}^+(\mathbf{T}_\mathbf{m}(\mathbf{G}_\mathbf{m}^+(\mathcal{A}))) = (\mathbf{c}_\mathbf{m}^+(\mathbf{T}_\mathbf{m}(G^+(A))) = (\frac{G^+(A) \times \mathbf{N}}{\sim})^+, \preceq_m, +, \smile, \mathbf{0}, \mathbf{u}).$$

Consider now the function

$$\eta_A^+ : G^+(A) \longrightarrow \mathbf{c}_\mathbf{m}^+(\mathbf{T}_\mathbf{m}(G^+(A)))$$

defined as follows: for each $\mathbf{x} = (x_0, x_1, \ldots, x_n) \in G^+(A)$,

(10.1) $$\eta_A^+(\mathbf{x}) \overset{def.}{=} \widehat{(\mathbf{x}, 0)};$$

in particular, for each $(x) \in G^+(A)$,

(10.2) $$\eta_A^+((x)) \overset{def.}{=} \widehat{((x), 0)}.$$

Note that the function η_A^+ is a particular case of the function η_u^+ (see Chapter 6) for the particular strong m_0-ME positive cone $\mathcal{M}_u^+ = \mathbf{G}_\mathbf{m}^+(\mathcal{A})$. Consequently, we have the following result.

Theorem 10.2.1 *(See ([10], Proposition 2.4.2 (iii)))*
The function η_A^+ is an isomorphism of strong m_0-ME positive cones.

Proof. By Propositions 6.1.1, 6.1.2. □

We have the situation from the following Figure.

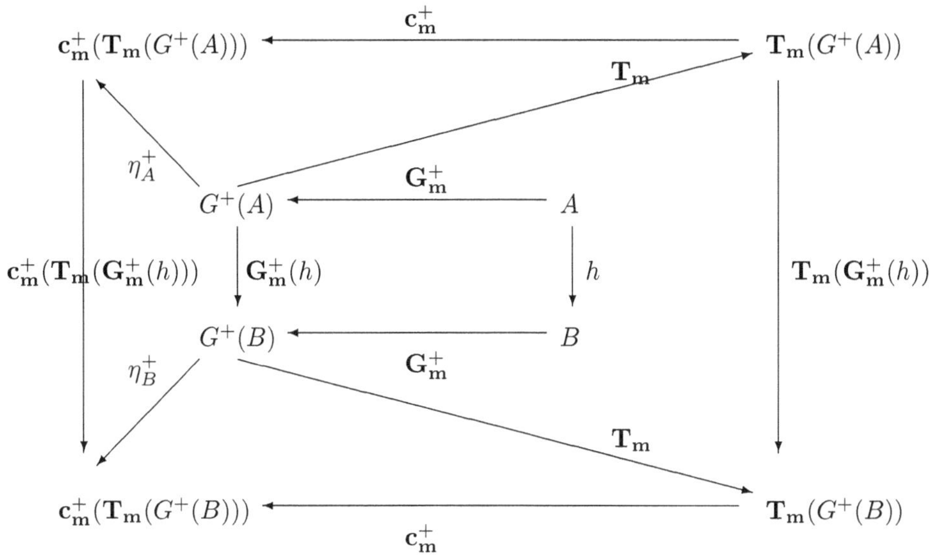

10.3 The isomorphism φ_A between $\mathcal{A}$ and $\mathbf{U_m}(\mathbf{c_m^+}(\mathbf{T_m}(\mathbf{G_m^+}(\mathcal{A}))))$

For each right-XY algebra

$$\mathcal{A} = (A, \oplus, ^-, 0_A), \qquad \textit{with} \quad 1_A \stackrel{def.}{=} 0_A^-,$$

consider its associated strong $\mathrm{m_0}$-ME positive cone $\mathbf{G_m^+}(\mathcal{A})$ (by Theorem 7.1.77) obtained by $\mathbf{G_m^+}$

$$\mathbf{G_m^+}(\mathcal{A}) = (G^+(A), \leq_m, +, \smile, 0_{G^+(A)} = (0_A), u = u_{G^+(A)} = (1_A)),$$

then, consider the strong $\mathrm{m_0}$-ME structure $\mathbf{T_m}(\mathbf{G_m^+}(\mathcal{A}))$ (by Theorem 5.1.27) obtained by $\mathbf{T_m}$

$$\mathbf{T_m}(\mathbf{G_m^+}(\mathcal{A})) = (\mathbf{T_m}(G^+(A)) = \frac{G^+(A) \times \mathbf{N}}{\sim}, \preceq_m, +, \smile, \mathbf{0}, \mathbf{u}),$$

where, $u = u_{G^+(A)}$, $\quad \mathbf{0} = \overbrace{(0_{G^+(A)}, 0)}$, by (5.2), $\quad \mathbf{u} = \overbrace{(u, 0)}$, by (5.3);
then, consider the strong $\mathrm{m_0}$-ME positive cone $\mathbf{c_m^+}(\mathbf{T_m}(\mathbf{G_m^+}(\mathcal{A})))$ obtained by $\mathbf{c_m^+}$

$$\mathbf{c_m^+}(\mathbf{T_m}(\mathbf{G_m^+}(\mathcal{A}))) = (\mathbf{c_m^+}(\mathbf{T_m}(G^+(A))) = (\frac{G^+(A) \times \mathbf{N}}{\sim})^+, \preceq_m, +, \smile, \mathbf{0}, \mathbf{u}),$$

where:

$$(\frac{G^+(A) \times \mathbf{N}}{\sim})^+$$

$$= \{\overbrace{(x + mu_{G^+(A)}, n)} \mid \overbrace{(x + mu_{G^+(A)}, n)} \succeq_m \mathbf{0}, \; s.t. \; (Nu^{+1R}), \; (S^{+1R}) \; hold\}$$

$$= \{\overbrace{(x + mu_{G^+(A)}, 0)} \mid x \in G^+(A), m \in \mathbf{N},\ s.t.\ (Nu^{+1R}),\ (S^{+1R})\ hold\}$$

$$= \{\overbrace{(x, 0)} \mid x \in G^+(A)\};$$

and finally consider the right-XY algebra $\mathbf{U_m}(\mathbf{c^+_m}(\mathbf{T_m}(\mathbf{G^+_m}(\mathcal{A}))))$ obtained by $\mathbf{U_m}$

$$\mathbf{U_m}(\mathbf{c^+_m}(\mathbf{T_m}(\mathbf{G^+_m}(\mathcal{A})))) = (\mathbf{U_m}(\mathbf{c^+_m}(\mathbf{T_m}(G^+(A)))) = \mathbf{U_m}((\frac{G^+(A) \times \mathbf{N}}{\sim})^+)$$

$$= [\mathbf{0}, \mathbf{u}], \oplus, \neg, \mathbf{0}),$$

with the unit interval

$$[\mathbf{0}, \mathbf{u}] \stackrel{def.}{=} \{\overbrace{(\mathbf{a}, 0)} \mid \mathbf{0} \preceq_m \overbrace{(\mathbf{a}, 0)} \preceq_m \mathbf{u}\}.$$

Consider now the function

$$\varphi_A : A \longrightarrow \mathbf{U_m}(\mathbf{c^+_m}(\mathbf{T_m}(G^+(A)))) = [\mathbf{0}, \mathbf{u}]$$

defined as follows: for each $x \in A$,

$$\varphi_A(x) \stackrel{def.}{=} \overbrace{((x), 0)}. \tag{10.3}$$

Then, we have the following important result:

Theorem 10.3.1 *(See the analogous ([10], Theorem 2.4.5))*
The function φ_A is an isomorphism of right-XY algebras, i.e. a bijective homomorphism of XY algebras.

Proof. • The function φ_A is obviously well defined.
• The function φ_A is bijective. Indeed,

- *injectivity*: If $\overbrace{((a), 0)} = \overbrace{((b), 0)}$, i.e. $(a) = (b)$ in $G^+(A)$, then obviously $a = b$ in A.

- *surjectivity*: Let $\overbrace{(\mathbf{a}, 0)} \in [\mathbf{0}, \mathbf{u}]$; we have:

$\mathbf{0} \preceq_m \overbrace{(\mathbf{a}, 0)} \preceq_m \mathbf{u}$
$\Longleftrightarrow \overbrace{(0_{G^+(A)}, 0)} \preceq_m \overbrace{(\mathbf{a}, 0)} \preceq_m \overbrace{(u_{G^+(A)}, 0)}$
$\Longleftrightarrow 0_{G^+(A)} \leq_m \mathbf{a} \leq_m u_{G^+(A)}$
$\Longleftrightarrow \mathbf{a} = (a)$, by Proposition 7.1.56.

Hence, there exists $a \in A$, such that $\varphi_A(a) = \overbrace{((a), 0)} = \overbrace{(\mathbf{a}, 0)}$.
• The function φ_A is a homomophism of right-XY algebras. Indeed,
(H1): $\varphi_A(a \oplus b) = \varphi_A(a) \oplus \varphi_A(b)$. Indeed,

$\varphi_A(a \oplus b) = \overbrace{((a \oplus b), 0)} \stackrel{(7.15)}{=} \overbrace{(\inf_m(u_{G^+(A)}, (a) + (b)), 0)}$

and

$$\begin{aligned}
\varphi_A(a) \oplus \varphi_A(b) &= \overbrace{((a),0)} \oplus \overbrace{((b),0)} \\
&= \inf_m(\mathbf{u}, \overbrace{((a),0)} + \overbrace{((b),0)}) \\
&= \inf_m(\mathbf{u}, \overbrace{((a)+(b),0)}) \\
&= \inf_m(\overbrace{(u_{G^+(A)},0)}, \overbrace{((a)+(b),0)}) \\
&\overset{(5.7)}{=} \overbrace{(\inf_m(u_{G^+(A)} + 0u_{G^+(A)}, (a)+(b)+0u_{G^+(A)}), 0+0)} \\
&= \overbrace{(\inf_m(u_{G^+(A)}, (a)+(b)), 0)}.
\end{aligned}$$

(H2): $\varphi_A(a^-) = \neg\varphi_A(a)$. Indeed,
$\varphi_A(a^-) = \overbrace{((a^-),0)} = \overbrace{(\smile (a),0)} = \smile \overbrace{((a),0)} = \smile \varphi_A(a) = \neg\varphi_A(a)$.

(H3): $\varphi_A(0) = \mathbf{0}$. Indeed,
$\varphi_A(0) = \varphi_A(0_A) = \overbrace{((0_A),0)} = \overbrace{(0_{G^+(A)},0)} = \mathbf{0}$. □

Hence, we have (see the previous Figure in the previous subsection 10.2):

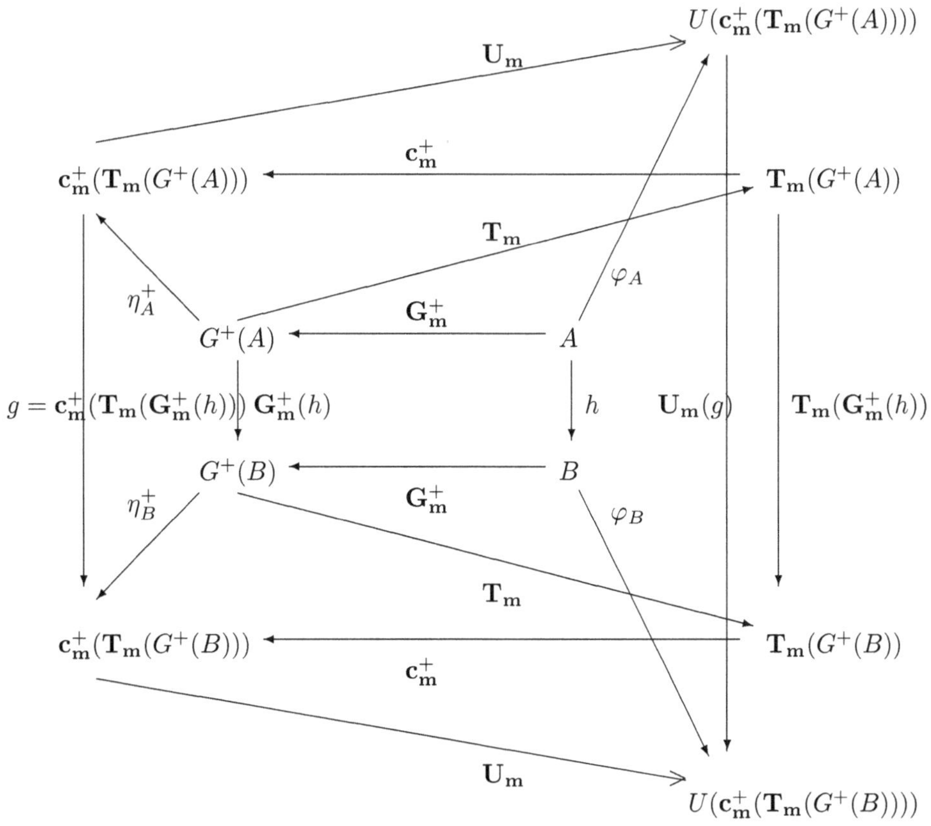

10.4 Future work

Resolve Open problems 7.1.45.

If (see Remark 7.1.78) there is at least an example of cancellative proper XY algebra $\mathcal{A}$ whose $G^+(A)$ verifies the cancellative property (C), then its strong m_0-ME positive cone $(G^+(A), \leq_m, +, \smile, 0_{G^+(A)}, u_{G^+(A)})$ verifies (C), and hence another construction of the functor $\mathbf{T_m}$ can be made, similar to that of $\mathbf{T}$ in MV algebras case (see Subsection 1.3).

The right-pseudo-MV algebras were introduced in 1999 [14], [15], as non-commutative generalizations $(A, \oplus, ^-, ^\sim, 0, 1)$ of right-MV algebras $(A, \oplus, ^-, 0)$ and, in [12], a generalization of Mundici's equivalence to the non-commutative case is presented. We have introduced and studied in 2024 [22] a new framework, containing, among many other (left-) algebras, the groups and the m_1-pseudo-ME algebras and the pseudo-MV algebras, m-pseudo-MEL algebras and the m-pseudo-BCK algebras; a group $(G, \cdot, ^{-1}, 1)$ coincides with a m_1-pseudo-ME algebra verifying the property (m_1-pRe) ([22], Theorem 10.1.26) and a left-pseudo-MV algebra $(A, \odot, ^-, ^\sim, 0, 1)$ is an involutive left-m-pseudo-MEL algebra verifying (sum) and the property ($\wedge_m$-comm) ([22], Def. 15.3.15), where (sum) is $(x^- \odot y^-)^\sim = (x^\sim \odot y^\sim)^-$ and ($\wedge_m$-comm) is $y \odot (x^- \odot y)^\sim = x \odot (y^- \odot x)^\sim$. Thus, an immediate further work could be to generalize the notions and the results from this book to the non-commutative case, in the new non-commutative framework centered on m-pseudo-BCK algebras from [22], and connect them to [12].

Chapter 11

Appendix

11.1 Appendix A

Recall (Remarks 2.1.13) that in the diagram of a hierarchy of classes of algebras, we shall represent:

- *reflexive* algebras by ◯

- *antisymmetric* algebras by ∘

- *transitive* algebras by •

- *reflexive* and *antisymmetric* algebras by ⊚

- *reflexive* and *transitive* algebras by ⊙

- *ordered* algebras by ●

- any other algebra by □

Recall, in the following two Figures 11.1 and 11.2, the connections between m_0-ME algebras and commutative groups, between right-m-MEL algebras and (involutive) right-m-BCK algebras from [21], [22].

Recall also, in the following Figure 11.3, the connections between involutive m_0-ME algebras and commutative groups, between involutive right-m-MEL algebras and (involutive) right-m-BCK algebras from [22].

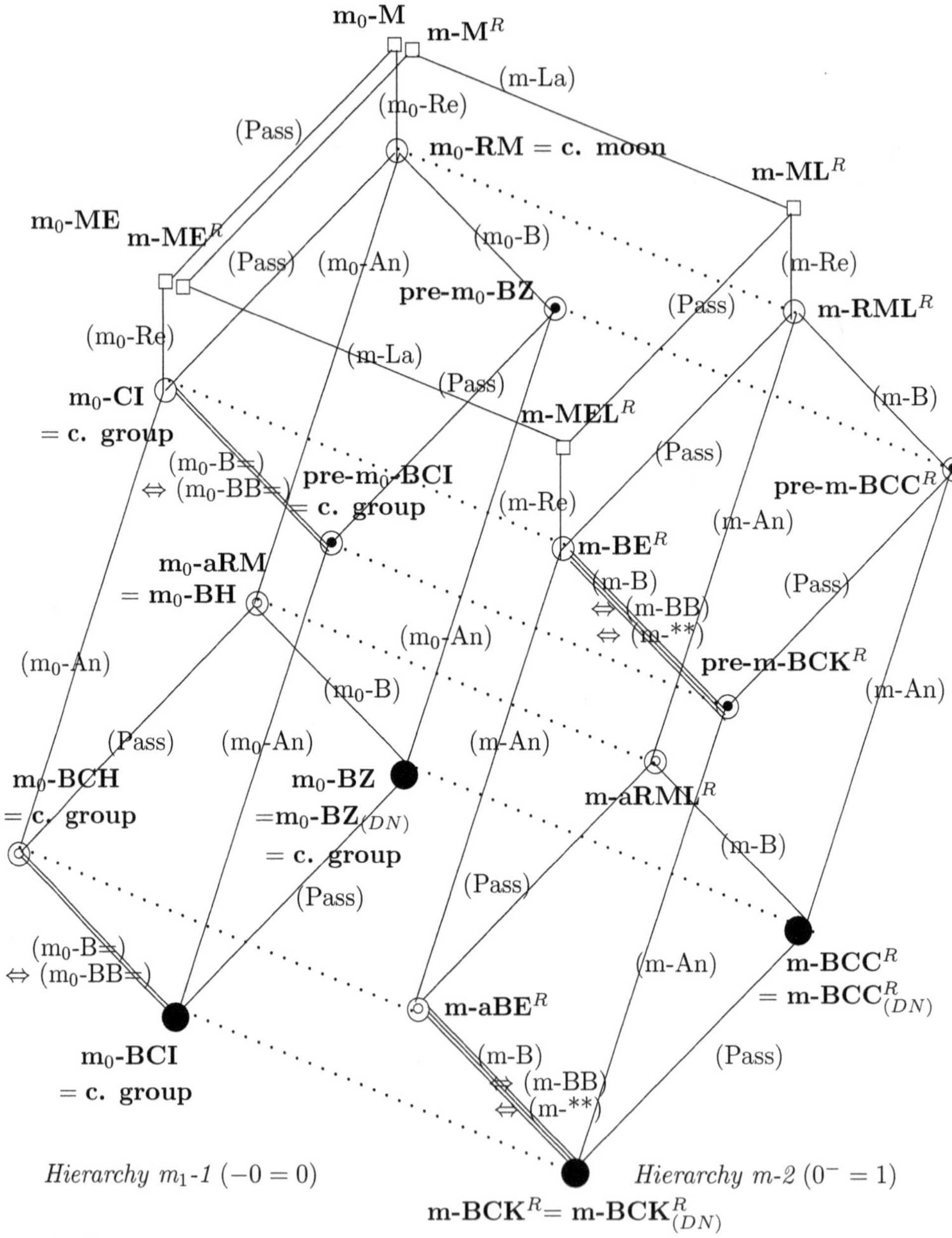

Figure 11.1: The "Big m-map": Hierarchies m_0-1 and m-2 of unital commutative right-magmas with additional operation (see ([21], Figure 8.1) and ([22], Figure 10.4) for the non-commutative left-case). We have omitted the superscript ^{R} for properties

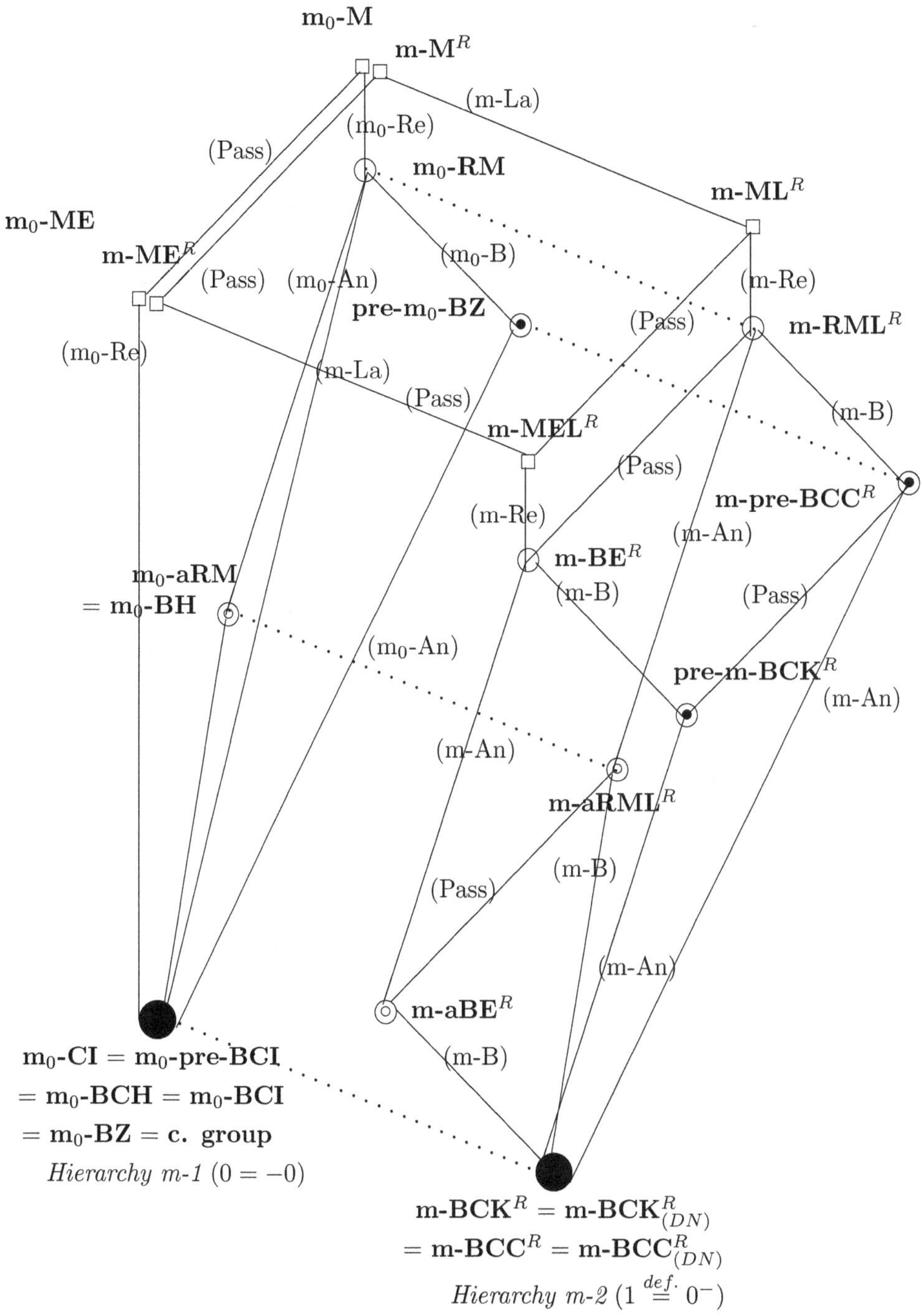

Figure 11.2: The better "Big $\mathrm{m_0}$-m-map": Hierarchies m-1 and m-2 and $\mathbf{m\text{-}M}^R$, $\mathbf{m\text{-}ME}^R$ (see ([22], Figure 10.5) for the non-commutative left-case). We have omitted the superscript ^{R} for properties

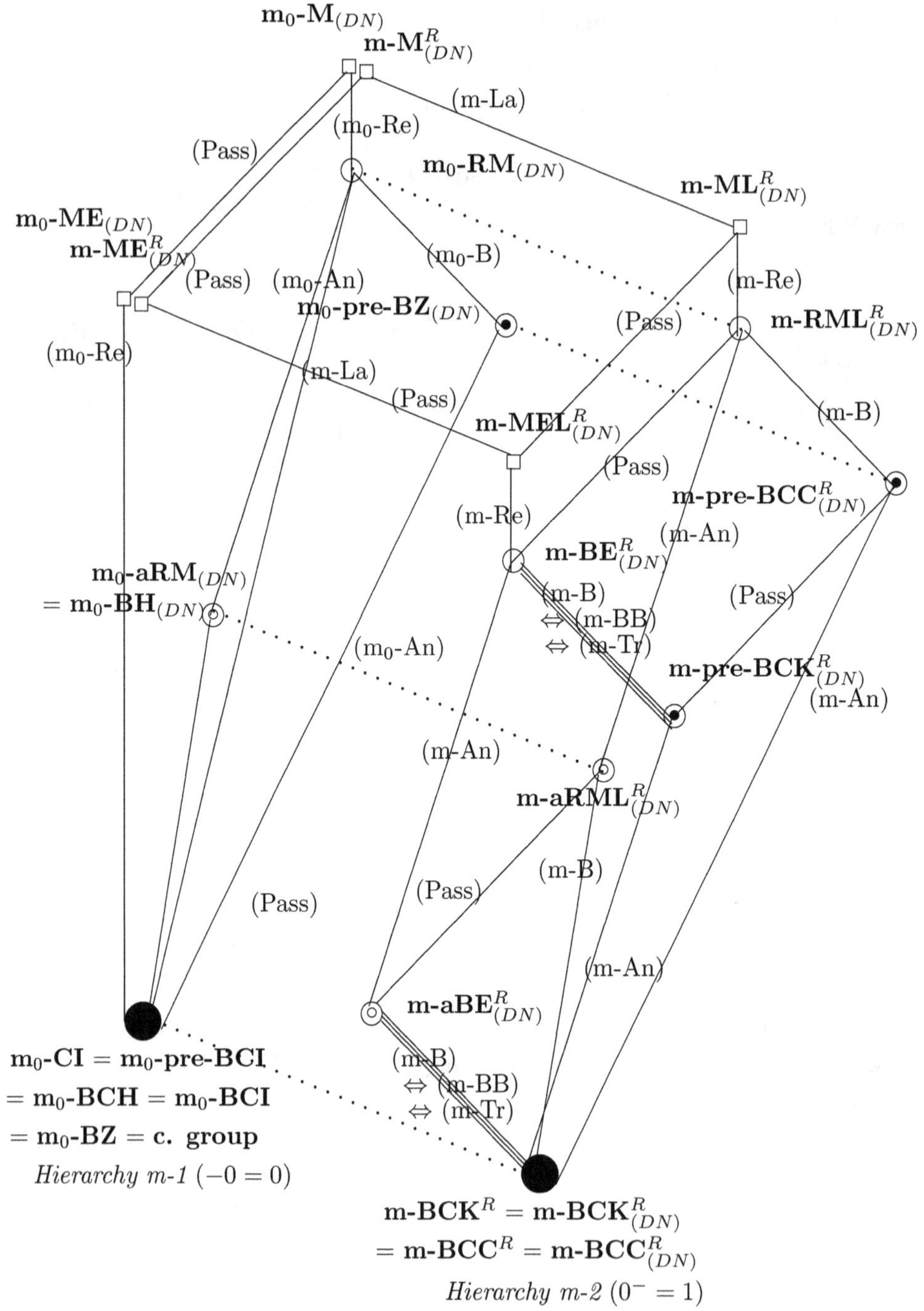

Figure 11.3: The "involutive Big m_0-m-map" of involutive right-algebras from Hierarchies m-1 and m-2 and $\mathbf{m\text{-}M}^R_{(DN)}$, $\mathbf{m\text{-}ME}^R_{(DN)}$ (see ([22], Figure 13.1) for the non-commutative left-case). We have omitted the superscript ^{R} for properties

11.2 Appendix B

11.2.1 The 'hybrid' proof of Proposition 2.2.60

Proposition Let $\mathcal{A} = (A, \oplus, ^-, 0)$ be a right-MV algebra. Then, the property (XX) holds, where:
(XX) $\quad (x \oplus y) \odot ((x \odot y) \oplus z) = (y \oplus z) \odot ((y \odot z) \oplus x).$

Proof. (Based on a proof by PROVER9, Length of proof was 92; the length of the expanded proof was 280.)

Let $\mathcal{A} = (A, \oplus, ^-, 0)$ be a right-MV algebra; then (SU), (Scomm), (Sass), (m-La^R), (DN), ($\vee_m$-comm) ($x \vee_m^M y = y \vee_m^M x$), equivalently ($M_\vee$) ($(x \oplus y) \odot ((x \odot y) \oplus y^-) = x$), (Z) ($(x \oplus y) \oplus (x \odot y) = x \oplus y$) and (m-Re^R) hold; dually, (PU), (Pcomm), (Pass), (m-La) and (m-Re) hold too, where:

$$x \odot y \overset{def.}{=} (x^- \oplus y^-)^- \tag{11.1}$$

and

$$x \oplus y \overset{def.}{=} (x^- \odot y^-)^-. \tag{11.2}$$

Hence, we immediately obtain:

$$(x \odot y)^- = x^- \oplus y^-, \tag{11.3}$$

$$(x \oplus y^-)^- = x^- \odot y, \tag{11.4}$$

$$(x \oplus y)^- = x^- \odot y^-. \tag{11.5}$$

Note that ($\vee_m$-comm) means $y \oplus (y \oplus x^-)^- = x \oplus (x \oplus y^-)^-$, hence, by (11.1), $y \oplus (y^- \odot x) = x \oplus (x^- \odot y)$, which, by (Pcomm), becomes:

$$x \oplus (x^- \odot y) = y \oplus (x \odot y^-). \tag{11.6}$$

Note also that (Z), by (Sass), becomes:

$$x \oplus (y \oplus (x \odot y)) = x \oplus y. \tag{11.7}$$

We have to prove that (XX) holds, i.e.

$$(c1 \oplus c2) \odot ((c1 \odot c2) \oplus c3) = (c2 \oplus c3) \odot ((c2 \odot c3) \oplus c1). \tag{11.8}$$

To prove this, the following properties are immediate:

$$x \oplus (y \oplus z) = y \oplus (x \oplus z), \tag{11.9}$$

$$x \oplus (x^- \oplus y) = 1, \tag{11.10}$$

(11.11) $$(x \odot y) \oplus (x^- \oplus y^-) = 1,$$

(11.12) $$x \oplus (y \oplus x^-) = 1,$$

(11.13) $$x \oplus (y \oplus (y \oplus x)^-) = 1,$$

(11.14) $$x \oplus (y \oplus (z \oplus (x^- \odot y^-))) = 1,$$

and, dually,

(11.15) $$x \odot (y \odot z) = y \odot (x \odot z),$$

(11.16) $$x \odot (x^- \odot y) = 0,$$

(11.17) $$x \odot (y \odot (x^- \oplus y^-)) = 0,$$

(11.18) $$x \odot (y \odot (z \odot x^-)) = 0,$$

(11.19) $$(x \oplus y) \odot (y^- \odot x^-) = 0.$$

Now, in (M$_\vee$) ($(x \oplus y) \odot ((x \odot y) \oplus y^-) = x$), interchange x with y to obtain:
$(y \oplus x) \odot ((y \odot x) \oplus x^-) = y$,
which, by (Scomm), becomes:

(11.20) $$(x \oplus y) \odot ((y \odot x) \oplus x^-) = y.$$

Then, from (M$_\vee$), by (Scomm), we obtain:

(11.21) $$(x \oplus y) \odot (y^- \oplus (x \odot y)) = x.$$

Then, in (M$_\vee$) again, take $X := x \oplus y$, $Y := z$, to obtain:
$((x \oplus y) \oplus z) \odot (((x \oplus y) \odot z) \oplus z^-) = x \oplus y$,
which, by (Sass), becomes:

(11.22) $$(x \oplus (y \oplus z)) \odot (((x \oplus y) \odot z) \oplus z^-) = x \oplus y.$$

Then, in (M$_\vee$) again, take $Y := y^-$, to obtain, by (DN):
$(x \oplus y^-) \odot ((x \odot y^-) \oplus y) = x$,
which, by (Scomm), becomes:

(11.23) $$(x \oplus y^-) \odot (y \oplus (x \odot y^-)) = x.$$

Then, from (M$_\vee$) again, by (Pcomm), we obtain:

(11.24) $$(x \oplus y) \odot ((y \odot x) \oplus y^-) = x.$$

Now, in (Pass) ($(x \odot y) \odot z = x \odot (y \odot z)$), take $X := x \oplus z$, $Y := (x \odot z) \oplus z^-$, $Z := y$, to obtain:
$((x \oplus z) \odot [(x \odot z) \oplus z^-]) \odot y = (x \oplus z) \odot ([(x \odot z) \oplus z^-] \odot y)$,
which, by ($M_\vee$), becomes:
$x \odot y = (x \oplus z) \odot ([(x \odot z) \oplus z^-] \odot y)$, i.e.

$$(x \oplus y) \odot (((x \odot y) \oplus y^-) \odot z) = x \odot z. \tag{11.25}$$

Now, from (11.25), by (Scomm) and (Pcomm), we obtain:

$$(x \oplus y) \odot (z \odot ((x \odot y) \oplus y^-)) = z \odot x. \tag{11.26}$$

Then, in ($M_\vee$) again, take $X := x \oplus y$, $Y := (x \odot y) \oplus y^-$, to obtain:
$((x \oplus y) \oplus [(x \odot y) \oplus y^-]) \odot ((x \oplus y) \odot [(x \odot y) \oplus y^-]) \oplus [(x \odot y) \oplus y^-]^-) = x \oplus y$,
which, by ($M_\vee$), becomes:
$((x \oplus y) \oplus ((x \odot y) \oplus y^-)) \odot (x \oplus [(x \odot y) \oplus y^-]^-) = x \oplus y$,
which, by (11.9), becomes:
$((x \odot y) \oplus ((x \oplus y) \oplus y^-)) \odot (x \oplus [(x \odot y) \oplus y^-]^-) = x \oplus y$,
which, by (Sass), becomes:
$((x \odot y) \oplus (x \oplus (y \oplus y^-))) \odot (x \oplus [(x \odot y) \oplus y^-]^-) = x \oplus y$,
which, by (m-Re^R), (m-La^R) twice, becomes:
$1 \odot (x \oplus [(x \odot y) \oplus y^-]^-) = x \oplus y$,
which, by (PU), becomes:
$x \oplus [(x \odot y) \oplus y^-]^- = x \oplus y$,
which, by (11.4), becomes:
$x \oplus [(x \odot y)^- \odot y] = x \oplus y$, hence
$x \oplus [(x^- \oplus y^-) \odot y] = x \oplus y$,
which, by (Pcomm), becomes:

$$x \oplus (y \odot (x^- \oplus y^-)) = x \oplus y. \tag{11.27}$$

Then, in ($M_\vee$) again, take $Y := y \oplus x^-$, to obtain:
$(x \oplus (y \oplus x^-)) \odot ((x \odot (y \oplus x^-)) \oplus (y \oplus x^-)^-) = x$, which, by (11.12) and (PU), becomes:
$(x \odot (y \oplus x^-)) \oplus (y \oplus x^-)^- = x$,
which, by (11.4), becomes:

$$(x \odot (y \oplus x^-)) \oplus (y^- \odot x) = x. \tag{11.28}$$

Now, in (11.10) ($x \oplus (x^- \oplus z) = 1$), take $X := x \oplus y^-$, to obtain:
$(x \oplus y^-) \oplus ((x \oplus y^-)^- \oplus z) = 1$,
which, by (11.4), becomes:
$(x \oplus y^-) \oplus ((x^- \odot y) \oplus z) = 1$,
which, by (Sass), becomes:

$$x \oplus (y^- \oplus ((x^- \odot y) \oplus z)) = 1. \tag{11.29}$$

Dually, in (11.16) ($x \odot (x^- \odot z) = 0$), take also $X := x \oplus y^-$, to obtain:
$(x \oplus y^-) \odot ((x \oplus y^-)^- \odot z) = 0$,

which, by (11.4), becomes:
$(x \oplus y^-) \odot ((x^- \odot y) \odot z) = 0$,
which, by (Pass), becomes:

$$(x \oplus y^-) \odot (x^- \odot (y \odot z)) = 0. \tag{11.30}$$

Now, from (11.13), by (11.5), we obtain:

$$x \oplus (y \oplus (y^- \odot x^-)) = 1. \tag{11.31}$$

Then, in (11.31), take $Y := y \odot z$, to obtain:
$x \oplus ((y \odot z) \oplus ((y \odot z)^- \odot x^-)) = 1$,
which, by (11.3), becomes:

$$x \oplus ((y \odot z) \oplus ((y^- \oplus z^-) \odot x^-)) = 1. \tag{11.32}$$

Now, in (Pass) $((x \odot y) \odot z = x \odot (y \odot z))$, take $X := z \oplus x$, $Y := (x \odot z) \oplus z^-$, $Z := y$, to obtain:
$((z \oplus x) \odot [(x \odot z) \oplus z^-]) \odot y = (z \oplus x) \odot ([(x \odot z) \oplus z^-] \odot y)$,
which, by (11.20), becomes:
$x \odot y = (z \oplus x) \odot (((x \odot z) \oplus z^-) \odot y)$, i.e.

$$(x \oplus y) \odot ([(y \odot x) \oplus x^-] \odot z) = y \odot z, \tag{11.33}$$

which, by (Pcomm) twice, becomes:

$$(x \oplus y) \odot (z \odot [(y \odot x) \oplus x^-]) = z \odot y. \tag{11.34}$$

Now, in (11.24) $((x \oplus y) \odot ((y \odot x) \oplus y^-) = x)$, take $X := (x \odot y) \oplus y^-$, $Y := x \oplus y$, to obtain:
$([(x \odot y) \oplus y^-] \oplus (x \oplus y)) \odot ((x \oplus y \odot [(x \odot y) \oplus y^-]) \oplus (x \oplus y)^-) = (x \odot y) \oplus y^-$,
which, by ($M_\vee$), becomes:
$([(x \odot y) \oplus y^-] \oplus (x \oplus y)) \odot (x \oplus (x \oplus y)^-) = (x \odot y) \oplus y^-$,
which, by (Scomm), becomes:
$((x \oplus y) \oplus ((x \odot y) \oplus y^-)) \odot (x \oplus (x \oplus y)^-) = (x \odot y) \oplus y^-$,
which, by (11.9), becomes:
$((x \odot y) \oplus ((x \oplus y) \oplus y^-)) \odot (x \oplus (x \oplus y)^-) = (x \odot y) \oplus y^-$,
which, by (Sass), (m-Re^R), (m-La^R), becomes:
$1 \odot (x \oplus (x \oplus y)^-) = (x \odot y) \oplus y^-$,
which, by (PU), becomes:
$x \oplus (x \oplus y)^- = (x \odot y) \oplus y^-$,
which, by definition of $\odot$, becomes:
$x \oplus (x^- \odot y^-) = (x \odot y) \oplus y^-$, hence:

$$(x \odot y) \oplus y^- = x \oplus (x^- \odot y^-). \tag{11.35}$$

Then, in (11.24) again, take $X := x \odot (y^- \oplus x^-)$, to obtain:
$([x \odot (y^- \oplus x^-)] \oplus y) \odot ((y \odot [x \odot (y^- \oplus x^-)]) \oplus y^-) = x \odot (y^- \oplus x^-)$,

which, by (11.17), becomes:
$([x \odot (y^- \oplus x^-)] \oplus y) \odot (0 \oplus y^-) = x \odot (y^- \oplus x^-)$,
which, by (SU) and (Scomm), becomes:
$(y \oplus [x \odot (y^- \oplus x^-)]) \odot y^- = x \odot (y^- \oplus x^-)$,
which, by (11.27), becomes:
$(y \oplus x) \odot y^- = x \odot (y^- \oplus x^-)$, i.e.

$$x \odot (y^- \oplus x^-) = (y \oplus x) \odot y^-. \tag{11.36}$$

Then, in (11.24) again, take $X := ((x \oplus y) \odot z) \oplus z^-$, $Y := x \oplus (y \oplus z)$, to obtain:
(a) $([((x \oplus y) \odot z) \oplus z^-] \oplus [x \oplus (y \oplus z)])\odot$
$(([x \oplus (y \oplus z)] \odot [((x \oplus y) \odot z) \oplus z^-]) \oplus [x \oplus (y \oplus z)]^-) = ((x \oplus y) \odot z) \oplus z^-$;
but, in (a), the part $[x \oplus (y \oplus z)] \odot [((x \oplus y) \odot z) \oplus z^-]$ equals $x \oplus y$, by (11.22); then, (a) becomes:
$([((x \oplus y) \odot z) \oplus z^-] \oplus [x \oplus (y \oplus z)])\odot ((x \oplus y) \oplus [x \oplus (y \oplus z)]^-) = ((x \oplus y) \odot z) \oplus z^-$,
which, by (Scomm), becomes:
$([x \oplus (y \oplus z)] \oplus [((x \oplus y) \odot z) \oplus z^-])\odot ((x \oplus y) \oplus [x \oplus (y \oplus z)]^-) = ((x \oplus y) \odot z) \oplus z^-$,
which, by (11.9), becomes:
$(((x \oplus y) \odot z) \oplus [(x \oplus (y \oplus z)) \oplus z^-]) \odot ((x \oplus y) \oplus (x \oplus (y \oplus z))^-) = ((x \oplus y) \odot z) \oplus z^-$,
which, by (Scomm), (Sass), (m-Re^R), (m-La^R), becomes:
$1 \odot ((x \oplus y) \oplus (x \oplus (y \oplus z))^-) = ((x \oplus y) \odot z) \oplus z^-$,
which, by (PU), becomes:
$(x \oplus y) \oplus (x \oplus (y \oplus z))^- = ((x \oplus y) \odot z) \oplus z^-$,
which, by (Sass), becomes:
$x \oplus (y \oplus (x \oplus (y \oplus z))^-) = ((x \oplus y) \odot z) \oplus z^-$,
which, by definition of $\odot$, becomes:
$x \oplus (y \oplus (x^- \odot (y^- \odot z^-))) = ((x \oplus y) \odot z) \oplus z^-$, i.e.

$$((x \oplus y) \odot z) \oplus z^- = x \oplus (y \oplus (x^- \odot (y^- \odot z^-))). \tag{11.37}$$

Now, in (11.25), take $Z := (((x \odot y) \odot y^-) \oplus y^=) \odot z$, to obtain:
(b) $(x \oplus y) \odot [((x \odot y) \oplus y^-) \odot [(((x \odot y) \odot y^-) \oplus y^=) \odot z]] = x \odot [(((x \odot y) \odot y^-) \oplus y^=) \odot z]$;
but, in (b), the part $((x \odot y) \oplus y^-) \odot [(((x \odot y) \odot y^-) \oplus y^=) \odot z]$ equals $(x \odot y) \odot z$, by (11.25); then, (b) becomes, by (DN):
$(x \oplus y) \odot [(x \odot y) \odot z] = x \odot [(((x \odot y) \odot y^-) \oplus y) \odot z]$,
which, by (Pass) on left side, becomes:
$(x \oplus y) \odot [x \odot (y \odot z)] = x \odot [(((x \odot y) \odot y^-) \oplus y) \odot z]$,
which, by (11.15) on left side, becomes:
$x \odot [(x \oplus y) \odot (y \odot z)] = x \odot [(((x \odot y) \odot y^-) \oplus y) \odot z]$,
which, by (11.15) again on left side, becomes:
$x \odot [y \odot ((x \oplus y) \odot z)] = x \odot [(((x \odot y) \odot y^-) \oplus y) \odot z]$,
which, by (Pass) on right side, becomes:
$x \odot [y \odot ((x \oplus y) \odot z)] = x \odot [((x \odot (y \odot y^-)) \oplus y) \odot z]$,
which, by (m-Re), (m-La), (SU) on right side, becomes:

$$x \odot [y \odot ((x \oplus y) \odot z)] = x \odot [y \odot z]. \tag{11.38}$$

Now, in (Pass) $((x \odot y) \odot z = x \odot (y \odot z))$, take $X := y \oplus z$, $Y := z^- \odot y^-$, $Z := x$, to obtain:

$((y \oplus z) \odot (z^- \odot y^-)) \odot x = (y \oplus z) \odot ((z^- \odot y^-) \odot x)$,
which, by (11.19), becomes:
$0 \odot x = (y \oplus z) \odot ((z^- \odot y^-) \odot x)$,
which, by (m-La), becomes:
$(y \oplus z) \odot ((z^- \odot y^-) \odot x) = 0$,
which, by (Pass), becomes:
$(y \oplus z) \odot (z^- \odot (y^- \odot x)) = 0$, i.e.

(11.39) $$(x \oplus y) \odot (y^- \odot (x^- \odot z)) = 0.$$

Now, in (11.28), take $Y := y^-$, to obtain, by (DN):
$(x \odot (y^- \oplus x^-)) \oplus (y \odot x) = x$,
which, by (11.36), becomes:
$((y \oplus x) \odot y^-) \oplus (y \odot x) = x$,
which, by (Scomm), becomes:

(11.40) $$(x \odot y) \oplus ((x \oplus y) \odot x^-) = y.$$

Now, in (11.30) $((x \oplus y^-) \odot (x^- \odot (y \odot z)) = 0)$, take $Y := y \oplus z^-$, $Z := z \oplus (y \odot z^-)$, to obtain:
$(x \oplus (y \oplus z^-)^-) \odot (x^- \odot ((y \oplus z^-) \odot [z \oplus (y \odot z^-)])) = 0$,
which, by (11.23), becomes:
$(x \oplus (y \oplus z^-)^-) \odot (x^- \odot y) = 0$,
which, by (11.5) and (DN), becomes:

(11.41) $$(x \oplus (y^- \odot z)) \odot (x^- \odot y) = 0.$$

Now, from (11.14) $(x \oplus (y \oplus (z \oplus (x^- \odot y^-))) = 1)$, by (Pcomm), we obtain:

(11.42) $$x \oplus (y \oplus (z \oplus (y^- \odot x^-))) = 1.$$

Now, in (11.29) $(x \oplus (y^- \oplus ((x^- \odot y) \oplus z)) = 1)$, take $Z := z \odot (x^- \odot y)^-$, to obtain:
(c) $x \oplus (y^- \oplus ((x^- \odot y) \oplus [z \odot (x^- \odot y)^-])) = 1$;
but, in (c), the part $(x^- \odot y) \oplus [z \odot (x^- \odot y)^-]$ equals $z \oplus (z^- \odot (x^- \odot y))$, by (11.6) for $Y := x^- \odot y$; then, (c) becomes:

(11.43) $$x \oplus (y^- \oplus (z \oplus (z^- \odot (x^- \odot y)))) = 1.$$

Now, in (11.26) $((x \oplus y) \odot (z \odot ((x \odot y) \oplus y^-)) = z \odot x)$, take $X := x \oplus y$, $Y := y^- \odot (x^- \odot z)$, $Z := u$, to obtain:
$((x \oplus y) \oplus [y^- \odot (x^- \odot z)]) \odot (u \odot (((x \oplus y) \odot [y^- \odot (x^- \odot z)]) \oplus [y^- \odot (x^- \odot z)]^-)) = u \odot (x \oplus y)$,
which, by (11.39), becomes:
$((x \oplus y) \oplus [y^- \odot (x^- \odot z)]) \odot (u \odot (0 \oplus [y^- \odot (x^- \odot z)]^-)) = u \odot (x \oplus y)$,
which, by (SU), becomes:
$((x \oplus y) \oplus [y^- \odot (x^- \odot z)]) \odot (u \odot [y^- \odot (x^- \odot z)]^-) = u \odot (x \oplus y)$,
which, by (Sass), becomes:

$(x \oplus (y \oplus [y^- \odot (x^- \odot z)])) \odot (u \odot [y^- \odot (x^- \odot z)]^-) = u \odot (x \oplus y)$,
which, by (11.3) and (DN), becomes:

$$(x \oplus (y \oplus [y^- \odot (x^- \odot z)])) \odot (u \odot [y \oplus (x \oplus z^-)]) = u \odot (x \oplus y). \quad (11.44)$$

Now, in (11.41) ($(x \oplus (y^- \odot z)) \odot (x^- \odot y) = 0$), take $Y := y^-$, to obtain, by (DN):

$$(x \oplus (y \odot z)) \odot (x^- \odot y^-) = 0. \quad (11.45)$$

Dually, we have:

$$(x \odot (y \oplus z)) \oplus (x^- \oplus y^-) = 1. \quad (11.46)$$

And, from (11.46), by (Scomm), we obtain:

$$(x \odot (y \oplus z)) \oplus (y^- \oplus x^-) = 1. \quad (11.47)$$

Now, in (11.33) ($(x \oplus y) \odot ([(y \odot x) \oplus x^-] \odot z) = y \odot z$), take $X := y \odot x$, $Y := (y \oplus x) \odot y^-$, to obtain:
$((y \odot x) \oplus [(y \oplus x) \odot y^-]) \odot ([([(y \oplus x) \odot y^-] \odot (y \odot x)) \oplus (y \odot x)^-] \odot z) = [(y \oplus x) \odot y^-] \odot z$,
which, by (11.40), becomes:
$x \odot ([([(y \oplus x) \odot y^-] \odot (y \odot x)) \oplus (y \odot x)^-] \odot z) = [(y \oplus x) \odot y^-] \odot z$,
which, by (Pcomm), becomes:
$x \odot ([((y \odot x) \odot [(y \oplus x) \odot y^-]) \oplus (y \odot x)^-] \odot z) = [(y \oplus x) \odot y^-] \odot z$,
which, by (Pass), becomes:
$x \odot ([(y \odot (x \odot [(y \oplus x) \odot y^-])) \oplus (y \odot x)^-] \odot z) = [(y \oplus x) \odot y^-] \odot z$,
which, by (11.18), becomes:
$x \odot ([0 \oplus (y \odot x)^-] \odot z) = [(y \oplus x) \odot y^-] \odot z$,
which, by (SU) and (11.3), becomes:
$x \odot ((y^- \oplus x^-) \odot z) = [(y \oplus x) \odot y^-] \odot z$,
which, by (Sass), becomes:

$$x \odot ((y^- \oplus x^-) \odot z) = (y \oplus x) \odot [y^- \odot z]. \quad (11.48)$$

Now, in (11.20) ($(x \oplus y) \odot ((y \odot x) \oplus x^-) = y$), take $X := z$, $Y := x \oplus (y \oplus (x^- \odot z^-))$, to obtain:
$(z \oplus [x \oplus (y \oplus (x^- \odot z^-))]) \odot (([x \oplus (y \oplus (x^- \odot z^-))] \odot z) \oplus z^-) = x \oplus (y \oplus (x^- \odot z^-))$,
which, by (11.42), becomes:
$1 \odot (([x \oplus (y \oplus (x^- \odot z^-))] \odot z) \oplus z^-) = x \oplus (y \oplus (x^- \odot z^-))$,
which, by (PU), becomes:
$([x \oplus (y \oplus (x^- \odot z^-))] \odot z) \oplus z^- = x \oplus (y \oplus (x^- \odot z^-))$,
which, by (Scomm), becomes:
$z^- \oplus ([x \oplus (y \oplus (x^- \odot z^-))] \odot z) = x \oplus (y \oplus (x^- \odot z^-))$,
which, by (Pcomm), becomes:

$$z^- \oplus (z \odot [x \oplus (y \oplus (x^- \odot z^-))]) = x \oplus (y \oplus (x^- \odot z^-)). \quad (11.49)$$

Now, in ($M_\vee$) ($(x \oplus y) \odot ((x \odot y) \oplus y^-) = x$) again, take $X := x \oplus (y \odot z)$, $Y := x^- \odot y^-$, to obtain:

$([x \oplus (y \odot z)] \oplus (x^- \odot y^-)) \odot (([x \oplus (y \odot z)] \odot (x^- \odot y^-)) \oplus (x^- \odot y^-)^-) = x \oplus (y \odot z)$,
which, by (11.45), becomes:
$([x \oplus (y \odot z)] \oplus (x^- \odot y^-)) \odot (0 \oplus (x^- \odot y^-)^-) = x \oplus (y \odot z)$,
which, by (SU), becomes:
$([x \oplus (y \odot z)] \oplus (x^- \odot y^-)) \odot (x^- \odot y^-)^- = x \oplus (y \odot z)$,
which, by (Sass), becomes:
$(x \oplus [(y \odot z) \oplus (x^- \odot y^-)]) \odot (x^- \odot y^-)^- = x \oplus (y \odot z)$,
which, by (11.3), becomes:
$(x \oplus [(y \odot z) \oplus (x^- \odot y^-)]) \odot (x \oplus y) = x \oplus (y \odot z)$,
which, by (Pcomm), becomes:

(11.50) $$(x \oplus y) \odot (x \oplus [(y \odot z) \oplus (x^- \odot y^-)]) = x \oplus (y \odot z).$$

Now, in (11.34) $((x \oplus y) \odot (z \odot [(y \odot x) \oplus x^-]) = z \odot y)$, take $X := x^- \odot y^-$, $Y := x \oplus (y \odot z)$, $Z := u$, to obtain:
$((x^- \odot y^-) \oplus [x \oplus (y \odot z)]) \odot (u \odot [([x \oplus (y \odot z)] \odot (x^- \odot y^-)) \oplus (x^- \odot y^-)^-]) = u \odot [x \oplus (y \odot z)]$,
which, by (11.45), becomes:
$((x^- \odot y^-) \oplus [x \oplus (y \odot z)]) \odot (u \odot [0 \oplus (x^- \odot y^-)^-]) = u \odot [x \oplus (y \odot z)]$,
which, by (SU), becomes:
$((x^- \odot y^-) \oplus [x \oplus (y \odot z)]) \odot (u \odot (x^- \odot y^-)^-) = u \odot [x \oplus (y \odot z)]$,
which, by (11.9), becomes:
$(x \oplus [(x^- \odot y^-) \oplus (y \odot z)]) \odot (u \odot (x^- \odot y^-)^-) = u \odot [x \oplus (y \odot z)]$,
which, by (11.3), becomes:

(11.51) $$(x \oplus [(x^- \odot y^-) \oplus (y \odot z)]) \odot (u \odot (x \oplus y)) = u \odot [x \oplus (y \odot z)].$$

Now, in (11.24) $((x \oplus y) \odot ((y \odot x) \oplus y^-) = x)$ again, take $X := y \odot (x \oplus z)$, $Y := x^- \oplus y^-$, to obtain:
$([y \odot (x \oplus z)] \oplus (x^- \oplus y^-)) \odot (((x^- \oplus y^-) \odot [y \odot (x \oplus z)]) \oplus (x^- \oplus y^-)^-) = y \odot (x \oplus z)$,
which, by (11.47), becomes:
$1 \odot (((x^- \oplus y^-) \odot [y \odot (x \oplus z)]) \oplus (x^- \oplus y^-)^-) = y \odot (x \oplus z)$,
which, by (PU), becomes:
$((x^- \oplus y^-) \odot [y \odot (x \oplus z)]) \oplus (x^- \oplus y^-)^- = y \odot (x \oplus z)$,
which, by (11.15), becomes:
(d) $(y \odot [(x^- \oplus y^-) \odot (x \oplus z)]) \oplus (x^- \oplus y^-)^- = y \odot (x \oplus z)$;
but, in (d), the part $y \odot [(x^- \oplus y^-) \odot (x \oplus z)]$ equals $(x \oplus y) \odot (x^- \odot (x \oplus z))$, by (11.48); then, (d) becomes:
$((x \oplus y) \odot (x^- \odot (x \oplus z))) \oplus (x^- \oplus y^-)^- = y \odot (x \oplus z)$,
which, by (11.5), becomes:
$((x \oplus y) \odot (x^- \odot (x \oplus z))) \oplus (x \odot y) = y \odot (x \oplus z)$,
which, by (Scomm), becomes:

(11.52) $$(x \odot y) \oplus ((x \oplus y) \odot (x^- \odot (x \oplus z))) = y \odot (x \oplus z).$$

Now, in (Sass) $((x \oplus y) \oplus z = x \oplus (y \oplus z))$, take $X := x \odot y$, $Y := y^-$, to obtain:
$((x \odot y) \oplus y^-) \oplus z = (x \odot y) \oplus (y^- \oplus z)$,

which, by (11.35), becomes:
$(x \oplus (x^- \odot y^-)) \oplus z = (x \odot y) \oplus (y^- \oplus z)$,
which, by (Sass), becomes:

$$x \oplus ((x^- \odot y^-) \oplus z) = (x \odot y) \oplus (y^- \oplus z). \tag{11.53}$$

Then, from (11.51) $((x \oplus [(x^- \odot y^-) \oplus (y \odot z)]) \odot (u \odot (x \oplus y)) = u \odot [x \oplus (y \odot z)])$, by (11.53), we obtain:

$$((x \odot y) \oplus (y^- \oplus (y \odot z))) \odot (u \odot (x \oplus y)) = u \odot [x \oplus (y \odot z)]. \tag{11.54}$$

Now, in (11.38) $(x \odot [y \odot ((x \oplus y) \odot z)] = x \odot [y \odot z])$, take $Y := y \oplus z$, $Z := (y \odot z) \oplus z^-$, to obtain:
$x \odot [(y \oplus z) \odot ((x \oplus (y \oplus z)) \odot [(y \odot z) \oplus z^-])] = x \odot [(y \oplus z) \odot [(y \odot z) \oplus z^-]]$,
which, by ($M_\vee$), becomes:
(e) $x \odot [(y \oplus z) \odot ((x \oplus (y \oplus z)) \odot [(y \odot z) \oplus z^-])] = x \odot y$;
but, in (e) the part $(y \oplus z) \odot ((x \oplus (y \oplus z)) \odot [(y \odot z) \oplus z^-])$ equals $(x \oplus (y \oplus z)) \odot y$, by (11.26); then, (e) becomes:
$x \odot [(x \oplus (y \oplus z)) \odot y] = x \odot y$,
which, by (Pcomm), becomes:

$$x \odot [y \odot (x \oplus (y \oplus z))] = x \odot y. \tag{11.55}$$

Now, by (11.7) $(x \oplus (y \oplus (x \odot y)) = x \oplus y)$,
$(x \oplus y) \oplus z = (x \oplus (y \oplus (x \odot y))) \oplus z \overset{(Sass)}{=} x \oplus ((y \oplus (x \odot y)) \oplus z)$,
hence, we have:
$(x \oplus y) \oplus z = x \oplus ((y \oplus (x \odot y)) \oplus z)$,
which, by (Sass) twice, becomes:
$x \oplus (y \oplus z) = x \oplus (y \oplus ((x \odot y) \oplus z))$, i.e.

$$x \oplus (y \oplus ((x \odot y) \oplus z)) = x \oplus (y \oplus z). \tag{11.56}$$

Finally, in (11.55), take $Z := x^- \odot (y^- \odot z^-)$, to obtain:
(f) $x \odot [y \odot (x \oplus (y \oplus [x^- \odot (y^- \odot z^-)]))] = x \odot y$;
but, in (f), the part $x \oplus (y \oplus [x^- \odot (y^- \odot z^-)])$ equals $((x \oplus y) \odot z) \oplus z^-$, by (11.37); then, (f) becomes:
$x \odot [y \odot (((x \oplus y) \odot z) \oplus z^-)] = x \odot y$,
which, for $Z := z^-$, becomes, by (DN):
$x \odot [y \odot (((x \oplus y) \odot z^-) \oplus z)] = x \odot y$,
which, by (Scomm), becomes:

$$x \odot [y \odot (z \oplus ((x \oplus y) \odot z^-))] = x \odot y. \tag{11.57}$$

We are ready now to prove (11.8).
Indeed, (keeping the numbering from the PROVER9 proof), we have:
$(c1 \oplus c2) \odot ((c1 \odot c2) \oplus c3)$
$= (c1 \oplus c2) \odot (c3 \oplus (c1 \odot c2))$, by (Scomm),

$235 = (c3 \oplus ((c1 \odot c2) \oplus ((c1 \odot c2)^- \odot (c3^- \odot x)))) \odot ((c1 \oplus c2) \odot ((c1 \odot c2) \oplus (c3 \oplus x^-)))$, by (11.44),
$236 = (c3 \oplus ((c1 \odot c2) \oplus ((c1^- \oplus c2^-) \odot (c3^- \odot x)))) \odot ((c1 \oplus c2) \odot ((c1 \odot c2) \oplus (c3 \oplus x^-)))$, by (11.3),
$237 = (c3 \oplus ((c1 \odot c2) \oplus ((c1^- \oplus c2^-) \odot (c3^- \odot x)))) \odot ((c1 \oplus c2) \odot (c3 \oplus ((c1 \odot c2) \oplus x^-)))$, by (11.9),
$238 = (c1 \oplus c2) \odot ((c3 \oplus ((c1 \odot c2) \oplus ((c1^- \oplus c2^-) \odot (c3^- \odot x)))) \odot (c3 \oplus ((c1 \odot c2) \oplus x^-)))$, by (11.15),
$239 = (c1 \oplus c2) \odot ((c3 \oplus ((c1 \odot c2) \oplus ((c1^- \oplus c2^-) \odot c3^-))) \odot (c3 \oplus ((c1 \odot c2) \oplus (x \oplus (((c1^- \oplus c2^-) \oplus c3^-) \odot x^-))^-)))$, by (11.57),
$240 = (c1 \oplus c2) \odot (1 \odot (c3 \oplus ((c1 \odot c2) \oplus (x \oplus (((c1^- \oplus c2^-) \oplus c3^-) \odot x^-))^-)))$, by (11.32),
$241 = (c1 \oplus c2) \odot (1 \odot (c3 \oplus ((c1 \odot c2) \oplus (x \oplus ((c1^- \oplus (c2^- \oplus c3^-)) \odot x^-))^-)))$, by (Sass),
$242 = (c1 \oplus c2) \odot (1 \odot (c3 \oplus ((c1 \odot c2) \oplus (x^- \odot ((c1^- \oplus (c2^- \oplus c3^-)) \odot x^-)^-))))$, by (11.5),
$243 = (c1 \oplus c2) \odot (1 \odot (c3 \oplus ((c1 \odot c2) \oplus (x^- \odot ((c1^- \oplus (c2^- \oplus c3^-))^- \oplus x^=)))))$, by (11.3),
$244 = (c1 \oplus c2) \odot (1 \odot (c3 \oplus ((c1 \odot c2) \oplus (x^- \odot ((-c1^- \odot (c2^- \oplus c3^-)^-) \oplus x^=)))))$, by (11.5),
$245 = (c1 \oplus c2) \odot (1 \odot (c3 \oplus ((c1 \odot c2) \oplus (x^- \odot ((c1 \odot (c2^- \oplus c3^-)^-) \oplus x^=)))))$, by (DN),
$246 = (c1 \oplus c2) \odot (1 \odot (c3 \oplus ((c1 \odot c2) \oplus (x^- \odot ((c1 \odot (c2^= \odot c3^=)) \oplus x^=)))))$, by (11.5),
$247 = (c1 \oplus c2) \odot (1 \odot (c3 \oplus ((c1 \odot c2) \oplus (x^- \odot ((c1 \odot (c2 \odot c3^=)) \oplus x^=)))))$, by (DN),
$248 = (c1 \oplus c2) \odot (1 \odot (c3 \oplus ((c1 \odot c2) \oplus (x^- \odot ((c1 \odot (c2 \odot c3)) \oplus x^=)))))$, by (DN),
$249 = (c1 \oplus c2) \odot (1 \odot (c3 \oplus ((c1 \odot c2) \oplus (x^- \odot ((c1 \odot (c2 \odot c3)) \oplus x)))))$, by (DN),
$250 = (c1 \oplus c2) \odot (1 \odot (c3 \oplus ((c1 \odot c2) \oplus (((c1 \odot (c2 \odot c3)) \oplus x) \odot x^-))))$, by (Pcomm),
$251 = (c1 \oplus c2) \odot (c3 \oplus ((c1 \odot c2) \oplus (((c1 \odot (c2 \odot c3)) \oplus x) \odot x^-)))$, by (PU),
$252 = (c1 \oplus c2) \odot (c3 \oplus ((c1 \odot c2) \oplus (1 \odot (x^- \oplus (y \oplus (y^- \odot ((c1 \odot (c2 \odot c3))^- \odot x))))^-)))$, by (11.43),
$253 = (c1 \oplus c2) \odot (c3 \oplus ((c1 \odot c2) \oplus (1 \odot (x^- \oplus (y \oplus (y^- \odot ((c1^- \oplus -(c2 \odot c3)) \odot x))))^-)))$, by (11.3),
$254 = (c1 \oplus c2) \odot (c3 \oplus ((c1 \odot c2) \oplus (1 \odot (x^- \oplus (y \oplus (y^- \odot ((c1^- \oplus (c2^- \oplus c3^-)) \odot x))))^-)))$, by (11.3),
$255 = (c1 \oplus c2) \odot (c3 \oplus ((c1 \odot c2) \oplus (1 \odot (x^= \odot (y \oplus (y^- \odot ((c1^- \oplus (c2^- \oplus c3^-)) \odot x)))^-))))$, by (11.5),
$256 = (c1 \oplus c2) \odot (c3 \oplus ((c1 \odot c2) \oplus (1 \odot (x \odot (y \oplus (y^- \odot ((c1^- \oplus (c2^- \oplus c3^-)) \odot x)))^-))))$, by (DN),
$257 = (c1 \oplus c2) \odot (c3 \oplus ((c1 \odot c2) \oplus (1 \odot (x \odot (y^- \odot (y^- \odot ((c1^- \oplus (c2^- \oplus c3^-)) \odot x))^-)))))$, by (11.5),
$258 = (c1 \oplus c2) \odot (c3 \oplus ((c1 \odot c2) \oplus (1 \odot (x \odot (y^- \odot (y^= \oplus ((c1^- \oplus (c2^- \oplus c3^-)) \odot x)^-))))))$, by (11.3),
$259 = (c1 \oplus c2) \odot (c3 \oplus ((c1 \odot c2) \oplus (1 \odot (x \odot (y^- \odot (y \oplus ((c1^- \oplus (c2^- \oplus c3^-)) \odot x)^-))))))$, by (DN),
$260 = (c1 \oplus c2) \odot (c3 \oplus ((c1 \odot c2) \oplus (1 \odot (x \odot (y^- \odot (y \oplus ((c1^- \oplus (c2^- \oplus c3^-))^- \oplus x^-)))))))$,

by (11.3),
$261 = (c1 \oplus c2) \odot (c3 \oplus ((c1 \odot c2) \oplus (1 \odot (x \odot (y^- \odot (y \oplus ((c1^{=} \odot (c2^- \oplus c3^-)^-)^-) \oplus x^-)))))))$,
by (11.5),
$262 = (c1 \oplus c2) \odot (c3 \oplus ((c1 \odot c2) \oplus (1 \odot (x \odot (y^- \odot (y \oplus ((c1 \odot (c2^- \oplus c3^-)^-)^-) \oplus x^-)))))))$,
by (DN),
$263 = (c1 \oplus c2) \odot (c3 \oplus ((c1 \odot c2) \oplus (1 \odot (x \odot (y^- \odot (y \oplus ((c1 \odot (c2^{=} \odot c3^{=})) \oplus x^-)))))))$,
by (11.5),
$264 = (c1 \oplus c2) \odot (c3 \oplus ((c1 \odot c2) \oplus (1 \odot (x \odot (y^- \odot (y \oplus ((c1 \odot (c2 \odot c3^{=})) \oplus x^-)))))))$,
by (DN),
$265 = (c1 \oplus c2) \odot (c3 \oplus ((c1 \odot c2) \oplus (1 \odot (x \odot (y^- \odot (y \oplus ((c1 \odot (c2 \odot c3)) \oplus x^-)))))))$,
by (DN),
$266 = (c1 \oplus c2) \odot (c3 \oplus ((c1 \odot c2) \oplus (x \odot (y^- \odot (y \oplus ((c1 \odot (c2 \odot c3)) \oplus x^-))))))$, by (PU),
$267 = (c1 \oplus c2) \odot (c3 \oplus (c2 \odot (c1 \oplus ((c1 \odot (c2 \odot c3)) \oplus (c1 \oplus c2)^-))))$, by (11.52),
$268 = (c1 \oplus c2) \odot (c3 \oplus (c2 \odot (c1 \oplus ((c1 \odot (c2 \odot c3)) \oplus (c1^- \odot c2^-)))))$, by (11.5),
$269 = ((c3 \odot c2) \oplus (c2^- \oplus (c2 \odot (c1 \oplus ((c1 \odot (c2 \odot c3)) \oplus (c1^- \odot c2^-)))))) \odot ((c1 \oplus c2) \odot (c3 \oplus c2))$, by (11.54),
$270 = ((c2 \odot c3) \oplus (c2^- \oplus (c2 \odot (c1 \oplus ((c1 \odot (c2 \odot c3)) \oplus (c1^- \odot c2^-)))))) \odot ((c1 \oplus c2) \odot (c3 \oplus c2))$, by (Pcomm),
$271 = ((c2 \odot c3) \oplus (c1 \oplus ((c1 \odot (c2 \odot c3)) \oplus (c1^- \odot c2^-)))) \odot ((c1 \oplus c2) \odot (c3 \oplus c2))$,
by (11.49),
$272 = (c1 \oplus ((c2 \odot c3) \oplus ((c1 \odot (c2 \odot c3)) \oplus (c1^- \odot c2^-)))) \odot ((c1 \oplus c2) \odot (c3 \oplus c2))$,
by (11.9),
$273 = (c1 \oplus ((c2 \odot c3) \oplus (c1^- \odot c2^-))) \odot ((c1 \oplus c2) \odot (c3 \oplus c2))$, by (11.56),
$274 = (c1 \oplus ((c2 \odot c3) \oplus (c1^- \odot c2^-))) \odot ((c1 \oplus c2) \odot (c2 \oplus c3))$, by (Scomm),
$275 = ((c1 \oplus c2) \odot (c2 \oplus c3)) \odot (c1 \oplus ((c2 \odot c3) \oplus (c1^- \odot c2^-)))$, by (Pcomm),
$276 = (c1 \oplus c2) \odot ((c2 \oplus c3) \odot (c1 \oplus ((c2 \odot c3) \oplus (c1^- \odot c2^-))))$, by (Pass),
$277 = (c2 \oplus c3) \odot ((c1 \oplus c2) \odot (c1 \oplus ((c2 \odot c3) \oplus (c1^- \odot c2^-))))$, by (11.15),
$278 = (c2 \oplus c3) \odot (c1 \oplus (c2 \odot c3))$, by (11.50),
$279 = (c2 \oplus c3) \odot ((c2 \odot c3) \oplus c1)$, by (Scomm).

Thus, (11.8) holds. □

11.3 Appendix C

11.3.1 Example of program and PROVER9 proof

Proposition. (See Proposition 3.4.4)

Let $(G, \vee, \wedge, +, -, 0, u)$ be a commutative l-group withe strong unit u. For all $0 \leq a, b \leq u$, we have: $0 \leq u \wedge (a + b)$, i.e. $0 \wedge (u \wedge (a + b)) = 0$.

The program we have written for PROVER9 and the expanded renumbered proof made by PROVER9 (Length is 17, Length of expanded proof is 24) are the following, where, in the program, u is denoted by 1, $\wedge$ is denoted by $*$, $\vee$ is denoted by v, $\Longleftrightarrow$ is denoted by $< - >$; in the proof by PROVER9, $! =$ means $\neq$ and $|$ means 'or'.

$x + 0 = x.$
$x + y = y + x.$
$(x + y) + z = x + (y + z).$
$x + (-x) = 0.$

$x * x = x.$
$x * y = y * x.$
$(x * y) * z = x * (y * z).$
$x * (x\, v\, y) = x.$

$x\, v\, x = x.$
$x\, v\, y = y\, v\, x.$
$(x\, v\, y)\, v\, z = x\, v\, (y\, v\, z).$
$x\, v\, (x * y) = x.$

$x * y = x <-> x\, v\, y = y.$

$x * y = x <-> (x + z) * (y + z) = x + z.$

$z + (x * y) = (z + x) * (z + y).$
$z + (x\, v\, y) = (z + x)\, v\, (z + y).$
$-(x * y) = (-x)\, v\, (-y).$
$-(x\, v\, y) = (-x) * (-y).$
$z\, v\, (x * y) = (z\, v\, x) * (z\, v\, y).$
$z * (x\, v\, y) = (z * x)\, v\, (z * y).$

$0 * 1 = 0.$

$0 * a = 0.$
$0 * b = 0.$
$a * 1 = a.$
$b * 1 = b.$

$(a\, v\, -a)\, v\, 1 = 1.$
$(b\, v\, -b)\, v\, 1 = 1.$
Goals :
$0 * (1 * (a + b)) = 0.$
================================== prooftrans ==========
Prover9 (32) version Dec-2007, Dec 2007.
Process 11084 was started by Afrodita on LAPTOP-3U6OSV86,
Tue Jul 22 15:37:51 2025
The command was "/cygdrive/c/Program Files (x86)/Prover9-Mace4/bin-win32/prover9".
================================== end of head =========

================================== end of input =========

=============================== PROOF ===========

——— Comments from original proof ———
Proof 1 at 0.03 (+ 0.03) seconds.
Length of proof is 17.
Level of proof is 4.
Maximum clause weight is 16.
Given clauses 89.

1 x * y = x < − > (x + z) * (y + z) = x + z # label(non-clause). [assumption].
2 0 * (1 * (a + b)) = 0 # label(non-clause) # label(goal). [goal].
3 x + 0 = x. [assumption].
4 x + y = y + x. [assumption].
5 (x * y) * z = x * (y * z). [assumption].
6 x * y != x | (x + z) * (y + z) = x + z. [clausify(1)].
7 0 * 1 = 0. [assumption].
8 0 * a = 0. [assumption].
9 0 * b = 0. [assumption].
10 0 * (1 * (a + b)) != 0. [deny(2)].
11 0 + x = x. [para(4(a,1),3(a,1))].
12 0 * x = 0 * (1 * x). [para(7(a,1),5(a,1,1))].
13 0 * (1 * x) = 0 * x. [copy(12),flip(a)].
14 0 * (a + b) != 0. [para(13(a,1),10(a,1))].
15 0 * x = 0 * (a * x). [para(8(a,1),5(a,1,1))].
16 0 * (a * x) = 0 * x. [copy(15),flip(a)].
17 (0 + x) * (b + x) = 0 + x. [resolve(6,a,9,a)].
18 x * (b + x) = 0 + x. [para(11(a,1),17(a,1,1))].
19 x * (b + x) = x. [para(11(a,1),18(a,2))].
20 0 * a = 0 * (b + a). [para(19(a,1),16(a,1,2))].
21 0 = 0 * (b + a). [para(8(a,1),20(a,1))].
22 0 = 0 * (a + b). [para(4(a,1),21(a,2,2))].
23 0 * (a + b) = 0. [copy(22),flip(a)].
24 $F. [resolve(23,a,14,a)].

=============================== end of proof =======

11.3.2 The 'hybrid' proof of Proposition 3.4.6

Proposition Let $\mathcal{G}_u = (G, \vee, \wedge, +, -, 0_G, u = u_G)$ be a commutative l-group with strong unit $u = u_G$. Then, we have:
($s4^R$) (ii) for all $0_G \leq a, b \leq u_G$,

$$u_G \wedge (u_G + -(u_G \wedge ((u_G + -a) + (u_G + -b)))) = u_G + -(u_G \wedge ((u_G + -a) + (u_G + -b))).$$

Proof. (By PROVER9, Length of proof was 99, in 18.02 seconds; the renumbered expanded proof had the Length 344.)

In order to simplify the writting (less space on a line), we shall replace 0_G by 0 and u_G by 1.

Hence, we have to prove that, for all $0 \leq a, b \leq 1$,

$$1 \wedge (1 + -(1 \wedge ((1 + -a) + (1 + -b)))) = 1 + -(1 \wedge ((1 + -a) + (1 + -b))). \tag{11.58}$$

The particular hypotheses used by PROVER9 (among the given ones - see (α)) are:

$$0 \wedge 1 = 0 \quad (i.e.\ 0 \leq 1), \tag{11.59}$$

$$a \wedge 1 = a \quad (i.e.\ a \leq 1). \tag{11.60}$$

$$b \wedge 1 = b \quad (i.e.\ b \leq 1), \tag{11.61}$$

hence, by (Wcomm):

$$1 \wedge a = a, \tag{11.62}$$

$$1 \wedge b = b. \tag{11.63}$$

The following properties are obvious, by (SU), (Scomm), (Sass), (m_0-Re), (NegS) and (Wcomm) ($x \wedge y = y \wedge x$), (Wass) ($(x \wedge y) \wedge z = x \wedge (y \wedge z)$):

$$x + (y + z) = y + (x + z), \tag{11.64}$$

$$x + (-x + y) = y, \tag{11.65}$$

$$x + (y + (-x + z)) = y + z, \tag{11.66}$$

$$-x + (y + x) = y, \tag{11.67}$$

$$x + (y + -x) = y, \tag{11.68}$$

$$x \wedge (y \wedge z) = y \wedge (x \wedge z), \tag{11.69}$$

$$x \wedge (y \wedge x) = y \wedge x, \tag{11.70}$$

$$x \vee y = -(-x \wedge -y). \tag{11.71}$$

Now, in (g3) ($(x+y)\wedge(x+z) = x+(y\wedge z)$), take first $Y := 0$, to obtain, by (SU):

$$x \wedge (x+z) = x + (0 \wedge z). \tag{11.72}$$

Then, in (g3) again, interchange x with y, to obtain, by (Scomm):

$$(x+y)\wedge(y+z) = y + (x\wedge z). \tag{11.73}$$

Then, from (g3) again, by (Sxcomm), we obtain:

$$(x+y)\wedge(z+x) = x + (y\wedge z). \tag{11.74}$$

Then, in (g3), take $Z := -x$, to obtain, by (m_0-Re): $(x+y)\wedge 0 = x + (y \wedge -x)$, hence, by (Wcomm):

$$0 \wedge (x+y) = x + (y\wedge -x). \tag{11.75}$$

Finally, in (g3), take $Z := -x+z$, to obtain:
$(x+y)\wedge(x+(-x+z)) = x + (y\wedge(-x+z))$,
which, by (11.65), becomes:

$$(x+y)\wedge z = x + (y \wedge (-x+z)). \tag{11.76}$$

Then, in (11.72), take first $Z := -x+y$, to obtain:
$x\wedge(x+(-x+y)) = x + (0\wedge(-x+y))$,
which, by (11.65), becomes:

$$x \wedge y = x + (0\wedge(-x+y)). \tag{11.77}$$

Then, in (11.72) again, take $X := -x$, $Z := y+x$, to obtain:
$-x\wedge(-x+(y+x)) = -x + (0\wedge(y+x))$,
which, by (11.67), becomes:

$$-x\wedge y = -x + (0 \wedge (y+x)). \tag{11.78}$$

Then, from (11.73), by (11.76) for $X := x$, $Y := y$, $Z := y+z$, we obtain:

$$x + (y\wedge(-x+(y+z))) = y + (x\wedge z). \tag{11.79}$$

Then, in (11.75), take $Y := 0$, to obtain, by (SU):
(*) $0\wedge x = x + (0\wedge -x)$;
then, in (*), take $X := -x$, to obtain, by (DN):

$$0\wedge -x = -x + (0\wedge x). \tag{11.80}$$

Now, in (11.75) again, interchange x with y, to obtain:
$0\wedge(y+x) = y + (x\wedge -y)$,
which, by (Scomm), becomes:

$$0\wedge(x+y) = y + (x\wedge -y). \tag{11.81}$$

Now, in (11.75) again, take $X := -x$, to obtain, by (DN):

$$0 \wedge (-x + y) = -x + (y \wedge x). \tag{11.82}$$

Now, in (Cp) ($x \leq y \Longleftrightarrow x + z \leq y + z$), i.e. in

$$x \wedge y = x \Longleftrightarrow (x + z) \wedge (y + z) = x + z, \tag{11.83}$$

take first $Z := -x + z$, to obtain:
$x \wedge y = x \Longleftrightarrow (x + (-x + z)) \wedge (y + (-x + z)) = x + (-x + z)$,
which, by (11.65), becomes:

$$x \wedge y = x \Longleftrightarrow z \wedge (y + (-x + z)) = z. \tag{11.84}$$

Then, from (11.83), by (Scomm), we obtain:
$x \wedge y = x \Longleftrightarrow (z + x) \wedge (y + z) = x + z$,
which, by (11.74), becomes:

$$x \wedge y = x \Longleftrightarrow z + (x \wedge y) = x + z. \tag{11.85}$$

Then, from (11.83), by (Scomm) again, we obtain:

$$x \wedge y = x \Longleftrightarrow (x + z) \wedge (y + z) = z + x. \tag{11.86}$$

Finally, in (11.83), take $X := 0$, $Y := 1$, to obtain, by (SU):
$0 \wedge 1 = 0 \Longleftrightarrow z \wedge (1 + z) = z$,
which, since $0 \wedge 1 = 0$ is true, by the hypothesis (11.59), we obtain:

$$z \wedge (1 + z) = z. \tag{11.87}$$

Now, in (11.84), take $Z := z \wedge x$, to obtain:
$x \wedge y = x \Longleftrightarrow (z \wedge x) \wedge (y + (-x + (z \wedge x))) = z \wedge x$,
which, by (DN), becomes:
$x \wedge y = x \Longleftrightarrow (z \wedge x) \wedge (y + (-x + (z \wedge -(-x)))) = z \wedge x$,
which, by (11.75) for $X := -x$, becomes:
$x \wedge y = x \Longleftrightarrow (z \wedge x) \wedge (y + (0 \wedge (-x + z))) = z \wedge x$,
which, by (Wass), becomes:

$$x \wedge y = x \Longleftrightarrow z \wedge (x \wedge (y + (0 \wedge (-x + z)))) = z \wedge x. \tag{11.88}$$

Then, in (11.85), take first $Z := -x + z$, to obtain:
$x \wedge y = x \Longleftrightarrow (-x + z) + (x \wedge y) = x + (-x + z)$,
which, by (11.65), becomes:
$x \wedge y = x \Longleftrightarrow (-x + z) + (x \wedge y) = z$,
which, by (Sass), becomes:

$$x \wedge y = x \Longleftrightarrow -x + (z + (x \wedge y)) = z. \tag{11.89}$$

Then, in (11.85), take $Y := y \wedge x$, to obtain:
$x \wedge (y \wedge x) = x \Longleftrightarrow z + (x \wedge (y \wedge x)) = x + z$,

which, by (11.70) twice, becomes:
$y \wedge x = x \Longleftrightarrow z + (y \wedge x) = x + z$,
which, by interchanging x with y, becomes:

$$x \wedge y = y \Longleftrightarrow z + (x \wedge y) = y + z. \tag{11.90}$$

Now, from (11.62) ($1 \wedge a = a$), by (11.77), we obtain:
$1 + (0 \wedge (-1 + a)) = a$,
which, by (Scomm), becomes:
$1 + (0 \wedge (a + -1)) = a$,
which, by (11.75), becomes:
$1 + (a + (-1 \wedge -a)) = a$,
which, by (11.78), becomes:
$1 + (a + (-1 + (0 \wedge (-a + 1)))) = a$,
which, by (Scomm), becomes:
$1 + (a + (-1 + (0 \wedge (1 + -a)))) = a$,
which, by (11.66), becomes:
(**) $a + (0 \wedge (1 + -a)) = a$;
then, in (11.85), take $X := 0$, $Y := 1 + -a$, $Z := a$, to obtain, by (SU):
$0 \wedge (1 + -a) = 0 \Longleftrightarrow a + (0 \wedge (1 + -a)) = a$,
which, by (**), implies:
(@) $0 \wedge (1 + -a) = 0$;
then, in (11.88), take $X := a$, $Y := x$, $Z := 1$, to obtain:
$a \wedge x = a \Longleftrightarrow 1 \wedge (a \wedge (x + (0 \wedge (-a + 1)))) = 1 \wedge a$,
which, by (Scomm) and (11.62), becomes:
$a \wedge x = a \Longleftrightarrow 1 \wedge (a \wedge (x + (0 \wedge (1 + -a)))) = a$,
which, by (@) and (SU), becomes:

$$a \wedge x = a \Longleftrightarrow 1 \wedge (a \wedge x) = a. \tag{11.91}$$

Similarly, from (11.63) ($1 \wedge b = b$), by (11.77), we obtain:
$1 + (0 \wedge (-1 + b)) = b$,
which, by (Scomm), becomes:
$1 + (0 \wedge (b + -1)) = b$,
which, by (11.75), becomes:
$1 + (b + (-1 \wedge -b)) = b$,
which, by (11.78), becomes:
$1 + (b + (-1 + (0 \wedge (-b + 1)))) = b$,
which, by (Scomm), becomes:
$1 + (b + (-1 + (0 \wedge (1 + -b)))) = b$,
which, by (11.66), becomes:
(**') $b + (0 \wedge (1 + -b)) = b$;
then, in (11.85), take $X := 0$, $Y := 1 + -b$, $Z := b$, to obtain, by (SU):
$0 \wedge (1 + -b) = 0 \Longleftrightarrow b + (0 \wedge (1 + -b)) = b$,
which, by (**'), implies:
(@') $0 \wedge (1 + -b) = 0$;
then, in (11.69), take $Z := y + z$, to obtain:

$x \wedge (y \wedge (y + z)) = y \wedge (x \wedge (y + z))$,
which, by (11.72), becomes:
$x \wedge (y + (0 \wedge z)) = y \wedge (x \wedge (y + z))$,
which, by interchanging x with y, becomes:
$y \wedge (x + (0 \wedge z)) = x \wedge (y \wedge (x + z))$,
which, for $Z := 1 + -b$, becomes:
$y \wedge (x + (0 \wedge (1 + -b))) = x \wedge (y \wedge (x + (1 + -b)))$,
which, by (@'), becomes, by (SU):
$y \wedge x = x \wedge (y \wedge (x + (1 + -b)))$,
hence:

(11.92) $$x \wedge (y \wedge (x + (1 + -b))) = y \wedge x.$$

Now, from (11.86), by (11.76) for $X := x$, $Y := z$, $Z := y + z$, we obtain:

(11.93) $$x \wedge y = x \Longleftrightarrow x + (z \wedge (-x + (y + z))) = z + x.$$

Now, in (11.87), take $Z := -1 + x$, to obtain:
$(-1 + x) \wedge (1 + (-1 + x)) = -1 + x$,
which, by (11.65), becomes:
$(-1 + x) \wedge x = -1 + x$,
hence, by (Wcomm):

(11.94) $$x \wedge (-1 + x) = -1 + x.$$

Now, in (Wass) ($(x \wedge y) \wedge z = x \wedge (y \wedge z)$), take first $X := 0$, $Y := 1$, to obtain:
$(0 \wedge 1) \wedge z = 0 \wedge (1 \wedge z)$,
which, by (11.59), becomes:
$0 \wedge z = 0 \wedge (1 \wedge z)$,
hence:

(11.95) $$0 \wedge (1 \wedge z) = 0 \wedge z.$$

Then, in (Wass) again, take $Y := 1 + x$, $Z := y$, to obtain:
$(x \wedge (1 + x)) \wedge y = x \wedge ((1 + x) \wedge y)$,
which, by (11.87), becomes:
$x \wedge y = x \wedge ((1 + x) \wedge y)$,
hence:

(11.96) $$x \wedge ((1 + x) \wedge y) = x \wedge y.$$

Then, in (Wass) again, take $Z := 1 + (x \wedge y)$, to obtain:
$(x \wedge y) \wedge (1 + (x \wedge y)) = x \wedge (y \wedge (1 + (x \wedge y)))$,
which, by (11.87) for $Z := x \wedge y$, becomes:
$x \wedge y = x \wedge (y \wedge (1 + (x \wedge y)))$,
hence:

(11.97) $$x \wedge (y \wedge (1 + (x \wedge y))) = x \wedge y.$$

Now, from (11.95), by (Wcomm), we obtain:

$$0 \wedge (z \wedge 1) = 0 \wedge z. \tag{11.98}$$

From (11.96), by (11.76), we obtain:
(***) $x \wedge (1 + (x \wedge (-1 + y))) = x \wedge y$;
then, in (***), take $Y := 1 + y$, to obtain:
$x \wedge (1 + (x \wedge (-1 + (1 + y)))) = x \wedge (1 + y)$,
which, by (11.65), becomes:

$$x \wedge (1 + (x \wedge y)) = x \wedge (1 + y). \tag{11.99}$$

Then, in (11.97), take $Y := 1$, to obtain:
$x \wedge (1 \wedge (1 + (x \wedge 1))) = x \wedge 1$,
which, by (11.72), becomes:
$x \wedge (1 + (0 \wedge (x \wedge 1))) = x \wedge 1$,
which, by (11.98), becomes:
(****) $x \wedge (1 + (0 \wedge x)) = x \wedge 1$;
then, in (****), take $X := -x$, to obtain:
$-x \wedge (1 + (0 \wedge -x)) = -x \wedge 1$,
which, by (11.80), becomes:
$-x \wedge (1 + (-x + (0 \wedge x))) = -x \wedge 1$,
which, by (11.78), becomes:
$-x + (0 \wedge ((1 + (-x + (0 \wedge x))) + x)) = -x \wedge 1$,
which, by (Scomm), becomes:
$-x + (0 \wedge (x + (1 + (-x + (0 \wedge x))))) = -x \wedge 1$,
which, by (11.66), becomes:
$-x + (0 \wedge (1 + (0 \wedge x))) = -x \wedge 1$,
which, by (11.99), becomes:
$-x + (0 \wedge (1 + x)) = -x \wedge 1$,
which, by (Wcomm), becomes:

$$1 \wedge -x = -x + (0 \wedge (1 + x)). \tag{11.100}$$

Finally, from (g6) $(x \vee (y \wedge z) = (x \vee y) \wedge (x \vee z))$, by (11.71) three times, we obtain:
$-(-x \wedge -(y \wedge z)) = -(-x \wedge -y) \wedge -(-x \wedge -z)$,
hence:
$-(-x \wedge -y) \wedge -(-x \wedge -z) = -(-x \wedge -(y \wedge z))$,

and from this identity we obtain the following identities (copying the corresponding lines from PROVER9 proof):
90 $-(-x + (0 \wedge (-y + x))) \wedge -(-x \wedge -z) = -(-x \wedge -(y \wedge z))$, by (11.78),
91 $(-(0 \wedge (-y + x)) + --x) \wedge -(-x \wedge -z) = -(-x \wedge -(y \wedge z))$, by (NegS),
92 $(-(0 \wedge (-y + x)) + x) \wedge -(-x \wedge -z) = -(-x \wedge -(y \wedge z))$, by (DN),
93 $(x + -(0 \wedge (-y + x))) \wedge -(-x \wedge -z) = -(-x \wedge -(y \wedge z))$, by (Scomm),
94 $(x + -(0 \wedge (-y + x))) \wedge -(-x + (0 \wedge (-z + x))) = -(-x \wedge -(y \wedge z))$, by (11.78),

95 $(x+-(0\wedge(-y+x)))\wedge(-(0\wedge(-z+x))+--x)=-(-x\wedge-(y\wedge z))$, by (NegS),
96 $(x+-(0\wedge(-y+x)))\wedge(-(0\wedge(-z+x))+x)=-(-x\wedge-(y\wedge z))$, by (DN),
97 $(x+-(0\wedge(-y+x)))\wedge(x+-(0\wedge(-z+x)))=-(-x\wedge-(y\wedge z))$, by (Scomm),
98 $x+(-(0\wedge(-y+x))\wedge(-x+(x+-(0\wedge(-z+x)))))=-(-x\wedge-(y\wedge z))$, by (11.76),
99 $x+(-(0\wedge(-y+x))\wedge(x+(-x+-(0\wedge(-z+x)))))=-(-x\wedge-(y\wedge z))$, by (11.64),
100 $x+(-(0\wedge(-y+x))\wedge-(0\wedge(-z+x)))=-(-x\wedge-(y\wedge z))$, by (11.65),
101 $x+(-(0\wedge(-y+x))+(0\wedge(-(0\wedge(-z+x))+(0\wedge(-y+x)))))=-(-x\wedge-(y\wedge z))$, by (11.78),
102 $x+(-(0\wedge(-y+x))+(0\wedge(-(0\wedge(-z+x))+(0\wedge(-y+x)))))=-(-x+(0\wedge(-(y\wedge z)+x)))$, by (11.78),
103 $x+(-(0\wedge(-y+x))+(0\wedge(-(0\wedge(-z+x))+(0\wedge(-y+x)))))=-(0\wedge(-(y\wedge z)+x))+--x$, by (NegS),
104 $x+(-(0\wedge(-y+x))+(0\wedge(-(0\wedge(-z+x))+(0\wedge(-y+x)))))=-(0\wedge(-(y\wedge z)+x))+x$, by (DN),
105 $x+(-(0\wedge(-y+x))+(0\wedge(-(0\wedge(-z+x))+(0\wedge(-y+x)))))=x+-(0\wedge(-(y\wedge z)+x))$, by (Scomm),
106 $x+-(0\wedge(-(y\wedge z)+x))=x+(-(0\wedge(-y+x))+(0\wedge(-(0\wedge(-z+x))+(0\wedge(-y+x)))))$, copy 105,

157 $0+-(0\wedge(-(1\wedge x)+0))=0+(-(-1+0)+(0\wedge(-(0\wedge(-x+0))+(0\wedge(-1+0)))))$, by (11.94),
158 $0+-(0\wedge(0+-(1\wedge x)))=0+(-(-1+0)+(0\wedge(-(0\wedge(-x+0))+(0\wedge(-1+0)))))$, by (Scomm),
159 $0+-(0\wedge-(1\wedge x))=0+(-(-1+0)+(0\wedge(-(0\wedge(-x+0))+(0\wedge(-1+0)))))$, by (SU),
160 $0+-(-(1\wedge x)+(0\wedge(1\wedge x)))=0+(-(-1+0)+(0\wedge(-(0\wedge(-x+0))+(0\wedge(-1+0)))))$, by (11.80),
161 $0+-(-(1\wedge x)+(0\wedge x))=0+(-(-1+0)+(0\wedge(-(0\wedge(-x+0))+(0\wedge(-1+0)))))$, by (11.95),
162 $0+-((0\wedge x)+-(1\wedge x))=0+(-(-1+0)+(0\wedge(-(0\wedge(-x+0))+(0\wedge(-1+0)))))$, by (Scomm),
163 $0+(--(1\wedge x)+-(0\wedge x))=0+(-(-1+0)+(0\wedge(-(0\wedge(-x+0))+(0\wedge(-1+0)))))$, by (NegS),
164 $0+((1\wedge x)+-(0\wedge x))=0+(-(-1+0)+(0\wedge(-(0\wedge(-x+0))+(0\wedge(-1+0)))))$, by (DN),
165 $0+(-(0\wedge x)+(1\wedge x))=0+(-(-1+0)+(0\wedge(-(0\wedge(-x+0))+(0\wedge(-1+0)))))$, by (Scomm),
166 $-(0\wedge x)+(1\wedge x)=0+(-(-1+0)+(0\wedge(-(0\wedge(-x+0))+(0\wedge(-1+0)))))$, by (SU),
167 $-(0\wedge x)+(1\wedge x)=0+(-(0+-1)+(0\wedge(-(0\wedge(-x+0))+(0\wedge(-1+0)))))$, by (Scomm),
168 $-(0\wedge x)+(1\wedge x)=0+(--1+(0\wedge(-(0\wedge(-x+0))+(0\wedge(-1+0)))))$, by (SU),
169 $-(0\wedge x)+(1\wedge x)=0+(1+(0\wedge(-(0\wedge(-x+0))+(0\wedge(-1+0)))))$, by (DN),

170 $-(0 \wedge x) + (1 \wedge x) = 0 + (1 + (0 \wedge (-(0 \wedge (0 + -x)) + (0 \wedge (-1 + 0)))))$, by (Scomm),
171 $-(0 \wedge x) + (1 \wedge x) = 0 + (1 + (0 \wedge (-(0 \wedge -x) + (0 \wedge (-1 + 0)))))$, by (SU),
172 $-(0 \wedge x) + (1 \wedge x) = 0 + (1 + (0 \wedge (-(-x + (0 \wedge x)) + (0 \wedge (-1 + 0)))))$, by (11.80),
173 $-(0 \wedge x) + (1 \wedge x) = 0 + (1 + (0 \wedge ((-(0 \wedge x) + - - x) + (0 \wedge (-1 + 0)))))$, by (NegS),
174 $-(0 \wedge x) + (1 \wedge x) = 0 + (1 + (0 \wedge ((-(0 \wedge x) + x) + (0 \wedge (-1 + 0)))))$, by (DN),
175 $-(0 \wedge x) + (1 \wedge x) = 0 + (1 + (0 \wedge ((x + -(0 \wedge x)) + (0 \wedge (-1 + 0)))))$, by (Scomm),
176 $-(0 \wedge x) + (1 \wedge x) = 0 + (1 + (0 \wedge ((x + -(0 \wedge x)) + (0 \wedge (0 + -1)))))$, by (Scomm),
177 $-(0 \wedge x) + (1 \wedge x) = 0 + (1 + (0 \wedge ((x + -(0 \wedge x)) + (0 \wedge -1))))$, by (SU),
178 $-(0 \wedge x) + (1 \wedge x) = 0 + (1 + (0 \wedge ((x + -(0 \wedge x)) + (-1 + (0 \wedge 1)))))$, by (11.80),
179 $-(0 \wedge x) + (1 \wedge x) = 0 + (1 + (0 \wedge ((x + -(0 \wedge x)) + (-1 + 0))))$, by (11.59),
180 $-(0 \wedge x) + (1 \wedge x) = 0 + (1 + (0 \wedge ((x + -(0 \wedge x)) + (0 + -1))))$, by (Scomm),
181 $-(0 \wedge x) + (1 \wedge x) = 0 + (1 + (0 \wedge ((x + -(0 \wedge x)) + -1)))$, by (SU),
182 $-(0 \wedge x) + (1 \wedge x) = 0 + (1 + (0 \wedge (-1 + (x + -(0 \wedge x)))))$, by (Scomm),
183 $-(0 \wedge x) + (1 \wedge x) = 1 + (0 \wedge (-1 + (x + -(0 \wedge x))))$, by (SU),
hence we have:

$$-(0 \wedge x) + (1 \wedge x) = 1 + (0 \wedge (-1 + (x + -(0 \wedge x)))). \tag{11.101}$$

We are ready now to prove (11.58). Indeed, we have the following equivalent identities (copying the corresponding lines from the PROVER9 proof):

$1 \wedge (1 + -(1 \wedge ((1 + -a) + (1 + -b)))) = 1 + -(1 \wedge ((1 + -a) + (1 + -b)))$
25 $\Longleftrightarrow 1 \wedge (1 + -(1 \wedge (1 + (-a + (1 + -b))))) = 1 + -(1 \wedge ((1 + -a) + (1 + -b)))$, by (Sass),
26 $\Longleftrightarrow 1 \wedge (1 + -(1 \wedge (1 + (-a + (1 + -b))))) = 1 + -(1 \wedge (1 + (-a + (1 + -b))))$, by (Sass),

30 $\Longleftrightarrow 1 \wedge (1 + -(1 \wedge (1 + (1 + (-a + -b))))) = 1 + -(1 \wedge (1 + (-a + (1 + -b))))$, by (11.64),
31 $\Longleftrightarrow 1 \wedge (1 + -(1 \wedge (1 + (1 + (-a + -b))))) = 1 + -(1 \wedge (1 + (1 + (-a + -b))))$, by (11.64),
44 $\Longleftrightarrow 1 \wedge (1 + -(1 + (0 \wedge (1 + (-a + -b))))) = 1 + -(1 \wedge (1 + (1 + (-a + -b))))$, by (11.72),
45 $\Longleftrightarrow 1 \wedge (1 + (-(0 \wedge (1 + (-a + -b))) + -1)) = 1 + -(1 \wedge (1 + (1 + (-a + -b))))$, by (NegS),
46 $\Longleftrightarrow 1 \wedge (1 + (-1 + -(0 \wedge (1 + (-a + -b))))) = 1 + -(1 \wedge (1 + (1 + (-a + -b))))$, by (Scomm),
47 $\Longleftrightarrow 1 \wedge -(0 \wedge (1 + (-a + -b))) = 1 + -(1 \wedge (1 + (1 + (-a + -b))))$, by (11.65),
48 $\Longleftrightarrow 1 \wedge -(0 \wedge (1 + (-a + -b))) = 1 + -(1 + (0 \wedge (1 + (-a + -b))))$, by (11.72),
49 $\Longleftrightarrow 1 \wedge -(0 \wedge (1 + (-a + -b))) = 1 + (-(0 \wedge (1 + (-a + -b))) + -1)$, by (NegS),
50 $\Longleftrightarrow 1 \wedge -(0 \wedge (1 + (-a + -b))) = 1 + (-1 + -(0 \wedge (1 + (-a + -b))))$, by (Scomm),
51 $\Longleftrightarrow 1 \wedge -(0 \wedge (1 + (-a + -b))) = -(0 \wedge (1 + (-a + -b)))$, by (11.65),

133 $\Longleftrightarrow -(0 \wedge (1 + (-a + -b))) + (0 \wedge (1 + (0 \wedge (1 + (-a + -b))))) = -(0 \wedge$

$(1+(-a+-b)))$, by (11.100),
134 $\Longleftrightarrow -(0\wedge(1+(-a+-b)))+(0\wedge(1+(1+(-a+-b)))) = -(0\wedge(1+(-a+-b)))$, by (11.99),

184 $\Longleftrightarrow -(0\wedge(1+(-a+-b)))+(0\wedge(1+(1+(-a+-b)))) = -((-a+-b)+(1\wedge -(-a+-b)))$, by (11.81),
185 $\Longleftrightarrow -(0\wedge(1+(-a+-b)))+(0\wedge(1+(1+(-a+-b)))) = -((-a+-b)+(1\wedge(--b+--a)))$, by (NegS),
186 $\Longleftrightarrow -(0\wedge(1+(-a+-b)))+(0\wedge(1+(1+(-a+-b)))) = -((-a+-b)+(1\wedge(b+--a)))$, by (DN),
187 $\Longleftrightarrow -(0\wedge(1+(-a+-b)))+(0\wedge(1+(1+(-a+-b)))) = -((-a+-b)+(1\wedge(b+a)))$, by (DN),
188 $\Longleftrightarrow -(0\wedge(1+(-a+-b)))+(0\wedge(1+(1+(-a+-b)))) = -((-a+-b)+(1\wedge(a+b)))$, by (Scomm),
189 $\Longleftrightarrow -(0\wedge(1+(-a+-b)))+(0\wedge(1+(1+(-a+-b)))) = -(-a+(-b+(1\wedge(a+b))))$, by (Sass),
190 $\Longleftrightarrow -(0\wedge(1+(-a+-b)))+(0\wedge(1+(1+(-a+-b)))) = -(-b+(1\wedge(a+b)))+--a$, by (NegS),
191 $\Longleftrightarrow -(0\wedge(1+(-a+-b)))+(0\wedge(1+(1+(-a+-b)))) = (-(1\wedge(a+b))+--b)+--a$, by (NegS),
192 $\Longleftrightarrow -(0\wedge(1+(-a+-b)))+(0\wedge(1+(1+(-a+-b)))) = (-(1\wedge(a+b))+b)+--a$, by (DN),
193 $\Longleftrightarrow -(0\wedge(1+(-a+-b)))+(0\wedge(1+(1+(-a+-b)))) = (b+-(1\wedge(a+b)))+--a$, by (Scomm),
194 $\Longleftrightarrow -(0\wedge(1+(-a+-b)))+(0\wedge(1+(1+(-a+-b)))) = (b+-(1\wedge(a+b)))+a$, by (DN),
195 $\Longleftrightarrow -(0\wedge(1+(-a+-b)))+(0\wedge(1+(1+(-a+-b)))) = a+(b+-(1\wedge(a+b)))$, by (Scomm),
196 $\Longleftrightarrow a+(b+-(1\wedge(a+b))) = -(0\wedge(1+(-a+-b)))+(0\wedge(1+(1+(-a+-b))))$, copy 195,

197 $\Longleftrightarrow a+(b+-(1+(0\wedge(-1+(a+b))))) = -(0\wedge(1+(-a+-b)))+(0\wedge(1+(1+(-a+-b))))$, by (11.77),
198 $\Longleftrightarrow a+(b+-(1+(0\wedge((a+b)+-1)))) = -(0\wedge(1+(-a+-b)))+(0\wedge(1+(1+(-a+-b))))$, by (Scomm),
199 $\Longleftrightarrow a+(b+-(1+(0\wedge(a+(b+-1))))) = -(0\wedge(1+(-a+-b)))+(0\wedge(1+(1+(-a+-b))))$, by (Sass),
200 $\Longleftrightarrow a+(b+(-(0\wedge(a+(b+-1)))+-1)) = -(0\wedge(1+(-a+-b)))+(0\wedge(1+(1+(-a+-b))))$, by (NegS),
201 $\Longleftrightarrow a+(b+(-1+-(0\wedge(a+(b+-1))))) = -(0\wedge(1+(-a+-b)))+(0\wedge(1+(1+(-a+-b))))$, by (Scomm),
202 $\Longleftrightarrow -(0\wedge(1+(-a+-b)))+(0\wedge(1+(1+(-a+-b)))) = a+(b+(-1+-(0\wedge(a+(b+-1)))))$, copy 201,

203 $\Longleftrightarrow -(0\wedge(1+(-a+-b)))+(0\wedge(1+(1+(-a+-b)))) = a+(b+(-1+-((b+-1)+(a\wedge-(b+-1)))))$, by (11.81),

204 $\Longleftrightarrow -(0 \wedge (1 + (-a + -b))) + (0 \wedge (1 + (1 + (-a + -b)))) = a + (b + (-1 + -((b + -1) + (a \wedge (- - 1 + -b)))))$, by (NegS),
205 $\Longleftrightarrow -(0 \wedge (1 + (-a + -b))) + (0 \wedge (1 + (1 + (-a + -b)))) = a + (b + (-1 + -((b + -1) + (a \wedge (1 + -b)))))$, by (DN),
206 $\Longleftrightarrow -(0 \wedge (1 + (-a + -b))) + (0 \wedge (1 + (1 + (-a + -b)))) = a + (b + (-1 + -(b + (-1 + (a \wedge (1 + -b))))))$, by (Sass),
207 $\Longleftrightarrow -(0 \wedge (1 + (-a + -b))) + (0 \wedge (1 + (1 + (-a + -b)))) = a + (b + (-1 + (-(-1 + (a \wedge (1 + -b))) + -b)))$, by (NegS),
208 $\Longleftrightarrow -(0 \wedge (1 + (-a + -b))) + (0 \wedge (1 + (1 + (-a + -b)))) = a + (b + (-1 + ((-(a \wedge (1 + -b)) + - - 1) + -b)))$, by (NegS),
209 $\Longleftrightarrow -(0 \wedge (1 + (-a + -b))) + (0 \wedge (1 + (1 + (-a + -b)))) = a + (b + (-1 + ((-(a \wedge (1 + -b)) + 1) + -b)))$, by (DN),
210 $\Longleftrightarrow -(0 \wedge (1 + (-a + -b))) + (0 \wedge (1 + (1 + (-a + -b)))) = a + (b + (-1 + ((1 + -(a \wedge (1 + -b))) + -b)))$, by (Scomm),
211 $\Longleftrightarrow -(0 \wedge (1 + (-a + -b))) + (0 \wedge (1 + (1 + (-a + -b)))) = a + (b + (-1 + (-b + (1 + -(a \wedge (1 + -b))))))$, by (Scomm),
212 $\Longleftrightarrow -(0 \wedge (1 + (-a + -b))) + (0 \wedge (1 + (1 + (-a + -b)))) = a + (b + (-1 + (1 + (-b + -(a \wedge (1 + -b))))))$, by (11.64),
213 $\Longleftrightarrow -(0 \wedge (1 + (-a + -b))) + (0 \wedge (1 + (1 + (-a + -b)))) = a + (b + (1 + (-1 + (-b + -(a \wedge (1 + -b))))))$, by (11.64),
214 $\Longleftrightarrow -(0 \wedge (1 + (-a + -b))) + (0 \wedge (1 + (1 + (-a + -b)))) = a + (b + (-b + -(a \wedge (1 + -b))))$, by (11.65),
215 $\Longleftrightarrow -(0 \wedge (1 + (-a + -b))) + (0 \wedge (1 + (1 + (-a + -b)))) = a + -(a \wedge (1 + -b))$, by (11.65),
216 $\Longleftrightarrow a + -(a \wedge (1 + -b)) = -(0 \wedge (1 + (-a + -b))) + (0 \wedge (1 + (1 + (-a + -b))))$, copy 215,

217 $\Longleftrightarrow a + -(a \wedge (1 + -b)) = -((-a + -b) + (1 \wedge -(-a + -b))) + (0 \wedge (1 + (1 + (-a + -b))))$, by (11.81),
218 $\Longleftrightarrow a + -(a \wedge (1 + -b)) = -((-a + -b) + (1 \wedge (- - b + - - a))) + (0 \wedge (1 + (1 + (-a + -b))))$, by (NegS),
219 $\Longleftrightarrow a + -(a \wedge (1 + -b)) = -((-a + -b) + (1 \wedge (b + - - a))) + (0 \wedge (1 + (1 + (-a + -b))))$, by (DN),
220 $\Longleftrightarrow a + -(a \wedge (1 + -b)) = -((-a + -b) + (1 \wedge (b + a))) + (0 \wedge (1 + (1 + (-a + -b))))$, by (DN),
221 $\Longleftrightarrow a + -(a \wedge (1 + -b)) = -((-a + -b) + (1 \wedge (a + b))) + (0 \wedge (1 + (1 + (-a + -b))))$, by (Scomm),
222 $\Longleftrightarrow a + -(a \wedge (1 + -b)) = -(-a + (-b + (1 \wedge (a + b)))) + (0 \wedge (1 + (1 + (-a + -b))))$, by (Sass),
223 $\Longleftrightarrow a + -(a \wedge (1 + -b)) = (-(-b + (1 \wedge (a + b))) + - - a) + (0 \wedge (1 + (1 + (-a + -b))))$, by (NegS),
224 $\Longleftrightarrow a + -(a \wedge (1 + -b)) = ((-(1 \wedge (a + b)) + - - b) + - - a) + (0 \wedge (1 + (1 + (-a + -b))))$, by (NegS),
225 $\Longleftrightarrow a + -(a \wedge (1 + -b)) = ((-(1 \wedge (a + b)) + b) + - - a) + (0 \wedge (1 + (1 + (-a + -b))))$, by (DN),
226 $\Longleftrightarrow a + -(a \wedge (1 + -b)) = ((b + -(1 \wedge (a + b))) + - - a) + (0 \wedge (1 + (1 + (-a + -b))))$,

by (Scomm),
227 $\Longleftrightarrow a+-(a\wedge(1+-b))=((b+-(1\wedge(a+b)))+a)+(0\wedge(1+(1+(-a+-b))))$, by (DN),
228 $\Longleftrightarrow a+-(a\wedge(1+-b))=(a+(b+-(1\wedge(a+b))))+(0\wedge(1+(1+(-a+-b))))$, by (Scomm),
229 $\Longleftrightarrow a+-(a\wedge(1+-b))=(0\wedge(1+(1+(-a+-b))))+(a+(b+-(1\wedge(a+b))))$, by (Scomm),
230 $\Longleftrightarrow a+-(a\wedge(1+-b))=a+((0\wedge(1+(1+(-a+-b))))+(b+-(1\wedge(a+b))))$, by (11.64),
231 $\Longleftrightarrow a+-(a\wedge(1+-b))=a+(b+((0\wedge(1+(1+(-a+-b))))+-(1\wedge(a+b))))$, by (11.64),
232 $\Longleftrightarrow a+-(a\wedge(1+-b))=a+(b+(((1+(-a+-b))+(1\wedge-(1+(-a+-b))))+-(1\wedge(a+b))))$, by (11.81),
233 $\Longleftrightarrow a+-(a\wedge(1+-b))=a+(b+(((1+(-a+-b))+(1\wedge(-(-a+-b)+-1)))+-(1\wedge(a+b))))$, by (NegS),
234 $\Longleftrightarrow a+-(a\wedge(1+-b))=a+(b+(((1+(-a+-b))+(1\wedge((--b+--a)+-1)))+-(1\wedge(a+b))))$, by (NegS),
235 $\Longleftrightarrow a+-(a\wedge(1+-b))=a+(b+(((1+(-a+-b))+(1\wedge((b+--a)+-1)))+-(1\wedge(a+b))))$, by (DN),
236 $\Longleftrightarrow a+-(a\wedge(1+-b))=a+(b+(((1+(-a+-b))+(1\wedge((b+a)+-1)))+-(1\wedge(a+b))))$, by (DN),
237 $\Longleftrightarrow a+-(a\wedge(1+-b))=a+(b+(((1+(-a+-b))+(1\wedge((a+b)+-1)))+-(1\wedge(a+b))))$, by (Scomm),
238 $\Longleftrightarrow a+-(a\wedge(1+-b))=a+(b+(((1+(-a+-b))+(1\wedge(a+(b+-1))))+-(1\wedge(a+b))))$, by (Sass),
239 $\Longleftrightarrow a+-(a\wedge(1+-b))=a+(b+((1+((-a+-b)+(1\wedge(a+(b+-1)))))+-(1\wedge(a+b))))$, by (Sass),
240 $\Longleftrightarrow a+-(a\wedge(1+-b))=a+(b+((1+(-a+(-b+(1\wedge(a+(b+-1))))))+-(1\wedge(a+b))))$, by (Sass),
241 $\Longleftrightarrow a+-(a\wedge(1+-b))=a+(b+(-(1\wedge(a+b))+(1+(-a+(-b+(1\wedge(a+(b+-1))))))))$, by (Scomm),
242 $\Longleftrightarrow a+-(a\wedge(1+-b))=a+(b+(1+(-(1\wedge(a+b))+(-a+(-b+(1\wedge(a+(b+-1))))))))$, by (11.64),
243 $\Longleftrightarrow a+-(a\wedge(1+-b))=a+(b+(1+(-a+(-(1\wedge(a+b))+(-b+(1\wedge(a+(b+-1))))))))$, by (11.64),
244 $\Longleftrightarrow a+-(a\wedge(1+-b))=a+(b+(1+(-a+(-b+(-(1\wedge(a+b))+(1\wedge(a+(b+-1))))))))$, by (11.64),
245 $\Longleftrightarrow a+-(a\wedge(1+-b))=a+(1+(b+(-a+(-b+(-(1\wedge(a+b))+(1\wedge(a+(b+-1))))))))$, by (11.64),
246 $\Longleftrightarrow a+-(a\wedge(1+-b))=a+(1+(-a+(-(1\wedge(a+b))+(1\wedge(a+(b+-1))))))$, by (11.66),
247 $\Longleftrightarrow a+-(a\wedge(1+-b))=1+(a+(-a+(-(1\wedge(a+b))+(1\wedge(a+(b+-1))))))$, by (11.64),
248 $\Longleftrightarrow a+-(a\wedge(1+-b))=1+(-(1\wedge(a+b))+(1\wedge(a+(b+-1))))$, by (11.65),
249 $\Longleftrightarrow a+-(a\wedge(1+-b))=1+(-(1+(0\wedge(-1+(a+b))))+(1\wedge(a+(b+-1))))$, by (11.77),

250 $\Longleftrightarrow a+-(a\wedge(1+-b)) = 1+(-(1+(0\wedge((a+b)+-1)))+(1\wedge(a+(b+-1))))$, by (Scomm),
251 $\Longleftrightarrow a+-(a\wedge(1+-b)) = 1+(-(1+(0\wedge(a+(b+-1))))+(1\wedge(a+(b+-1))))$, by (Sass),
252 $\Longleftrightarrow a+-(a\wedge(1+-b)) = 1+((-(0\wedge(a+(b+-1)))+-1)+(1\wedge(a+(b+-1))))$, by (NegS),
253 $\Longleftrightarrow a+-(a\wedge(1+-b)) = 1+((-1+-(0\wedge(a+(b+-1))))+(1\wedge(a+(b+-1))))$, by (Scomm),
254 $\Longleftrightarrow a+-(a\wedge(1+-b)) = 1+(-1+(-(0\wedge(a+(b+-1)))+(1\wedge(a+(b+-1)))))$, by (Sass),
255 $\Longleftrightarrow a+-(a\wedge(1+-b)) = 1+(-1+(1+(0\wedge(-1+((a+(b+-1))+-(0\wedge(a+(b+-1))))))))$, by (11.101),
256 $\Longleftrightarrow a+-(a\wedge(1+-b)) = 1+(-1+(1+(0\wedge(-1+(a+((b+-1)+-(0\wedge(a+(b+-1)))))))))$, by (Sass),
257 $\Longleftrightarrow a+-(a\wedge(1+-b)) = 1+(-1+(1+(0\wedge(-1+(a+(b+(-1+-(0\wedge(a+(b+-1))))))))))$, by (Sass),
258 $\Longleftrightarrow a+-(a\wedge(1+-b)) = 1+(-1+(1+(0\wedge(a+(-1+(b+(-1+-(0\wedge(a+(b+-1))))))))))$, by (11.64),
259 $\Longleftrightarrow a+-(a\wedge(1+-b)) = 1+(-1+(1+(0\wedge(a+(b+(-1+(-1+-(0\wedge(a+(b+-1))))))))))$, by (11.64),
260 $\Longleftrightarrow a+-(a\wedge(1+-b)) = 1+(1+(-1+(0\wedge(a+(b+(-1+(-1+-(0\wedge(a+(b+-1))))))))))$, by (11.64),
261 $\Longleftrightarrow a+-(a\wedge(1+-b)) = 1+(0\wedge(a+(b+(-1+(-1+-(0\wedge(a+(b+-1))))))))$, by (11.65),
262 $\Longleftrightarrow a+-(a\wedge(1+-b)) = 1+(0\wedge(a+(b+(-1+(-1+-((b+-1)+(a\wedge-(b+-1))))))))$, by (11.81),
263 $\Longleftrightarrow a+-(a\wedge(1+-b)) = 1+(0\wedge(a+(b+(-1+(-1+-((b+-1)+(a\wedge(--1+-b))))))))$, by (NegS),
264 $\Longleftrightarrow a+-(a\wedge(1+-b)) = 1+(0\wedge(a+(b+(-1+(-1+-((b+-1)+(a\wedge(1+-b))))))))$, by (DN),
265 $\Longleftrightarrow a+-(a\wedge(1+-b)) = 1+(0\wedge(a+(b+(-1+(-1+-(b+(-1+(a\wedge(1+-b)))))))))$, by (Sass),
266 $\Longleftrightarrow a+-(a\wedge(1+-b)) = 1+(0\wedge(a+(b+(-1+(-1+(-(-1+(a\wedge(1+-b)))+-b))))))$, by (NegS),
267 $\Longleftrightarrow a+-(a\wedge(1+-b)) = 1+(0\wedge(a+(b+(-1+(-1+((-(a\wedge(1+-b))+--1)+-b))))))$, by (NegS),
268 $\Longleftrightarrow a+-(a\wedge(1+-b)) = 1+(0\wedge(a+(b+(-1+(-1+((-(a\wedge(1+-b))+1)+-b))))))$, by (DN),
269 $\Longleftrightarrow a+-(a\wedge(1+-b)) = 1+(0\wedge(a+(b+(-1+(-1+((1+-(a\wedge(1+-b)))+-b))))))$, by (Scomm),
270 $\Longleftrightarrow a+-(a\wedge(1+-b)) = 1+(0\wedge(a+(b+(-1+(-1+(-b+(1+-(a\wedge(1+-b)))))))))$, by (Scomm),
271 $\Longleftrightarrow a+-(a\wedge(1+-b)) = 1+(0\wedge(a+(b+(-1+(-1+(1+(-b+-(a\wedge(1+-b)))))))))$, by (11.64),
272 $\Longleftrightarrow a+-(a\wedge(1+-b)) = 1+(0\wedge(a+(b+(-1+(1+(-1+(-b+-(a\wedge(1+-b)))))))))$, by (11.64),

273 $\Longleftrightarrow a + -(a \wedge (1 + -b)) = 1 + (0 \wedge (a + (b + (-1 + (-b + -(a \wedge (1 + -b)))))))$, by (11.65),
274 $\Longleftrightarrow a + -(a \wedge (1 + -b)) = 1 + (0 \wedge (a + (-1 + -(a \wedge (1 + -b)))))$, by (11.66),
275 $\Longleftrightarrow 1 + (0 \wedge (a + (-1 + -(a \wedge (1 + -b))))) = a + -(a \wedge (1 + -b))$, copy 274,

276 $\Longleftrightarrow 1 \wedge (1 + (a + (-1 + -(a \wedge (1 + -b))))) = a + -(a \wedge (1 + -b))$, by (11.72),
277 $\Longleftrightarrow 1 \wedge (a + -(a \wedge (1 + -b))) = a + -(a \wedge (1 + -b))$, by (11.66),
278 $\Longleftrightarrow x + (1 \wedge (a + -(a \wedge (1 + -b)))) = (a + -(a \wedge (1 + -b))) + x$, by (11.90),
279 $\Longleftrightarrow x + (1 \wedge (a + -(a \wedge (1 + -b)))) = a + (-(a \wedge (1 + -b)) + x)$, by (Sass),
280 $\Longleftrightarrow a + (-(a \wedge (1 + -b)) + x) = x + (1 \wedge (a + -(a \wedge (1 + -b))))$, copy 279,

281 $\Longleftrightarrow a+(-(a\wedge(1+-b))+-(a+-(a\wedge(1+-b)))) = 0\wedge(-(a+-(a\wedge(1+-b)))+1)$, by (11.82),
282 $\Longleftrightarrow a + (-(a \wedge (1 + -b)) + (- - (a \wedge (1 + -b)) + -a)) = 0 \wedge (-(a + -(a \wedge (1 + -b))) + 1)$, by (NegS),
283 $\Longleftrightarrow a+(-(a\wedge(1+-b))+((a\wedge(1+-b))+-a)) = 0\wedge(-(a+-(a\wedge(1+-b)))+1)$, by (DN),
284 $\Longleftrightarrow a+(-(a\wedge(1+-b))+(-a+(a\wedge(1+-b)))) = 0\wedge(-(a+-(a\wedge(1+-b)))+1)$, by (Scomm),
285 $\Longleftrightarrow a+((-a+(a\wedge(1+-b)))+-(a\wedge(1+-b))) = 0\wedge(-(a+-(a\wedge(1+-b)))+1)$, by (Scomm),
286 $\Longleftrightarrow a+(-a+((a\wedge(1+-b))+-(a\wedge(1+-b)))) = 0\wedge(-(a+-(a\wedge(1+-b)))+1)$, by (Sass),
287 $\Longleftrightarrow a + (-a + 0) = 0 \wedge (-(a + -(a \wedge (1 + -b))) + 1)$, by ($\mathrm{m}_0$-Re),
288 $\Longleftrightarrow a + (0 + -a) = 0 \wedge (-(a + -(a \wedge (1 + -b))) + 1)$, by (Scomm),
289 $\Longleftrightarrow a + -a = 0 \wedge (-(a + -(a \wedge (1 + -b))) + 1)$, by (SU),
290 $\Longleftrightarrow 0 = 0 \wedge (-(a + -(a \wedge (1 + -b))) + 1)$, by ($\mathrm{m}_0$-Re),
291 $\Longleftrightarrow 0 = 0 \wedge ((- - (a \wedge (1 + -b)) + -a) + 1)$, by (NegS),
292 $\Longleftrightarrow 0 = 0 \wedge (((a \wedge (1 + -b)) + -a) + 1)$, by (DN),
293 $\Longleftrightarrow 0 = 0 \wedge ((-a + (a \wedge (1 + -b))) + 1)$, by (Scomm),
294 $\Longleftrightarrow 0 = 0 \wedge (1 + (-a + (a \wedge (1 + -b))))$, by (Scomm),
295 $\Longleftrightarrow 0 \wedge (1 + (-a + (a \wedge (1 + -b)))) = 0$, copy 294,

296 $\Longleftrightarrow 0 \wedge (1 + (a + (-a \wedge (-a + (-a + (1 + -b)))))) = 0$, by (11.79),
297 $\Longleftrightarrow 0 \wedge (1 + (a + (-a \wedge (-a + (1 + (-a + -b)))))) = 0$, by (11.64),
298 $\Longleftrightarrow 0 \wedge (1 + (a + (-a \wedge (1 + (-a + (-a + -b)))))) = 0$, by (11.64),
299 $\Longleftrightarrow 0 \wedge (1 + (a + (-a + (0 \wedge ((1 + (-a + (-a + -b))) + a))))) = 0$, by (11.78),
300 $\Longleftrightarrow 0 \wedge (1 + (a + (-a + (0 \wedge (a + (1 + (-a + (-a + -b)))))))) = 0$, by (Scomm),
301 $\Longleftrightarrow 0 \wedge (1 + (a + (-a + (0 \wedge (1 + (a + (-a + (-a + -b)))))))) = 0$, by (11.64),
302 $\Longleftrightarrow 0 \wedge (1 + (a + (-a + (0 \wedge (1 + (-a + -b)))))) = 0$, by (11.66),
303 $\Longleftrightarrow 0 \wedge (1 + (0 \wedge (1 + (-a + -b)))) = 0$, by (11.65),
304 $\Longleftrightarrow 0 \wedge (1 + (1 + (-a + -b))) = 0$, by (11.99),
305 $\Longleftrightarrow 0 + (x \wedge (-0 + ((1 + (1 + (-a + -b))) + x))) = x + 0$, by (11.93),
306 $\Longleftrightarrow 0 + (x \wedge (0 + ((1 + (1 + (-a + -b))) + x))) = x + 0$, by (N0),
307 $\Longleftrightarrow 0 + (x \wedge (0 + (1 + ((1 + (-a + -b)) + x)))) = x + 0$, by (Sass),
308 $\Longleftrightarrow 0 + (x \wedge (0 + (1 + (1 + ((-a + -b) + x))))) = x + 0$, by (Sass),

309 $\Longleftrightarrow 0+(x \wedge (0+(1+(1+(-a+(-b+x)))))) = x+0$, by (Sass),
310 $\Longleftrightarrow 0+(x \wedge (1+(1+(-a+(-b+x))))) = x+0$, by (SU),
311 $\Longleftrightarrow x \wedge (1+(1+(-a+(-b+x)))) = x+0$, by (SU),
312 $\Longleftrightarrow x \wedge (1+(1+(-a+(-b+x)))) = x$, by (SU),
313 $\Longleftrightarrow -x+(y+(x \wedge (1+(1+(-a+(-b+x)))))) = y$, by (11.89),
314 $\Longleftrightarrow -x+(x+(y \wedge (-x+(y+(1+(1+(-a+(-b+x)))))))) = y$, by (11.79),
315 $\Longleftrightarrow x+(-x+(y \wedge (-x+(y+(1+(1+(-a+(-b+x)))))))) = y$, by (11.64),
316 $\Longleftrightarrow x \wedge (-y+(x+(1+(1+(-a+(-b+y)))))) = x$, by (11.65),
317 $\Longleftrightarrow x \wedge (-(--b+y)+(x+(1+(1+(-a+y))))) = x$, by (11.65),
318 $\Longleftrightarrow x \wedge (-(b+y)+(x+(1+(1+(-a+y))))) = x$, by (DN),
319 $\Longleftrightarrow x \wedge ((-y+-b)+(x+(1+(1+(-a+y))))) = x$, by (NegS),
320 $\Longleftrightarrow x \wedge (-y+(-b+(x+(1+(1+(-a+y)))))) = x$, by (Sass),

327 $\Longleftrightarrow x \wedge (---a+(-b+(x+(1+(1+0))))) = x$, by ($\mathrm{m_0}$-Re),
328 $\Longleftrightarrow x \wedge (-a+(-b+(x+(1+(1+0))))) = x$, by (DN),
329 $\Longleftrightarrow x \wedge (-a+(-b+(x+(1+(0+1))))) = x$, by (Scomm),
330 $\Longleftrightarrow x \wedge (-a+(-b+(x+(1+1)))) = x$, by (SU),
331 $\Longleftrightarrow 1 \wedge (a \wedge (-a+(-b+(a+(1+1))))) = a$, by (11.91),
332 $\Longleftrightarrow 1 \wedge (a \wedge (-a+(-b+(1+(a+1))))) = a$, by (11.64),
333 $\Longleftrightarrow 1 \wedge (a \wedge (-a+(-b+(1+(1+a))))) = a$, by (Scomm),
334 $\Longleftrightarrow 1 \wedge (a \wedge (-a+((1+(1+a))+-b))) = a$, by (Scomm),
335 $\Longleftrightarrow 1 \wedge (a \wedge (-a+(1+((1+a)+-b)))) = a$, by (Sass),
336 $\Longleftrightarrow 1 \wedge (a \wedge (-a+(1+(1+(a+-b))))) = a$, by (Sass),
337 $\Longleftrightarrow 1 \wedge (a \wedge (1+(-a+(1+(a+-b))))) = a$, by (11.64),
338 $\Longleftrightarrow 1 \wedge (a \wedge (1+(1+(-a+(a+-b))))) = a$, by (11.64),
339 $\Longleftrightarrow 1 \wedge (a \wedge (1+(1+(a+(-a+-b))))) = a$, by (11.64),
340 $\Longleftrightarrow 1 \wedge (a \wedge (1+(1+-b))) = a$, by (11.65),
341 $\Longleftrightarrow a \wedge 1 = a$, by (11.92),
343 $\Longleftrightarrow a = a$, by (11.60),
that is true. Thus,
$1 \wedge (1+-(1 \wedge ((1+-a)+(1+-b)))) = 1+-(1 \wedge ((1+-a)+(1+-b)))$
is true, hence (11.58) holds. □

11.3.3 The 'hybrid' proof of Proposition 3.4.7

Proposition Let $\mathcal{G}_u = (G, \vee, \wedge, +, -, 0_G, u = u_G)$ be a commutative l-group with strong unit $u = u_G$, with $x \leq y \Longleftrightarrow x \wedge y = x \Longleftrightarrow x \vee y = y$. Then, the following property holds (see Remark 3.4.3):
(XX_u^R) for $0_G \leq a, b, c \leq u_G$,
$u_G + -\inf(u_G, [(u_G + -\inf(u_G, a+b)) + (u_G + -\inf(u_G, [c + (u_G + -\inf(u_G, (u_G + -a) + (u_G + -b)))]))]) =$
$u_G + -\inf(u_G, [(u_G + -\inf(u_G, b+c)) + (u_G + -\inf(u_G, [a + (u_G + -\inf(u_G, (u_G + -b) + (u_G + -c)))]))])$, i.e.

$u_G + -(u_G \wedge [(u_G + -(u_G \wedge (a+b))) + (u_G + -(u_G \wedge [c + (u_G + -(u_G \wedge ((u_G + -a) + (u_G + -b))))]))]) =$

$u_G + -(u_G \wedge [(u_G + -(u_G \wedge (b+c))) + (u_G + -(u_G \wedge [a + (u_G + -(u_G \wedge ((u_G + -b) + (u_G + -c))))]))])$.

Proof. (By PROVER9, Length of proof was 79, in 2.33 seconds; the renumbered expanded proof had the Length 358.)

In order to simplify the writting (less space on a line), we shall replace 0_G by 0 and u_G by 1.

Hence, we have to prove that, for all $0 \leq a, b, c \leq 1$:

(11.102)

$$1+-(1\wedge[(1+-(1\wedge(a+b)))+(1+-(1\wedge[c+(1+-(1\wedge((1+-a)+(1+-b))))]))]) =$$

$$1+-(1\wedge[(1+-(1\wedge(b+c)))+(1+-(1\wedge[a+(1+-(1\wedge((1+-b)+(1+-c))))]))]).$$

The particular hypotheses used by PROVER9 (among the given ones - see (α)) are:

$$0 \wedge 1 = 0 \quad (i.e.\ 0 \leq 1), \tag{11.103}$$

$$a \wedge 1 = a \quad (i.e.\ a \leq 1), \tag{11.104}$$

$$c \wedge 1 = c \quad (i.e.\ c \leq 1), \tag{11.105}$$

hence, by (Wcomm),

$$1 \wedge a = a \quad (i.e.\ a \leq 1), \tag{11.106}$$

$$1 \wedge c = c \quad (i.e.\ c \leq 1). \tag{11.107}$$

The following properties are obvious, by (SU), (Scomm), (Sass), (m_0-Re), (NegS) and (Wcomm) ($x \wedge y = y \wedge x$), (Wass) ($(x \wedge y) \wedge z = x \wedge (y \wedge z)$):

$$x + (y+z) = y + (x+z), \tag{11.108}$$

$$x + (-x+y) = y, \tag{11.109}$$

$$x + (y + -x) = y, \tag{11.110}$$

$$-x + (y+x) = y, \tag{11.111}$$

$$x + (y + (-x+z)) = y+z, \tag{11.112}$$

$$x + (y + (z + -x)) = y+z, \tag{11.113}$$

$$x + (y + (z + (-x+u))) = (y+z)+u, \tag{11.114}$$

$$x + (y + (z + (-x + u))) = y + (z + u). \tag{11.115}$$

Now, in (g3) ($(x + y) \wedge (x + z) = x + (y \wedge z)$), take first $Y := 0$, to obtain, by (SU):

$$x \wedge (x + z) = x + (0 \wedge z). \tag{11.116}$$

Then, from (g3) again, by interchanging x with y, we obtain:
$(y + x) \wedge (y + z) = y + (x \wedge z)$,
which, by (Scomm), becomes:

$$(x + y) \wedge (y + z) = y + (x \wedge z). \tag{11.117}$$

Then, from (g3) again, by (Scomm), we obtain:

$$(x + y) \wedge (z + x) = x + (y \wedge z). \tag{11.118}$$

Then, in (g3) again, take $Z := -x$, to obtain, by (m_0-Re):
$(x + y) \wedge 0 = x + (y \wedge -x)$,
which, by (Wcomm), becomes:

$$0 \wedge (x + y) = x + (y \wedge -x). \tag{11.119}$$

Finally, in (g3) again, take $Z := -x + z$, to obtain:
$(x + y) \wedge (x + (-x + z)) = x + (y \wedge (-x + z))$,
which, by (11.109), becomes:

$$(x + y) \wedge z = x + (y \wedge (-x + z)). \tag{11.120}$$

Now, in (Sass) ($(x + y) + z = x + (y + z)$), take $Y := 0 \wedge y$, to obtain:
$(x + (0 \wedge y)) + z = x + ((0 \wedge y) + z)$,
which, by (11.116), becomes:

$$(x \wedge (x + y)) + z = x + ((0 \wedge y) + z). \tag{11.121}$$

Then, in (11.108) ($x + (y + z) = y + (x + z)$), take $Z := 0 \wedge z$, to obtain:
$x + (y + (0 \wedge z)) = y + (x + (0 \wedge z))$,
by (11.116), becomes:

$$x + (y \wedge (y + z)) = y + (x + (0 \wedge z)). \tag{11.122}$$

Then, in (11.116), take $Z := y + z$, to obtain:
$x \wedge (x + (y + z)) = x + (0 \wedge (y + z))$,
by (11.108), becomes:

$$x \wedge (y + (x + z)) = x + (0 \wedge (y + z)). \tag{11.123}$$

Then, in (11.116) again, take $Z := -x + y$, to obtain:
$x \wedge (x + (-x + y)) = x + (0 \wedge (-x + y))$,
which, by (11.109), becomes:

$$x \wedge y = x + (0 \wedge (-x + y)). \tag{11.124}$$

Then, in (11.116) again, take $X := -x$, $Z := y + x$, to obtain:
$-x \wedge (-x + (y + x)) = -x + (0 \wedge (y + x))$,
which, by (11.111), becomes:

$$-x \wedge y = -x + (0 \wedge (y + x)). \tag{11.125}$$

Now, from (11.117) ($(x + y) \wedge (y + z) = y + (x \wedge z)$), by (11.120) for $Z := y + z$, we obtain:

$$x + (y \wedge (-x + (y + z))) = y + (x \wedge z). \tag{11.126}$$

Now, in (11.119) ($0 \wedge (x + y) = x + (y \wedge -x)$), take $Y := 0$, to obtain, by (SU):
$0 \wedge x = x + (0 \wedge -x)$,
which, for $X := -x$, becomes, by (DN):

$$0 \wedge -x = -x + (0 \wedge x). \tag{11.127}$$

Now, from (11.106) ($1 \wedge a = a$), by (11.124), we obtain:
$1 + (0 \wedge (-1 + a)) = a$,
which, by (Scomm), becomes:
$1 + (0 \wedge (a + -1)) = a$,
which, by (11.119) for $X := a$, $Y := -1$, becomes:
$1 + (a + (-1 \wedge -a)) = a$,
which, by (11.125), becomes:
$1 + (a + (-1 + (0 \wedge (-a + 1)))) = a$,
which, by (Scomm), becomes:
$1 + (a + (-1 + (0 \wedge (1 + -a)))) = a$,
which, by (11.112) for $Z := 0 \wedge (1 + -a)$, becomes:

$$a + (0 \wedge (1 + -a)) = a. \tag{11.128}$$

Similarly, from (11.107) ($1 \wedge c = c$), by (11.124), we obtain:
$1 + (0 \wedge (-1 + c)) = c$,
which, by (Scomm), becomes:
$1 + (0 \wedge (c + -1)) = c$,
which, by (11.119) for $X := c$, $Y := -1$, becomes:
$1 + (c + (-1 \wedge -c)) = c$,
which, by (11.125), becomes:
$1 + (c + (-1 + (0 \wedge (-c + 1)))) = c$,
which, by (Scomm), becomes:
$1 + (c + (-1 + (0 \wedge (1 + -c)))) = c$,
which, by (11.112) for $Z := 0 \wedge (1 + -c)$, becomes:

$$c + (0 \wedge (1 + -c)) = c. \tag{11.129}$$

Now, in (g2) ($(x + y) \vee (x + z) = x + (y \vee z)$), take $Z := -x + z$, to obtain:
$(x + y) \vee (x + (-x + z)) = x + (y \vee (-x + z))$,
which, by (11.109), becomes:

$$(x + y) \vee z = x + (y \vee (-x + z)). \tag{11.130}$$

Now, in (g5) ($-x \vee -y = -(x \wedge y)$), take $Y := -y$, to obtain, by (DN):

$$-x \vee y = -(x \wedge -y). \tag{11.131}$$

Then, in (g5) again, take $X := x + y$, $Y := x + z$, to obtain:
$-(x + y) \vee -(x + z) = -((x + y) \wedge (x + z))$,
which, by (g3), becomes:
$-(x + y) \vee -(x + z) = -(x + (y \wedge z))$,
which by (NegS) three times, becomes:
$(-y + -x) \vee (-z + -x) = -(y \wedge z) + -x$,
which, by (11.130) for $Z := -z + -x$ and by (DN), becomes:
(*) $-y + (-x \vee (y + (-z + -x))) = -(y \wedge z) + -x$;
but, in (*) the part $-x \vee (y + (-z + -x))$ equals, by (11.131), $-(x \wedge -(y + (-z + -x)))$,
which equals, by (NegS), $-(x \wedge (-(-z + -x) + -y))$,
which equals, by (NegS) and (DN), $-(x \wedge ((x + z) + -y))$;
then, (*) becomes:
$-y + -(x \wedge ((x + z) + -y)) = -(y \wedge z) + -x$,
which, by (Sass), becomes:
$-y + -(x \wedge (x + (z + -y))) = -(y \wedge z) + -x$,
hence, we have:

$$-(x \wedge y) + -z = -x + -(z \wedge (z + (y + -x))). \tag{11.132}$$

Now, (Cp) ($x \leq y \iff x + z \leq y + z$), i.e. $x \wedge y = x \iff (x + z) \wedge (y + z) = x + z$, becomes, by (Scomm):
$x \wedge y = x \iff (z + x) \wedge (y + z) = x + z$,
which, by (11.118), becomes:
(**) $x \wedge y = x \iff z + (x \wedge y) = x + z$;
then, in (**), take $X := 0$, $Y := 1 + -a$, $Z := a$, to obtain, by (SU):
(**') $0 \wedge (1 + -a) = 0 \iff a + (0 \wedge (1 + -a)) = a$;
but, in (**'), the part $a + (0 \wedge (1 + -a)) = a$ is always true, by (11.128); it follows that we obtain:

$$0 \wedge (1 + -a) = 0. \tag{11.133}$$

Finally, in (Wass) ($(x \wedge y) \wedge z = x \wedge (y \wedge z)$), take $X := 0$, $Y := 1$, to obtain:
$(0 \wedge 1) \wedge z = 0 \wedge (1 \wedge z)$,
which, by (11.103), becomes:
$0 \wedge z = 0 \wedge (1 \wedge z)$,
hence, we have:

$$0 \wedge (1 \wedge x) = 0 \wedge x. \tag{11.134}$$

We are ready now to prove (11.102). Indeed, we have the following equivalent identities (copying the corresponding lines from the PROVER9 proof):

24 $1 + -(1 \wedge ((1 + -(1 \wedge (b + c))) + (1 + -(1 \wedge (a + (1 + -(1 \wedge ((1 + -b) + (1 + -c))))))))) = 1 + -(1 \wedge ((1 + -(1 \wedge (a + b))) + (1 + -(1 \wedge (c + (1 + -(1 \wedge ((1 + -a) + (1 + -b)))))))))$

25 $\Longleftrightarrow 1 + -(1 \wedge ((1 + -(1 \wedge (b + c))) + (1 + -(1 \wedge (a + (1 + -(1 \wedge (1 + (-b + (1 + -c)))))))))) = 1 + -(1 \wedge ((1 + -(1 \wedge (a + b))) + (1 + -(1 \wedge (c + (1 + -(1 \wedge ((1 + -a) + (1 + -b)))))))))$, by (Sass),

26 $\Longleftrightarrow 1 + -(1 \wedge (1 + (-(1 \wedge (b + c)) + (1 + -(1 \wedge (a + (1 + -(1 \wedge (1 + (-b + (1 + -c))))))))))) = 1 + -(1 \wedge ((1 + -(1 \wedge (a + b))) + (1 + -(1 \wedge (c + (1 + -(1 \wedge ((1 + -a) + (1 + -b)))))))))$, by (Sass),

27 $\Longleftrightarrow 1 + -(1 \wedge (1 + (-(1 \wedge (b + c)) + (1 + -(1 \wedge (a + (1 + -(1 \wedge (1 + (-b + (1 + -c))))))))))) = 1 + -(1 \wedge ((1 + -(1 \wedge (a + b))) + (1 + -(1 \wedge (c + (1 + -(1 \wedge (1 + (-a + (1 + -b))))))))))$, by (Sass),

28 $\Longleftrightarrow 1 + -(1 \wedge (1 + (-(1 \wedge (b + c)) + (1 + -(1 \wedge (a + (1 + -(1 \wedge (1 + (-b + (1 + -c))))))))))) = 1 + -(1 \wedge (1 + (-(1 \wedge (a + b)) + (1 + -(1 \wedge (c + (1 + -(1 \wedge (1 + (-a + (1 + -b)))))))))))$, by (Sass)

32 $\Longleftrightarrow 1 + -(1 \wedge (1 + (-(1 \wedge (b + c)) + (1 + -(1 \wedge (a + (1 + -(1 \wedge (1 + (1 + (-b + -c))))))))))) = 1 + -(1 \wedge (1 + (-(1 \wedge (a + b)) + (1 + -(1 \wedge (c + (1 + -(1 \wedge (1 + (-a + (1 + -b)))))))))))$, by (11.108),

33 $\Longleftrightarrow 1 + -(1 \wedge (1 + (-(1 \wedge (b + c)) + (1 + -(1 \wedge (1 + (a + -(1 \wedge (1 + (1 + (-b + -c))))))))))) = 1 + -(1 \wedge (1 + (-(1 \wedge (a + b)) + (1 + -(1 \wedge (c + (1 + -(1 \wedge (1 + (-a + (1 + -b)))))))))))$, by (11.108),

34 $\Longleftrightarrow 1 + -(1 \wedge (1 + (1 + (-(1 \wedge (b + c)) + -(1 \wedge (1 + (a + -(1 \wedge (1 + (1 + (-b + -c))))))))))) = 1 + -(1 \wedge (1 + (-(1 \wedge (a + b)) + (1 + -(1 \wedge (c + (1 + -(1 \wedge (1 + (-a + (1 + -b)))))))))))$, by (11.108),

35 $\Longleftrightarrow 1 + -(1 \wedge (1 + (1 + (-(1 \wedge (b + c)) + -(1 \wedge (1 + (a + -(1 \wedge (1 + (1 + (-b + -c))))))))))) = 1 + -(1 \wedge (1 + (-(1 \wedge (a + b)) + (1 + -(1 \wedge (c + (1 + -(1 \wedge (1 + (1 + (-a + -b)))))))))))$, by (11.108),

36 $\Longleftrightarrow 1 + -(1 \wedge (1 + (1 + (-(1 \wedge (b + c)) + -(1 \wedge (1 + (a + -(1 \wedge (1 + (1 + (-b + -c))))))))))) = 1 + -(1 \wedge (1 + (-(1 \wedge (a + b)) + (1 + -(1 \wedge (1 + (c + -(1 \wedge (1 + (1 + (-a + -b)))))))))))$, by (11.108),

37 $\Longleftrightarrow 1 + -(1 \wedge (1 + (1 + (-(1 \wedge (b + c)) + -(1 \wedge (1 + (a + -(1 \wedge (1 + (1 + (-b + -c))))))))))) = 1 + -(1 \wedge (1 + (1 + (-(1 \wedge (a + b)) + -(1 \wedge (1 + (c + -(1 \wedge (1 + (1 + (-a + -b)))))))))))$, by (11.108),

46 $\Longleftrightarrow 1 + -(1 \wedge (1 + (1 + (-(1 \wedge (b + c)) + -(1 \wedge (1 + (a + -(1 + (0 \wedge (1 + (-b + -c))))))))))) = 1 + -(1 \wedge (1 + (1 + (-(1 \wedge (a + b)) + -(1 \wedge (1 + (c + -(1 \wedge (1 + (1 + (-a + -b)))))))))))$, by (11.116),

47 $\Longleftrightarrow 1 + -(1 \wedge (1 + (1 + (-(1 \wedge (b + c)) + -(1 \wedge (1 + (a + (-(0 \wedge (1 + (-b + -c))) + -1)))))))) = 1 + -(1 \wedge (1 + (1 + (-(1 \wedge (a + b)) + -(1 \wedge (1 + (c + -(1 \wedge (1 + (1 + (-a + -b)))))))))))$, by (NegS),

48 $\Longleftrightarrow 1 + -(1 \wedge (1 + (1 + (-(1 \wedge (b + c)) + -(1 \wedge (1 + (a + (-1 + -(0 \wedge (1 + (-b + -c))))))))))) = 1 + -(1 \wedge (1 + (1 + (-(1 \wedge (a + b)) + -(1 \wedge (1 + (c + -(1 \wedge (1 + (1 + (-a + -b)))))))))))$, by (Scomm),

49 $\Longleftrightarrow 1 + -(1 \wedge (1 + (1 + (-(1 \wedge (b + c)) + -(1 + (0 \wedge (a + (-1 + -(0 \wedge (1 + (-b + -c))))))))))) = 1 + -(1 \wedge (1 + (1 + (-(1 \wedge (a + b)) + -(1 \wedge (1 + (c + -(1 \wedge (1 + (1 +$

$(-a+-b)))))))))))$, by (11.116),
$50 \iff 1+-(1\wedge(1+(1+(-(1\wedge(b+c))+(-(0\wedge(a+(-1+-(0\wedge(1+(-b+-c))))))+-1)))))=1+-(1\wedge(1+(1+(-(1\wedge(a+b))+-(1\wedge(1+(c+-(1\wedge(1+(1+(-a+-b)))))))))))$, by (NegS),
$51 \iff 1+-(1\wedge(1+(1+(-(1\wedge(b+c))+(-1+-(0\wedge(a+(-1+-(0\wedge(1+(-b+-c)))))))))))=1+-(1\wedge(1+(1+(-(1\wedge(a+b))+-(1\wedge(1+(c+-(1\wedge(1+(1+(-a+-b)))))))))))$, by (Scomm),
$52 \iff 1+-(1\wedge(1+(1+((-1+-(0\wedge(a+(-1+-(0\wedge(1+(-b+-c)))))))+-(1\wedge(b+c))))))=1+-(1\wedge(1+(1+(-(1\wedge(a+b))+-(1\wedge(1+(c+-(1\wedge(1+(1+(-a+-b)))))))))))$, by (Scomm),
$53 \iff 1+-(1\wedge(1+(1+(-1+(-(0\wedge(a+(-1+-(0\wedge(1+(-b+-c))))))+-(1\wedge(b+c)))))))=1+-(1\wedge(1+(1+(-(1\wedge(a+b))+-(1\wedge(1+(c+-(1\wedge(1+(1+(-a+-b)))))))))))$, by (Sass),
$54 \iff 1+-(1\wedge(1+(-(0\wedge(a+(-1+-(0\wedge(1+(-b+-c))))))+-(1\wedge(b+c)))))=1+-(1\wedge(1+(1+(-(1\wedge(a+b))+-(1\wedge(1+(c+-(1\wedge(1+(1+(-a+-b)))))))))))$, by (11.109),
$55 \iff 1+-(1+(0\wedge(-(0\wedge(a+(-1+-(0\wedge(1+(-b+-c))))))+-(1\wedge(b+c)))))=1+-(1\wedge(1+(1+(-(1\wedge(a+b))+-(1\wedge(1+(c+-(1\wedge(1+(1+(-a+-b)))))))))))$, by (11.116),
$56 \iff 1+(-(0\wedge(-(0\wedge(a+(-1+-(0\wedge(1+(-b+-c))))))+-(1\wedge(b+c))))+-1)=1+-(1\wedge(1+(1+(-(1\wedge(a+b))+-(1\wedge(1+(c+-(1\wedge(1+(1+(-a+-b)))))))))))$, by (NegS),
$57 \iff 1+(-1+-(0\wedge(-(0\wedge(a+(-1+-(0\wedge(1+(-b+-c))))))+-(1\wedge(b+c)))))=1+-(1\wedge(1+(1+(-(1\wedge(a+b))+-(1\wedge(1+(c+-(1\wedge(1+(1+(-a+-b)))))))))))$, by (Scomm),
$58 \iff -(0\wedge(-(0\wedge(a+(-1+-(0\wedge(1+(-b+-c))))))+-(1\wedge(b+c))))=1+-(1\wedge(1+(1+(-(1\wedge(a+b))+-(1\wedge(1+(c+-(1\wedge(1+(1+(-a+-b)))))))))))$, by (11.109),
$59 \iff -(0\wedge(-(0\wedge(a+(-1+-(0\wedge(1+(-b+-c))))))+-(1\wedge(b+c))))=1+-(1\wedge(1+(1+(-(1\wedge(a+b))+-(1\wedge(1+(c+-(1+(0\wedge(1+(-a+-b)))))))))))$, by (11.116),
$60 \iff -(0\wedge(-(0\wedge(a+(-1+-(0\wedge(1+(-b+-c))))))+-(1\wedge(b+c))))=1+-(1\wedge(1+(1+(-(1\wedge(a+b))+-(1\wedge(1+(c+(-(0\wedge(1+(-a+-b)))+-1))))))))$, by (NegS),
$61 \iff -(0\wedge(-(0\wedge(a+(-1+-(0\wedge(1+(-b+-c))))))+-(1\wedge(b+c))))=1+-(1\wedge(1+(1+(-(1\wedge(a+b))+-(1\wedge(1+(c+(-1+-(0\wedge(1+(-a+-b)))))))))))$, by (Scomm),
$62 \iff -(0\wedge(-(0\wedge(a+(-1+-(0\wedge(1+(-b+-c))))))+-(1\wedge(b+c))))=1+-(1\wedge(1+(1+(-(1\wedge(a+b))+-(1+(0\wedge(c+(-1+-(0\wedge(1+(-a+-b)))))))))))$, by (11.116),
$63 \iff -(0\wedge(-(0\wedge(a+(-1+-(0\wedge(1+(-b+-c))))))+-(1\wedge(b+c))))=1+-(1\wedge(1+(1+(-(1\wedge(a+b))+(-(0\wedge(c+(-1+-(0\wedge(1+(-a+-b))))))+-1)))))$, by (NegS),
$64 \iff -(0\wedge(-(0\wedge(a+(-1+-(0\wedge(1+(-b+-c))))))+-(1\wedge(b+c))))=1+-(1\wedge(1+(1+(-(1\wedge(a+b))+(-1+-(0\wedge(c+(-1+-(0\wedge(1+(-a+-b)))))))))))$, by (Scomm),

65 $\Longleftrightarrow -(0 \wedge (-(0 \wedge (a + (-1 + -(0 \wedge (1 + (-b + -c)))))) + -(1 \wedge (b + c)))) = 1+-(1\wedge(1+(1+((-1+-(0\wedge(c+(-1+-(0\wedge(1+(-a+-b)))))))+-(1\wedge(a+b))))))$, by (Scomm),
66 $\Longleftrightarrow -(0 \wedge (-(0 \wedge (a + (-1 + -(0 \wedge (1 + (-b + -c)))))) + -(1 \wedge (b + c)))) = 1+-(1\wedge(1+(1+(-1+(-(0\wedge(c+(-1+-(0\wedge(1+(-a+-b))))))+-(1\wedge(a+b)))))))$, by (Sass),
67 $\Longleftrightarrow -(0 \wedge (-(0 \wedge (a + (-1 + -(0 \wedge (1 + (-b + -c)))))) + -(1 \wedge (b + c)))) = 1 + -(1 \wedge (1 + (-(0 \wedge (c + (-1 + -(0 \wedge (1 + (-a + -b)))))) + -(1 \wedge (a + b)))))$, by (11.109),
68 $\Longleftrightarrow -(0 \wedge (-(0 \wedge (a + (-1 + -(0 \wedge (1 + (-b + -c)))))) + -(1 \wedge (b + c)))) = 1 + -(1 + (0 \wedge (-(0 \wedge (c + (-1 + -(0 \wedge (1 + (-a + -b)))))) + -(1 \wedge (a + b)))))$, by (11.116),
69 $\Longleftrightarrow -(0 \wedge (-(0 \wedge (a + (-1 + -(0 \wedge (1 + (-b + -c)))))) + -(1 \wedge (b + c)))) = 1 + (-(0 \wedge (-(0 \wedge (c + (-1 + -(0 \wedge (1 + (-a + -b)))))) + -(1 \wedge (a + b)))) + -1)$, by (NegS),
70 $\Longleftrightarrow -(0 \wedge (-(0 \wedge (a + (-1 + -(0 \wedge (1 + (-b + -c)))))) + -(1 \wedge (b + c)))) = 1 + (-1 + -(0 \wedge (-(0 \wedge (c + (-1 + -(0 \wedge (1 + (-a + -b)))))) + -(1 \wedge (a + b)))))$, by (Scomm),
71 $\Longleftrightarrow -(0 \wedge (-(0 \wedge (a + (-1 + -(0 \wedge (1 + (-b + -c)))))) + -(1 \wedge (b + c)))) = -(0 \wedge (-(0 \wedge (c + (-1 + -(0 \wedge (1 + (-a + -b)))))) + -(1 \wedge (a + b))))$, by (11.109),

111 $\Longleftrightarrow -(0 \wedge (-0 + -((1 \wedge (b+c)) \wedge ((1 \wedge (b+c)) + ((a + (-1 + -(0 \wedge (1 + (-b + -c))))) + -0))))) = -(0\wedge(-(0\wedge(c+(-1+-(0\wedge(1+(-a+-b))))))+-(1\wedge(a+b))))$, by (11.132),
112 $\Longleftrightarrow -(0 \wedge (0 + -((1 \wedge (b+c)) \wedge ((1 \wedge (b+c)) + ((a + (-1 + -(0 \wedge (1 + (-b + -c))))) + -0))))) = -(0\wedge(-(0\wedge(c+(-1+-(0\wedge(1+(-a+-b))))))+-(1\wedge(a+b))))$, by (N0),
113 $\Longleftrightarrow -(0 \wedge (0 + -((1 \wedge (b+c)) \wedge ((1 \wedge (b+c)) + ((a + (-1 + -(0 \wedge (1 + (-b + -c))))) + 0))))) = -(0\wedge(-(0\wedge(c+(-1+-(0\wedge(1+(-a+-b))))))+-(1\wedge(a+b))))$, by (N0),
114 $\Longleftrightarrow -(0 \wedge (0 + -((1 \wedge (b+c)) \wedge ((1 \wedge (b+c)) + (0 + (a + (-1 + -(0 \wedge (1 + (-b + -c)))))))))) = -(0 \wedge (-(0 \wedge (c + (-1 + -(0 \wedge (1 + (-a + -b)))))) + -(1 \wedge (a + b))))$, by (Scomm),
115 $\Longleftrightarrow -(0\wedge(0+-((1\wedge(b+c))\wedge((1\wedge(b+c))+(a+(-1+-(0\wedge(1+(-b+-c))))))))) = -(0 \wedge (-(0 \wedge (c + (-1 + -(0 \wedge (1 + (-a + -b)))))) + -(1 \wedge (a + b))))$, by (SU),
116 $\Longleftrightarrow -(0\wedge(0+-((1\wedge(b+c))\wedge((a+(-1+-(0\wedge(1+(-b+-c)))))+(1\wedge(b+c)))))) = -(0 \wedge (-(0 \wedge (c + (-1 + -(0 \wedge (1 + (-a + -b)))))) + -(1 \wedge (a + b))))$, by (Scomm),
117 $\Longleftrightarrow -(0\wedge(0+-((1\wedge(b+c))\wedge(a+((-1+-(0\wedge(1+(-b+-c))))+(1\wedge(b+c))))))) = -(0 \wedge (-(0 \wedge (c + (-1 + -(0 \wedge (1 + (-a + -b)))))) + -(1 \wedge (a + b))))$, by (Sass),
118 $\Longleftrightarrow -(0\wedge(0+-((1\wedge(b+c))\wedge(a+(-1+(-(0\wedge(1+(-b+-c)))+(1\wedge(b+c)))))))) = -(0 \wedge (-(0 \wedge (c + (-1 + -(0 \wedge (1 + (-a + -b)))))) + -(1 \wedge (a + b))))$, by (Sass),
119 $\Longleftrightarrow -(0\wedge(0+-(1\wedge((b+c)\wedge(a+(-1+(-(0\wedge(1+(-b+-c)))+(1\wedge(b+c))))))))) = -(0 \wedge (-(0 \wedge (c + (-1 + -(0 \wedge (1 + (-a + -b)))))) + -(1 \wedge (a + b))))$, by (Wass),
120 $\Longleftrightarrow -(0\wedge(0+-(1\wedge(b+(c\wedge(-b+(a+(-1+(-(0\wedge(1+(-b+-c)))+(1\wedge(b+c)))))))))))) = -(0 \wedge (-(0 \wedge (c + (-1 + -(0 \wedge (1 + (-a + -b)))))) + -(1 \wedge (a + b))))$, by (11.120),

121 $\iff -(0\wedge(0+-(1\wedge(b+(c\wedge(a+(-b+(-1+(-(0\wedge(1+(-b+-c)))+(1\wedge(b+c)))))))))) = -(0\wedge(-(0\wedge(c+(-1+-(0\wedge(1+(-a+-b))))))+-(1\wedge(a+b))))$, by (11.108),

122 $\iff -(0\wedge(0+-(1\wedge(b+(c\wedge(a+(-1+(-b+(-(0\wedge(1+(-b+-c)))+(1\wedge(b+c)))))))))) = -(0\wedge(-(0\wedge(c+(-1+-(0\wedge(1+(-a+-b))))))+-(1\wedge(a+b))))$, by (11.108),

123 $\iff -(0\wedge-(1\wedge(b+(c\wedge(a+(-1+(-b+(-(0\wedge(1+(-b+-c)))+(1\wedge(b+c))))))))) = -(0\wedge(-(0\wedge(c+(-1+-(0\wedge(1+(-a+-b))))))+-(1\wedge(a+b))))$, by (SU),

124 $\iff -(-(1\wedge(b+(c\wedge(a+(-1+(-b+(-(0\wedge(1+(-b+-c)))+(1\wedge(b+c))))))))+(0\wedge(1\wedge(b+(c\wedge(a+(-1+(-b+(-(0\wedge(1+(-b+-c)))+(1\wedge(b+c)))))))))) = -(0\wedge(-(0\wedge(c+(-1+-(0\wedge(1+(-a+-b))))))+-(1\wedge(a+b))))$, by (11.127),

125 $\iff -(-(1\wedge(b+(c\wedge(a+(-1+(-b+(-(0\wedge(1+(-b+-c)))+(1\wedge(b+c))))))))+(0\wedge(b+(c\wedge(a+(-1+(-b+(-(0\wedge(1+(-b+-c)))+(1\wedge(b+c))))))))) = -(0\wedge(-(0\wedge(c+(-1+-(0\wedge(1+(-a+-b))))))+-(1\wedge(a+b))))$, by (11.134),

126 $\iff -((0\wedge(b+(c\wedge(a+(-1+(-b+(-(0\wedge(1+(-b+-c)))+(1\wedge(b+c))))))))+-(1\wedge(b+(c\wedge(a+(-1+(-b+(-(0\wedge(1+(-b+-c)))+(1\wedge(b+c))))))))) = -(0\wedge(-(0\wedge(c+(-1+-(0\wedge(1+(-a+-b))))))+-(1\wedge(a+b))))$, by (Scomm),

127 $\iff --(1\wedge(b+(c\wedge(a+(-1+(-b+(-(0\wedge(1+(-b+-c)))+(1\wedge(b+c))))))))+-(0\wedge(b+(c\wedge(a+(-1+(-b+(-(0\wedge(1+(-b+-c)))+(1\wedge(b+c)))))))) = -(0\wedge(-(0\wedge(c+(-1+-(0\wedge(1+(-a+-b))))))+-(1\wedge(a+b))))$, by (NegS),

128 $\iff (1\wedge(b+(c\wedge(a+(-1+(-b+(-(0\wedge(1+(-b+-c)))+(1\wedge(b+c))))))))+-(0\wedge(b+(c\wedge(a+(-1+(-b+(-(0\wedge(1+(-b+-c)))+(1\wedge(b+c)))))))) = -(0\wedge(-(0\wedge(c+(-1+-(0\wedge(1+(-a+-b))))))+-(1\wedge(a+b))))$, by (DN),

129 $\iff -(0\wedge(b+(c\wedge(a+(-1+(-b+(-(0\wedge(1+(-b+-c)))+(1\wedge(b+c))))))))+(1\wedge(b+(c\wedge(a+(-1+(-b+(-(0\wedge(1+(-b+-c)))+(1\wedge(b+c)))))))) = -(0\wedge(-(0\wedge(c+(-1+-(0\wedge(1+(-a+-b))))))+-(1\wedge(a+b))))$, by (Scomm),

130 $\iff -(0\wedge(b+(c\wedge(a+(-1+(-b+(1+(-(0\wedge(1+(-b+-c)))\wedge(-1+(-(0\wedge(1+(-b+-c)))+(b+c))))))))))+(1\wedge(b+(c\wedge(a+(-1+(-b+(-(0\wedge(1+(-b+-c)))+(1\wedge(b+c))))))))) = -(0\wedge(-(0\wedge(c+(-1+-(0\wedge(1+(-a+-b))))))+-(1\wedge(a+b))))$, by (11.126),

131 $\iff -(0\wedge(b+(c\wedge(a+(-1+(-b+(1+(-(0\wedge(1+(-b+-c)))\wedge(-1+((b+c)+-(0\wedge(1+(-b+-c))))))))))))+(1\wedge(b+(c\wedge(a+(-1+(-b+(-(0\wedge(1+(-b+-c)))+(1\wedge(b+c))))))))) = -(0\wedge(-(0\wedge(c+(-1+-(0\wedge(1+(-a+-b))))))+-(1\wedge(a+b))))$, by (Scomm),

132 $\iff -(0\wedge(b+(c\wedge(a+(-1+(-b+(1+(-(0\wedge(1+(-b+-c)))\wedge(-1+(b+(c+-(0\wedge(1+(-b+-c)))))))))))))+(1\wedge(b+(c\wedge(a+(-1+(-b+(-(0\wedge(1+(-b+-c)))+(1\wedge(b+c))))))))) = -(0\wedge(-(0\wedge(c+(-1+-(0\wedge(1+(-a+-b))))))+-(1\wedge(a+b))))$, by (Sass),

133 $\iff -(0\wedge(b+(c\wedge(a+(-1+(-b+(1+(-(0\wedge(1+(-b+-c)))\wedge(b+(-1+(c+-(0\wedge(1+(-b+-c)))))))))))))+(1\wedge(b+(c\wedge(a+(-1+(-b+(-(0\wedge(1+(-b+-c)))+(1\wedge(b+c))))))))) = -(0\wedge(-(0\wedge(c+(-1+-(0\wedge(1+(-a+-b))))))+-(1\wedge(a+b))))$, by (11.108),

134 $\iff -(0\wedge(b+(c\wedge(a+(-1+(-b+(1+(-(0\wedge(1+(-b+-c)))\wedge(b+(c+(-1+-(0\wedge(1+(-b+-c)))))))))))))+(1\wedge(b+(c\wedge(a+(-1+(-b+(-(0\wedge(1+(-b+-c)))+(1\wedge(b+c))))))))) = -(0\wedge(-(0\wedge(c+(-1+-(0\wedge(1+(-a+-b))))))+-(1\wedge(a+b))))$,

by (11.108),

135 $\iff -(0 \wedge (b + (c \wedge (a + (-1 + (-b + (1 + (-(0 \wedge (1 + (-b + -c))) + (0 \wedge ((b + (c + (-1 + -(0 \wedge (1 + (-b + -c)))))) + (0 \wedge (1 + (-b + -c)))))))))))) + (1 \wedge (b + (c \wedge (a + (-1 + (-b + (-(0 \wedge (1 + (-b + -c))) + (1 \wedge (b + c))))))))) = -(0 \wedge (-(0 \wedge (c + (-1 + -(0 \wedge (1 + (-a + -b)))))) + -(1 \wedge (a + b))))$, by (11.125),

136 $\iff -(0 \wedge (b + (c \wedge (a + (-1 + (-b + (1 + (-(0 \wedge (1 + (-b + -c))) + (0 \wedge ((0 \wedge (1 + (-b + -c))) + (b + (c + (-1 + -(0 \wedge (1 + (-b + -c)))))))))))))))) + (1 \wedge (b + (c \wedge (a + (-1 + (-b + (-(0 \wedge (1 + (-b + -c))) + (1 \wedge (b + c))))))))) = -(0 \wedge (-(0 \wedge (c + (-1 + -(0 \wedge (1 + (-a + -b)))))) + -(1 \wedge (a + b))))$, by (Scomm),

137 $\iff -(0 \wedge (b + (c \wedge (a + (-1 + (-b + (1 + (-(0 \wedge (1 + (-b + -c))) + (0 \wedge (b + ((0 \wedge (1 + (-b + -c))) + (c + (-1 + -(0 \wedge (1 + (-b + -c)))))))))))))))) + (1 \wedge (b + (c \wedge (a + (-1 + (-b + (-(0 \wedge (1 + (-b + -c))) + (1 \wedge (b + c))))))))) = -(0 \wedge (-(0 \wedge (c + (-1 + -(0 \wedge (1 + (-a + -b)))))) + -(1 \wedge (a + b))))$, by (11.108),

138 $\iff -(0 \wedge (b + (c \wedge (a + (-1 + (-b + (1 + (-(0 \wedge (1 + (-b + -c))) + (0 \wedge (b + (c + ((0 \wedge (1 + (-b + -c))) + (-1 + -(0 \wedge (1 + (-b + -c)))))))))))))))) + (1 \wedge (b + (c \wedge (a + (-1 + (-b + (-(0 \wedge (1 + (-b + -c))) + (1 \wedge (b + c))))))))) = -(0 \wedge (-(0 \wedge (c + (-1 + -(0 \wedge (1 + (-a + -b)))))) + -(1 \wedge (a + b))))$, by (11.108),

139 $\iff -(0 \wedge (b + (c \wedge (a + (-1 + (-b + (1 + (-(0 \wedge (1 + (-b + -c))) + (0 \wedge (b + (c + (-1 + ((0 \wedge (1 + (-b + -c))) + -(0 \wedge (1 + (-b + -c)))))))))))))))) + (1 \wedge (b + (c \wedge (a + (-1 + (-b + (-(0 \wedge (1 + (-b + -c))) + (1 \wedge (b + c))))))))) = -(0 \wedge (-(0 \wedge (c + (-1 + -(0 \wedge (1 + (-a + -b)))))) + -(1 \wedge (a + b))))$, by (11.108),

140 $\iff -(0 \wedge (b + (c \wedge (a + (-1 + (-b + (1 + (-(0 \wedge (1 + (-b + -c))) + (0 \wedge (b + (c + (-1 + 0)))))))))))) + (1 \wedge (b + (c \wedge (a + (-1 + (-b + (-(0 \wedge (1 + (-b + -c))) + (1 \wedge (b + c))))))))) = -(0 \wedge (-(0 \wedge (c + (-1 + -(0 \wedge (1 + (-a + -b)))))) + -(1 \wedge (a + b))))$, by ($m_0$-Re),

141 $\iff -(0 \wedge (b + (c \wedge (a + (-1 + (-b + (1 + (-(0 \wedge (1 + (-b + -c))) + (0 \wedge (b + (c + (0 + -1)))))))))))) + (1 \wedge (b + (c \wedge (a + (-1 + (-b + (-(0 \wedge (1 + (-b + -c))) + (1 \wedge (b + c))))))))) = -(0 \wedge (-(0 \wedge (c + (-1 + -(0 \wedge (1 + (-a + -b)))))) + -(1 \wedge (a + b))))$, by (Scomm),

142 $\iff -(0 \wedge (b + (c \wedge (a + (-1 + (-b + (1 + (-(0 \wedge (1 + (-b + -c))) + (0 \wedge (b + (c + -1))))))))))) + (1 \wedge (b + (c \wedge (a + (-1 + (-b + (-(0 \wedge (1 + (-b + -c))) + (1 \wedge (b + c))))))))) = -(0 \wedge (-(0 \wedge (c + (-1 + -(0 \wedge (1 + (-a + -b)))))) + -(1 \wedge (a + b))))$, by (SU),

143 $\iff -(0 \wedge (b + (c \wedge (a + (-1 + (-b + (1 + ((0 \wedge (b + (c + -1))) + -(0 \wedge (1 + (-b + -c))))))))))) + (1 \wedge (b + (c \wedge (a + (-1 + (-b + (-(0 \wedge (1 + (-b + -c))) + (1 \wedge (b + c))))))))) = -(0 \wedge (-(0 \wedge (c + (-1 + -(0 \wedge (1 + (-a + -b)))))) + -(1 \wedge (a + b))))$, by (Scomm),

144 $\iff -(0 \wedge (b + (c \wedge (a + (-1 + (1 + (-b + ((0 \wedge (b + (c + -1))) + -(0 \wedge (1 + (-b + -c))))))))))) + (1 \wedge (b + (c \wedge (a + (-1 + (-b + (-(0 \wedge (1 + (-b + -c))) + (1 \wedge (b + c))))))))) = -(0 \wedge (-(0 \wedge (c + (-1 + -(0 \wedge (1 + (-a + -b)))))) + -(1 \wedge (a + b))))$, by (11.108),

145 $\iff -(0 \wedge (b + (c \wedge (a + (1 + (-1 + (-b + ((0 \wedge (b + (c + -1))) + -(0 \wedge (1 + (-b + -c))))))))))) + (1 \wedge (b + (c \wedge (a + (-1 + (-b + (-(0 \wedge (1 + (-b + -c))) + (1 \wedge (b + c))))))))) = -(0 \wedge (-(0 \wedge (c + (-1 + -(0 \wedge (1 + (-a + -b)))))) + -(1 \wedge (a + b))))$, by (11.108),

146 $\iff -(0 \wedge (b + (c \wedge (a + (-b + ((0 \wedge (b + (c + -1))) + -(0 \wedge (1 + (-b + -c))))))))) + (1 \wedge (b + (c \wedge (a + (-1 + (-b + (-(0 \wedge (1 + (-b + -c))) + (1 \wedge (b + c))))))))) = -(0 \wedge (-(0 \wedge (c + (-1 + -(0 \wedge (1 + (-a + -b)))))) + -(1 \wedge (a + b))))$, by (11.109),

147 $\iff -(0 \wedge (b + (c \wedge (a + (-b + ((b + ((c + -1) \wedge -b)) + -(0 \wedge (1 + (-b + -c))))))))) + (1 \wedge (b + (c \wedge (a + (-1 + (-b + (-(0 \wedge (1 + (-b + -c))) + (1 \wedge (b + c))))))))) = -(0 \wedge (-(0 \wedge (c + (-1 + -(0 \wedge (1 + (-a + -b)))))) + -(1 \wedge (a + b))))$, by (11.119),

148 $\Longleftrightarrow -(0 \wedge (b+(c \wedge (a+(-b+((b+(c+(-1 \wedge (-c+-b))))+-(0 \wedge (1+(-b+-c)))))))))+(1 \wedge (b+(c \wedge (a+(-1+(-b+(-(0 \wedge (1+(-b+-c)))+(1 \wedge (b+c)))))))))) = -(0 \wedge (-(0 \wedge (c+(-1+-(0 \wedge (1+(-a+-b))))))+-(1 \wedge (a+b))))$, by (11.120),
149 $\Longleftrightarrow -(0 \wedge (b+(c \wedge (a+(-b+((b+(c+(-1 \wedge (-b+-c))))+-(0 \wedge (1+(-b+-c)))))))))+(1 \wedge (b+(c \wedge (a+(-1+(-b+(-(0 \wedge (1+(-b+-c)))+(1 \wedge (b+c)))))))))) = -(0 \wedge (-(0 \wedge (c+(-1+-(0 \wedge (1+(-a+-b))))))+-(1 \wedge (a+b))))$, by (Scomm),
150 $\Longleftrightarrow -(0 \wedge (b+(c \wedge (a+(-b+((b+(c+(-1+(0 \wedge ((-b+-c)+1)))))+-(0 \wedge (1+(-b+-c)))))))))+(1 \wedge (b+(c \wedge (a+(-1+(-b+(-(0 \wedge (1+(-b+-c)))+(1 \wedge (b+c)))))))))) = -(0 \wedge (-(0 \wedge (c+(-1+-(0 \wedge (1+(-a+-b))))))+-(1 \wedge (a+b))))$, by (11.125),
151 $\Longleftrightarrow -(0 \wedge (b+(c \wedge (a+(-b+((b+(c+(-1+(0 \wedge (1+(-b+-c))))))+-(0 \wedge (1+(-b+-c)))))))))+(1 \wedge (b+(c \wedge (a+(-1+(-b+(-(0 \wedge (1+(-b+-c)))+(1 \wedge (b+c)))))))))) = -(0 \wedge (-(0 \wedge (c+(-1+-(0 \wedge (1+(-a+-b))))))+-(1 \wedge (a+b))))$, by (Scomm),
152 $\Longleftrightarrow -(0 \wedge (b+(c \wedge (a+(-b+(b+((c+(-1+(0 \wedge (1+(-b+-c)))))+-(0 \wedge (1+(-b+-c))))))))))+(1 \wedge (b+(c \wedge (a+(-1+(-b+(-(0 \wedge (1+(-b+-c)))+(1 \wedge (b+c)))))))))) = -(0 \wedge (-(0 \wedge (c+(-1+-(0 \wedge (1+(-a+-b))))))+-(1 \wedge (a+b))))$, by (Sass),
153 $\Longleftrightarrow -(0 \wedge (b+(c \wedge (a+(-b+(b+(c+((-1+(0 \wedge (1+(-b+-c))))+-(0 \wedge (1+(-b+-c)))))))))))+(1 \wedge (b+(c \wedge (a+(-1+(-b+(-(0 \wedge (1+(-b+-c)))+(1 \wedge (b+c)))))))))) = -(0 \wedge (-(0 \wedge (c+(-1+-(0 \wedge (1+(-a+-b))))))+-(1 \wedge (a+b))))$, by (Sass),
154 $\Longleftrightarrow -(0 \wedge (b+(c \wedge (a+(-b+(b+(c+(-1+((0 \wedge (1+(-b+-c)))+-(0 \wedge (1+(-b+-c))))))))))))+(1 \wedge (b+(c \wedge (a+(-1+(-b+(-(0 \wedge (1+(-b+-c)))+(1 \wedge (b+c)))))))))) = -(0 \wedge (-(0 \wedge (c+(-1+-(0 \wedge (1+(-a+-b))))))+-(1 \wedge (a+b))))$, by (Sass),
155 $\Longleftrightarrow -(0 \wedge (b+(c \wedge (a+(-b+(b+(c+(-1+0))))))))+(1 \wedge (b+(c \wedge (a+(-1+(-b+(-(0 \wedge (1+(-b+-c)))+(1 \wedge (b+c)))))))))) = -(0 \wedge (-(0 \wedge (c+(-1+-(0 \wedge (1+(-a+-b))))))+-(1 \wedge (a+b))))$, by ($m_0$-Re),
156 $\Longleftrightarrow -(0 \wedge (b+(c \wedge (a+(-b+(b+(c+(0+-1))))))))+(1 \wedge (b+(c \wedge (a+(-1+(-b+(-(0 \wedge (1+(-b+-c)))+(1 \wedge (b+c)))))))))) = -(0 \wedge (-(0 \wedge (c+(-1+-(0 \wedge (1+(-a+-b))))))+-(1 \wedge (a+b))))$, by (Scomm),
157 $\Longleftrightarrow -(0 \wedge (b+(c \wedge (a+(-b+(b+(c+-1)))))))+(1 \wedge (b+(c \wedge (a+(-1+(-b+(-(0 \wedge (1+(-b+-c)))+(1 \wedge (b+c)))))))))) = -(0 \wedge (-(0 \wedge (c+(-1+-(0 \wedge (1+(-a+-b))))))+-(1 \wedge (a+b))))$, by (SU),
158 $\Longleftrightarrow -(0 \wedge (b+(c \wedge (a+((b+(c+-1))+-b)))))+(1 \wedge (b+(c \wedge (a+(-1+(-b+(-(0 \wedge (1+(-b+-c)))+(1 \wedge (b+c)))))))))) = -(0 \wedge (-(0 \wedge (c+(-1+-(0 \wedge (1+(-a+-b))))))+-(1 \wedge (a+b))))$, by (Scomm),
159 $\Longleftrightarrow -(0 \wedge (b+(c \wedge (a+(b+((c+-1)+-b))))))+(1 \wedge (b+(c \wedge (a+(-1+(-b+(-(0 \wedge (1+(-b+-c)))+(1 \wedge (b+c)))))))))) = -(0 \wedge (-(0 \wedge (c+(-1+-(0 \wedge (1+(-a+-b))))))+-(1 \wedge (a+b))))$, by (Sass),
160 $\Longleftrightarrow -(0 \wedge (b+(c \wedge (a+(b+(c+(-1+-b)))))))+(1 \wedge (b+(c \wedge (a+(-1+(-b+(-(0 \wedge (1+(-b+-c)))+(1 \wedge (b+c)))))))))) = -(0 \wedge (-(0 \wedge (c+(-1+-(0 \wedge (1+(-a+-b))))))+-(1 \wedge (a+b))))$, by (Sass),
161 $\Longleftrightarrow -(0 \wedge (b+(c \wedge (a+(c+-1)))))+(1 \wedge (b+(c \wedge (a+(-1+(-b+(-(0 \wedge (1+(-b+-c)))+(1 \wedge (b+c)))))))))) = -(0 \wedge (-(0 \wedge (c+(-1+-(0 \wedge (1+(-a+-b))))))+-(1 \wedge (a+b))))$, by (11.113),
162 $\Longleftrightarrow -(0 \wedge (b+(c \wedge (a+(c+-1)))))+(1 \wedge (b+(c \wedge (a+(-1+(-b+(1+(-(0 \wedge (1+(-b+-c))) \wedge (-1+(-(0 \wedge (1+(-b+-c)))+(b+c))))))))))) = -(0 \wedge (-(0 \wedge (c+(-1+-(0 \wedge (1+(-a+-b))))))+-(1 \wedge (a+b))))$, by (11.126),
163 $\Longleftrightarrow w-(0 \wedge (b+(c \wedge (a+(c+-1)))))+(1 \wedge (b+(c \wedge (a+(-1+(-b+$

$(1+(-(0\wedge(1+(-b+-c)))\wedge(-1+((b+c)+-(0\wedge(1+(-b+-c)))))))))))) =$
$-(0\wedge(-(0\wedge(c+(-1+-(0\wedge(1+(-a+-b))))))+-(1\wedge(a+b))))$, by (Scomm),
164 $\iff -(0\wedge(b+(c\wedge(a+(c+-1)))))+(1\wedge(b+(c\wedge(a+(-1+(-b+(1+$
$(-(0\wedge(1+(-b+-c)))\wedge(-1+(b+(c+-(0\wedge(1+(-b+-c)))))))))))) =$
$-(0\wedge(-(0\wedge(c+(-1+-(0\wedge(1+(-a+-b))))))+-(1\wedge(a+b))))$, by (Sass),
165 $\iff -(0\wedge(b+(c\wedge(a+(c+-1)))))+(1\wedge(b+(c\wedge(a+(-1+(-b+(1+$
$(-(0\wedge(1+(-b+-c)))\wedge(b+(-1+(c+-(0\wedge(1+(-b+-c)))))))))))) =$
$-(0\wedge(-(0\wedge(c+(-1+-(0\wedge(1+(-a+-b))))))+-(1\wedge(a+b))))$, by (11.108),
166 $\iff -(0\wedge(b+(c\wedge(a+(c+-1)))))+(1\wedge(b+(c\wedge(a+(-1+(-b+(1+$
$(-(0\wedge(1+(-b+-c)))\wedge(b+(c+(-1+-(0\wedge(1+(-b+-c)))))))))))) =$
$-(0\wedge(-(0\wedge(c+(-1+-(0\wedge(1+(-a+-b))))))+-(1\wedge(a+b))))$, by (11.108),
167 $\iff -(0\wedge(b+(c\wedge(a+(c+-1)))))+(1\wedge(b+(c\wedge(a+(-1+(-b+(1+(-(0\wedge$
$(1+(-b+-c)))+(0\wedge((b+(c+(-1+-(0\wedge(1+(-b+-c))))))+(0\wedge(1+(-b+$
$-c))))))))))))) = -(0\wedge(-(0\wedge(c+(-1+-(0\wedge(1+(-a+-b))))))+-(1\wedge(a+b))))$,
by (11.125),
168 $\iff -(0\wedge(b+(c\wedge(a+(c+-1)))))+(1\wedge(b+(c\wedge(a+(-1+(-b+(1+(-(0\wedge$
$(1+(-b+-c)))+(0\wedge((0\wedge(1+(-b+-c)))+(b+(c+(-1+-(0\wedge(1+(-b+$
$-c)))))))))))))))) = -(0\wedge(-(0\wedge(c+(-1+-(0\wedge(1+(-a+-b))))))+-(1\wedge(a+b))))$,
by (Scomm),
169 $\iff -(0\wedge(b+(c\wedge(a+(c+-1)))))+(1\wedge(b+(c\wedge(a+(-1+(-b+(1+(-(0\wedge$
$(1+(-b+-c)))+(0\wedge(b+((0\wedge(1+(-b+-c)))+(c+(-1+-(0\wedge(1+(-b+$
$-c)))))))))))))))) = -(0\wedge(-(0\wedge(c+(-1+-(0\wedge(1+(-a+-b))))))+-(1\wedge(a+b))))$,
by (11.108),
170 $\iff -(0\wedge(b+(c\wedge(a+(c+-1)))))+(1\wedge(b+(c\wedge(a+(-1+(-b+(1+(-(0\wedge$
$(1+(-b+-c)))+(0\wedge(b+(c+((0\wedge(1+(-b+-c)))+(-1+-(0\wedge(1+(-b+$
$-c)))))))))))))))) = -(0\wedge(-(0\wedge(c+(-1+-(0\wedge(1+(-a+-b))))))+-(1\wedge(a+b))))$,
by (11.108),
171 $\iff -(0\wedge(b+(c\wedge(a+(c+-1)))))+(1\wedge(b+(c\wedge(a+(-1+(-b+(1+(-(0\wedge$
$(1+(-b+-c)))+(0\wedge(b+(c+(-1+((0\wedge(1+(-b+-c)))+-(0\wedge(1+(-b+$
$-c)))))))))))))))) = -(0\wedge(-(0\wedge(c+(-1+-(0\wedge(1+(-a+-b))))))+-(1\wedge(a+b))))$,
by (11.108),
172 $\iff -(0\wedge(b+(c\wedge(a+(c+-1)))))+(1\wedge(b+(c\wedge(a+(-1+(-b+(1+$
$(-(0\wedge(1+(-b+-c)))+(0\wedge(b+(c+(-1+0)))))))))))) = -(0\wedge(-(0\wedge(c+$
$(-1+-(0\wedge(1+(-a+-b))))))+-(1\wedge(a+b))))$, by ($m_0$-Re),
173 $\iff -(0\wedge(b+(c\wedge(a+(c+-1)))))+(1\wedge(b+(c\wedge(a+(-1+(-b+(1+$
$(-(0\wedge(1+(-b+-c)))+(0\wedge(b+(c+(0+-1)))))))))))) = -(0\wedge(-(0\wedge(c+$
$(-1+-(0\wedge(1+(-a+-b))))))+-(1\wedge(a+b))))$, by (Scomm),
174 $\iff -(0\wedge(b+(c\wedge(a+(c+-1)))))+(1\wedge(b+(c\wedge(a+(-1+(-b+(1+$
$(-(0\wedge(1+(-b+-c)))+(0\wedge(b+(c+-1))))))))))) = -(0\wedge(-(0\wedge(c+(-1+$
$-(0\wedge(1+(-a+-b))))))+-(1\wedge(a+b))))$, by (SU),
175 $\iff -(0\wedge(b+(c\wedge(a+(c+-1)))))+(1\wedge(b+(c\wedge(a+(-1+(-b+(1+((0\wedge$
$(b+(c+-1)))+-(0\wedge(1+(-b+-c)))))))))))) = -(0\wedge(-(0\wedge(c+(-1+-(0\wedge$
$(1+(-a+-b))))))+-(1\wedge(a+b))))$, by (Scomm),
176 $\iff -(0\wedge(b+(c\wedge(a+(c+-1)))))+(1\wedge(b+(c\wedge(a+(-1+(1+(-b+((0\wedge$
$(b+(c+-1)))+-(0\wedge(1+(-b+-c)))))))))))) = -(0\wedge(-(0\wedge(c+(-1+-(0\wedge$
$(1+(-a+-b))))))+-(1\wedge(a+b))))$, by (11.108),

177 $\iff -(0\wedge(b+(c\wedge(a+(c+-1)))))+(1\wedge(b+(c\wedge(a+(1+(-1+(-b+((0\wedge(b+(c+-1)))+-(0\wedge(1+(-b+-c)))))))))))=-(0\wedge(-(0\wedge(c+(-1+-(0\wedge(1+(-a+-b))))))+-(1\wedge(a+b))))$, by (11.108),
178 $\iff -(0\wedge(b+(c\wedge(a+(c+-1)))))+(1\wedge(b+(c\wedge(a+(-b+((0\wedge(b+(c+-1)))+-(0\wedge(1+(-b+-c)))))))))=-(0\wedge(-(0\wedge(c+(-1+-(0\wedge(1+(-a+-b))))))+-(1\wedge(a+b))))$, by (11.109),
179 $\iff -(0\wedge(b+(c\wedge(a+(c+-1)))))+(1\wedge(b+(c\wedge(a+(-b+((b+((c+-1)\wedge-b))+-(0\wedge(1+(-b+-c)))))))))=-(0\wedge(-(0\wedge(c+(-1+-(0\wedge(1+(-a+-b))))))+-(1\wedge(a+b))))$, by (11.119),
180 $\iff -(0\wedge(b+(c\wedge(a+(c+-1)))))+(1\wedge(b+(c\wedge(a+(-b+((b+(c+(-1\wedge(-c+-b))))+-(0\wedge(1+(-b+-c)))))))))=-(0\wedge(-(0\wedge(c+(-1+-(0\wedge(1+(-a+-b))))))+-(1\wedge(a+b))))$, by (11.120),
181 $\iff -(0\wedge(b+(c\wedge(a+(c+-1)))))+(1\wedge(b+(c\wedge(a+(-b+((b+(c+(-1\wedge(-b+-c))))+-(0\wedge(1+(-b+-c)))))))))=-(0\wedge(-(0\wedge(c+(-1+-(0\wedge(1+(-a+-b))))))+-(1\wedge(a+b))))$, by (Scomm)
182 $\iff -(0\wedge(b+(c\wedge(a+(c+-1)))))+(1\wedge(b+(c\wedge(a+(-b+((b+(c+(-1+(0\wedge((-b+-c)+1)))))+-(0\wedge(1+(-b+-c)))))))))=-(0\wedge(-(0\wedge(c+(-1+-(0\wedge(1+(-a+-b))))))+-(1\wedge(a+b))))$, by (11.125),
183 $\iff -(0\wedge(b+(c\wedge(a+(c+-1)))))+(1\wedge(b+(c\wedge(a+(-b+((b+(c+(-1+(0\wedge(1+(-b+-c))))))+-(0\wedge(1+(-b+-c)))))))))=-(0\wedge(-(0\wedge(c+(-1+-(0\wedge(1+(-a+-b))))))+-(1\wedge(a+b))))$, by (Scomm),
184 $\iff -(0\wedge(b+(c\wedge(a+(c+-1)))))+(1\wedge(b+(c\wedge(a+(-b+(b+((c+(-1+(0\wedge(1+(-b+-c)))))+-(0\wedge(1+(-b+-c))))))))))=-(0\wedge(-(0\wedge(c+(-1+-(0\wedge(1+(-a+-b))))))+-(1\wedge(a+b))))$, by (Sass),
185 $\iff -(0\wedge(b+(c\wedge(a+(c+-1)))))+(1\wedge(b+(c\wedge(a+(-b+(b+(c+((-1+(0\wedge(1+(-b+-c))))+-(0\wedge(1+(-b+-c))))))))))))=-(0\wedge(-(0\wedge(c+(-1+-(0\wedge(1+(-a+-b))))))+-(1\wedge(a+b))))$, by (Sass),
186 $\iff -(0\wedge(b+(c\wedge(a+(c+-1)))))+(1\wedge(b+(c\wedge(a+(-b+(b+(c+(-1+((0\wedge(1+(-b+-c)))+-(0\wedge(1+(-b+-c))))))))))))=-(0\wedge(-(0\wedge(c+(-1+-(0\wedge(1+(-a+-b))))))+-(1\wedge(a+b))))$, by (Sass),
187 $\iff -(0\wedge(b+(c\wedge(a+(c+-1)))))+(1\wedge(b+(c\wedge(a+(-b+(b+(c+(-1+0))))))))=-(0\wedge(-(0\wedge(c+(-1+-(0\wedge(1+(-a+-b))))))+-(1\wedge(a+b))))$, by ($\mathrm{m_0}$-Re),
188 $\iff -(0\wedge(b+(c\wedge(a+(c+-1)))))+(1\wedge(b+(c\wedge(a+(-b+(b+(c+(0+-1))))))))=-(0\wedge(-(0\wedge(c+(-1+-(0\wedge(1+(-a+-b))))))+-(1\wedge(a+b))))$, by (Scomm),
189 $\iff -(0\wedge(b+(c\wedge(a+(c+-1)))))+(1\wedge(b+(c\wedge(a+(-b+(b+(c+-1)))))))=-(0\wedge(-(0\wedge(c+(-1+-(0\wedge(1+(-a+-b))))))+-(1\wedge(a+b))))$, by (SU),
190 $\iff -(0\wedge(b+(c\wedge(a+(c+-1)))))+(1\wedge(b+(c\wedge(a+((b+(c+-1))+-b)))))=-(0\wedge(-(0\wedge(c+(-1+-(0\wedge(1+(-a+-b))))))+-(1\wedge(a+b))))$, by (Scomm),
191 $\iff -(0\wedge(b+(c\wedge(a+(c+-1)))))+(1\wedge(b+(c\wedge(a+(b+((c+-1)+-b))))))=-(0\wedge(-(0\wedge(c+(-1+-(0\wedge(1+(-a+-b))))))+-(1\wedge(a+b))))$, by (Sass),
192 $\iff -(0\wedge(b+(c\wedge(a+(c+-1)))))+(1\wedge(b+(c\wedge(a+(b+(c+(-1+-b)))))))=-(0\wedge(-(0\wedge(c+(-1+-(0\wedge(1+(-a+-b))))))+-(1\wedge(a+b))))$, by (Sass),
193 $\iff -(0\wedge(b+(c\wedge(a+(c+-1)))))+(1\wedge(b+(c\wedge(a+(c+-1)))))=-(0\wedge(-(0\wedge(c+(-1+-(0\wedge(1+(-a+-b))))))+-(1\wedge(a+b))))$, by (11.113),
194 $\iff -(0\wedge(b+(c\wedge(a+(c+-1)))))+(1\wedge(b+(c\wedge(a+(c+-1)))))=-(0\wedge(-0+-((1\wedge(a+b))\wedge((1\wedge(a+b))+((c+(-1+-(0\wedge(1+(-a+-b)))))+-0))))),$

by (11.132),
195 $\iff -(0 \wedge (b + (c \wedge (a + (c + -1))))) + (1 \wedge (b + (c \wedge (a + (c + -1))))) = -(0\wedge(0+-((1\wedge(a+b))\wedge((1\wedge(a+b))+((c+(-1+-(0\wedge(1+(-a+-b)))))+-0)))))$,
by (N0),
196 $\iff -(0 \wedge (b + (c \wedge (a + (c + -1))))) + (1 \wedge (b + (c \wedge (a + (c + -1))))) = -(0\wedge(0+-((1\wedge(a+b))\wedge((1\wedge(a+b))+((c+(-1+-(0\wedge(1+(-a+-b)))))+0)))))$,
by (N0),
197 $\iff -(0 \wedge (b + (c \wedge (a + (c + -1))))) + (1 \wedge (b + (c \wedge (a + (c + -1))))) = -(0\wedge(0+-((1\wedge(a+b))\wedge((1\wedge(a+b))+(0+(c+(-1+-(0\wedge(1+(-a+-b))))))))))$,
by (Scomm),
198 $\iff -(0 \wedge (b + (c \wedge (a + (c + -1))))) + (1 \wedge (b + (c \wedge (a + (c + -1))))) = -(0\wedge(0+-((1\wedge(a+b))\wedge((1\wedge(a+b))+(c+(-1+-(0\wedge(1+(-a+-b)))))))))$,
by (SU),
199 $\iff -(0 \wedge (b + (c \wedge (a + (c + -1))))) + (1 \wedge (b + (c \wedge (a + (c + -1))))) = -(0\wedge(0+-((1\wedge(a+b))\wedge((c+(-1+-(0\wedge(1+(-a+-b)))))+(1\wedge(a+b))))))$,
by (Scomm),
200 $\iff -(0 \wedge (b + (c \wedge (a + (c + -1))))) + (1 \wedge (b + (c \wedge (a + (c + -1))))) = -(0\wedge(0+-((1\wedge(a+b))\wedge(c+((-1+-(0\wedge(1+(-a+-b))))+(1\wedge(a+b)))))))$,
by (Sass),
201 $\iff -(0 \wedge (b + (c \wedge (a + (c + -1))))) + (1 \wedge (b + (c \wedge (a + (c + -1))))) = -(0\wedge(0+-((1\wedge(a+b))\wedge(c+(-1+(-(0\wedge(1+(-a+-b)))+(1\wedge(a+b))))))))$,
by (Sass),
202 $\iff -(0 \wedge (b + (c \wedge (a + (c + -1))))) + (1 \wedge (b + (c \wedge (a + (c + -1))))) = -(0\wedge(0+-(1\wedge((a+b)\wedge(c+(-1+(-(0\wedge(1+(-a+-b)))+(1\wedge(a+b))))))))))$,
by (Wass),
203 $\iff -(0 \wedge (b + (c \wedge (a + (c + -1))))) + (1 \wedge (b + (c \wedge (a + (c + -1))))) = -(0\wedge(0+-(1\wedge(a+(b\wedge(-a+(c+(-1+(-(0\wedge(1+(-a+-b)))+(1\wedge(a+b)))))))))))$,
by (11.120),
204 $\iff -(0 \wedge (b + (c \wedge (a + (c + -1))))) + (1 \wedge (b + (c \wedge (a + (c + -1))))) = -(0\wedge(0+-(1\wedge(a+(b\wedge(c+(-a+(-1+(-(0\wedge(1+(-a+-b)))+(1\wedge(a+b)))))))))))$,
by (11.108),
205 $\iff -(0 \wedge (b + (c \wedge (a + (c + -1))))) + (1 \wedge (b + (c \wedge (a + (c + -1))))) = -(0\wedge(0+-(1\wedge(a+(b\wedge(c+(-1+(-a+(-(0\wedge(1+(-a+-b)))+(1\wedge(a+b)))))))))))$,
by (11.108),
206 $\iff -(0 \wedge (b + (c \wedge (a + (c + -1))))) + (1 \wedge (b + (c \wedge (a + (c + -1))))) = -(0\wedge-(1\wedge(a+(b\wedge(c+(-1+(-a+(-(0\wedge(1+(-a+-b)))+(1\wedge(a+b))))))))))$,
by (SU),
207 $\iff -(0 \wedge (b + (c \wedge (a + (c + -1))))) + (1 \wedge (b + (c \wedge (a + (c + -1))))) = -(-(1\wedge(a+(b\wedge(c+(-1+(-a+(-(0\wedge(1+(-a+-b)))+(1\wedge(a+b)))))))))+ (0\wedge(1\wedge(a+(b\wedge(c+(-1+(-a+(-(0\wedge(1+(-a+-b)))+(1\wedge(a+b)))))))))))$,
by (11.127),
208 $\iff -(0 \wedge (b + (c \wedge (a + (c + -1))))) + (1 \wedge (b + (c \wedge (a + (c + -1))))) = -(-(1\wedge(a+(b\wedge(c+(-1+(-a+(-(0\wedge(1+(-a+-b)))+(1\wedge(a+b)))))))))+ (0\wedge(a+(b\wedge(c+(-1+(-a+(-(0\wedge(1+(-a+-b)))+(1\wedge(a+b))))))))))$, by (11.134),
209 $\iff -(0 \wedge (b + (c \wedge (a + (c + -1))))) + (1 \wedge (b + (c \wedge (a + (c + -1))))) =$

$-((0 \wedge (a + (b \wedge (c + (-1 + (-a + (-(0 \wedge (1 + (-a + -b))) + (1 \wedge (a + b)))))))) + -(1 \wedge (a + (b \wedge (c + (-1 + (-a + (-(0 \wedge (1 + (-a + -b))) + (1 \wedge (a + b)))))))))$, by (Scomm),

210 $\iff -(0 \wedge (b + (c \wedge (a + (c + -1))))) + (1 \wedge (b + (c \wedge (a + (c + -1))))) = - - (1 \wedge (a + (b \wedge (c + (-1 + (-a + (-(0 \wedge (1 + (-a + -b))) + (1 \wedge (a + b)))))))) + -(0 \wedge (a + (b \wedge (c + (-1 + (-a + (-(0 \wedge (1 + (-a + -b))) + (1 \wedge (a + b))))))))$, by (NegS),

211 $\iff -(0 \wedge (b + (c \wedge (a + (c + -1))))) + (1 \wedge (b + (c \wedge (a + (c + -1))))) = (1 \wedge (a + (b \wedge (c + (-1 + (-a + (-(0 \wedge (1 + (-a + -b))) + (1 \wedge (a + b)))))))) + -(0 \wedge (a + (b \wedge (c + (-1 + (-a + (-(0 \wedge (1 + (-a + -b))) + (1 \wedge (a + b))))))))$, by (DN),

212 $\iff -(0 \wedge (b + (c \wedge (a + (c + -1))))) + (1 \wedge (b + (c \wedge (a + (c + -1))))) = -(0 \wedge (a + (b \wedge (c + (-1 + (-a + (-(0 \wedge (1 + (-a + -b))) + (1 \wedge (a + b)))))))) + (1 \wedge (a + (b \wedge (c + (-1 + (-a + (-(0 \wedge (1 + (-a + -b))) + (1 \wedge (a + b))))))))$, by (Scomm),

213 $\iff -(0 \wedge (b + (c \wedge (a + (c + -1))))) + (1 \wedge (b + (c \wedge (a + (c + -1))))) = -(0 \wedge (a + (b \wedge (c + (-1 + (-a + (1 + (-(0 \wedge (1 + (-a + -b))) \wedge (-1 + (-(0 \wedge (1 + (-a + -b))) + (a + b))))))))))) + (1 \wedge (a + (b \wedge (c + (-1 + (-a + (-(0 \wedge (1 + (-a + -b))) + (1 \wedge (a + b))))))))$, by (11.126),

214 $\iff -(0 \wedge (b + (c \wedge (a + (c + -1))))) + (1 \wedge (b + (c \wedge (a + (c + -1))))) = -(0 \wedge (a + (b \wedge (c + (-1 + (-a + (1 + (-(0 \wedge (1 + (-a + -b))) \wedge (-1 + ((a + b) + -(0 \wedge (1 + (-a + -b)))))))))))) + (1 \wedge (a + (b \wedge (c + (-1 + (-a + (-(0 \wedge (1 + (-a + -b))) + (1 \wedge (a + b))))))))$, by (Scomm),

215 $\iff -(0 \wedge (b + (c \wedge (a + (c + -1))))) + (1 \wedge (b + (c \wedge (a + (c + -1))))) = -(0 \wedge (a + (b \wedge (c + (-1 + (-a + (1 + (-(0 \wedge (1 + (-a + -b))) \wedge (-1 + (a + (b + -(0 \wedge (1 + (-a + -b))))))))))))) + (1 \wedge (a + (b \wedge (c + (-1 + (-a + (-(0 \wedge (1 + (-a + -b))) + (1 \wedge (a + b))))))))$, by (Sass),

216 $\iff -(0 \wedge (b + (c \wedge (a + (c + -1))))) + (1 \wedge (b + (c \wedge (a + (c + -1))))) = -(0 \wedge (a + (b \wedge (c + (-1 + (-a + (1 + (-(0 \wedge (1 + (-a + -b))) \wedge (a + (-1 + (b + -(0 \wedge (1 + (-a + -b))))))))))))) + (1 \wedge (a + (b \wedge (c + (-1 + (-a + (-(0 \wedge (1 + (-a + -b))) + (1 \wedge (a + b))))))))$, by (11.108),

217 $\iff -(0 \wedge (b + (c \wedge (a + (c + -1))))) + (1 \wedge (b + (c \wedge (a + (c + -1))))) = -(0 \wedge (a + (b \wedge (c + (-1 + (-a + (1 + (-(0 \wedge (1 + (-a + -b))) \wedge (a + (b + (-1 + -(0 \wedge (1 + (-a + -b))))))))))))) + (1 \wedge (a + (b \wedge (c + (-1 + (-a + (-(0 \wedge (1 + (-a + -b))) + (1 \wedge (a + b))))))))$, by (11.108),

218 $\iff -(0 \wedge (b + (c \wedge (a + (c + -1))))) + (1 \wedge (b + (c \wedge (a + (c + -1))))) = -(0 \wedge (a + (b \wedge (c + (-1 + (-a + (1 + (-(0 \wedge (1 + (-a + -b))) + (0 \wedge ((a + (b + (-1 + -(0 \wedge (1 + (-a + -b)))))) + (0 \wedge (1 + (-a + -b)))))))))))) + (1 \wedge (a + (b \wedge (c + (-1 + (-a + (-(0 \wedge (1 + (-a + -b))) + (1 \wedge (a + b))))))))$, by (11.125),

219 $\iff -(0 \wedge (b + (c \wedge (a + (c + -1))))) + (1 \wedge (b + (c \wedge (a + (c + -1))))) = -(0 \wedge (a + (b \wedge (c + (-1 + (-a + (1 + (-(0 \wedge (1 + (-a + -b))) + (0 \wedge ((0 \wedge (1 + (-a + -b))) + (a + (b + (-1 + -(0 \wedge (1 + (-a + -b))))))))))))))) + (1 \wedge (a + (b \wedge (c + (-1 + (-a + (-(0 \wedge (1 + (-a + -b))) + (1 \wedge (a + b))))))))$, by (Scomm),

220 $\iff -(0 \wedge (b + (c \wedge (a + (c + -1))))) + (1 \wedge (b + (c \wedge (a + (c + -1))))) = -(0 \wedge (a + (b \wedge (c + (-1 + (-a + (1 + (-(0 \wedge (1 + (-a + -b))) + (0 \wedge (a + ((0 \wedge (1 + (-a + -b))) + (b + (-1 + -(0 \wedge (1 + (-a + -b))))))))))))))) + (1 \wedge (a + (b \wedge (c + (-1 + (-a + (-(0 \wedge (1 + (-a + -b))) + (1 \wedge (a + b))))))))$, by (11.108),

221 $\iff -(0 \wedge (b + (c \wedge (a + (c + -1))))) + (1 \wedge (b + (c \wedge (a + (c + -1))))) = -(0 \wedge (a + (b \wedge (c + (-1 + (-a + (1 + (-(0 \wedge (1 + (-a + -b))) + (0 \wedge (a + (b + ((0 \wedge (1 + (-a + -b))) + (-1 + -(0 \wedge (1 + (-a + -b)))))))))))))))) + (1 \wedge (a + (b \wedge (c + (-1 + (-a + (-(0 \wedge (1 + (-a + -b))) + (1 \wedge (a + b)))))))))$, by (11.108),
222 $\iff -(0 \wedge (b + (c \wedge (a + (c + -1))))) + (1 \wedge (b + (c \wedge (a + (c + -1))))) = -(0 \wedge (a + (b \wedge (c + (-1 + (-a + (1 + (-(0 \wedge (1 + (-a + -b))) + (0 \wedge (a + (b + (-1 + ((0 \wedge (1 + (-a + -b))) + -(0 \wedge (1 + (-a + -b)))))))))))))))) + (1 \wedge (a + (b \wedge (c + (-1 + (-a + (-(0 \wedge (1 + (-a + -b))) + (1 \wedge (a + b)))))))))$, by (11.108),
223 $\iff -(0 \wedge (b + (c \wedge (a + (c + -1))))) + (1 \wedge (b + (c \wedge (a + (c + -1))))) = -(0 \wedge (a + (b \wedge (c + (-1 + (-a + (1 + (-(0 \wedge (1 + (-a + -b))) + (0 \wedge (a + (b + (-1 + 0)))))))))))) + (1 \wedge (a + (b \wedge (c + (-1 + (-a + (-(0 \wedge (1 + (-a + -b))) + (1 \wedge (a + b)))))))))$, by ($m_0$-Re),
224 $\iff -(0 \wedge (b + (c \wedge (a + (c + -1))))) + (1 \wedge (b + (c \wedge (a + (c + -1))))) = -(0 \wedge (a + (b \wedge (c + (-1 + (-a + (1 + (-(0 \wedge (1 + (-a + -b))) + (0 \wedge (a + (b + (0 + -1)))))))))))) + (1 \wedge (a + (b \wedge (c + (-1 + (-a + (-(0 \wedge (1 + (-a + -b))) + (1 \wedge (a + b)))))))))$, by (Scomm),
225 $\iff -(0 \wedge (b + (c \wedge (a + (c + -1))))) + (1 \wedge (b + (c \wedge (a + (c + -1))))) = -(0 \wedge (a + (b \wedge (c + (-1 + (-a + (1 + (-(0 \wedge (1 + (-a + -b))) + (0 \wedge (a + (b + -1))))))))))) + (1 \wedge (a + (b \wedge (c + (-1 + (-a + (-(0 \wedge (1 + (-a + -b))) + (1 \wedge (a + b)))))))))$, by (SU),
226 $\iff -(0 \wedge (b + (c \wedge (a + (c + -1))))) + (1 \wedge (b + (c \wedge (a + (c + -1))))) = -(0 \wedge (a + (b \wedge (c + (-1 + (-a + (1 + ((0 \wedge (a + (b + -1))) + -(0 \wedge (1 + (-a + -b))))))))))) + (1 \wedge (a + (b \wedge (c + (-1 + (-a + (-(0 \wedge (1 + (-a + -b))) + (1 \wedge (a + b)))))))))$, by (Scomm),
227 $\iff -(0 \wedge (b + (c \wedge (a + (c + -1))))) + (1 \wedge (b + (c \wedge (a + (c + -1))))) = -(0 \wedge (a + (b \wedge (c + (-1 + (1 + (-a + ((0 \wedge (a + (b + -1))) + -(0 \wedge (1 + (-a + -b))))))))))) + (1 \wedge (a + (b \wedge (c + (-1 + (-a + (-(0 \wedge (1 + (-a + -b))) + (1 \wedge (a + b)))))))))$, by (11.108),
228 $\iff -(0 \wedge (b + (c \wedge (a + (c + -1))))) + (1 \wedge (b + (c \wedge (a + (c + -1))))) = -(0 \wedge (a + (b \wedge (c + (1 + (-1 + (-a + ((0 \wedge (a + (b + -1))) + -(0 \wedge (1 + (-a + -b))))))))))) + (1 \wedge (a + (b \wedge (c + (-1 + (-a + (-(0 \wedge (1 + (-a + -b))) + (1 \wedge (a + b)))))))))$, by (11.108),
229 $\iff -(0 \wedge (b + (c \wedge (a + (c + -1))))) + (1 \wedge (b + (c \wedge (a + (c + -1))))) = -(0 \wedge (a + (b \wedge (c + (-a + ((0 \wedge (a + (b + -1))) + -(0 \wedge (1 + (-a + -b))))))))) + (1 \wedge (a + (b \wedge (c + (-1 + (-a + (-(0 \wedge (1 + (-a + -b))) + (1 \wedge (a + b)))))))))$, by (11.109),
230 $\iff -(0 \wedge (b + (c \wedge (a + (c + -1))))) + (1 \wedge (b + (c \wedge (a + (c + -1))))) = -(0 \wedge (a + (b \wedge (c + (-a + ((a + ((b + -1) \wedge -a)) + -(0 \wedge (1 + (-a + -b))))))))) + (1 \wedge (a + (b \wedge (c + (-1 + (-a + (-(0 \wedge (1 + (-a + -b))) + (1 \wedge (a + b)))))))))$, by (11.119),
231 $\iff -(0 \wedge (b + (c \wedge (a + (c + -1))))) + (1 \wedge (b + (c \wedge (a + (c + -1))))) = -(0 \wedge (a + (b \wedge (c + (-a + ((a + (b + (-1 \wedge (-b + -a)))) + -(0 \wedge (1 + (-a + -b))))))))) + (1 \wedge (a + (b \wedge (c + (-1 + (-a + (-(0 \wedge (1 + (-a + -b))) + (1 \wedge (a + b)))))))))$, by (11.120),
232 $\iff -(0 \wedge (b + (c \wedge (a + (c + -1))))) + (1 \wedge (b + (c \wedge (a + (c + -1))))) = -(0 \wedge (a + (b \wedge (c + (-a + ((a + (b + (-1 \wedge (-a + -b)))) + -(0 \wedge (1 + (-a + -b))))))))) +$

$(1 \wedge (a + (b \wedge (c + (-1 + (-a + (-(0 \wedge (1 + (-a + -b))) + (1 \wedge (a + b))))))))))$, by (Scomm),

233 $\Longleftrightarrow -(0 \wedge (b + (c \wedge (a + (c + -1))))) + (1 \wedge (b + (c \wedge (a + (c + -1))))) = -(0 \wedge (a + (b \wedge (c + (-a + ((a + (b + (-1 + (0 \wedge ((-a + -b) + 1))))) + -(0 \wedge (1 + (-a + -b))))))))) + (1 \wedge (a + (b \wedge (c + (-1 + (-a + (-(0 \wedge (1 + (-a + -b))) + (1 \wedge (a + b))))))))))$, by (11.125),

234 $\Longleftrightarrow -(0 \wedge (b + (c \wedge (a + (c + -1))))) + (1 \wedge (b + (c \wedge (a + (c + -1))))) = -(0 \wedge (a + (b \wedge (c + (-a + ((a + (b + (-1 + (0 \wedge (1 + (-a + -b)))))) + -(0 \wedge (1 + (-a + -b))))))))) + (1 \wedge (a + (b \wedge (c + (-1 + (-a + (-(0 \wedge (1 + (-a + -b))) + (1 \wedge (a + b))))))))))$, by (Scomm),

235 $\Longleftrightarrow -(0 \wedge (b + (c \wedge (a + (c + -1))))) + (1 \wedge (b + (c \wedge (a + (c + -1))))) = -(0 \wedge (a + (b \wedge (c + (-a + (a + ((b + (-1 + (0 \wedge (1 + (-a + -b))))) + -(0 \wedge (1 + (-a + -b)))))))))) + (1 \wedge (a + (b \wedge (c + (-1 + (-a + (-(0 \wedge (1 + (-a + -b))) + (1 \wedge (a + b))))))))))$, by (Sass),

236 $\Longleftrightarrow -(0 \wedge (b + (c \wedge (a + (c + -1))))) + (1 \wedge (b + (c \wedge (a + (c + -1))))) = -(0 \wedge (a + (b \wedge (c + (-a + (a + (b + ((-1 + (0 \wedge (1 + (-a + -b)))) + -(0 \wedge (1 + (-a + -b))))))))))) + (1 \wedge (a + (b \wedge (c + (-1 + (-a + (-(0 \wedge (1 + (-a + -b))) + (1 \wedge (a + b))))))))))$, by (Sass),

237 $\Longleftrightarrow -(0 \wedge (b + (c \wedge (a + (c + -1))))) + (1 \wedge (b + (c \wedge (a + (c + -1))))) = -(0 \wedge (a + (b \wedge (c + (-a + (a + (b + (-1 + ((0 \wedge (1 + (-a + -b))) + -(0 \wedge (1 + (-a + -b)))))))))))) + (1 \wedge (a + (b \wedge (c + (-1 + (-a + (-(0 \wedge (1 + (-a + -b))) + (1 \wedge (a + b))))))))))$, by (Sass),

238 $\Longleftrightarrow -(0 \wedge (b + (c \wedge (a + (c + -1))))) + (1 \wedge (b + (c \wedge (a + (c + -1))))) = -(0 \wedge (a + (b \wedge (c + (-a + (a + (b + (-1 + 0)))))))) + (1 \wedge (a + (b \wedge (c + (-1 + (-a + (-(0 \wedge (1 + (-a + -b))) + (1 \wedge (a + b))))))))))$, by ($m_0$-Re),

239 $\Longleftrightarrow -(0 \wedge (b + (c \wedge (a + (c + -1))))) + (1 \wedge (b + (c \wedge (a + (c + -1))))) = -(0 \wedge (a + (b \wedge (c + (-a + (a + (b + (0 + -1)))))))) + (1 \wedge (a + (b \wedge (c + (-1 + (-a + (-(0 \wedge (1 + (-a + -b))) + (1 \wedge (a + b))))))))))$, by (Scomm),

240 $\Longleftrightarrow -(0 \wedge (b + (c \wedge (a + (c + -1))))) + (1 \wedge (b + (c \wedge (a + (c + -1))))) = -(0 \wedge (a + (b \wedge (c + (-a + (a + (b + -1))))))) + (1 \wedge (a + (b \wedge (c + (-1 + (-a + (-(0 \wedge (1 + (-a + -b))) + (1 \wedge (a + b))))))))))$, by (SU),

241 $\Longleftrightarrow -(0 \wedge (b + (c \wedge (a + (c + -1))))) + (1 \wedge (b + (c \wedge (a + (c + -1))))) = -(0 \wedge (a + (b \wedge (c + ((a + (b + -1)) + -a))))) + (1 \wedge (a + (b \wedge (c + (-1 + (-a + (-(0 \wedge (1 + (-a + -b))) + (1 \wedge (a + b))))))))))$, by (Scomm),

242 $\Longleftrightarrow -(0 \wedge (b + (c \wedge (a + (c + -1))))) + (1 \wedge (b + (c \wedge (a + (c + -1))))) = -(0 \wedge (a + (b \wedge (c + (a + ((b + -1) + -a)))))) + (1 \wedge (a + (b \wedge (c + (-1 + (-a + (-(0 \wedge (1 + (-a + -b))) + (1 \wedge (a + b))))))))))$, by (Sass),

243 $\Longleftrightarrow -(0 \wedge (b + (c \wedge (a + (c + -1))))) + (1 \wedge (b + (c \wedge (a + (c + -1))))) = -(0 \wedge (a + (b \wedge (c + (a + (b + (-1 + -a))))))) + (1 \wedge (a + (b \wedge (c + (-1 + (-a + (-(0 \wedge (1 + (-a + -b))) + (1 \wedge (a + b))))))))))$, by (Sass),

244 $\Longleftrightarrow -(0 \wedge (b + (c \wedge (a + (c + -1))))) + (1 \wedge (b + (c \wedge (a + (c + -1))))) = -(0 \wedge (a + (b \wedge (c + (b + -1))))) + (1 \wedge (a + (b \wedge (c + (-1 + (-a + (-(0 \wedge (1 + (-a + -b))) + (1 \wedge (a + b))))))))))$, by (11.113),

245 $\Longleftrightarrow -(0 \wedge (b + (c \wedge (a + (c + -1))))) + (1 \wedge (b + (c \wedge (a + (c + -1))))) = -(0 \wedge (a + (b \wedge (b + (c + -1))))) + (1 \wedge (a + (b \wedge (c + (-1 + (-a + (-(0 \wedge (1 + (-a + -b))) + (1 \wedge (a + b))))))))))$, by (11.108),

246 $\iff -(0\wedge(b+(c\wedge(a+(c+-1)))))+(1\wedge(b+(c\wedge(a+(c+-1))))) = -(0\wedge(a+(b+(0\wedge(c+-1)))))+(1\wedge(a+(b\wedge(c+(-1+(-a+(-(0\wedge(1+(-a+-b)))+(1\wedge(a+b)))))))))$, by (11.116),

247 $\iff -(0\wedge(b+(c\wedge(a+(c+-1)))))+(1\wedge(b+(c\wedge(a+(c+-1))))) = -(0\wedge(a+(b+(0\wedge(c+-1)))))+(1\wedge(a+(b\wedge(c+(-1+(-a+(1+(-(0\wedge(1+(-a+-b)))\wedge(-1+(-(0\wedge(1+(-a+-b)))+(a+b)))))))))))$, by (11.126),

248 $\iff -(0\wedge(b+(c\wedge(a+(c+-1)))))+(1\wedge(b+(c\wedge(a+(c+-1))))) = -(0\wedge(a+(b+(0\wedge(c+-1)))))+(1\wedge(a+(b\wedge(c+(-1+(-a+(1+(-(0\wedge(1+(-a+-b)))\wedge(-1+((a+b)+-(0\wedge(1+(-a+-b)))))))))))))$, by (Scomm),

249 $\iff -(0\wedge(b+(c\wedge(a+(c+-1)))))+(1\wedge(b+(c\wedge(a+(c+-1))))) = -(0\wedge(a+(b+(0\wedge(c+-1)))))+(1\wedge(a+(b\wedge(c+(-1+(-a+(1+(-(0\wedge(1+(-a+-b)))\wedge(-1+(a+(b+-(0\wedge(1+(-a+-b))))))))))))))$, by (Sass),

250 $\iff -(0\wedge(b+(c\wedge(a+(c+-1)))))+(1\wedge(b+(c\wedge(a+(c+-1))))) = -(0\wedge(a+(b+(0\wedge(c+-1)))))+(1\wedge(a+(b\wedge(c+(-1+(-a+(1+(-(0\wedge(1+(-a+-b)))\wedge(a+(-1+(b+-(0\wedge(1+(-a+-b))))))))))))))$, by (11.108),

251 $\iff -(0\wedge(b+(c\wedge(a+(c+-1)))))+(1\wedge(b+(c\wedge(a+(c+-1))))) = -(0\wedge(a+(b+(0\wedge(c+-1)))))+(1\wedge(a+(b\wedge(c+(-1+(-a+(1+(-(0\wedge(1+(-a+-b)))\wedge(a+(b+(-1+-(0\wedge(1+(-a+-b))))))))))))))$, by (11.108),

252 $\iff -(0\wedge(b+(c\wedge(a+(c+-1)))))+(1\wedge(b+(c\wedge(a+(c+-1))))) = -(0\wedge(a+(b+(0\wedge(c+-1)))))+(1\wedge(a+(b\wedge(c+(-1+(-a+(1+(-(0\wedge(1+(-a+-b)))+(0\wedge((a+(b+(-1+-(0\wedge(1+(-a+-b)))))))+(0\wedge(1+(-a+-b))))))))))))$, by (11.125),

253 $\iff -(0\wedge(b+(c\wedge(a+(c+-1)))))+(1\wedge(b+(c\wedge(a+(c+-1))))) = -(0\wedge(a+(b+(0\wedge(c+-1)))))+(1\wedge(a+(b\wedge(c+(-1+(-a+(1+(-(0\wedge(1+(-a+-b)))+(0\wedge((0\wedge(1+(-a+-b)))+(a+(b+(-1+-(0\wedge(1+(-a+-b))))))))))))))))$, by (Scomm),

254 $\iff -(0\wedge(b+(c\wedge(a+(c+-1)))))+(1\wedge(b+(c\wedge(a+(c+-1))))) = -(0\wedge(a+(b+(0\wedge(c+-1)))))+(1\wedge(a+(b\wedge(c+(-1+(-a+(1+(-(0\wedge(1+(-a+-b)))+(0\wedge(a+((0\wedge(1+(-a+-b)))+(b+(-1+-(0\wedge(1+(-a+-b))))))))))))))))$, by (11.108),

255 $\iff -(0\wedge(b+(c\wedge(a+(c+-1)))))+(1\wedge(b+(c\wedge(a+(c+-1))))) = -(0\wedge(a+(b+(0\wedge(c+-1)))))+(1\wedge(a+(b\wedge(c+(-1+(-a+(1+(-(0\wedge(1+(-a+-b)))+(0\wedge(a+(b+((0\wedge(1+(-a+-b)))+(-1+-(0\wedge(1+(-a+-b))))))))))))))))$, by (11.108),

256 $\iff -(0\wedge(b+(c\wedge(a+(c+-1)))))+(1\wedge(b+(c\wedge(a+(c+-1))))) = -(0\wedge(a+(b+(0\wedge(c+-1)))))+(1\wedge(a+(b\wedge(c+(-1+(-a+(1+(-(0\wedge(1+(-a+-b)))+(0\wedge(a+(b+(-1+((0\wedge(1+(-a+-b)))+-(0\wedge(1+(-a+-b))))))))))))))))$, by (11.108),

257 $\iff -(0\wedge(b+(c\wedge(a+(c+-1)))))+(1\wedge(b+(c\wedge(a+(c+-1))))) = -(0\wedge(a+(b+(0\wedge(c+-1)))))+(1\wedge(a+(b\wedge(c+(-1+(-a+(1+(-(0\wedge(1+(-a+-b)))+(0\wedge(a+(b+(-1+0))))))))))))$, by ($m_0$-Re),

258 $\iff -(0\wedge(b+(c\wedge(a+(c+-1)))))+(1\wedge(b+(c\wedge(a+(c+-1))))) = -(0\wedge(a+(b+(0\wedge(c+-1)))))+(1\wedge(a+(b\wedge(c+(-1+(-a+(1+(-(0\wedge(1+(-a+-b)))+(0\wedge(a+(b+(0+-1))))))))))))$, by (Scomm),

259 $\iff -(0\wedge(b+(c\wedge(a+(c+-1)))))+(1\wedge(b+(c\wedge(a+(c+-1))))) = -(0\wedge(a+(b+(0\wedge(c+-1)))))+(1\wedge(a+(b\wedge(c+(-1+(-a+(1+(-(0\wedge(1+$

$(-a+-b))) + (0 \wedge (a+(b+-1))))))))))$, by (SU),

260 $\iff -(0 \wedge (b+(c \wedge (a+(c+-1))))) + (1 \wedge (b+(c \wedge (a+(c+-1))))) = -(0 \wedge (a+(b+(0 \wedge (c+-1))))) + (1 \wedge (a+(b \wedge (c+(-1+(-a+(1+((0 \wedge (a+(b+-1))) + -(0 \wedge (1+(-a+-b)))))))))))$, by (Scomm),

261 $\iff -(0 \wedge (b+(c \wedge (a+(c+-1))))) + (1 \wedge (b+(c \wedge (a+(c+-1))))) = -(0 \wedge (a+(b+(0 \wedge (c+-1))))) + (1 \wedge (a+(b \wedge (c+(-1+(1+(-a+((0 \wedge (a+(b+-1))) + -(0 \wedge (1+(-a+-b)))))))))))$, by (11.108),

262 $\iff -(0 \wedge (b+(c \wedge (a+(c+-1))))) + (1 \wedge (b+(c \wedge (a+(c+-1))))) = -(0 \wedge (a+(b+(0 \wedge (c+-1))))) + (1 \wedge (a+(b \wedge (c+(1+(-1+(-a+((0 \wedge (a+(b+-1))) + -(0 \wedge (1+(-a+-b)))))))))))$, by (11.108),

263 $\iff -(0\wedge(b+(c\wedge(a+(c+-1)))))+(1\wedge(b+(c\wedge(a+(c+-1))))) = -(0\wedge(a+(b+(0\wedge(c+-1)))))+(1\wedge(a+(b\wedge(c+(-a+((0\wedge(a+(b+-1)))+-(0\wedge(1+(-a+-b)))))))))$, by (11.109),

264 $\iff -(0 \wedge (b+(c \wedge (a+(c+-1))))) + (1 \wedge (b+(c \wedge (a+(c+-1))))) = -(0 \wedge (a+(b+(0 \wedge (c+-1))))) + (1 \wedge (a+(b \wedge (c+(-a+((a+((b+-1) \wedge -a)) + -(0 \wedge (1+(-a+-b)))))))))$, by (11.119),

265 $\iff -(0 \wedge (b+(c \wedge (a+(c+-1))))) + (1 \wedge (b+(c \wedge (a+(c+-1))))) = -(0 \wedge (a+(b+(0 \wedge (c+-1))))) + (1 \wedge (a+(b \wedge (c+(-a+((a+(b+(-1 \wedge (-b+-a)))) + -(0 \wedge (1+(-a+-b)))))))))$, by (11.120),

266 $\iff -(0 \wedge (b+(c \wedge (a+(c+-1))))) + (1 \wedge (b+(c \wedge (a+(c+-1))))) = -(0 \wedge (a+(b+(0 \wedge (c+-1))))) + (1 \wedge (a+(b \wedge (c+(-a+((a+(b+(-1 \wedge (-a+-b)))) + -(0 \wedge (1+(-a+-b)))))))))$, by (Scomm),

267 $\iff -(0 \wedge (b+(c \wedge (a+(c+-1))))) + (1 \wedge (b+(c \wedge (a+(c+-1))))) = -(0 \wedge (a+(b+(0 \wedge (c+-1))))) + (1 \wedge (a+(b \wedge (c+(-a+((a+(b+(-1+(0 \wedge ((-a+-b)+1))))) + -(0 \wedge (1+(-a+-b)))))))))$, by (11.125),

268 $\iff -(0 \wedge (b+(c \wedge (a+(c+-1))))) + (1 \wedge (b+(c \wedge (a+(c+-1))))) = -(0 \wedge (a+(b+(0 \wedge (c+-1))))) + (1 \wedge (a+(b \wedge (c+(-a+((a+(b+(-1+(0 \wedge (1+(-a+-b)))))) + -(0 \wedge (1+(-a+-b)))))))))$, by (Scomm),

269 $\iff -(0 \wedge (b+(c \wedge (a+(c+-1))))) + (1 \wedge (b+(c \wedge (a+(c+-1))))) = -(0 \wedge (a+(b+(0 \wedge (c+-1))))) + (1 \wedge (a+(b \wedge (c+(-a+(a+((b+(-1+(0 \wedge (1+(-a+-b))))) + -(0 \wedge (1+(-a+-b))))))))))$, by (Sass),

270 $\iff -(0 \wedge (b+(c \wedge (a+(c+-1))))) + (1 \wedge (b+(c \wedge (a+(c+-1))))) = -(0 \wedge (a+(b+(0 \wedge (c+-1))))) + (1 \wedge (a+(b \wedge (c+(-a+(a+(b+((-1+(0 \wedge (1+(-a+-b)))) + -(0 \wedge (1+(-a+-b))))))))))$, by (Sass),

271 $\iff -(0 \wedge (b+(c \wedge (a+(c+-1))))) + (1 \wedge (b+(c \wedge (a+(c+-1))))) = -(0 \wedge (a+(b+(0 \wedge (c+-1))))) + (1 \wedge (a+(b \wedge (c+(-a+(a+(b+(-1+((0 \wedge (1+(-a+-b))) + -(0 \wedge (1+(-a+-b)))))))))))$, by (Sass),

272 $\iff -(0 \wedge (b+(c \wedge (a+(c+-1))))) + (1 \wedge (b+(c \wedge (a+(c+-1))))) = -(0 \wedge (a+(b+(0 \wedge (c+-1))))) + (1 \wedge (a+(b \wedge (c+(-a+(a+(b+(-1+0))))))))$, by ($m_0$-Re),

273 $\iff -(0 \wedge (b+(c \wedge (a+(c+-1))))) + (1 \wedge (b+(c \wedge (a+(c+-1))))) = -(0 \wedge (a+(b+(0 \wedge (c+-1))))) + (1 \wedge (a+(b \wedge (c+(-a+(a+(b+(0+-1))))))))$, by (Scomm),

274 $\iff -(0 \wedge (b+(c \wedge (a+(c+-1))))) + (1 \wedge (b+(c \wedge (a+(c+-1))))) = -(0 \wedge (a+(b+(0 \wedge (c+-1))))) + (1 \wedge (a+(b \wedge (c+(-a+(a+(b+-1)))))))$, by (SU),

275 $\iff -(0 \wedge (b + (c \wedge (a + (c + -1))))) + (1 \wedge (b + (c \wedge (a + (c + -1))))) = -(0 \wedge (a + (b + (0 \wedge (c + -1))))) + (1 \wedge (a + (b \wedge (c + ((a + (b + -1)) + -a)))))$, by (Scomm),
276 $\iff -(0 \wedge (b + (c \wedge (a + (c + -1))))) + (1 \wedge (b + (c \wedge (a + (c + -1))))) = -(0 \wedge (a + (b + (0 \wedge (c + -1))))) + (1 \wedge (a + (b \wedge (c + (a + ((b + -1) + -a))))))$, by (Sass),
277 $\iff -(0 \wedge (b + (c \wedge (a + (c + -1))))) + (1 \wedge (b + (c \wedge (a + (c + -1))))) = -(0 \wedge (a + (b + (0 \wedge (c + -1))))) + (1 \wedge (a + (b \wedge (c + (a + (b + (-1 + -a)))))))$, by (Sass),
278 $\iff -(0 \wedge (b + (c \wedge (a + (c + -1))))) + (1 \wedge (b + (c \wedge (a + (c + -1))))) = -(0 \wedge (a + (b + (0 \wedge (c + -1))))) + (1 \wedge (a + (b \wedge (c + (b + -1)))))$, by (11.113),
279 $\iff -(0 \wedge (b + (c \wedge (a + (c + -1))))) + (1 \wedge (b + (c \wedge (a + (c + -1))))) = -(0 \wedge (a + (b + (0 \wedge (c + -1))))) + (1 \wedge (a + (b \wedge (b + (c + -1)))))$, by (11.108),
280 $\iff -(0 \wedge (b + (c \wedge (a + (c + -1))))) + (1 \wedge (b + (c \wedge (a + (c + -1))))) = -(0 \wedge (a + (b + (0 \wedge (c + -1))))) + (1 \wedge (a + (b + (0 \wedge (c + -1)))))$, by (11.116),
281 $\iff 1+(-(0\wedge(b+(c\wedge(a+(c+-1)))))\wedge(-1+(-(0\wedge(b+(c\wedge(a+(c+-1)))))+(b+(c\wedge(a+(c+-1))))))) = -(0\wedge(a+(b+(0\wedge(c+-1)))))+(1\wedge(a+(b+(0\wedge(c+-1)))))$, by (11.126),
282 $\iff 1+(-(0\wedge(b+(c\wedge(a+(c+-1)))))\wedge(-1+((b+(c\wedge(a+(c+-1))))+-(0\wedge(b+(c\wedge(a+(c+-1)))))))) = -(0\wedge(a+(b+(0\wedge(c+-1)))))+(1\wedge(a+(b+(0\wedge(c+-1)))))$, by (Scomm),
283 $\iff 1+(-(0\wedge(b+(c\wedge(a+(c+-1)))))\wedge(-1+(b+((c\wedge(a+(c+-1)))+-(0\wedge(b+(c\wedge(a+(c+-1))))))))) = -(0\wedge(a+(b+(0\wedge(c+-1)))))+(1\wedge(a+(b+(0\wedge(c+-1)))))$, by (Sass),
284 $\iff 1+(-(0\wedge(b+(c\wedge(a+(c+-1)))))\wedge(b+(-1+((c\wedge(a+(c+-1)))+-(0\wedge(b+(c\wedge(a+(c+-1))))))))) = -(0\wedge(a+(b+(0\wedge(c+-1)))))+(1\wedge(a+(b+(0\wedge(c+-1)))))$, by (11.108),
285 $\iff 1 + (-(0 \wedge (b + (c \wedge (a + (c + -1))))) + (0 \wedge ((b + (-1 + ((c \wedge (a + (c + -1))) + -(0 \wedge (b + (c \wedge (a + (c + -1)))))))) + (0 \wedge (b + (c \wedge (a + (c + -1)))))))) = -(0 \wedge (a + (b + (0 \wedge (c + -1))))) + (1 \wedge (a + (b + (0 \wedge (c + -1)))))$, by (11.125),
286 $\iff 1 + (-(0 \wedge (b + (c \wedge (a + (c + -1))))) + (0 \wedge ((0 \wedge (b + (c \wedge (a + (c + -1))))) + (b + (-1 + ((c \wedge (a + (c + -1))) + -(0 \wedge (b + (c \wedge (a + (c + -1)))))))))) = -(0 \wedge (a + (b + (0 \wedge (c + -1))))) + (1 \wedge (a + (b + (0 \wedge (c + -1)))))$, by (Scomm),
287 $\iff 1 + (-(0 \wedge (b + (c \wedge (a + (c + -1))))) + (0 \wedge (b + ((0 \wedge (b + (c \wedge (a + (c + -1))))) + (-1 + ((c \wedge (a + (c + -1))) + -(0 \wedge (b + (c \wedge (a + (c + -1)))))))))) = -(0 \wedge (a + (b + (0 \wedge (c + -1))))) + (1 \wedge (a + (b + (0 \wedge (c + -1)))))$, by (11.108),
288 $\iff 1 + (-(0 \wedge (b + (c \wedge (a + (c + -1))))) + (0 \wedge (b + (-1 + ((0 \wedge (b + (c \wedge (a + (c + -1))))) + ((c \wedge (a + (c + -1))) + -(0 \wedge (b + (c \wedge (a + (c + -1)))))))))) = -(0 \wedge (a + (b + (0 \wedge (c + -1))))) + (1 \wedge (a + (b + (0 \wedge (c + -1)))))$, by (11.108),
289 $\iff 1 + (-(0 \wedge (b + (c \wedge (a + (c + -1))))) + (0 \wedge (b + (-1 + ((c \wedge (a + (c + -1))) + ((0 \wedge (b + (c \wedge (a + (c + -1))))) + -(0 \wedge (b + (c \wedge (a + (c + -1)))))))))) = -(0 \wedge (a + (b + (0 \wedge (c + -1))))) + (1 \wedge (a + (b + (0 \wedge (c + -1)))))$, by (11.108),
290 $\iff 1+(-(0\wedge(b+(c\wedge(a+(c+-1)))))+(0\wedge(b+(-1+((c\wedge(a+(c+-1)))+0))))) = -(0 \wedge (a + (b + (0 \wedge (c + -1))))) + (1 \wedge (a + (b + (0 \wedge (c + -1)))))$, by ($\mathrm{m}_0$-Re),
291 $\iff 1+(-(0\wedge(b+(c\wedge(a+(c+-1)))))+(0\wedge(b+(-1+(0+(c\wedge(a+(c+-1)))))))) = -(0 \wedge (a + (b + (0 \wedge (c + -1))))) + (1 \wedge (a + (b + (0 \wedge (c + -1)))))$, by (Scomm),

292 $\Longleftrightarrow 1+(-(0\wedge(b+(c\wedge(a+(c+-1)))))+(0\wedge(b+(-1+(c\wedge(a+(c+-1))))))) = -(0\wedge(a+(b+(0\wedge(c+-1)))))+(1\wedge(a+(b+(0\wedge(c+-1)))))$, by (SU),
293 $\Longleftrightarrow 1+(-(0\wedge(b+(c\wedge(a+(c+-1)))))+(0\wedge(b+(c+(-1\wedge(-c+(-1+(a+(c+-1))))))))) = -(0\wedge(a+(b+(0\wedge(c+-1)))))+(1\wedge(a+(b+(0\wedge(c+-1)))))$, by (11.126),
294 $\Longleftrightarrow 1+(-(0\wedge(b+(c\wedge(a+(c+-1)))))+(0\wedge(b+(c+(-1\wedge(-c+(a+(-1+(c+-1))))))))) = -(0\wedge(a+(b+(0\wedge(c+-1)))))+(1\wedge(a+(b+(0\wedge(c+-1)))))$, by (11.108),
295 $\Longleftrightarrow 1+(-(0\wedge(b+(c\wedge(a+(c+-1)))))+(0\wedge(b+(c+(-1\wedge(-c+(a+(c+(-1+-1))))))))) = -(0\wedge(a+(b+(0\wedge(c+-1)))))+(1\wedge(a+(b+(0\wedge(c+-1)))))$, by (11.108),
296 $\Longleftrightarrow 1+(-(0\wedge(b+(c\wedge(a+(c+-1)))))+(0\wedge(b+(c+(-1\wedge((a+(c+(-1+-1)))+-c)))))) = -(0\wedge(a+(b+(0\wedge(c+-1)))))+(1\wedge(a+(b+(0\wedge(c+-1)))))$, by (Scomm),
297 $\Longleftrightarrow 1+(-(0\wedge(b+(c\wedge(a+(c+-1)))))+(0\wedge(b+(c+(-1\wedge(a+((c+(-1+-1))+-c))))))) = -(0\wedge(a+(b+(0\wedge(c+-1)))))+(1\wedge(a+(b+(0\wedge(c+-1)))))$, by (Sass),
298 $\Longleftrightarrow 1+(-(0\wedge(b+(c\wedge(a+(c+-1)))))+(0\wedge(b+(c+(-1\wedge(a+(c+((-1+-1)+-c)))))))) = -(0\wedge(a+(b+(0\wedge(c+-1)))))+(1\wedge(a+(b+(0\wedge(c+-1)))))$, by (Sass),
299 $\Longleftrightarrow 1+(-(0\wedge(b+(c\wedge(a+(c+-1)))))+(0\wedge(b+(c+(-1\wedge(a+(c+(-1+(-1+-c))))))))) = -(0\wedge(a+(b+(0\wedge(c+-1)))))+(1\wedge(a+(b+(0\wedge(c+-1)))))$, by (Sass),
300 $\Longleftrightarrow 1+(-(0\wedge(b+(c\wedge(a+(c+-1)))))+(0\wedge(b+(c+(-1\wedge(a+(-1+-1))))))) = -(0\wedge(a+(b+(0\wedge(c+-1)))))+(1\wedge(a+(b+(0\wedge(c+-1)))))$, by (11.113),
301 $\Longleftrightarrow 1+(-(0\wedge(b+(c\wedge(a+(c+-1)))))+(0\wedge(b+(c+(-1+(0\wedge((a+(-1+-1))+1))))))) = -(0\wedge(a+(b+(0\wedge(c+-1)))))+(1\wedge(a+(b+(0\wedge(c+-1)))))$, by (11.125),
302 $\Longleftrightarrow 1+(-(0\wedge(b+(c\wedge(a+(c+-1)))))+(0\wedge(b+(c+(-1+(0\wedge(1+(a+(-1+-1))))))))) = -(0\wedge(a+(b+(0\wedge(c+-1)))))+(1\wedge(a+(b+(0\wedge(c+-1)))))$, by (Scomm),
303 $\Longleftrightarrow 1+(-(0\wedge(b+(c\wedge(a+(c+-1)))))+(0\wedge(b+(c+(-1+(0\wedge(a+-1))))))) = -(0\wedge(a+(b+(0\wedge(c+-1)))))+(1\wedge(a+(b+(0\wedge(c+-1)))))$, by (11.113),
304 $\Longleftrightarrow 1+((0\wedge(b+(c+(-1+(0\wedge(a+-1))))))+-(0\wedge(b+(c\wedge(a+(c+-1)))))) = -(0\wedge(a+(b+(0\wedge(c+-1)))))+(1\wedge(a+(b+(0\wedge(c+-1)))))$, by (Scomm),
305 $\Longleftrightarrow 1+((0\wedge(b+(c+(-1+(0\wedge(a+-1))))))+-(0\wedge(b+(c\wedge(a+(c+-1)))))) = -(0\wedge(a+(b+(c+(-1\wedge-c)))))+(1\wedge(a+(b+(0\wedge(c+-1)))))$, by (11.119),
306 $\Longleftrightarrow 1+((0\wedge(b+(c+(-1+(0\wedge(a+-1))))))+-(0\wedge(b+(c\wedge(a+(c+-1)))))) = -(0\wedge(a+(b+(c+(-1+(0\wedge(-c+1)))))))+(1\wedge(a+(b+(0\wedge(c+-1)))))$, by (11.125),
307 $\Longleftrightarrow 1+((0\wedge(b+(c+(-1+(0\wedge(a+-1))))))+-(0\wedge(b+(c\wedge(a+(c+-1)))))) = -(0\wedge(a+(b+(c+(-1+(0\wedge(1+-c)))))))+(1\wedge(a+(b+(0\wedge(c+-1)))))$, by (Scomm),
308 $\Longleftrightarrow (1\wedge(1+(b+(c+(-1+(0\wedge(a+-1)))))))+-(0\wedge(b+(c\wedge(a+(c+-1))))) = -(0\wedge(a+(b+(c+(-1+(0\wedge(1+-c)))))))+(1\wedge(a+(b+(0\wedge(c+-1)))))$, by (11.121),

309 $\iff (1 \wedge (b + (c + (0 \wedge (a + -1))))) + -(0 \wedge (b + (c \wedge (a + (c + -1))))) = -(0 \wedge (a + (b + (c + (-1 + (0 \wedge (1 + -c))))))) + (1 \wedge (a + (b + (0 \wedge (c + -1)))))$, by (11.115),
310 $\iff (1 \wedge (b + (c + (0 \wedge (a + -1))))) + -(0 \wedge (b + (c \wedge (a + (c + -1))))) = -(0 \wedge (a + (b + (-1 + (c \wedge (c + (1 + -c))))))) + (1 \wedge (a + (b + (0 \wedge (c + -1)))))$, by (11.122),
311 $\iff (1 \wedge (b + (c + (0 \wedge (a + -1))))) + -(0 \wedge (b + (c \wedge (a + (c + -1))))) = -(0 \wedge (a + (b + (-1 + (c \wedge (1 + (c + -c))))))) + (1 \wedge (a + (b + (0 \wedge (c + -1)))))$, by (11.108),
312 $\iff (1 \wedge (b + (c + (0 \wedge (a + -1))))) + -(0 \wedge (b + (c \wedge (a + (c + -1))))) = -(0 \wedge (a + (b + (-1 + (c \wedge (1 + 0)))))) + (1 \wedge (a + (b + (0 \wedge (c + -1)))))$, by ($m_0$-Re),
313 $\iff (1 \wedge (b + (c + (0 \wedge (a + -1))))) + -(0 \wedge (b + (c \wedge (a + (c + -1))))) = -(0 \wedge (a + (b + (-1 + (c \wedge (0 + 1)))))) + (1 \wedge (a + (b + (0 \wedge (c + -1)))))$, by (Scomm),
314 $\iff (1 \wedge (b + (c + (0 \wedge (a + -1))))) + -(0 \wedge (b + (c \wedge (a + (c + -1))))) = -(0 \wedge (a + (b + (-1 + (c \wedge 1))))) + (1 \wedge (a + (b + (0 \wedge (c + -1)))))$, by (SU),
315 $\iff (1 \wedge (b + (c + (0 \wedge (a + -1))))) + -(0 \wedge (b + (c \wedge (a + (c + -1))))) = -(0 \wedge (a + (b + (-1 + (1 \wedge c))))) + (1 \wedge (a + (b + (0 \wedge (c + -1)))))$, by (Wcomm),
316 $\iff (1 \wedge (b + (c + (0 \wedge (a + -1))))) + -(0 \wedge (b + (c \wedge (a + (c + -1))))) = -(0 \wedge (a + (b + (-1 + c)))) + (1 \wedge (a + (b + (0 \wedge (c + -1)))))$, by (11.107),
317 $\iff (1 \wedge (b + (c + (0 \wedge (a + -1))))) + -(0 \wedge (b + (c \wedge (a + (c + -1))))) = -(0 \wedge (a + (b + (c + -1)))) + (1 \wedge (a + (b + (0 \wedge (c + -1)))))$, by (Scomm),

322 $\iff (1 \wedge (b + (c + (0 \wedge (a + -1))))) + -(0 \wedge (b + (c + (0 \wedge (a + -1))))) = -(0 \wedge (a + (b + (c + -1)))) + (1 \wedge (a + (b + (0 \wedge (c + -1)))))$, by (11.123),
323 $\iff -(0 \wedge (b + (c + (0 \wedge (a + -1))))) + (1 \wedge (b + (c + (0 \wedge (a + -1))))) = -(0 \wedge (a + (b + (c + -1)))) + (1 \wedge (a + (b + (0 \wedge (c + -1)))))$, by (Scomm),
324 $\iff -(0 \wedge (a + (b + (c + -1)))) + (1 \wedge (a + (b + (0 \wedge (c + -1))))) = -(0 \wedge (b + (c + (0 \wedge (a + -1))))) + (1 \wedge (b + (c + (0 \wedge (a + -1)))))$, copy 323

336 $\iff -(0 \wedge (a + (b + (c + -1)))) + (1 \wedge (a + (b + (0 \wedge (c + -1))))) = -(0 \wedge (b + (c + (a + (-1 \wedge -a))))) + (1 \wedge (b + (c + (0 \wedge (a + -1)))))$, by (11.119),
337 $\iff -(0 \wedge (a + (b + (c + -1)))) + (1 \wedge (a + (b + (0 \wedge (c + -1))))) = -(0 \wedge (b + (c + (a + (-1 + (0 \wedge (-a + 1))))))) + (1 \wedge (b + (c + (0 \wedge (a + -1)))))$, by (11.125),
338 $\iff -(0 \wedge (a + (b + (c + -1)))) + (1 \wedge (a + (b + (0 \wedge (c + -1))))) = -(0 \wedge (b + (c + (a + (-1 + (0 \wedge (1 + -a))))))) + (1 \wedge (b + (c + (0 \wedge (a + -1)))))$, by (Scomm),
339 $\iff -(0 \wedge (a + (b + (c + -1)))) + (1 \wedge (a + (b + (0 \wedge (c + -1))))) = -(0 \wedge (b + (c + (a + (-1 + 0))))) + (1 \wedge (b + (c + (0 \wedge (a + -1)))))$, by (11.133),
340 $\iff -(0 \wedge (a + (b + (c + -1)))) + (1 \wedge (a + (b + (0 \wedge (c + -1))))) = -(0 \wedge (b + (c + (a + (0 + -1))))) + (1 \wedge (b + (c + (0 \wedge (a + -1)))))$, by (Scomm),
341 $\iff -(0 \wedge (a + (b + (c + -1)))) + (1 \wedge (a + (b + (0 \wedge (c + -1))))) = -(0 \wedge (b + (c + (a + -1)))) + (1 \wedge (b + (c + (0 \wedge (a + -1)))))$, by (SU),
342 $\iff -(0 \wedge (a + (b + (c + -1)))) + (1 \wedge (a + (b + (0 \wedge (c + -1))))) = -(0 \wedge (b + (a + (c + -1)))) + (1 \wedge (b + (c + (0 \wedge (a + -1)))))$, by (11.108),
343 $\iff -(0 \wedge (a + (b + (c + -1)))) + (1 \wedge (a + (b + (0 \wedge (c + -1))))) = -(0 \wedge (a + (b + (c + -1)))) + (1 \wedge (b + (c + (0 \wedge (a + -1)))))$, by (11.108),
344 $\iff -(0 \wedge (a + (b + (c + -1)))) + (1 \wedge (a + (b + (0 \wedge (c + -1))))) = -(0 \wedge (a +$

$(b+(c+-1)))) + (1 \wedge (b+(c+(a+(-1 \wedge -a)))))$, by (11.119),
345 $\iff -(0 \wedge (a+(b+(c+-1)))) + (1 \wedge (a+(b+(0 \wedge (c+-1))))) = -(0 \wedge (a+ (b+(c+-1)))) + (1 \wedge (b+(c+(a+(-1+(0 \wedge (-a+1)))))))$, by (11.125),
346 $\iff -(0 \wedge (a+(b+(c+-1)))) + (1 \wedge (a+(b+(0 \wedge (c+-1))))) = -(0 \wedge (a+ (b+(c+-1)))) + (1 \wedge (b+(c+(a+(-1+(0 \wedge (1+-a)))))))$, by (Scomm),
347 $\iff -(0 \wedge (a+(b+(c+-1)))) + (1 \wedge (a+(b+(0 \wedge (c+-1))))) = -(0 \wedge (a+ (b+(c+-1)))) + (1 \wedge (b+(c+(a+(-1+0)))))$, by (11.133),
348 $\iff -(0 \wedge (a+(b+(c+-1)))) + (1 \wedge (a+(b+(0 \wedge (c+-1))))) = -(0 \wedge (a+ (b+(c+-1)))) + (1 \wedge (b+(c+(a+(0+-1)))))$, by (Scomm),
349 $\iff -(0 \wedge (a+(b+(c+-1)))) + (1 \wedge (a+(b+(0 \wedge (c+-1))))) = -(0 \wedge (a+ (b+(c+-1)))) + (1 \wedge (b+(c+(a+-1))))$, by (SU),
350 $\iff -(0 \wedge (a+(b+(c+-1)))) + (1 \wedge (a+(b+(0 \wedge (c+-1))))) = -(0 \wedge (a+ (b+(c+-1)))) + (1 \wedge (b+(a+(c+-1))))$, by (11.108),
351 $\iff -(0 \wedge (a+(b+(c+-1)))) + (1 \wedge (a+(b+(0 \wedge (c+-1))))) = -(0 \wedge (a+ (b+(c+-1)))) + (1 \wedge (a+(b+(c+-1))))$, by (11.108),
352 $\iff -(0 \wedge (a+(b+(c+-1)))) + (1 \wedge (a+(b+(c+(-1 \wedge -c))))) = -(0 \wedge (a+(b+(c+-1)))) + (1 \wedge (a+(b+(c+-1))))$, by (11.119),
353 $\iff -(0 \wedge (a+(b+(c+-1)))) + (1 \wedge (a+(b+(c+(-1+(0 \wedge (-c+1))))))) = -(0 \wedge (a+(b+(c+-1)))) + (1 \wedge (a+(b+(c+-1))))$, by (11.125),
354 $\iff -(0 \wedge (a+(b+(c+-1)))) + (1 \wedge (a+(b+(c+(-1+(0 \wedge (1+-c))))))) = -(0 \wedge (a+(b+(c+-1)))) + (1 \wedge (a+(b+(c+-1))))$, by (Scomm),
355 $\iff -(0 \wedge (a+(b+(c+-1)))) + (1 \wedge (a+(b+(-1+(c+(0 \wedge (1+-c))))))) = -(0 \wedge (a+(b+(c+-1)))) + (1 \wedge (a+(b+(c+-1))))$, by (11.108),
356 $\iff -(0 \wedge (a+(b+(c+-1)))) + (1 \wedge (a+(b+(-1+c)))) = -(0 \wedge (a+(b+ (c+-1)))) + (1 \wedge (a+(b+(c+-1))))$, by (11.129),
357 $\iff -(0 \wedge (a+(b+(c+-1)))) + (1 \wedge (a+(b+(c+-1)))) = -(0 \wedge (a+(b+ (c+-1)))) + (1 \wedge (a+(b+(c+-1))))$, by (Scomm).

This last identity, 357, being always true, it follows that the first identity, 24, is also true, hence (11.102) holds. □

11.3.4 The proof of Proposition 3.4.10

Proposition Let $\mathcal{G}_u = (G, \vee, \wedge, +, -, 0_G, u = u_G)$ be a commutative l-group with strong unit $u = u_G$, with $x \leq y \iff x \wedge y = x \iff x \vee y = y$.
Then, the property (YY_u^R) holds (see Remark 3.4.3):
(YY_u^R) for all $0_G \leq a, b, c \leq u_G$, if $\inf(u_G, a+b) = a$, then
$u_G + -\inf(u_G, (u_G + -a) + (u_G + -\inf(u_G, b+c))) =$
$\inf(u_G, b + (u_G + -\inf(u_G, (u_G + -a) + (u_G + -c))))$, i.e.

for all $0_G \leq a, b, c \leq u_G$, if $u_G \wedge (a+b) = a$, then
$u_G + -(u_G \wedge ((u_G + -a) + (u_G + -(u_G \wedge (b+c))))) =$
$u_G \wedge (b + (u_G + -(u_G \wedge ((u_G + -a) + (u_G + -c)))))$.

Proof. (Michael Kinyon, by PROVER9 on Linux, Length of proof was 121, in 2.76 seconds; the expanded proof had the Length 284.)

In order to simplify the writting (less space on a line), we shall replace 0_G by 0 and u_G by 1.

Hence, we have to prove that, for all $0 \leq a, b, c \leq 1$, if $1 \wedge (a+b) = a$, then: $1+-(1\wedge((1+-a)+(1+-(1\wedge(b+c))))) = 1\wedge(b+(1+-(1\wedge((1+-a)+(1+-c)))))$.

In order to shorten the expanded proof (which on PROVER9 on Windows had the length of about 700), Michael Kinyon introduced a new negation, x', defined by:

$$x' \overset{def.}{=} 1 + -x. \tag{11.135}$$

It follows that $x'' = 1 + -x' = 1 + -(1 + -x) \overset{(NegS)}{=} 1 + (- - x + -1) \overset{(DN)}{=} 1 + (x + -1) \overset{(m_0-Re)}{=} x$, hence, we have:

$$x'' = x. \tag{11.136}$$

It follows also that $-x' = -(1 + -x) \overset{(NegS)}{=} - - x + -1 \overset{(DN)}{=} x + -1$, hence, we have:

$$-x' = x + -1. \tag{11.137}$$

Hence, we have to prove that, for all $0 \leq a, b, c \leq 1$, if $1 \wedge (a+b) = a$, then: $(1 \wedge (a' + (1 \wedge (b+c))'))' = 1 \wedge (b + (1 \wedge (a' + c'))')$.

In order to shorten the proof, Michael Kinyon denoted the left part of the identity by $T1$ and the right part by $T2$:
$T1 = (1 \wedge (a' + (1 \wedge (b+c))'))'$ and $T2 = 1 \wedge (b + (1 \wedge (a' + c'))')$, i.e. we have:

$$(1 \wedge (a' + (1 \wedge (b+c))'))' = T1, \tag{11.138}$$

$$1 \wedge (b + (1 \wedge (a' + c'))') = T2. \tag{11.139}$$

Hence, we have to prove that, for all $0 \leq a, b, c \leq 1$, if $1 \wedge (a+b) = a$, then $T1 = T2$.

The are the particular hypotheses used by PROVER9 (among the given ones - see (α)):

$$0 \wedge a = 0, \tag{11.140}$$

$$0 \wedge c = 0, \tag{11.141}$$

$$a \wedge 1 = a, \tag{11.142}$$

$$c \wedge 1 = c, \tag{11.143}$$

hence,

$$a \wedge 0 = 0, \tag{11.144}$$

$$c \wedge 0 = 0. \tag{11.145}$$

The following properties are obvious, by (SU), (Scomm), (Sass), (m_0-Re), (NegS) and (Wcomm) ($x \wedge y = y \wedge x$):

$$x + (y + z) = y + (x + z), \tag{11.146}$$

$$x + (-x + y) = y, \tag{11.147}$$

$$x + (-x + -y) = -y, \tag{11.148}$$

$$-x + (x + -y) = -y, \tag{11.149}$$

$$x + (y + -x) = y; \tag{11.150}$$

$$-x + -y = -(x + y), \tag{11.151}$$

$$-(x + -y) = -x + y, \tag{11.152}$$

$$-(-x + y) = x + -y, \tag{11.153}$$

$$-x + -(-x + y) = -y \tag{11.154}$$

$$x + -(x + y) = -y, \tag{11.155}$$

$$x + -(y + x) = -y; \tag{11.156}$$

$$x + x' = 1, \tag{11.157}$$

$$x + 1 = (-x)', \tag{11.158}$$

$$1 + x = (-x)', \tag{11.159}$$

$$1 + (-x + y) = x' + y; \tag{11.160}$$

$$x \wedge (y \wedge x) = y \wedge x, \tag{11.161}$$

$$x \vee (y \wedge x) = x. \tag{11.162}$$

Suppose that

$$1 \wedge (a + b) = a. \tag{11.163}$$

We have to prove that $T1 = T2$. Indeed,

first, in (g4) $(-x \wedge -y = -(x \vee y))$, take $X := 0$, to obtain, by (N0):

$$0 \wedge -x = -(0 \vee x). \tag{11.164}$$

Then, in (g5) $(-x \vee -y = -(x \wedge y))$, take $X := 0$, to obtain, by (N0):

$$0 \vee -x = -(0 \wedge x). \tag{11.165}$$

Then, in (g5) also, take $Y := -y$, to obtain, by (DN):
$-x \vee y = -(x \wedge -y)$, hence:

$$-(x \wedge -y) = -x \vee y. \tag{11.166}$$

Now, in (g3) $((x + y) \wedge (x + z) = x + (y \wedge z))$, take first $Y := 0$, to obtain, by (SU):

$$x \wedge (x + z) = x + (0 \wedge z). \tag{11.167}$$

Then, in (g3) again, interchange x with y, to obtain, by (Scomm):

$$(x + y) \wedge (y + z) = y + (x \wedge z). \tag{11.168}$$

Now, in (11.168), take $Y := -x$, to obtain, by (m_0-Re):

$$0 \wedge (-x + z) = -x + (x \wedge z). \tag{11.169}$$

Then, in (g3) again, take $Z := -x$, to obtain, by (m_0-Re),
$(x + y) \wedge 0 = x + (y \wedge -x)$, hence, by (Wcomm),
$0 \wedge (x + y) = x + (y \wedge -x)$, i.e.

$$x + (y \wedge -x) = 0 \wedge (x + y). \tag{11.170}$$

Then, in (g3) again, take $Y := c$, $Z := 0$, to obtain:
$(x + c) \wedge (x + 0) = x + (c \wedge 0)$, which, by (11.145), becomes, by (SU):
$(x + c) \wedge x = x$, which, by (Wcomm), becomes:
$x \wedge (x + c) = x$, which, by (Scomm), becomes:

$$x \wedge (c + x) = x. \tag{11.171}$$

Then, in (g3) again, take $Y := x'$, to obtain:
$(x + x') \wedge (x + z) = x + (x' \wedge z)$, which, by (11.157), becomes:
$1 \wedge (x + z) = x + (x' \wedge z)$, i.e.

$$x + (x' \wedge y) = 1 \wedge (x + y). \tag{11.172}$$

Then, in (g3) again, take $Z := z + -x$, to obtain:
$(x + y) \wedge (x + (z + -x)) = x + (y \wedge (z + -x))$, which, by (11.150), becomes:
$(x + y) \wedge z = x + (y \wedge (z + -x))$, i.e.

$$x + (y \wedge (z + -x)) = (x + y) \wedge z. \tag{11.173}$$

Now, in (11.167), take $Z := -(x + y)$, to obtain:
$x \wedge (x + -(x + y)) = x + (0 \wedge -(x + y))$, which, by (11.155), becomes:
$x \wedge -y = x + (0 \wedge -(x + y))$, which, by (11.164), becomes:
$x \wedge -y = x + -(0 \vee (x + y))$, i.e.

$$x + -(0 \vee (x + y)) = x \wedge -y. \tag{11.174}$$

Then, from (11.167), by (Scomm), we obtain:

$$x \wedge (z + x) = x + (0 \wedge z). \tag{11.175}$$

Then, in (11.175), take $X := -x$, $Z := x$, to obtain, by (m_0-Re):
$-x \wedge 0 = -x + (0 \wedge x)$, which, by (Wcomm), becomes:
$0 \wedge -x = -x + (0 \wedge x)$, which, by (11.164), becomes:
$-(0 \vee x) = -x + (0 \wedge x)$,i.e.

$$-x + (0 \wedge x) = -(0 \vee x). \tag{11.176}$$

Now, in (11.159) ($1 + x = (-x)'$), take $X := 0 \wedge x$, to obtain:
$1 + (0 \wedge x) = (-(0 \wedge x))'$, which, by (11.167), becomes:
$1 \wedge (1 + x) = (-(0 \wedge x))'$, which, by (11.159), becomes:

$$1 \wedge (-x)' = (-(0 \wedge x))'. \tag{11.177}$$

Then, in (11.177), take $X := -x$, to obtain, by (DN):
$1 \wedge x' = (-(0 \wedge -x))'$, which, by (11.164), becomes:
$1 \wedge x' = (- - (0 \vee x))'$, which, by (DN), becomes:

$$1 \wedge x' = (0 \vee x)', \tag{11.178}$$

which, for $X := x'$, becomes:
$1 \wedge x'' = (0 \vee x')'$, which, by (11.136), becomes:
$1 \wedge x = (0 \vee x')'$, i.e.

$$(0 \vee x')' = 1 \wedge x. \tag{11.179}$$

Now, in (g2) ($(x + y) \vee (x + z) = x + (y \vee z)$), take $Y := -x$, to obtain, by (m_0-Re):
$0 \vee (x + y) = x + (-x \vee y)$, i.e.

$$x + (-x \vee y) = 0 \vee (x + y). \tag{11.180}$$

Then, in (g2) again, take $Y := x'$, to obtain:
$(x + x') \vee (x + z) = x + (x' \vee z)$, which, by (11.157), becomes:
$1 \vee (x + z) = x + (x' \vee z)$, i.e.

$$x + (x' \vee y) = 1 \vee (x + y). \tag{11.181}$$

Then, in (g2) again, take $Z := x'$, to obtain:
$(x + y) \vee (x + x') = x + (y \vee x')$, which, by (11.157), becomes:
$(x + y) \vee 1 = x + (y \vee x')$, which, by (Vcomm), becomes:
$1 \vee (x + y) = x + (y \vee x')$, i.e.

$$x + (y \vee x') = 1 \vee (x + y). \tag{11.182}$$

Then, in (11.146) $(x + (y + z) = y + (x + z))$, take $Z := z \vee y'$, to obtain:
$x + (y + (z \vee y')) = y + (x + (z \vee y'))$, which, by (11.182), becomes:
$x + (1 \vee (y + z)) = y + (x + (z \vee y'))$, i.e.

$$x + (y + (z \vee x')) = y + (1 \vee (x + z)). \tag{11.183}$$

Now, in (11.160), take $X := -x + y$, to obtain, by (11.159),
$(-(-x + y))' = x' + y$, hence, by (11.153):

$$(x + -y)' = x' + y. \tag{11.184}$$

Then, in (11.184), take $Y := -y$, to obtain, by (DN):
$(x + y)' = x' + -y$, i.e.

$$x' + -y = (x + y)', \tag{11.185}$$

which, by (Scomm), becomes:
$-y + x' = (x + y)'$, i.e.

$$-x + y' = (y + x)'. \tag{11.186}$$

Now, in (Sass) $((x + y) + z = x + (y + z))$, take $Z := -1$, to obtain:
$(x + y) + -1 = x + (y + -1)$, hence, by (11.137) for $X := x + y$:
$-(x + y)' = x + (y + -1)$, which, by (11.137) again, becomes:
$-(x + y)' = x + -y'$, i.e.

$$x + -y' = -(x + y)'. \tag{11.187}$$

Then, in (Sass) again, take $X := 1$, to obtain:
$(1 + y) + z = 1 + (y + z)$, which, by (11.159), becomes:
$(-y)' + z = 1 + (y + z)$, which, by (11.159) again, becomes:

$$(-y)' + z = (-(y + z))'. \tag{11.188}$$

Then, in (Sass) again, take $Y := 1$, to obtain:
$(x + 1) + z = x + (1 + z)$, which, by (11.159), becomes:

$(x+1)+z = x+(-z)'$, which, by (11.158), becomes:
$(-x)'+z = x+(-z)'$, which, by (11.188), becomes:
$(-(x+z))' = x+(-z)'$, i.e.

$$x+(-y)' = (-(x+y))'. \tag{11.189}$$

Then, in (Sass) again, take $Y := x'$, to obtain:
$(x+x')+z = x+(x'+z)$, which, by (11.157), becomes:
$1+z = x+(x'+z)$, which, by (11.159), becomes:
$(-z)' = x+(x'+z)$, i.e.

$$x+(x'+y) = (-y)'. \tag{11.190}$$

Now, in (11.190) again, take $Y := -(y+x')$, to obtain:
$x+(x'+-(y+x')) = (--(y+x'))'$, which, by (11.156) and (DN), becomes:
$x+-y = (y+x')'$, i.e.

$$(x+y')' = y+-x. \tag{11.191}$$

Then, in (11.190), take $Y := -y$, to obtain, by (DN):
$x+(x'+-y) = y'$, which, by (11.185), becomes:

$$x+(x+y)' = y'. \tag{11.192}$$

Then, in (11.192), take $Y := x' \vee y$, to obtain:
$x+(x+(x' \vee y))' = (x' \vee y)'$, which, by (11.181), becomes:

$$x+(1 \vee (x+y))' = (x' \vee y)'. \tag{11.193}$$

Now, in (Wdis) $((x \wedge y) \vee (x \wedge z) = x \wedge (y \vee z))$, take $X := a$, $Y := 0$, to obtain:
$(a \wedge 0) \vee (a \wedge z) = a \wedge (0 \vee z)$, which, by (11.144), becomes:
$0 \vee (a \wedge x) = a \wedge (0 \vee x)$, i.e.

$$a \wedge (0 \vee x) = 0 \vee (a \wedge x). \tag{11.194}$$

Now, in (Wass) $((x \wedge y) \wedge z = x \wedge (y \wedge z))$, take $X := a$, $Y := 1$, to obtain:
$(a \wedge 1) \wedge z = a \wedge (1 \wedge z)$, which, by (11.142), becomes:
$a \wedge z = a \wedge (1 \wedge z)$, i.e.

$$a \wedge (1 \wedge x) = a \wedge x. \tag{11.195}$$

Then, in (Wass) again, take $X := 1$, $Y := a+b$, to obtain:
$(1 \wedge (a+b) \wedge z = 1 \wedge ((a+b) \wedge z)$, which, by (11.163), becomes:
$a \wedge x = 1 \wedge ((a+b) \wedge x)$, i.e.

$$1 \wedge ((a+b) \wedge x) = a \wedge x. \tag{11.196}$$

Then, in (Wass) again, take $Y := c+x$, to obtain:
$(x \wedge (c+x)) \wedge z = x \wedge ((c+x) \wedge z)$, which, by (11.171), becomes:
$x \wedge y = x \wedge ((c+x) \wedge y)$, i.e.

$$x \wedge ((c+x) \wedge y) = x \wedge y. \tag{11.197}$$

Now, in (11.197), take $Y := c + y$, to obtain:
$x \wedge ((c + x) \wedge (c + y)) = x \wedge (c + y)$, which, by (g3), becomes:

$$x \wedge (c + (x \wedge y)) = x \wedge (c + y). \tag{11.198}$$

Then, in (11.198), take$X := 1$, $Y := a + b$, to obtain:
$1 \wedge (c + (1 \wedge (a + b))) = 1 \wedge (c + (a + b))$, which, by (11.163), becomes:
$1 \wedge (c + a) = 1 \wedge (c + (a + b))$, which, by (Scomm), becomes:
$1 \wedge (a + c) = 1 \wedge (c + (a + b))$, which, by (11.146), becomes:
$1 \wedge (a + c) = 1 \wedge (a + (c + b))$, which, by (Scomm), becomes:
$1 \wedge (a + c) = 1 \wedge (a + (b + c))$, i.e.

$$1 \wedge (a + (b + c)) = 1 \wedge (a + c). \tag{11.199}$$

Now, in (11.195), take $X := x'$, to obtain:
$a \wedge (1 \wedge x') = a \wedge x'$, which, by (11.178), becomes:

$$a \wedge (0 \vee x)' = a \wedge x'. \tag{11.200}$$

Then, in (11.136) ($x'' = x$), take $X := 0 \vee x'$, to obtain:
$(0 \vee x')'' = 0 \vee x'$, which, by (11.179), becomes:
$(1 \wedge x)' = 0 \vee x'$, i.e.

$$0 \vee x' = (1 \wedge x)', \tag{11.201}$$

which, for $X := y + x'$, becomes:
$0 \vee (y + x')' = (1 \wedge (y + x'))'$, which, by (11.191), becomes:
$0 \vee (x + -y) = (1 \wedge (y + x'))'$, i.e.

$$(1 \wedge (x + y'))' = 0 \vee (y + -x). \tag{11.202}$$

Then, in (11.194) ($a \wedge (0 \vee x) = 0 \vee (a \wedge x)$), take $X := x'$, to obtain:
$a \wedge (0 \vee x') = 0 \vee (a \wedge x')$, which, by (11.201), becomes:

$$a \wedge (1 \wedge x)' = 0 \vee (a \wedge x'). \tag{11.203}$$

Now, in (11.164), take $X := x + -y$, to obtain:
$0 \wedge -(x + -y) = -(0 \vee (x + -y))$, which, by (11.152), becomes:
$0 \wedge (-x + y) = -(0 \vee (x + -y))$, i.e.

$$-(0 \vee (x + -y)) = 0 \wedge (-x + y). \tag{11.204}$$

Now, in (11.176), take $X := x + -y$, to obtain:
$-(x + -y) + (0 \wedge (x + -y)) = -(0 \vee (x + -y))$, which, by (11.152), becomes:
$(-x + y) + (0 \wedge (x + -y)) = -(0 \vee (x + -y))$, which, by (Sass), becomes:
$-x + (y + (0 \wedge (x + -y))) = -(0 \vee (x + -y))$, which, by (11.173), becomes:
$-x + ((y + 0) \wedge x) = -(0 \vee (x + -y))$, which, by (SU), becomes:
$-x + (y \wedge x) = -(0 \vee (x + -y))$, which, by (11.204), becomes:
$-x + (y \wedge x) = 0 \wedge (-x + y)$, i.e.

$$0 \wedge (-x + y) = -x + (y \wedge x). \tag{11.205}$$

Then, in (11.205), take $X := x \wedge y$, to obtain:
$0 \wedge (-(x \wedge y) + y) = -(x \wedge y) + (y \wedge (x \wedge y))$, which, by (11.161), becomes:
$0 \wedge (-(x \wedge y) + y) = -(x \wedge y) + (x \wedge y)$, which, by (Scomm), becomes:
$0 \wedge (y + -(x \wedge y)) = -(x \wedge y) + (x \wedge y)$,which, by (Scomm) again, becomes:
$0 \wedge (y + -(x \wedge y)) = (x \wedge y) + -(x \wedge y)$, which, by ($\mathrm{m_0}$-Re), becomes:

$$0 \wedge (x + -(y \wedge x)) = 0. \tag{11.206}$$

Then, in (11.175) $(x \wedge (z + x) = x + (0 \wedge z))$, take $Z := y + -(z \wedge y)$, to obtain:
$x \wedge ((y + -(z \wedge y)) + x) = x + (0 \wedge (y + -(z \wedge y)))$, which, by (11.206), becomes:
$x \wedge ((y + -(z \wedge y)) + x) = x + 0$, which, by (SU), becomes:
$x \wedge ((y + -(z \wedge y)) + x) = x$, which, by (Sass), becomes:
$x \wedge (y + (-(z \wedge y) + x)) = x$, which, for $Y := 1$, $Z := c$, becomes:
$x \wedge (1 + (-(c \wedge 1) + x)) = x$, which, by (11.143), becomes:
$x \wedge (1 + (-c + x)) = x$, which, by (11.159), becomes:
$x \wedge (-(-c + x))' = x$, which, by (11.153), becomes:
$x \wedge (c + -x)' = x$, which, by (11.184), becomes:

$$x \wedge (c' + x) = x. \tag{11.207}$$

Then, in (11.162) $(x \vee (y \wedge x) = x)$, take $X := c' + x$, $Y := x$, to obtain:
$(c' + x) \vee (x \wedge (c' + x)) = c' + x$, which, by (11.207), becomes:
$(c' + x) \vee x = c' + x$, which, by (Vcomm), becomes:

$$x \vee (c' + x) = c' + x. \tag{11.208}$$

Now, in (11.194) $(a \wedge (0 \vee x) = 0 \vee (a \wedge x))$, take $X := c' + 0$, to obtain:
$a \wedge (0 \vee (c' + 0)) = 0 \vee (a \wedge (c' + 0))$, which, by (11.208), becomes:
$a \wedge (c' + 0) = 0 \vee (a \wedge (c' + 0))$, which, by (SU) twice, becomes:
$a \wedge c' = 0 \vee (a \wedge c')$, i.e.

$$0 \vee (a \wedge c') = a \wedge c'. \tag{11.209}$$

Now, in (11.159) $(1 + x = (-x)')$, take $X := x \wedge -1$, to obtain:
$1 + (x \wedge -1) = (-(x \wedge -1))'$, which, by (11.170), becomes:
$0 \wedge (1 + x) = (-(x \wedge -1))'$, which, by (11.159), becomes:
$0 \wedge (-x)' = (-(x \wedge -1))'$, which, by (11.166), becomes:
$0 \wedge (-x)' = (-x \vee 1)'$, which, by (Vcomm), becomes:

$$0 \wedge (-x)' = (1 \vee -x)'. \tag{11.210}$$

Then, in (11.210) $(0 \wedge (-x)' = (1 \vee -x)')$, take $X := -x$, to obtain, by (DN):

$$0 \wedge x' = (1 \vee x)'. \tag{11.211}$$

Now, in (11.170) $(x + (y \wedge -x) = 0 \wedge (x + y))$, take $X := x'$, $Y := -y$, to obtain:
$x' + (-y \wedge -x') = 0 \wedge (x' + -y)$, which, by (11.185), becomes:
$x' + (-y \wedge -x') = 0 \wedge (x + y)'$, which, by (g4), becomes:

$x' + -(y \vee x') = 0 \wedge (x+y)'$, which, by (11.185), becomes:
$(x + (y \vee x'))' = 0 \wedge (x+y)'$, which, by (11.182), becomes:
$(1 \vee (x+y))' = 0 \wedge (x+y)'$, i.e.

$$0 \wedge (x+y)' = (1 \vee (x+y))'. \tag{11.212}$$

Now, from (11.139) $(1 \wedge (b + (1 \wedge (a' + c'))') = T2)$, by (11.202), we obtain:
$1 \wedge (b + (0 \vee (c + -a'))) = T2$, which, by (11.187), becomes:
$1 \wedge (b + (0 \vee -(c+a)')) = T2$, which, by (Scomm), becomes:
$1 \wedge (b + (0 \vee -(a+c)')) = T2$, which, by (11.165), becomes:

$$1 \wedge (b + -(0 \wedge (a+c)')) = T2. \tag{11.213}$$

Then, from (11.213), by (11.212), we obtain:
$1 \wedge (b + -(1 \vee (a+c))') = T2$, which, by (11.187), becomes:

$$1 \wedge -(b + (1 \vee (a+c)))' = T2. \tag{11.214}$$

Now, in (11.174), take $Y := -y'$, to obtain, by (DN):
$x + -(0 \vee (x + -y')) = x \wedge y'$, which, by (11.187), becomes:
$x + -(0 \vee -(x+y)') = x \wedge y'$, which, by (11.165), becomes:
$x + - - (0 \wedge (x+y)') = x \wedge y'$, which, by (DN) and (11.211), becomes:
$x + (1 \vee (x+y))' = x \wedge y'$, which, by (11.193), becomes:

$$(x' \vee y)' = x \wedge y'. \tag{11.215}$$

Now, in (11.186) $(-x + y' = (y+x)')$, take $Y := y' \vee z$, to obtain:
$-x + (y' \vee z)' = ((y' \vee z) + x)'$, which, by (11.215), becomes:

$$-x + (y \wedge z') = ((y' \vee z) + x)'. \tag{11.216}$$

And, in (11.136) $(x'' = x)$, take $X := x' \vee y$, to obtain:
$(x' \vee y)'' = x' \vee y$, which, by (11.215), becomes:

$$(x \wedge y')' = x' \vee y. \tag{11.217}$$

Then, in (11.178) $(1 \wedge x' = (0 \vee x)')$, take $X := x \wedge y'$, to obtain:
$1 \wedge (x \wedge y')' = (0 \vee (x \wedge y'))'$, which, by (11.217), becomes:
$1 \wedge (x' \vee y) = (0 \vee (x \wedge y'))'$, i.e.

$$(0 \vee (x \wedge y'))' = 1 \wedge (x' \vee y). \tag{11.218}$$

Now, in (11.217), take $X := a$, $Y := 1 \wedge x$, to obtain:
$(a \wedge (1 \wedge x)')' = a' \vee (1 \wedge x)$, which, by (11.203), becomes:
$(0 \vee (a \wedge x'))' = a' \vee (1 \wedge x)$, which, by (11.218), becomes:

$$1 \wedge (a' \vee x) = a' \vee (1 \wedge x). \tag{11.219}$$

Now, in (11.169) $(0 \wedge (-x + z) = -x + (x \wedge z))$, take $X := a$, $Z := (0 \vee x)'$, to obtain:

$0 \wedge (-a + (0 \vee x)') = -a + (a \wedge (0 \vee x)')$, which, by (11.200), becomes:
$0 \wedge (-a + (0 \vee x)') = -a + (a \wedge x')$, which, by (11.186), becomes:
$0 \wedge ((0 \vee x) + a)' = -a + (a \wedge x')$, which, by (Scomm), becomes:
$0 \wedge (a + (0 \vee x))' = -a + (a \wedge x')$, which, by (11.211), becomes:
$(1 \vee (a + (0 \vee x)))' = -a + (a \wedge x')$, which, by (11.216), becomes:
$(1 \vee (a + (0 \vee x)))' = ((a' \vee x) + a)'$, which, by (Scomm), becomes:
$(1 \vee (a + (0 \vee x)))' = (a + (a' \vee x))'$, which, by (11.181), becomes:

(11.220) $$(1 \vee (a + (0 \vee x)))' = (1 \vee (a + x))'.$$

Now, in (11.180), take $Y := -y$, to obtain:
$x + (-x \vee -y) = 0 \vee (x + -y)$, which, by (g5), becomes:
$x + -(x \wedge y) = 0 \vee (x + -y)$, i.e.

(11.221) $$0 \vee (x + -y) = x + -(x \wedge y).$$

Then, in (11.221), take $X := a$, $Y := (0 \vee x)'$, to obtain:
$0 \vee (a + -(0 \vee x)') = a + -(a \wedge (0 \vee x)')$, which, by (11.200), becomes:
$0 \vee (a + -(0 \vee x)') = a + -(a \wedge x')$, which, by (11.187), becomes:
$0 \vee -(a + (0 \vee x))' = a + -(a \wedge x')$, which, by (11.165), becomes:
$-(0 \wedge (a + (0 \vee x))') = a + -(a \wedge x')$, which, by (11.211), becomes:

(11.222) $$-(1 \vee (a + (0 \vee x)))' = a + -(a \wedge x').$$

Now, from (11.222), by (11.220), we obtain:
$-(1 \vee (a + x))' = a + -(a \wedge x')$, i.e.

(11.223) $$a + -(a \wedge x') = -(1 \vee (a + x))'.$$

Now, in (11.136) ($x'' = x$), take $X := 1 \wedge (a' + (1 \wedge (b + c))')$, to obtain:
$(1 \wedge (a' + (1 \wedge (b + c))'))'' = 1 \wedge (a' + (1 \wedge (b + c))')$, hence, by (11.138):
$T1' = 1 \wedge (a' + (1 \wedge (b + c))')$, i.e.

(11.224) $$1 \wedge (a' + (1 \wedge (b + c))') = T1'.$$

Now, in (11.172) ($x + (x' \wedge y) = 1 \wedge (x + y)$), take $X := a'$, $Y := (1 \wedge (b + c))'$, to obtain:
$a' + (a'' \wedge (1 \wedge (b + c))') = 1 \wedge (a' + (1 \wedge (b + c))')$, which, by (11.224), becomes:
$a' + (a'' \wedge (1 \wedge (b + c))') = T1'$, which, by (11.136), becomes:
$a' + (a \wedge (1 \wedge (b + c))') = T1'$, which, by (11.203), becomes:

(11.225) $$a' + (0 \vee (a \wedge (b + c)')) = T1'.$$

Now, in (11.153) ($-(-x + y) = x + -y$), take $Y := y \wedge (z + - - x)$, to obtain:
$-(-x + (y \wedge (z + - - x))) = x + -(y \wedge (z + - - x))$, which, by (11.173), becomes:
$-((-x + y) \wedge z) = x + -(y \wedge (z + - - x))$, which, by (DN), becomes:
$-((-x + y) \wedge z) = x + -(y \wedge (z + x))$, i.e.

(11.226) $$x + -(y \wedge (z + x)) = -((-x + y) \wedge z).$$

Now, in (11.192) $(x + (x + y)' = y')$, take $X := a'$, $Y := 0 \vee (a \wedge (b + c)')$, to obtain:
$a' + (a' + (0 \vee (a \wedge (b + c)')))' = (0 \vee (a \wedge (b + c)'))'$, which, by (11.225), becomes:
$a' + T1'' = (0 \vee (a \wedge (b + c)'))'$, which, by (11.136), becomes:
$a' + T1 = (0 \vee (a \wedge (b + c)'))'$, which, by (Scomm), becomes:
$T1 + a' = (0 \vee (a \wedge (b + c)'))'$, which, by (11.218), becomes:
$T1 + a' = 1 \wedge (a' \vee (b + c))$, which, by (11.219), becomes:
$T1 + a' = a' \vee (1 \wedge (b + c))$, i.e.

(11.227) $$a' \vee (1 \wedge (b + c)) = T1 + a'.$$

Now, in (11.215) $((x' \vee y)' = x \wedge y')$, take $X := a$, $Y := 1 \wedge (b + c)$, to obtain:
$(a' \vee (1 \wedge (b + c)))' = a \wedge (1 \wedge (b + c))'$, which, by (11.227), becomes:
$(T1 + a')' = a \wedge (1 \wedge (b + c))'$, which, by (11.191), becomes:
$a + -T1 = a \wedge (1 \wedge (b + c))'$, which, by (11.203), becomes:
$a + -T1 = 0 \vee (a \wedge (b + c)')$, i.e.

(11.228) $$0 \vee (a \wedge (b + c)') = a + -T1.$$

Now, in (11.226) $(x + -(y \wedge (z + x)) = -((-x + y) \wedge z))$, take $X := b + c$, $Y := 1$, $Z := a$, to obtain:
$(b + c) + -(1 \wedge (a + (b + c))) = -((-(b + c) + 1) \wedge a)$, which, by (11.199), becomes:
$(b + c) + -(1 \wedge (a + c)) = -((-(b + c) + 1) \wedge a)$, which, by (Sass), becomes:
$b + (c + -(1 \wedge (a + c))) = -((-(b + c) + 1) \wedge a)$, which, by (11.226), becomes:
$b + -((-c + 1) \wedge a) = -((-(b + c) + 1) \wedge a)$, which, by (Scomm), becomes:
$b + -((1 + -c) \wedge a) = -((-(b + c) + 1) \wedge a)$, which, by (11.159), becomes:
$b + -((- - c)' \wedge a) = -((-(b + c) + 1) \wedge a)$, which, by (DN), becomes:
$b + -(c' \wedge a) = -((-(b + c) + 1) \wedge a)$, which, by (Wcomm), becomes:
$b + -(a \wedge c') = -((-(b + c) + 1) \wedge a)$, which, by (Scomm), becomes:
$b + -(a \wedge c') = -((1 + -(b + c)) \wedge a)$, which, by (11.159), becomes:
$b + -(a \wedge c') = -((- - (b + c))' \wedge a)$, which, by (DN), becomes:
$b + -(a \wedge c') = -((b + c)' \wedge a)$, which, by (Wcomm), becomes:

(11.229) $$b + -(a \wedge c') = -(a \wedge (b + c)').$$

Now, in (11.189) $(x + (-y)' = (-(x + y))')$, take $X := b$, $Y := -(a \wedge c')$, to obtain:
$b + (- - (a \wedge c'))' = (-(b + -(a \wedge c')))'$, which, by (11.229), becomes:
$b + (- - (a \wedge c'))' = (- - (a \wedge (b + c)'))'$, which, by (DN) on both sides, becomes:
$b + (a \wedge c')' = (a \wedge (b + c)')'$, which, by (11.217) on both sides, becomes:
$b + (a' \vee c) = a' \vee (b + c)$, which, by (Vcomm), becomes:
$b + (c \vee a') = a' \vee (b + c)$, i.e.

(11.230) $$a' \vee (b + c) = b + (c \vee a').$$

Now, in (11.181) $(x + (x' \vee y) = 1 \vee (x + y))$, take $X := a$, $Y := b + c$, to obtain:
$a + (a' \vee (b + c)) = 1 \vee (a + (b + c))$, which, by (11.230), becomes:

$a + (b + (c \vee a')) = 1 \vee (a + (b + c))$, which, by (11.183), becomes:
$b + (1 \vee (a + c)) = 1 \vee (a + (b + c))$, i.e.

$$(11.231) \qquad 1 \vee (a + (b + c)) = b + (1 \vee (a + c)).$$

Now, in (11.196) $(1 \wedge ((a + b) \wedge x) = a \wedge x)$, take $X := (a + b) + x$ to obtain:
$1 \wedge ((a + b) \wedge ((a + b) + x)) = a \wedge ((a + b) + x)$, which, by (11.167) for $X := a + b$, becomes:
$1 \wedge ((a + b) + (0 \wedge x)) = a \wedge ((a + b) + x)$, which, by (Sass) twice, becomes:
$1 \wedge (a + (b + (0 \wedge x))) = a \wedge (a + (b + x))$, which, by (11.167), becomes:

$$(11.232) \qquad 1 \wedge (a + (b + (0 \wedge x))) = a + (0 \wedge (b + x)).$$

Finally, in (11.232), take $X := -(a \wedge c')$ to obtain:
$1 \wedge (a + (b + (0 \wedge -(a \wedge c')))) = a + (0 \wedge (b + -(a \wedge c')))$, which, by (11.229), becomes:
$1 \wedge (a + (b + (0 \wedge -(a \wedge c')))) = a + (0 \wedge -(a \wedge (b + c)'))$, which, by (11.164), becomes:
$1 \wedge (a + (b + -(0 \vee (a \wedge c')))) = a + (0 \wedge -(a \wedge (b + c)'))$, which, by (11.209), becomes:
$1 \wedge (a + (b + -(a \wedge c'))) = a + (0 \wedge -(a \wedge (b + c)'))$, which, by (11.229), becomes:
$1 \wedge (a + -(a \wedge (b + c)')) = a + (0 \wedge -(a \wedge (b + c)'))$, which, by (11.223), becomes:
$1 \wedge -(1 \vee (a + (b + c)))' = a + (0 \wedge -(a \wedge (b + c)'))$, which, by (11.231), becomes:
$1 \wedge -(b + (1 \vee (a + c)))' = a + (0 \wedge -(a \wedge (b + c)'))$, which, by (11.214), becomes:
$T2 = a + (0 \wedge -(a \wedge (b + c)'))$, which, by (11.164), becomes:
$T2 = a + -(0 \vee (a \wedge (b + c)'))$, which, by (11.228), becomes:
$T2 = a + -(a + -T1)$, which, by (11.152), becomes:
$T2 = a + (-a + T1)$, which, by (11.147), becomes:
$T2 = T1$, i.e.
$T1 = T2$. □

11.3.5 The 'hybrid' proof of Proposition 3.4.14

Proposition Let $\mathcal{G}_u = (G, \vee, \wedge, +, -, 0_G, u = u_G)$ be a commutative l-group with strong unit $u = u_G$, with $x \leq y \iff x \wedge y = x \iff x \vee y = y$.
Then, the following property holds (see Remark 3.4.3):
(s5^R) (V^{10}) For all $0_G \leq a_0, a_1, b_0 \leq u_G$, if $u_G \wedge (a_0 + a_1) = a_0$, then:
$(a_0 + a_1) + b_0 = ((u_G \wedge (a_0 + b_0)) + (u_G + -(u_G \wedge ((u_G + -a_0) + (u_G + -(u_G \wedge (a_1 + b_0))))))) + (u_G + -(u_G \wedge ((u_G + -a_1) + (u_G + -b_0))))$.

Proof. (By Prover9, Length of proof was 54, in 1.83 seconds; the renumbered expanded proof had the Length 261.)

In order to simplify the writting (less space on a line), we shall replace a_0 by a, a_1 by b, b_0 by c, 0_G by 0 and u_G by 1.

Hence, we have to prove that, for all $0 \leq a, b, c \leq 1$, if $1 \wedge (a + b) = a$, then:
$(a + b) + c = ((1 \wedge (a + c)) + (1 + -(1 \wedge ((1 + -a) + (1 + -(1 \wedge (b + c))))))) + (1 + -(1 \wedge ((1 + -b) + (1 + -c))))$.

We shall denote the right side by $\mathbf{T_m}$:
$T \overset{notation}{=} ((1 \wedge (a + c)) + (1 + -(1 \wedge ((1 + -a) + (1 + -(1 \wedge (b + c))))))) + (1 + -(1 \wedge$

$((1 + -b) + (1 + -c))))$. So, we must prove that $T = (a + b) + c$.

The particular hypotheses used by PROVER9 (among the given ones - see (α)) are:

$$0 \wedge a = 0 \quad (i.e.\ 0 \leq a), \tag{11.233}$$

$$0 \wedge c = 0 \quad (i.e.\ 0 \leq c) \tag{11.234}$$

and

$$1 \wedge (a + b) = a. \tag{11.235}$$

The following properties are obvious, by (SU), (Scomm), (Sass), (m_0-Re), (NegS) and (Wcomm) ($x \wedge y = y \wedge x$), (Wass) ($(x \wedge y) \wedge z = x \wedge (y \wedge z)$):

$$x + (y + z) = y + (x + z), \tag{11.236}$$

$$x + (-x + y) = y, \tag{11.237}$$

$$x + (y + -x) = y, \tag{11.238}$$

$$x + (y + (-x + z)) = y + z, \tag{11.239}$$

$$x + (y + (z + -x)) = y + z, \tag{11.240}$$

$$x + (y + (z + (-x + u))) = y + (z + u), \tag{11.241}$$

$$x + (y + (z + (u + -x))) = y + (z + u), \tag{11.242}$$

$$x \wedge (y \wedge z) = y \wedge (x \wedge z). \tag{11.243}$$

Now, in (g3) ($(x + y) \wedge (x + z) = x + (y \wedge z)$), take $Y := 0$, to obtain, by (SU):

$$x \wedge (x + z) = x + (0 \wedge z). \tag{11.244}$$

Then, in (g3) again, take $Z := -x$, to obtain:
$(x + y) \wedge (x + -x) = x + (y \wedge -x)$,
hence, by (m_0-Re):

$$0 \wedge (x + y) = x + (y \wedge -x). \tag{11.245}$$

Again in (g3), take now $Z := -x + z$, to obtain:
$(x + y) \wedge (x + (-x + z)) = x + (y \wedge (-x + z))$,
which, by (11.237), becomes:

$$(x + y) \wedge z = x + (y \wedge (-x + z)). \tag{11.246}$$

Now, in (11.244), take $Z := a$, to obtain:
$x \wedge (x + a) = x + (0 \wedge a)$,
which, by (11.233) and (SU), (Scomm), becomes:

$$x \wedge (a + x) = x. \tag{11.247}$$

Then, in (11.244) again, take $Z := c$, to obtain, similarly, by (11.234):

$$x \wedge (c + x) = x. \tag{11.248}$$

Now, in (11.247), take $X := x + -a$, to obtain:
$(x + -a) \wedge (a + (x + -a)) = x + -a$,
which, by (11.238), becomes:
$(x + -a) \wedge x = x + -a$,
hence:

$$x \wedge (x + -a) = x + -a. \tag{11.249}$$

Now, in (11.243), take $Z := c + x$, to obtain:
$x \wedge (y \wedge (c + x)) = y \wedge (x \wedge (c + x))$,
which, by (11.248), becomes:

$$x \wedge (y \wedge (c + x)) = y \wedge x. \tag{11.250}$$

Now, in (Wass) ($(x \wedge y) \wedge z = x \wedge (y \wedge z)$), take $Z := (x \wedge y) + z$, to obtain:
$(x \wedge y) \wedge ((x \wedge y) + z) = x \wedge (y \wedge ((x \wedge y) + z))$,
which, by (11.244) for $X := x \wedge y$, becomes:
$(x \wedge y) + (0 \wedge z) = x \wedge (y \wedge ((x \wedge y) + z))$,
hence:

$$x \wedge (y \wedge ((x \wedge y) + z)) = (x \wedge y) + (0 \wedge z). \tag{11.251}$$

Now, in (11.244), take $Z := -x + y$, to obtain:
$x \wedge (x + (-x + y)) = x + (0 \wedge (-x + y))$,
which, by (11.237), becomes:

$$x \wedge y = x + (0 \wedge (-x + y)). \tag{11.252}$$

If now, in (11.252), we take $X := -x$, we obtain, by (DN), (Scomm):

$$-x \wedge y = -x + (0 \wedge (y + x)). \tag{11.253}$$

And, if in (11.252) again, we interchange x with y, we obtain, by (Wcomm):

$$x \wedge y = y + (0 \wedge (-y + x)). \tag{11.254}$$

Now, in (11.245), interchange x with y, to obtain, by (Scomm):

$$0 \wedge (x+y) = y + (x \wedge -y). \tag{11.255}$$

Then, in (11.245) again, take $Y := y+z$, to obtain:
$0 \wedge (x+(y+z)) = x + ((y+z) \wedge -x)$,
which, by (11.246), becomes:
$0 \wedge (x+(y+z)) = x + (y + (z \wedge (-y + -x)))$,
which, by (11.236), (Scomm), becomes:

$$0 \wedge (x+(y+z)) = y + (x + (z \wedge (-x + -y))). \tag{11.256}$$

Then, in (11.245) again, take $X := -x$, to obtain, by (DN):

$$0 \wedge (-x+y) = -x + (y \wedge x). \tag{11.257}$$

Now, in (11.250), take $Y := c + (0 \wedge x)$, to obtain:
$x \wedge ((c+(0 \wedge x)) \wedge (c+x)) = (c+(0 \wedge x)) \wedge x$,
which, by (Wass), becomes:
$(x \wedge (c+(0 \wedge x))) \wedge (c+x) = (c+(0 \wedge x)) \wedge x$,
which, by (11.244), becomes:
$(x \wedge (c \wedge (c+x))) \wedge (c+x) = (c+(0 \wedge x)) \wedge x$,
which, by (11.243), becomes:
$(c \wedge (x \wedge (c+x))) \wedge (c+x) = (c+(0 \wedge x)) \wedge x$,
which, by (11.248), becomes:
$(c \wedge x) \wedge (c+x) = (c+(0 \wedge x)) \wedge x$,
which, by (Wass) again, becomes:
$c \wedge (x \wedge (c+x)) = (c+(0 \wedge x)) \wedge x$,
which, by (11.248) again, becomes, by (Wcomm):

$$c \wedge x = x \wedge (c + (0 \wedge x)). \tag{11.258}$$

Now, in (11.252), take $Y := c + (0 \wedge x)$, to obtain:
$x \wedge (c+(0 \wedge x)) = x + (0 \wedge (-x + (c + (0 \wedge x))))$,
which, by above (11.258) on the left side, becomes:
$c \wedge x = x + (0 \wedge (-x + (c + (0 \wedge x))))$,
which, by (11.236) on the right side, becomes:

$$c \wedge x = x + (0 \wedge (c + (-x + (0 \wedge x)))). \tag{11.259}$$

Now, in (11.251), take $Y := x + -a$, $Z := y$, to obtain:
$x \wedge ((x+-a) \wedge ((x \wedge (x+-a)) + y)) = (x \wedge (x+-a)) + (0 \wedge y)$,
which, by (11.249) on both sides, becomes:
$x \wedge ((x+-a) \wedge ((x+-a)+y)) = (x+-a) + (0 \wedge y)$,
which, by (11.244) on the left side, for $X := x + -a$, becomes:
$x \wedge ((x+-a) + (0 \wedge y)) = (x+-a) + (0 \wedge y)$,
which, by (Sass) on both sides, becomes:

$$x \wedge (x + (-a + (0 \wedge y))) = x + (-a + (0 \wedge y)). \tag{11.260}$$

Finally, in (11.257), take $X := a+b$, $Y := 1$, to obtain:
$0 \wedge (-(a+b)+1) = -(a+b) + (1 \wedge (a+b))$,
which, by (11.235), becomes:
$0 \wedge (-(a+b)+1) = -(a+b) + a$,
which, by (NegS) on both sides, becomes:
$0 \wedge ((-b+-a)+1) = (-b+-a)+a$,
which, by (Scomm) on both sides, becomes:
$0 \wedge (1+(-b+-a)) = a+(-b+-a)$,
which, by (11.238), (Scomm), becomes:

$$0 \wedge (1+(-a+-b)) = -b. \tag{11.261}$$

Then, we have (by copying the corresponding lines from the expanded renumbered PROVER9 proof):

$T = ((1 \wedge (a+c)) + (1 + -(1 \wedge ((1+-a) + (1+-(1 \wedge (b+c))))))) + (1 + -(1 \wedge ((1+-b) + (1+-c))))$

$16 = ((1 \wedge (a+c)) + (1 + -(1 \wedge (1 + (-a + (1 + -(1 \wedge (b+c)))))))) + (1 + -(1 \wedge ((1+-b) + (1+-c))))$, by (Sass),
$17 = ((1 \wedge (a+c)) + (1 + -(1 \wedge (1 + (-a + (1 + -(1 \wedge (b+c)))))))) + (1 + -(1 \wedge (1 + (-b + (1+-c)))))$, by (Sass),
$18 = (1 + -(1 \wedge (1 + (-b + (1+-c))))) + ((1 \wedge (a+c)) + (1 + -(1 \wedge (1 + (-a + (1 + -(1 \wedge (b+c))))))))$, by (Scomm),
$19 = 1 + (-(1 \wedge (1 + (-b + (1+-c)))) + ((1 \wedge (a+c)) + (1 + -(1 \wedge (1 + (-a + (1 + -(1 \wedge (b+c)))))))))$, by (Sass),

$25 = 1 + (-(1 \wedge (1 + (1 + (-b+-c)))) + ((1 \wedge (a+c)) + (1 + -(1 \wedge (1 + (-a + (1 + -(1 \wedge (b+c)))))))))$, by (11.236),
$26 = 1 + (-(1 \wedge (1 + (1 + (-b+-c)))) + ((1 \wedge (a+c)) + (1 + -(1 \wedge (1 + (1 + (-a + -(1 \wedge (b+c)))))))))$, by (11.236),
$27 = 1 + (-(1 \wedge (1 + (1 + (-b+-c)))) + (1 + ((1 \wedge (a+c)) + -(1 \wedge (1 + (1 + (-a + -(1 \wedge (b+c)))))))))$, by (11.236),
$28 = 1 + (1 + (-(1 \wedge (1 + (1 + (-b+-c)))) + ((1 \wedge (a+c)) + -(1 \wedge (1 + (1 + (-a + -(1 \wedge (b+c)))))))))$, by (11.236),
$29 = 1 + (1 + ((1 \wedge (a+c)) + (-(1 \wedge (1 + (1 + (-b+-c)))) + -(1 \wedge (1 + (1 + (-a + -(1 \wedge (b+c)))))))))$, by (11.236),

$92 = 1 + (1 + ((1 \wedge (a+c)) + (-((1 + (1 + (-b+-c))) + (0 \wedge (-(1 + (1 + (-b+-c))) + 1))) + -(1 \wedge (1 + (1 + (-a + -(1 \wedge (b+c)))))))))$, by (11.254),
$93 = 1 + (1 + ((1 \wedge (a+c)) + (-((1 + (1 + (-b+-c))) + (0 \wedge ((-(1 + (-b+-c)) + -1) + 1))) + -(1 \wedge (1 + (1 + (-a + -(1 \wedge (b+c)))))))))$, by (NegS),
$94 = 1 + (1 + ((1 \wedge (a+c)) + (-((1 + (1 + (-b+-c))) + (0 \wedge (((-(-b+-c) + -1) + -1) + 1))) + -(1 \wedge (1 + (1 + (-a + -(1 \wedge (b+c)))))))))$, by (NegS),
$95 = 1 + (1 + ((1 \wedge (a+c)) + (-((1 + (1 + (-b+-c))) + (0 \wedge ((((--c+--b) + -1) + -1) + 1))) + -(1 \wedge (1 + (1 + (-a + -(1 \wedge (b+c)))))))))$, by (NegS),
$96 = 1 + (1 + ((1 \wedge (a+c)) + (-((1 + (1 + (-b+-c))) + (0 \wedge ((((c+--b) + -1) +$

$-1)+1)))+-(1\wedge(1+(1+(-a+-(1\wedge(b+c)))))))))$, by (DN),
$97=1+(1+((1\wedge(a+c))+(-((1+(1+(-b+-c)))+(0\wedge((((c+b)+-1)+$
$-1)+1)))+-(1\wedge(1+(1+(-a+-(1\wedge(b+c)))))))))$, by (DN),
$98=1+(1+((1\wedge(a+c))+(-((1+(1+(-b+-c)))+(0\wedge((((b+c)+-1)+$
$-1)+1)))+-(1\wedge(1+(1+(-a+-(1\wedge(b+c)))))))))$, by (Scomm),
$99=1+(1+((1\wedge(a+c))+(-((1+(1+(-b+-c)))+(0\wedge(((b+(c+-1))+$
$-1)+1)))+-(1\wedge(1+(1+(-a+-(1\wedge(b+c)))))))))$, by (Sass),
$100=1+(1+((1\wedge(a+c))+(-((1+(1+(-b+-c)))+(0\wedge((-1+(b+(c+$
$-1)))+1)))+-(1\wedge(1+(1+(-a+-(1\wedge(b+c)))))))))$, by (Scomm),
$101=1+(1+((1\wedge(a+c))+(-((1+(1+(-b+-c)))+(0\wedge((b+(-1+(c+$
$-1)))+1)))+-(1\wedge(1+(1+(-a+-(1\wedge(b+c)))))))))$, by (11.236),
$102=1+(1+((1\wedge(a+c))+(-((1+(1+(-b+-c)))+(0\wedge((b+(c+(-1+$
$-1)))+1)))+-(1\wedge(1+(1+(-a+-(1\wedge(b+c)))))))))$, by (11.236),
$103=1+(1+((1\wedge(a+c))+(-((1+(1+(-b+-c)))+(0\wedge(1+(b+(c+(-1+$
$-1))))))+-(1\wedge(1+(1+(-a+-(1\wedge(b+c)))))))))$, by (Scomm),
$104=1+(1+((1\wedge(a+c))+(-((1+(1+(-b+-c)))+(0\wedge(b+(c+-1))))+$
$-(1\wedge(1+(1+(-a+-(1\wedge(b+c)))))))))$, by (11.242),
$105=1+(1+((1\wedge(a+c))+(-(1+((1+(-b+-c))+(0\wedge(b+(c+-1)))))+$
$-(1\wedge(1+(1+(-a+-(1\wedge(b+c)))))))))$, by (Sass),
$106=1+(1+((1\wedge(a+c))+(-(1+(1+((-b+-c)+(0\wedge(b+(c+-1))))))+$
$-(1\wedge(1+(1+(-a+-(1\wedge(b+c)))))))))$, by (Sass),
$107=1+(1+((1\wedge(a+c))+(-(1+(1+(-b+(-c+(0\wedge(b+(c+-1)))))))+$
$-(1\wedge(1+(1+(-a+-(1\wedge(b+c)))))))))$, by (Sass),
$108=1+(1+((1\wedge(a+c))+((-(1+(-b+(-c+(0\wedge(b+(c+-1))))))+-1)+$
$-(1\wedge(1+(1+(-a+-(1\wedge(b+c)))))))))$, by (NegS),
$109=1+(1+((1\wedge(a+c))+(((-(-b+(-c+(0\wedge(b+(c+-1)))))+-1)+-1)+$
$-(1\wedge(1+(1+(-a+-(1\wedge(b+c)))))))))$, by (NegS),
$110=1+(1+((1\wedge(a+c))+((((-(-c+(0\wedge(b+(c+-1))))+--b)+-1)+$
$-1)+-(1\wedge(1+(1+(-a+-(1\wedge(b+c)))))))))$, by (NegS),
$111=1+(1+((1\wedge(a+c))+(((((-(0\wedge(b+(c+-1)))+--c)+--b)+-1)+$
$-1)+-(1\wedge(1+(1+(-a+-(1\wedge(b+c)))))))))$, by (NegS),
$112=1+(1+((1\wedge(a+c))+(((((-(0\wedge(b+(c+-1)))+c)+--b)+-1)+-1)+$
$-(1\wedge(1+(1+(-a+-(1\wedge(b+c)))))))))$, by (DN),
$113=1+(1+((1\wedge(a+c))+(((((c+-(0\wedge(b+(c+-1))))+--b)+-1)+-1)+$
$-(1\wedge(1+(1+(-a+-(1\wedge(b+c)))))))))$, by (Scomm),
$114=1+(1+((1\wedge(a+c))+(((((c+-(0\wedge(b+(c+-1))))+b)+-1)+-1)+$
$-(1\wedge(1+(1+(-a+-(1\wedge(b+c)))))))))$, by (DN),
$115=1+(1+((1\wedge(a+c))+((((b+(c+-(0\wedge(b+(c+-1)))))+-1)+-1)+$
$-(1\wedge(1+(1+(-a+-(1\wedge(b+c)))))))))$, by (Scomm),
$116=1+(1+((1\wedge(a+c))+(((-1+(b+(c+-(0\wedge(b+(c+-1))))))+-1)+$
$-(1\wedge(1+(1+(-a+-(1\wedge(b+c)))))))))$, by (Scomm),
$117=1+(1+((1\wedge(a+c))+(((b+(-1+(c+-(0\wedge(b+(c+-1))))))+-1)+$
$-(1\wedge(1+(1+(-a+-(1\wedge(b+c)))))))))$, by (11.236),
$118=1+(1+((1\wedge(a+c))+(((b+(c+(-1+-(0\wedge(b+(c+-1))))))+-1)+$
$-(1\wedge(1+(1+(-a+-(1\wedge(b+c)))))))))$, by (11.236),
$119=1+(1+((1\wedge(a+c))+((-1+(b+(c+(-1+-(0\wedge(b+(c+-1)))))))+$

$-(1 \wedge (1 + (1 + (-a + -(1 \wedge (b + c))))))))$, by (Scomm),
$120 = 1 + (1 + ((1 \wedge (a + c)) + ((b + (-1 + (c + (-1 + -(0 \wedge (b + (c + -1))))))) + -(1 \wedge (1 + (1 + (-a + -(1 \wedge (b + c))))))))$, by (11.236),
$121 = 1 + (1 + ((1 \wedge (a + c)) + ((b + (c + (-1 + (-1 + -(0 \wedge (b + (c + -1))))))) + -(1 \wedge (1 + (1 + (-a + -(1 \wedge (b + c))))))))$, by (11.236),

$122 = 1 + (1 + ((1 \wedge (a + c)) + (-(1 \wedge (1 + (1 + (-a + -(1 \wedge (b + c)))))) + (b + (c + (-1 + (-1 + -(0 \wedge (b + (c + -1))))))))))$, by (Scomm),
$123 = 1 + (1 + ((1 \wedge (a + c)) + (b + (-(1 \wedge (1 + (1 + (-a + -(1 \wedge (b + c)))))) + (c + (-1 + (-1 + -(0 \wedge (b + (c + -1))))))))))$, by (11.236),
$124 = 1 + (1 + ((1 \wedge (a + c)) + (b + (c + (-(1 \wedge (1 + (1 + (-a + -(1 \wedge (b + c)))))) + (-1 + (-1 + -(0 \wedge (b + (c + -1))))))))))$, by (11.236),
$125 = 1 + (1 + ((1 \wedge (a + c)) + (b + (c + (-1 + (-(1 \wedge (1 + (1 + (-a + -(1 \wedge (b + c)))))) + (-1 + -(0 \wedge (b + (c + -1))))))))))$, by (11.236),
$126 = 1 + (1 + ((1 \wedge (a + c)) + (b + (c + (-1 + (-1 + (-(1 \wedge (1 + (1 + (-a + -(1 \wedge (b + c)))))) + -(0 \wedge (b + (c + -1))))))))))$, by (11.236),
$127 = 1 + (1 + (b + ((1 \wedge (a + c)) + (c + (-1 + (-1 + (-(1 \wedge (1 + (1 + (-a + -(1 \wedge (b + c)))))) + -(0 \wedge (b + (c + -1))))))))))$, by (11.236),
$128 = 1 + (1 + (b + (c + ((1 \wedge (a + c)) + (-1 + (-1 + (-(1 \wedge (1 + (1 + (-a + -(1 \wedge (b + c)))))) + -(0 \wedge (b + (c + -1))))))))))$, by (11.236),
$129 = 1 + (1 + (b + (c + (-1 + ((1 \wedge (a + c)) + (-1 + (-(1 \wedge (1 + (1 + (-a + -(1 \wedge (b + c)))))) + -(0 \wedge (b + (c + -1))))))))))$, by (11.236),
$130 = 1 + (1 + (b + (c + (-1 + (-1 + ((1 \wedge (a + c)) + (-(1 \wedge (1 + (1 + (-a + -(1 \wedge (b + c)))))) + -(0 \wedge (b + (c + -1))))))))))$, by (11.236),
$131 = 1 + (b + (c + (-1 + ((1 \wedge (a + c)) + (-(1 \wedge (1 + (1 + (-a + -(1 \wedge (b + c)))))) + -(0 \wedge (b + (c + -1))))))))$, by (11.241),
$132 = b + (c + ((1 \wedge (a + c)) + (-(1 \wedge (1 + (1 + (-a + -(1 \wedge (b + c)))))) + -(0 \wedge (b + (c + -1))))))$, by (11.241),

$133 = b + (c + ((1 \wedge (a + c)) + (-(1 \wedge (1 + (1 + (-a + -(1 \wedge (b + c)))))) + -(b + ((c + -1) \wedge -b)))))$, by (11.245),
$134 = b + (c + ((1 \wedge (a + c)) + (-(1 \wedge (1 + (1 + (-a + -(1 \wedge (b + c)))))) + -(b + (c + (-1 \wedge (-c + -b)))))))$, by (11.246),
$135 = b + (c + ((1 \wedge (a + c)) + (-(1 \wedge (1 + (1 + (-a + -(1 \wedge (b + c)))))) + -(b + (c + (-1 \wedge (-b + -c)))))))$, by (Scomm),
$136 = b + (c + ((1 \wedge (a + c)) + (-(1 \wedge (1 + (1 + (-a + -(1 \wedge (b + c)))))) + -(b + (c + (-1 + (0 \wedge ((-b + -c) + 1))))))))$, by (11.253),
$137 = b + (c + ((1 \wedge (a + c)) + (-(1 \wedge (1 + (1 + (-a + -(1 \wedge (b + c)))))) + -(b + (c + (-1 + (0 \wedge (1 + (-b + -c)))))))))$, by (Scomm),
$138 = b + (c + ((1 \wedge (a + c)) + (-(1 \wedge (1 + (1 + (-a + -(1 \wedge (b + c)))))) + (-(c + (-1 + (0 \wedge (1 + (-b + -c))))) + -b))))$, by (NegS),
$139 = b + (c + ((1 \wedge (a + c)) + (-(1 \wedge (1 + (1 + (-a + -(1 \wedge (b + c)))))) + ((-(-1 + (0 \wedge (1 + (-b + -c)))) + -c) + -b))))$, by (NegS),
$140 = b + (c + ((1 \wedge (a + c)) + (-(1 \wedge (1 + (1 + (-a + -(1 \wedge (b + c)))))) + (((-(0 \wedge (1 + (-b + -c))) + --1) + -c) + -b))))$, by (NegS),
$141 = b + (c + ((1 \wedge (a + c)) + (-(1 \wedge (1 + (1 + (-a + -(1 \wedge (b + c)))))) + (((-(0 \wedge$

$(1 + (-b + -c))) + 1) + -c) + -b)))$, by (DN),
$142 = b + (c + ((1 \wedge (a + c)) + (-(1 \wedge (1 + (1 + (-a + -(1 \wedge (b + c)))))) + (((1 + -(0 \wedge (1 + (-b + -c)))) + -c) + -b))))$, by (Scomm),
$143 = b + (c + ((1 \wedge (a + c)) + (-(1 \wedge (1 + (1 + (-a + -(1 \wedge (b + c)))))) + ((-c + (1 + -(0 \wedge (1 + (-b + -c))))) + -b))))$, by (Scomm),
$144 = b + (c + ((1 \wedge (a + c)) + (-(1 \wedge (1 + (1 + (-a + -(1 \wedge (b + c)))))) + ((1 + (-c + -(0 \wedge (1 + (-b + -c))))) + -b))))$, by (11.236),
$145 = b + (c + ((1 \wedge (a + c)) + (-(1 \wedge (1 + (1 + (-a + -(1 \wedge (b + c)))))) + (-b + (1 + (-c + -(0 \wedge (1 + (-b + -c)))))))))$, by (Scomm),
$146 = b + (c + ((1 \wedge (a + c)) + (-(1 \wedge (1 + (1 + (-a + -(1 \wedge (b + c)))))) + (1 + (-b + (-c + -(0 \wedge (1 + (-b + -c)))))))))$, by (11.236),
$147 = b + (c + ((1 \wedge (a + c)) + (1 + (-(1 \wedge (1 + (1 + (-a + -(1 \wedge (b + c)))))) + (-b + (-c + -(0 \wedge (1 + (-b + -c)))))))))$, by (11.236),
$148 = b + (c + ((1 \wedge (a + c)) + (1 + (-b + (-(1 \wedge (1 + (1 + (-a + -(1 \wedge (b + c)))))) + (-c + -(0 \wedge (1 + (-b + -c)))))))))$, by (11.236),
$149 = b + (c + ((1 \wedge (a + c)) + (1 + (-b + (-c + (-(1 \wedge (1 + (1 + (-a + -(1 \wedge (b + c)))))) + -(0 \wedge (1 + (-b + -c)))))))))$, by (11.236),
$150 = b + (c + (1 + ((1 \wedge (a + c)) + (-b + (-c + (-(1 \wedge (1 + (1 + (-a + -(1 \wedge (b + c)))))) + -(0 \wedge (1 + (-b + -c)))))))))$, by (11.236),
$151 = b + (c + (1 + (-b + ((1 \wedge (a + c)) + (-c + (-(1 \wedge (1 + (1 + (-a + -(1 \wedge (b + c)))))) + -(0 \wedge (1 + (-b + -c)))))))))$, by (11.236),
$152 = b + (c + (1 + (-b + (-c + ((1 \wedge (a + c)) + (-(1 \wedge (1 + (1 + (-a + -(1 \wedge (b + c)))))) + -(0 \wedge (1 + (-b + -c)))))))))$, by (11.236),
$153 = b + (1 + (c + (-b + (-c + ((1 \wedge (a + c)) + (-(1 \wedge (1 + (1 + (-a + -(1 \wedge (b + c)))))) + -(0 \wedge (1 + (-b + -c)))))))))$, by (11.236),

$154 = b + (1 + (-b + ((1 \wedge (a + c)) + (-(1 \wedge (1 + (1 + (-a + -(1 \wedge (b + c)))))) + -(0 \wedge (1 + (-b + -c)))))))$, by (11.239),
$155 = 1 + (b + (-b + ((1 \wedge (a + c)) + (-(1 \wedge (1 + (1 + (-a + -(1 \wedge (b + c)))))) + -(0 \wedge (1 + (-b + -c)))))))$, by (11.236),
$156 = 1 + ((1 \wedge (a + c)) + (-(1 \wedge (1 + (1 + (-a + -(1 \wedge (b + c)))))) + -(0 \wedge (1 + (-b + -c)))))$, by (11.237),

$157 = 1 + ((1 \wedge (a + c)) + (-(1 \wedge (1 + (1 + (-a + -(1 \wedge (b + c)))))) + -((-b + -c) + (1 \wedge -(-b + -c)))))$, by (11.255),
$158 = 1 + ((1 \wedge (a + c)) + (-(1 \wedge (1 + (1 + (-a + -(1 \wedge (b + c)))))) + -((-b + -c) + (1 \wedge (- - c + - - b)))))$, by (NegS),
$159 = 1 + ((1 \wedge (a + c)) + (-(1 \wedge (1 + (1 + (-a + -(1 \wedge (b + c)))))) + -((-b + -c) + (1 \wedge (c + - - b)))))$, by (DN),
$160 = 1 + ((1 \wedge (a + c)) + (-(1 \wedge (1 + (1 + (-a + -(1 \wedge (b + c)))))) + -((-b + -c) + (1 \wedge (c + b)))))$, by (DN),
$161 = 1 + ((1 \wedge (a + c)) + (-(1 \wedge (1 + (1 + (-a + -(1 \wedge (b + c)))))) + -((-b + -c) + (1 \wedge (b + c)))))$, by (Scomm),
$162 = 1 + ((1 \wedge (a + c)) + (-(1 \wedge (1 + (1 + (-a + -(1 \wedge (b + c)))))) + -(-b + (-c + (1 \wedge (b + c))))))$, by (Sass),
$163 = 1 + ((1 \wedge (a + c)) + (-(1 \wedge (1 + (1 + (-a + -(1 \wedge (b + c)))))) + (-(-c + (1 \wedge$

$(b+c))) + --b))),$ by (NegS),
$164 = 1+((1\wedge(a+c))+(-(1\wedge(1+(1+(-a+-(1\wedge(b+c))))))+((-(1\wedge(b+c))+--c)+--b))),$ by (NegS),
$165 = 1+((1\wedge(a+c))+(-(1\wedge(1+(1+(-a+-(1\wedge(b+c))))))+((-(1\wedge(b+c))+c)+--b))),$ by (DN),
$166 = 1+((1\wedge(a+c))+(-(1\wedge(1+(1+(-a+-(1\wedge(b+c))))))+((c+-(1\wedge(b+c)))+--b))),$ by (Scomm),
$167 = 1+((1\wedge(a+c))+(-(1\wedge(1+(1+(-a+-(1\wedge(b+c))))))+((c+-(1\wedge(b+c)))+b))),$ by (DN),
$168 = 1+((1\wedge(a+c))+(-(1\wedge(1+(1+(-a+-(1\wedge(b+c))))))+(b+(c+-(1\wedge(b+c)))))),$ by (Scomm),
$169 = 1+((1\wedge(a+c))+((b+(c+-(1\wedge(b+c))))+-(1\wedge(1+(1+(-a+-(1\wedge(b+c)))))))),$ by (Scomm),
$170 = 1+((1\wedge(a+c))+(b+((c+-(1\wedge(b+c)))+-(1\wedge(1+(1+(-a+-(1\wedge(b+c)))))))),$ by (Sass),
$171 = 1+((1\wedge(a+c))+(b+(c+(-(1\wedge(b+c))+-(1\wedge(1+(1+(-a+-(1\wedge(b+c)))))))))),$ by (Sass),
$172 = 1+(b+((1\wedge(a+c))+(c+(-(1\wedge(b+c))+-(1\wedge(1+(1+(-a+-(1\wedge(b+c)))))))))),$ by (11.236),
$173 = 1+(b+(c+((1\wedge(a+c))+(-(1\wedge(b+c))+-(1\wedge(1+(1+(-a+-(1\wedge(b+c)))))))))),$ by (11.236),

$174 = 1+(b+(c+((1\wedge(a+c))+(-(1\wedge(b+c))+-((1+(1+(-a+-(1\wedge(b+c)))))+(0\wedge(-(1+(1+(-a+-(1\wedge(b+c)))))+1))))))),$ by (11.254),
$175 = 1+(b+(c+((1\wedge(a+c))+(-(1\wedge(b+c))+-((1+(1+(-a+-(1\wedge(b+c)))))+(0\wedge((-(1+(-a+-(1\wedge(b+c))))+-1)+1))))))),$ by (NegS),
$176 = 1+(b+(c+((1\wedge(a+c))+(-(1\wedge(b+c))+-((1+(1+(-a+-(1\wedge(b+c)))))+(0\wedge(((-(-a+-(1\wedge(b+c)))+-1)+-1)+1))))))),$ by (NegS),
$177 = 1+(b+(c+((1\wedge(a+c))+(-(1\wedge(b+c))+-((1+(1+(-a+-(1\wedge(b+c)))))+(0\wedge((((--(1\wedge(b+c))+--a)+-1)+-1)+1))))))),$ by (NegS),
$178 = 1+(b+(c+((1\wedge(a+c))+(-(1\wedge(b+c))+-((1+(1+(-a+-(1\wedge(b+c)))))+(0\wedge(((((1\wedge(b+c))+--a)+-1)+-1)+1))))))),$ by (DN),
$179 = 1+(b+(c+((1\wedge(a+c))+(-(1\wedge(b+c))+-((1+(1+(-a+-(1\wedge(b+c)))))+(0\wedge(((((1\wedge(b+c))+a)+-1)+-1)+1))))))),$ by (DN),
$180 = 1+(b+(c+((1\wedge(a+c))+(-(1\wedge(b+c))+-((1+(1+(-a+-(1\wedge(b+c)))))+(0\wedge((((a+(1\wedge(b+c)))+-1)+-1)+1))))))),$ by (Scomm),
$181 = 1+(b+(c+((1\wedge(a+c))+(-(1\wedge(b+c))+-((1+(1+(-a+-(1\wedge(b+c)))))+(0\wedge(((-1+(a+(1\wedge(b+c))))+-1)+1))))))),$ by (Scomm),
$182 = 1+(b+(c+((1\wedge(a+c))+(-(1\wedge(b+c))+-((1+(1+(-a+-(1\wedge(b+c)))))+(0\wedge(((a+(-1+(1\wedge(b+c))))+-1)+1))))))),$ by (11.236),
$183 = 1+(b+(c+((1\wedge(a+c))+(-(1\wedge(b+c))+-((1+(1+(-a+-(1\wedge(b+c)))))+(0\wedge((-1+(a+(-1+(1\wedge(b+c)))))+1))))))),$ by (Scomm),
$184 = 1+(b+(c+((1\wedge(a+c))+(-(1\wedge(b+c))+-((1+(1+(-a+-(1\wedge(b+c)))))+(0\wedge((a+(-1+(-1+(1\wedge(b+c)))))+1))))))),$ by (11.236),
$185 = 1+(b+(c+((1\wedge(a+c))+(-(1\wedge(b+c))+-((1+(1+(-a+-(1\wedge(b+c)))))+(0\wedge(1+(a+(-1+(-1+(1\wedge(b+c))))))))))))),$ by (Scomm),

$186 = 1 + (b + (c + ((1 \wedge (a + c)) + (-(1 \wedge (b + c)) + -((1 + (1 + (-a + -(1 \wedge (b + c))))) + (0 \wedge (a + (-1 + (1 \wedge (b + c))))))))))$, by (11.241),
$187 = 1 + (b + (c + ((1 \wedge (a + c)) + (-(1 \wedge (b + c)) + -(1 + ((1 + (-a + -(1 \wedge (b + c)))) + (0 \wedge (a + (-1 + (1 \wedge (b + c))))))))))))$, by (Sass),
$188 = 1 + (b + (c + ((1 \wedge (a + c)) + (-(1 \wedge (b + c)) + -(1 + (1 + ((-a + -(1 \wedge (b + c))) + (0 \wedge (a + (-1 + (1 \wedge (b + c)))))))))))))$, by (Sass),
$189 = 1 + (b + (c + ((1 \wedge (a + c)) + (-(1 \wedge (b + c)) + -(1 + (1 + (-a + (-(1 \wedge (b + c)) + (0 \wedge (a + (-1 + (1 \wedge (b + c))))))))))))))$, by (Sass),
$190 = 1 + (b + (c + ((1 \wedge (a + c)) + (-(1 \wedge (b + c)) + (-(1 + (-a + (-(1 \wedge (b + c)) + (0 \wedge (a + (-1 + (1 \wedge (b + c))))))))) + -1)))))$, by (NegS),
$191 = 1 + (b + (c + ((1 \wedge (a + c)) + (-(1 \wedge (b + c)) + ((-(-a + (-(1 \wedge (b + c)) + (0 \wedge (a + (-1 + (1 \wedge (b + c))))))) + -1) + -1)))))$, by (NegS),
$192 = 1 + (b + (c + ((1 \wedge (a + c)) + (-(1 \wedge (b + c)) + (((-(-(1 \wedge (b + c)) + (0 \wedge (a + (-1 + (1 \wedge (b + c)))))) + - - a) + -1) + -1)))))$, by (NegS),
$193 = 1 + (b + (c + ((1 \wedge (a + c)) + (-(1 \wedge (b + c)) + ((((-(0 \wedge (a + (-1 + (1 \wedge (b + c))))) + - - (1 \wedge (b + c))) + - - a) + -1) + -1)))))$, by(NegS),
$194 = 1 + (b + (c + ((1 \wedge (a + c)) + (-(1 \wedge (b + c)) + ((((-(0 \wedge (a + (-1 + (1 \wedge (b + c))))) + (1 \wedge (b + c))) + - - a) + -1) + -1)))))$, by (DN),
$195 = 1 + (b + (c + ((1 \wedge (a + c)) + (-(1 \wedge (b + c)) + (((((1 \wedge (b + c)) + -(0 \wedge (a + (-1 + (1 \wedge (b + c)))))) + - - a) + -1) + -1)))))$, by (Scomm),
$196 = 1 + (b + (c + ((1 \wedge (a + c)) + (-(1 \wedge (b + c)) + (((((1 \wedge (b + c)) + -(0 \wedge (a + (-1 + (1 \wedge (b + c)))))) + a) + -1) + -1)))))$, by (DN),
$197 = 1 + (b + (c + ((1 \wedge (a + c)) + (-(1 \wedge (b + c)) + (((a + ((1 \wedge (b + c)) + -(0 \wedge (a + (-1 + (1 \wedge (b + c))))))) + -1) + -1)))))$, by (Scomm),
$198 = 1 + (b + (c + ((1 \wedge (a + c)) + (-(1 \wedge (b + c)) + ((-1 + (a + ((1 \wedge (b + c)) + -(0 \wedge (a + (-1 + (1 \wedge (b + c)))))))) + -1)))))$, by (Scomm),
$199 = 1 + (b + (c + ((1 \wedge (a + c)) + (-(1 \wedge (b + c)) + ((a + (-1 + ((1 \wedge (b + c)) + -(0 \wedge (a + (-1 + (1 \wedge (b + c)))))))) + -1)))))$, by (11.236),
$200 = 1 + (b + (c + ((1 \wedge (a + c)) + (-(1 \wedge (b + c)) + (-1 + (a + (-1 + ((1 \wedge (b + c)) + -(0 \wedge (a + (-1 + (1 \wedge (b + c)))))))))))))$, by (Scomm),
$201 = 1 + (b + (c + ((1 \wedge (a + c)) + (-(1 \wedge (b + c)) + (a + (-1 + (-1 + ((1 \wedge (b + c)) + -(0 \wedge (a + (-1 + (1 \wedge (b + c)))))))))))))$, by (11.236),
$202 = 1 + (b + (c + ((1 \wedge (a + c)) + (a + (-(1 \wedge (b + c)) + (-1 + (-1 + ((1 \wedge (b + c)) + -(0 \wedge (a + (-1 + (1 \wedge (b + c)))))))))))))$, by (11.236),
$203 = 1 + (b + (c + ((1 \wedge (a + c)) + (a + (-1 + (-(1 \wedge (b + c)) + (-1 + ((1 \wedge (b + c)) + -(0 \wedge (a + (-1 + (1 \wedge (b + c)))))))))))))$, by (11.236),
$204 = 1 + (b + (c + ((1 \wedge (a + c)) + (a + (-1 + (-1 + (-(1 \wedge (b + c)) + ((1 \wedge (b + c)) + -(0 \wedge (a + (-1 + (1 \wedge (b + c)))))))))))))$, by (11.236),
$205 = 1 + (b + (c + ((1 \wedge (a + c)) + (a + (-1 + (-1 + ((1 \wedge (b + c)) + (-(1 \wedge (b + c)) + -(0 \wedge (a + (-1 + (1 \wedge (b + c)))))))))))))$, by (11.236),
$206 = 1+(b+(c+((1\wedge(a+c))+(a+(-1+(-1+-(0\wedge(a+(-1+(1\wedge(b+c)))))))))))$, by (11.237),
$207 = 1+(b+(c+(a+((1\wedge(a+c))+(-1+(-1+-(0\wedge(a+(-1+(1\wedge(b+c)))))))))))$, by (11.236),
$208 = 1+(b+(c+(a+(-1+((1\wedge(a+c))+(-1+-(0\wedge(a+(-1+(1\wedge(b+c)))))))))))$, by (11.236),

$209 = 1+(b+(c+(a+(-1+(-1+((1\wedge(a+c))+-(0\wedge(a+(-1+(1\wedge(b+c)))))))))))$, by (11.236),
$210 = 1+(b+(a+(c+(-1+(-1+((1\wedge(a+c))+-(0\wedge(a+(-1+(1\wedge(b+c)))))))))))$, by (11.236),
$211 = 1+(a+(b+(c+(-1+(-1+((1\wedge(a+c))+-(0\wedge(a+(-1+(1\wedge(b+c)))))))))))$, by (11.236),

$220 = 1+(a+(b+(c+(-1+(-1+((1\wedge(a+c))+-(-1+(a+((1\wedge(b+c))\wedge(-a+--1))))))))))$, by (11.256),
$221 = 1+(a+(b+(c+(-1+(-1+((1\wedge(a+c))+-(-1+(a+((1\wedge(b+c))\wedge(-a+1))))))))))$, by (DN),
$222 = 1+(a+(b+(c+(-1+(-1+((1\wedge(a+c))+-(-1+(a+((1\wedge(b+c))\wedge(1+-a))))))))))$, by (Scomm),
$223 = 1+(a+(b+(c+(-1+(-1+((1\wedge(a+c))+-(-1+(a+((1+-a)\wedge(1\wedge(b+c)))))))))))$, by (Wcomm),
$224 = 1+(a+(b+(c+(-1+(-1+((1\wedge(a+c))+-(-1+(a+(1\wedge((1+-a)\wedge(b+c)))))))))))$, by (11.243),
$225 = 1+(a+(b+(c+(-1+(-1+((1\wedge(a+c))+-(-1+(a+(1\wedge((b+c)\wedge(1+-a)))))))))))$, by (Wcomm),
$226 = 1+(a+(b+(c+(-1+(-1+((1\wedge(a+c))+-(-1+(a+(1\wedge(b+(c\wedge(-b+(1+-a)))))))))))))$, by (11.246),
$227 = 1+(a+(b+(c+(-1+(-1+((1\wedge(a+c))+-(-1+(a+(1\wedge(b+(c\wedge((1+-a)+-b))))))))))))$, by (Scomm),
$228 = 1+(a+(b+(c+(-1+(-1+((1\wedge(a+c))+-(-1+(a+(1\wedge(b+(c\wedge(1+(-a+-b)))))))))))))$, by (Sass),
$229 = 1+(a+(b+(c+(-1+(-1+((1\wedge(a+c))+-(-1+(a+(1\wedge(b+((1+(-a+-b))+(0\wedge(c+(-(1+(-a+-b))+(0\wedge(1+(-a+-b)))))))))))))))))$, by (11.259),
$230 = 1+(a+(b+(c+(-1+(-1+((1\wedge(a+c))+-(-1+(a+(1\wedge(b+((1+(-a+-b))+(0\wedge(c+((-(-a+-b)+-1)+(0\wedge(1+(-a+-b)))))))))))))))))$, by (NegS),
$231 = 1+(a+(b+(c+(-1+(-1+((1\wedge(a+c))+-(-1+(a+(1\wedge(b+((1+(-a+-b))+(0\wedge(c+(((--b+--a)+-1)+(0\wedge(1+(-a+-b)))))))))))))))))$, by (NegS),
$232 = 1+(a+(b+(c+(-1+(-1+((1\wedge(a+c))+-(-1+(a+(1\wedge(b+((1+(-a+-b))+(0\wedge(c+(((b+--a)+-1)+(0\wedge(1+(-a+-b)))))))))))))))))$, by (DN),
$233 = 1+(a+(b+(c+(-1+(-1+((1\wedge(a+c))+-(-1+(a+(1\wedge(b+((1+(-a+-b))+(0\wedge(c+(((b+a)+-1)+(0\wedge(1+(-a+-b)))))))))))))))))$, by (DN),
$234 = 1+(a+(b+(c+(-1+(-1+((1\wedge(a+c))+-(-1+(a+(1\wedge(b+((1+(-a+-b))+(0\wedge(c+(((a+b)+-1)+(0\wedge(1+(-a+-b)))))))))))))))))$, by (Scomm),
$235 = 1+(a+(b+(c+(-1+(-1+((1\wedge(a+c))+-(-1+(a+(1\wedge(b+((1+(-a+-b))+(0\wedge(c+((a+(b+-1))+(0\wedge(1+(-a+-b)))))))))))))))))$, by (Sass),
$236 = 1+(a+(b+(c+(-1+(-1+((1\wedge(a+c))+-(-1+(a+(1\wedge(b+((1+(-a+-b))+(0\wedge(c+((a+(b+-1))+-b))))))))))))))$, by (11.261),
$237 = 1+(a+(b+(c+(-1+(-1+((1\wedge(a+c))+-(-1+(a+(1\wedge(b+((1+$

$(-a + -b)) + (0 \wedge (c + (a + ((b + -1) + -b)))))))))))))))$, by (Sass),
$238 = 1 + (a + (b + (c + (-1 + (-1 + ((1 \wedge (a + c)) + -(-1 + (a + (1 \wedge (b + ((1 + (-a + -b)) + (0 \wedge (c + (a + (b + (-1 + -b))))))))))))))))$, by (Sass),
$239 = 1 + (a + (b + (c + (-1 + (-1 + ((1 \wedge (a + c)) + -(-1 + (a + (1 \wedge (b + ((1 + (-a + -b)) + (0 \wedge (c + (a + -1))))))))))))))$, by (11.238),
$240 = 1 + (a + (b + (c + (-1 + (-1 + ((1 \wedge (a + c)) + -(-1 + (a + (1 \wedge (b + ((1 + (-a + -b)) + (0 \wedge (a + (c + -1))))))))))))))$, by (11.236),
$241 = 1 + (a + (b + (c + (-1 + (-1 + ((1 \wedge (a + c)) + -(-1 + (a + (1 \wedge (b + (1 + ((-a + -b) + (0 \wedge (a + (c + -1)))))))))))))))$, by (Sass),
$242 = 1 + (a + (b + (c + (-1 + (-1 + ((1 \wedge (a + c)) + -(-1 + (a + (1 \wedge (b + (1 + (-a + (-b + (0 \wedge (a + (c + -1))))))))))))))))$, by (Sass),
$243 = 1 + (a + (b + (c + (-1 + (-1 + ((1 \wedge (a + c)) + -(-1 + (a + (1 \wedge (1 + (b + (-a + (-b + (0 \wedge (a + (c + -1))))))))))))))))$, by (11.236),
$244 = 1 + (a + (b + (c + (-1 + (-1 + ((1 \wedge (a + c)) + -(-1 + (a + (1 \wedge (1 + (-a + (0 \wedge (a + (c + -1))))))))))))))$, by (11.239),
$245 = 1 + (a + (b + (c + (-1 + (-1 + ((1 \wedge (a + c)) + -(-1 + (a + (1 + (-a + (0 \wedge (a + (c + -1)))))))))))))$, by (11.260),
$246 = 1 + (a + (b + (c + (-1 + (-1 + ((1 \wedge (a + c)) + -(-1 + (1 + (a + (-a + (0 \wedge (a + (c + -1)))))))))))))$, by (11.236),
$247 = 1 + (a + (b + (c + (-1 + (-1 + ((1 \wedge (a + c)) + -(-1 + (1 + (0 \wedge (a + (c + -1)))))))))))$, by (11.237),
$248 = 1 + (a + (b + (c + (-1 + (-1 + ((1 \wedge (a + c)) + -(-1 + (1 \wedge (1 + (a + (c + -1)))))))))))$, by (11.244),
$249 = 1 + (a + (b + (c + (-1 + (-1 + ((1 \wedge (a + c)) + -(-1 + (1 \wedge (a + c)))))))))$, by (11.240),
$250 = 1 + (a + (b + (c + (-1 + (-1 + ((1 \wedge (a + c)) + (-(1 \wedge (a + c)) + - - 1)))))))$, by (NegS),
$251 = 1 + (a + (b + (c + (-1 + (-1 + ((1 \wedge (a + c)) + (-(1 \wedge (a + c)) + 1)))))))$, by (DN),
$252 = 1 + (a + (b + (c + (-1 + (-1 + ((1 \wedge (a + c)) + (1 + -(1 \wedge (a + c)))))))))$, by (Scomm),
$253 = 1 + (a + (b + (c + (-1 + (-1 + (1 + ((1 \wedge (a + c)) + -(1 \wedge (a + c)))))))))$, by (11.236),
$254 = 1 + (a + (b + (c + (-1 + (-1 + (1 + 0))))))$, by ($m_0$-Re),
$255 = 1 + (a + (b + (c + (-1 + (-1 + 1)))))$, by (SU),
$256 = 1 + (a + (b + (c + (-1 + (1 + -1)))))$, by (Scomm),
$257 = 1 + (a + (b + (c + (-1 + 0))))$, by ($m_0$-Re),
$258 = 1 + (a + (b + (c + (0 + -1))))$, by (Scomm),
$259 = 1 + (a + (b + (c + -1)))$, by (SU),
$260 = a + (b + c)$, by (11.242),
$= (a + b) + c$, by (Sass). □

11.3.6 The 'hybrid' proof of Proposition 3.4.16

Proposition Let $\mathcal{G}_u = (G, \vee, \wedge, +, -, 0_G, u = u_G)$ be a commutative l-group with strong unit $u = u_G$.

Then, for all $0_G \leq a_0, a_1, b_0 \leq u_G$, we have (see $(s5^R)$):

$$(V^{00}) \implies (V^{10}).$$

Proof. (By PROVER9, Length of proof was 73, in 2.27 seconds; the renumbered expanded proof had the Length 327.)

Recall that (V^{00}) is: $a_0 + b_0 = (u_G \wedge (a_0 + b_0)) + (u_G + -(u_G \wedge ((u_G + -a_0) + (u_G + -b_0))))$
and
(V^{10}) is: if $u_G \wedge (a_0 + a_1) = a_0$, then:
$(a_0 + a_1) + b_0 = ((u_G \wedge (a_0 + b_0)) + (u_G + -(u_G \wedge ((u_G + -a_0) + (u_G + -(u_G \wedge (a_1 + b_0))))))) + (u_G + -(u_G \wedge ((u_G + -a_1) + (u_G + -b_0))))$.

In order to simplify the writting (less space on a line), we shall replace a_0 by a, a_1 by b, b_0 by c, 0_G by 0 and u_G by 1.

Hence, we have to prove that, for all $0 \leq a, b, c \leq 1$,
if (V^{00}) holds, i.e.

$$a + c = (1 \wedge (a + c)) + (1 + -(1 \wedge ((1 + -a) + (1 + -c)))), \tag{11.262}$$

then
(V^{10}) holds, i.e. if

$$1 \wedge (a + b) = a, \tag{11.263}$$

then:

$$(a + b) + c = \tag{11.264}$$

$$((1\wedge(a+c))+(1+-(1\wedge((1+-a)+(1+-(1\wedge(b+c)))))))+(1+-(1\wedge((1+-b)+(1+-c)))).$$

We shall denote the right side by H:
$H \overset{notation}{=} ((1 \wedge (a + c)) + (1 + -(1 \wedge ((1 + -a) + (1 + -(1 \wedge (b + c)))))))+ (1 + -(1 \wedge ((1 + -b) + (1 + -c))))$. So, we must prove that $H = (a + b) + c$.

The particular hypotheses used by PROVER9 (among the given ones - see (α)) are:

$$0 \wedge a = 0 \quad (i.e.\ 0 \leq a), \tag{11.265}$$

$$0 \wedge c = 0 \quad (i.e.\ 0 \leq c). \tag{11.266}$$

The following properties are obvious, by (SU), (Scomm), (Sass), (m_0-Re), (NegS) and (Wcomm) ($x \wedge y = y \wedge x$), (Wass) ($(x \wedge y) \wedge z = x \wedge (y \wedge z)$):

$$x + (y + z) = y + (x + z), \tag{11.267}$$

$$x + (-x + y) = y, \tag{11.268}$$

(11.269) $$x + (y + -x) = y,$$

(11.270) $$x + (y + (-x + z)) = y + z,$$

(11.271) $$x + (y + (z + (-x + u))) = y + (z + u),$$

(11.272) $$x + (y + (z + (u + -x))) = y + (z + u),$$

(11.273) $$x + -(-y + x) = y,$$

(11.274) $$x + -(x + y) = -y,$$

(11.275) $$x \wedge (y \wedge z) = y \wedge (x \wedge z).$$

Now, in (g3) ($(x + y) \wedge (x + z) = x + (y \wedge z)$), take $Y := 0$, to obtain, by (SU):

(11.276) $$x \wedge (x + z) = x + (0 \wedge z).$$

Then, in (g3) again, interchange x with y, to obtain:
$(y + x) \wedge (y + z) = y + (x \wedge z)$,
which, by (Scomm), becomes:

(11.277) $$(x + y) \wedge (y + z) = y + (x \wedge z).$$

Then, in (g3) again, take $Z := -x$, to obtain, by (m_0-Re):
$(x + y) \wedge 0 = x + (y \wedge -x)$,
which, by (Wcomm), becomes:

(11.278) $$0 \wedge (x + y) = x + (y \wedge -x).$$

Then, in (g3) ($(x + y) \wedge (x + z) = x + (y \wedge z)$) again, take $Z := -x + z$, to obtain:
$(x + y) \wedge (x + (-x + z)) = x + (y \wedge (-x + z))$,
which, by (11.268), becomes:

(11.279) $$(x + y) \wedge z = x + (y \wedge (-x + z)).$$

Now, in (Cp) ($x \leq y \Longleftrightarrow x + z \leq y + z$), i.e. in

(11.280) $$x \wedge y = x \Longleftrightarrow (x + z) \wedge (y + z) = x + z,$$

take first $X := 0$, $Y := a$, to obtain:
$0 \wedge a = 0 \Longleftrightarrow (0 + z) \wedge (a + z) = 0 + z$,
which, by (SU), becomes:

$0 \wedge a = 0 \Longleftrightarrow z \wedge (a+z) = z$;
but, since $0 \wedge a = 0$ is always true, by (11.265), it follows that we have:

$$z \wedge (a+z) = z; \tag{11.281}$$

then, second, take (in (11.280)) $X := 0$, $Y := c$, to obtain similarly, by (11.266):

$$z \wedge (c+z) = z. \tag{11.282}$$

Now, in (11.281), take $Z := -a + x$, to obtain:
$(-a+x) \wedge (a + (-a+x)) = -a + x$,
which, by (11.268), becomes:
$(-a+x) \wedge x = -a + x$,
which, by (Wcomm), becomes:

$$x \wedge (-a+x) = -a + x. \tag{11.283}$$

Now, in (11.279), take $Z := y + z$, to obtain:
$(x+y) \wedge (y+z) = x + (y \wedge (-x + (y+z)))$,
which, by (11.277), becomes:
$y + (x \wedge z) = x + (y \wedge (-x + (y+z)))$,
hence:

$$x + (y \wedge (-x + (y+z))) = y + (x \wedge z). \tag{11.284}$$

Now, in (11.275), take $Z := c + y$, to obtain:
$x \wedge (y \wedge (c+y)) = y \wedge (x \wedge (c+y))$,
which, by (11.282), becomes:

$$x \wedge y = y \wedge (x \wedge (c+y)). \tag{11.285}$$

Now, in (11.276), take $Z := -x + y$, to obtain:
$x \wedge (x + (-x+y)) = x + (0 \wedge (-x+y))$,
which, by (11.268), becomes:

$$x \wedge y = x + (0 \wedge (-x+y)). \tag{11.286}$$

Then, in (11.286), take $X := -x$, to obtain, by (DN):
$-x \wedge y = -x + (0 \wedge (x+y))$,
hence, by (Scomm):

$$-x \wedge y = -x + (0 \wedge (y+x)). \tag{11.287}$$

Then, in (11.286) again, interchange x with y to obtain, by (Wcomm):

$$x \wedge y = y + (0 \wedge (-y+x)). \tag{11.288}$$

Now, in (11.278), interchange x with y, to obtain:
$0 \wedge (y+x) = y + (x \wedge -y)$,
which, by (Scomm), becomes:

$$0 \wedge (x+y) = y + (x \wedge -y). \tag{11.289}$$

Then, in (11.278) again, take $X := y$, $Y := x + z$, to obtain:
$0 \wedge (y + (x + z)) = y + ((x + z) \wedge -y)$,
which, by (11.267) on the left side and by (11.279) on the right side, becomes:

$$0 \wedge (x + (y + z)) = y + (x + (z \wedge (-x + -y))). \tag{11.290}$$

Now, in (Wass) ($(x \wedge y) \wedge z = x \wedge (y \wedge z)$), take $Y := -a + x$, $Z := y$, to obtain:
$(x \wedge (-a + x)) \wedge y = x \wedge ((-a + x) \wedge y)$,
which, by (11.283) on left side, becomes:
$(-a + x) \wedge y = x \wedge ((-a + x) \wedge y)$,
which, by (11.279) on left side, becomes, by (DN):
$-a + (x \wedge (a + y)) = x \wedge ((-a + x) \wedge y)$,
which, by (11.279) on right side, becomes:
$-a + (x \wedge (a + y)) = x \wedge (-a + (x \wedge (-(-a) + y)))$,
which, by (DN), becomes:

$$x \wedge (-a + (x \wedge (a + y))) = -a + (x \wedge (a + y)). \tag{11.291}$$

Now, in (11.285), take $X := c$, to obtain:
$c \wedge y = y \wedge (c \wedge (c + y))$,
which, by (11.276), becomes:
(*) $c \wedge x = x \wedge (c + (0 \wedge x))$;
then, in (11.286), take $Y := c + (0 \wedge x)$, to obtain:
$x \wedge (c + (0 \wedge x)) = x + (0 \wedge (-x + (c + (0 \wedge x))))$,
which, by (*) on the left side, becomes:
$c \wedge x = x + (0 \wedge (-x + (c + (0 \wedge x))))$,
which, by (11.267), becomes:

$$c \wedge x = x + (0 \wedge (c + (-x + (0 \wedge x)))). \tag{11.292}$$

Finally, in (11.278), take $X := -x$, to obtain, by (DN):

$$0 \wedge (-x + y) = -x + (y \wedge x). \tag{11.293}$$

Then, in (11.293), take $X := a + b$, $Y := 1$, to obtain:
$0 \wedge (-(a + b) + 1) = -(a + b) + (1 \wedge (a + b))$,
which, by the hypothesis (11.263) of (V^{10}), becomes:
$0 \wedge (-(a + b) + 1) = -(a + b) + a$,
which, by (NegS), becomes:
$0 \wedge ((-b + -a) + 1) = (-b + -a) + a$,
which, by (Scomm) on both sides, becomes:
$0 \wedge (1 + (-b + -a)) = a + (-b + -a)$,
which, by (Scomm) and (11.269), becomes:

$$0 \wedge (1 + (-a + -b)) = -b. \tag{11.294}$$

Now, from (V^{00}), i.e. from (11.262), we obtain:
$(1 \wedge (a + c)) + (1 + -(1 \wedge (1 + (-a + (1 + -c))))) = a + c$, by (Sass),

$(1 \wedge (a+c)) + (1 + -(1 \wedge (1 + (1 + (-a + -c))))) = a + c$, by (11.267), and, by (11.267) again:

$$1 + ((1 \wedge (a+c)) + -(1 \wedge (1 + (1 + (-a + -c))))) = a + c. \tag{11.295}$$

Then, in (Sass) $((x+y)+z = x+(y+z))$, take $X := 1$, $Y := (1 \wedge (a+c)) + -(1 \wedge (1 + (1 + (-a + -c))))$, $Z := x$, to obtain:
$(1 + ((1 \wedge (a+c)) + -(1 \wedge (1 + (1 + (-a + -c)))))) + x$
$= 1 + (((1 \wedge (a+c)) + -(1 \wedge (1 + (1 + (-a + -c))))) + x)$,
which, by above (11.295), becomes:
$(a+c) + x = 1 + (((1 \wedge (a+c)) + -(1 \wedge (1 + (1 + (-a + -c))))) + x)$,
which, by (Sass) on both sides, becomes:
$a + (c+x) = 1 + ((1 \wedge (a+c)) + (-(1 \wedge (1 + (1 + (-a + -c)))) + x))$,
i.e.

$$1 + ((1 \wedge (a+c)) + (-(1 \wedge (1 + (1 + (-a + -c)))) + x)) = a + (c+x). \tag{11.296}$$

Then, in (11.289), take $X := 1$,
$Y := (1 \wedge (a+c)) + (-(1 \wedge (1 + (1 + (-a + -c)))) + x)$, to obtain:
$0 \wedge (1 + ((1 \wedge (a+c)) + (-(1 \wedge (1 + (1 + (-a + -c)))) + x))) =$
$((1 \wedge (a+c)) + (-(1 \wedge (1 + (1 + (-a + -c)))) + x)) +$
$(1 \wedge -((1 \wedge (a+c)) + (-(1 \wedge (1 + (1 + (-a + -c)))) + x)))$,
which, by (11.296) on left side, becomes:
$0 \wedge (a + (c+x)) =$
$((1 \wedge (a+c)) + (-(1 \wedge (1 + (1 + (-a + -c)))) + x)) +$
$(1 \wedge -((1 \wedge (a+c)) + (-(1 \wedge (1 + (1 + (-a + -c)))) + x)))$,
which, by (NegS), becomes:
$0 \wedge (a + (c+x)) =$
$((1 \wedge (a+c)) + (-(1 \wedge (1 + (1 + (-a + -c)))) + x)) +$
$(1 \wedge (-(-(1 \wedge (1 + (1 + (-a + -c)))) + x) + -(1 \wedge (a+c))))$,
which, by (NegS) again, becomes:
$0 \wedge (a + (c+x)) =$
$((1 \wedge (a+c)) + (-(1 \wedge (1 + (1 + (-a + -c)))) + x)) +$
$(1 \wedge ((-x + - - (1 \wedge (1 + (1 + (-a + -c))))) + -(1 \wedge (a+c))))$,
which, by (DN), becomes:
$0 \wedge (a + (c+x)) =$
$((1 \wedge (a+c)) + (-(1 \wedge (1 + (1 + (-a + -c)))) + x)) +$
$(1 \wedge ((-x + (1 \wedge (1 + (1 + (-a + -c))))) + -(1 \wedge (a+c))))$,
which, by (Scomm), becomes:
$0 \wedge (a + (c+x)) =$
$((1 \wedge (a+c)) + (-(1 \wedge (1 + (1 + (-a + -c)))) + x)) +$
$(1 \wedge (-(1 \wedge (a+c)) + (-x + (1 \wedge (1 + (1 + (-a + -c)))))))$,
which, by (Sass), becomes:
$0 \wedge (a + (c+x)) =$
$(1 \wedge (a+c)) + ((-(1 \wedge (1 + (1 + (-a + -c)))) + x) +$
$(1 \wedge (-(1 \wedge (a+c)) + (-x + (1 \wedge (1 + (1 + (-a + -c))))))))$,

which, by (Sass) again, becomes:

$$0 \wedge (a + (c + x)) = (1 \wedge (a + c)) + (-(1 \wedge (1 + (1 + (-a + -c))))+ (x + (1 \wedge (-(1 \wedge (a + c)) + (-x + (1 \wedge (1 + (1 + (-a + -c))))))))). \tag{11.297}$$

Then, in (11.274) $(x + -(x + y) = -y)$, take
$X := 1$, $Y := (1 \wedge (a + c)) + -(1 \wedge (1 + (1 + (-a + -c))))$, to obtain:
$1 + -(1 + ((1 \wedge (a + c)) + -(1 \wedge (1 + (1 + (-a + -c))))))$
$= -((1 \wedge (a + c)) + -(1 \wedge (1 + (1 + (-a + -c)))))$,
which, by (11.295) on the left side, becomes:
$1 + -(a + c) = -((1 \wedge (a + c)) + -(1 \wedge (1 + (1 + (-a + -c)))))$,
which, by (NegS) and (Scomm) on the left side, becomes:
$1 + (-a + -c) = -((1 \wedge (a + c)) + -(1 \wedge (1 + (1 + (-a + -c)))))$,
which, by (NegS) on the right side now, becomes:
$-(-(1 \wedge (1 + (1 + (-a + -c))))) + -(1 \wedge (a + c)) = 1 + (-a + -c)$,
which, by (DN) and (Scomm), becomes:

$$-(1 \wedge (a + c)) + (1 \wedge (1 + (1 + (-a + -c)))) = 1 + (-a + -c). \tag{11.298}$$

Finally, we shall prove that (11.264) holds, i.e. that $H = (a + b) + c$. Indeed,

$H = ((1 \wedge (a + c)) + (1 + -(1 \wedge ((1 + -a) + (1 + -(1 \wedge (b + c))))))) + (1 + -(1 \wedge ((1 + -b) + (1 + -c))))$ 22 $= ((1 \wedge (a + c)) + (1 + -(1 \wedge ((1 + -a) + (1 + -(1 \wedge (c + b))))))) + (1 + -(1 \wedge ((1 + -b) + (1 + -c))))$, by (Scomm),
23 $= ((1 \wedge (a + c)) + (1 + -(1 \wedge (1 + (-a + (1 + -(1 \wedge (c + b)))))))) + (1 + -(1 \wedge ((1 + -b) + (1 + -c))))$, by (Sass),
24 $= ((1 \wedge (a + c)) + (1 + -(1 \wedge (1 + (-a + (1 + -(1 \wedge (c + b)))))))) + (1 + -(1 \wedge ((1 + -c) + (1 + -b))))$, by (Scomm),
25 $= ((1 \wedge (a + c)) + (1 + -(1 \wedge (1 + (-a + (1 + -(1 \wedge (c + b)))))))) + (1 + -(1 \wedge (1 + (-c + (1 + -b)))))$, by (Sass),
26 $= (1 + -(1 \wedge (1 + (-c + (1 + -b))))) + ((1 \wedge (a + c)) + (1 + -(1 \wedge (1 + (-a + (1 + -(1 \wedge (c + b))))))))$, by (Scomm),
27 $= 1 + (-(1 \wedge (1 + (-c + (1 + -b)))) + ((1 \wedge (a + c)) + (1 + -(1 \wedge (1 + (-a + (1 + -(1 \wedge (c + b)))))))))$, by (Sass),

32 $= 1 + (-(1 \wedge (1 + (1 + (-c + -b)))) + ((1 \wedge (a + c)) + (1 + -(1 \wedge (1 + (-a + (1 + -(1 \wedge (c + b)))))))))$, by (11.267),
33 $= 1 + (-(1 \wedge (1 + (1 + (-c + -b)))) + ((1 \wedge (a + c)) + (1 + -(1 \wedge (1 + (1 + (-a + -(1 \wedge (c + b)))))))))$, by (11.267),
34 $= 1 + (-(1 \wedge (1 + (1 + (-c + -b)))) + (1 + ((1 \wedge (a + c)) + -(1 \wedge (1 + (1 + (-a + -(1 \wedge (c + b)))))))))$, by (11.267),
35 $= 1 + (1 + (-(1 \wedge (1 + (1 + (-c + -b)))) + ((1 \wedge (a + c)) + -(1 \wedge (1 + (1 + (-a + -(1 \wedge (c + b)))))))))$, by (11.267),
36 $= 1 + (1 + ((1 \wedge (a + c)) + (-(1 \wedge (1 + (1 + (-c + -b)))) + -(1 \wedge (1 + (1 + (-a + -(1 \wedge (c + b)))))))))$, by (11.267),

$133 = 1 + (1 + ((1 \wedge (a + c)) + (-((1 + (1 + (-c + -b))) + (0 \wedge (-(1 + (1 + (-c + -b))) + 1))) + -(1 \wedge (1 + (1 + (-a + -(1 \wedge (c + b))))))))))$, by (11.288),
$134 = 1 + (1 + ((1 \wedge (a + c)) + (-((1 + (1 + (-c + -b))) + (0 \wedge ((-(1 + (-c + -b)) + -1) + 1))) + -(1 \wedge (1 + (1 + (-a + -(1 \wedge (c + b))))))))))$, by (NegS),
$135 = 1 + (1 + ((1 \wedge (a + c)) + (-((1 + (1 + (-c + -b))) + (0 \wedge (((-(-c + -b) + -1) + -1) + 1))) + -(1 \wedge (1 + (1 + (-a + -(1 \wedge (c + b))))))))))$, by (NegS),
$136 = 1 + (1 + ((1 \wedge (a + c)) + (-((1 + (1 + (-c + -b))) + (0 \wedge ((((- - b + - - c) + -1) + -1) + 1))) + -(1 \wedge (1 + (1 + (-a + -(1 \wedge (c + b))))))))))$, by (NegS),
$137 = 1 + (1 + ((1 \wedge (a + c)) + (-((1 + (1 + (-c + -b))) + (0 \wedge ((((b + - - c) + -1) + -1) + 1))) + -(1 \wedge (1 + (1 + (-a + -(1 \wedge (c + b))))))))))$, by (DN),
$138 = 1 + (1 + ((1 \wedge (a + c)) + (-((1 + (1 + (-c + -b))) + (0 \wedge ((((b + c) + -1) + -1) + 1))) + -(1 \wedge (1 + (1 + (-a + -(1 \wedge (c + b))))))))))$, by (DN),
$139 = 1 + (1 + ((1 \wedge (a + c)) + (-((1 + (1 + (-c + -b))) + (0 \wedge ((((c + b) + -1) + -1) + 1))) + -(1 \wedge (1 + (1 + (-a + -(1 \wedge (c + b))))))))))$, by (Scomm),
$140 = 1 + (1 + ((1 \wedge (a + c)) + (-((1 + (1 + (-c + -b))) + (0 \wedge (((c + (b + -1)) + -1) + 1))) + -(1 \wedge (1 + (1 + (-a + -(1 \wedge (c + b))))))))))$, by (Sass),
$141 = 1 + (1 + ((1 \wedge (a + c)) + (-((1 + (1 + (-c + -b))) + (0 \wedge ((-1 + (c + (b + -1))) + 1))) + -(1 \wedge (1 + (1 + (-a + -(1 \wedge (c + b))))))))))$, by (Scomm),
$142 = 1 + (1 + ((1 \wedge (a + c)) + (-((1 + (1 + (-c + -b))) + (0 \wedge ((c + (-1 + (b + -1))) + 1))) + -(1 \wedge (1 + (1 + (-a + -(1 \wedge (c + b))))))))))$, by (11.267),
$143 = 1 + (1 + ((1 \wedge (a + c)) + (-((1 + (1 + (-c + -b))) + (0 \wedge ((c + (b + (-1 + -1))) + 1))) + -(1 \wedge (1 + (1 + (-a + -(1 \wedge (c + b))))))))))$, by (11.267),
$144 = 1 + (1 + ((1 \wedge (a + c)) + (-((1 + (1 + (-c + -b))) + (0 \wedge (1 + (c + (b + (-1 + -1))))))) + -(1 \wedge (1 + (1 + (-a + -(1 \wedge (c + b))))))))))$, by (Scomm),
$145 = 1 + (1 + ((1 \wedge (a + c)) + (-((1 + (1 + (-c + -b))) + (0 \wedge (c + (b + -1))))) + -(1 \wedge (1 + (1 + (-a + -(1 \wedge (c + b))))))))))$, by (11.272),
$146 = 1 + (1 + ((1 \wedge (a + c)) + (-(1 + ((1 + (-c + -b)) + (0 \wedge (c + (b + -1)))))) + -(1 \wedge (1 + (1 + (-a + -(1 \wedge (c + b))))))))))$, by (Sass),
$147 = 1 + (1 + ((1 \wedge (a + c)) + (-(1 + (1 + ((-c + -b) + (0 \wedge (c + (b + -1))))))) + -(1 \wedge (1 + (1 + (-a + -(1 \wedge (c + b))))))))))$, by (Sass),
$148 = 1 + (1 + ((1 \wedge (a + c)) + (-(1 + (1 + (-c + (-b + (0 \wedge (c + (b + -1)))))))) + -(1 \wedge (1 + (1 + (-a + -(1 \wedge (c + b))))))))))$, by (Sass),
$149 = 1 + (1 + ((1 \wedge (a + c)) + ((-(1 + (-c + (-b + (0 \wedge (c + (b + -1)))))) + -1) + -(1 \wedge (1 + (1 + (-a + -(1 \wedge (c + b))))))))))$, by (NegS),
$150 = 1 + (1 + ((1 \wedge (a + c)) + (((-(-c + (-b + (0 \wedge (c + (b + -1))))) + -1) + -1) + -(1 \wedge (1 + (1 + (-a + -(1 \wedge (c + b))))))))))$, by (NegS),
$151 = 1 + (1 + ((1 \wedge (a + c)) + ((((-(-b + (0 \wedge (c + (b + -1)))) + - - c) + -1) + -1) + -(1 \wedge (1 + (1 + (-a + -(1 \wedge (c + b))))))))))$, by (NegS),
$152 = 1 + (1 + ((1 \wedge (a + c)) + (((((-(0 \wedge (c + (b + -1))) + - - b) + - - c) + -1) + -1) + -(1 \wedge (1 + (1 + (-a + -(1 \wedge (c + b))))))))))$, by (NegS),
$153 = 1 + (1 + ((1 \wedge (a + c)) + (((((-(0 \wedge (c + (b + -1))) + b) + - - c) + -1) + -1) + -(1 \wedge (1 + (1 + (-a + -(1 \wedge (c + b))))))))))$, by (DN),
$154 = 1 + (1 + ((1 \wedge (a + c)) + (((((b + -(0 \wedge (c + (b + -1)))) + - - c) + -1) + -1) + -(1 \wedge (1 + (1 + (-a + -(1 \wedge (c + b))))))))))$, by (Scomm),
$155 = 1 + (1 + ((1 \wedge (a + c)) + (((((b + -(0 \wedge (c + (b + -1)))) + c) + -1) + -1) + -(1 \wedge (1 + (1 + (-a + -(1 \wedge (c + b))))))))))$, by (DN),

$156 = 1 + (1 + ((1 \wedge (a + c)) + ((((c + (b + -(0 \wedge (c + (b + -1))))) + -1) + -1) + -(1 \wedge (1 + (1 + (-a + -(1 \wedge (c + b))))))))))$, by (Scomm),
$157 = 1 + (1 + ((1 \wedge (a + c)) + (((-1 + (c + (b + -(0 \wedge (c + (b + -1)))))) + -1) + -(1 \wedge (1 + (1 + (-a + -(1 \wedge (c + b))))))))))$, by (Scomm),
$158 = 1 + (1 + ((1 \wedge (a + c)) + (((c + (-1 + (b + -(0 \wedge (c + (b + -1)))))) + -1) + -(1 \wedge (1 + (1 + (-a + -(1 \wedge (c + b))))))))))$, by (11.267),
$159 = 1 + (1 + ((1 \wedge (a + c)) + (((c + (b + (-1 + -(0 \wedge (c + (b + -1)))))) + -1) + -(1 \wedge (1 + (1 + (-a + -(1 \wedge (c + b))))))))))$, by (11.267),
$160 = 1 + (1 + ((1 \wedge (a + c)) + ((-1 + (c + (b + (-1 + -(0 \wedge (c + (b + -1))))))) + -(1 \wedge (1 + (1 + (-a + -(1 \wedge (c + b))))))))))$, by (Scomm),
$161 = 1 + (1 + ((1 \wedge (a + c)) + ((c + (-1 + (b + (-1 + -(0 \wedge (c + (b + -1))))))) + -(1 \wedge (1 + (1 + (-a + -(1 \wedge (c + b))))))))))$, by (11.267),
$162 = 1 + (1 + ((1 \wedge (a + c)) + ((c + (b + (-1 + (-1 + -(0 \wedge (c + (b + -1))))))) + -(1 \wedge (1 + (1 + (-a + -(1 \wedge (c + b))))))))))$, by (11.267),
$163 = 1 + (1 + ((1 \wedge (a + c)) + (-(1 \wedge (1 + (1 + (-a + -(1 \wedge (c + b)))))) + (c + (b + (-1 + (-1 + -(0 \wedge (c + (b + -1))))))))))))$, by (Scomm),
$164 = 1 + (1 + ((1 \wedge (a + c)) + (c + (-(1 \wedge (1 + (1 + (-a + -(1 \wedge (c + b)))))) + (b + (-1 + (-1 + -(0 \wedge (c + (b + -1))))))))))))$, by (11.267),
$165 = 1 + (1 + ((1 \wedge (a + c)) + (c + (b + (-(1 \wedge (1 + (1 + (-a + -(1 \wedge (c + b)))))) + (-1 + (-1 + -(0 \wedge (c + (b + -1))))))))))))$, by (11.267),
$166 = 1 + (1 + ((1 \wedge (a + c)) + (c + (b + (-1 + (-(1 \wedge (1 + (1 + (-a + -(1 \wedge (c + b)))))) + (-1 + -(0 \wedge (c + (b + -1))))))))))))$, by (11.267),
$167 = 1 + (1 + ((1 \wedge (a + c)) + (c + (b + (-1 + (-1 + (-(1 \wedge (1 + (1 + (-a + -(1 \wedge (c + b)))))) + -(0 \wedge (c + (b + -1))))))))))))$, by (11.267),
$168 = 1 + (1 + (c + ((1 \wedge (a + c)) + (b + (-1 + (-1 + (-(1 \wedge (1 + (1 + (-a + -(1 \wedge (c + b)))))) + -(0 \wedge (c + (b + -1))))))))))))$, by (11.267),
$169 = 1 + (1 + (c + (b + ((1 \wedge (a + c)) + (-1 + (-1 + (-(1 \wedge (1 + (1 + (-a + -(1 \wedge (c + b)))))) + -(0 \wedge (c + (b + -1))))))))))))$, by (11.267),
$170 = 1 + (1 + (c + (b + (-1 + ((1 \wedge (a + c)) + (-1 + (-(1 \wedge (1 + (1 + (-a + -(1 \wedge (c + b)))))) + -(0 \wedge (c + (b + -1))))))))))))$, by (11.267),
$171 = 1 + (1 + (c + (b + (-1 + (-1 + ((1 \wedge (a + c)) + (-(1 \wedge (1 + (1 + (-a + -(1 \wedge (c + b)))))) + -(0 \wedge (c + (b + -1))))))))))))$, by (11.267),
$172 = 1 + (c + (b + (-1 + ((1 \wedge (a + c)) + (-(1 \wedge (1 + (1 + (-a + -(1 \wedge (c + b)))))) + -(0 \wedge (c + (b + -1))))))))))$, by (11.271),
$173 = c + (b + ((1 \wedge (a + c)) + (-(1 \wedge (1 + (1 + (-a + -(1 \wedge (c + b)))))) + -(0 \wedge (c + (b + -1)))))))$, by (11.271),
$174 = c + (b + ((1 \wedge (a + c)) + (-(1 \wedge (1 + (1 + (-a + -(1 \wedge (c + b)))))) + -(c + ((b + -1) \wedge -c)))))$, by (11.278),
$175 = c + (b + ((1 \wedge (a + c)) + (-(1 \wedge (1 + (1 + (-a + -(1 \wedge (c + b)))))) + -(c + (b + (-1 \wedge (-b + -c)))))))$, by (11.279),
$176 = c + (b + ((1 \wedge (a + c)) + (-(1 \wedge (1 + (1 + (-a + -(1 \wedge (c + b)))))) + -(c + (b + (-1 \wedge (-c + -b)))))))$, by (Scomm),
$177 = c + (b + ((1 \wedge (a + c)) + (-(1 \wedge (1 + (1 + (-a + -(1 \wedge (c + b)))))) + -(c + (b + (-1 + (0 \wedge ((-c + -b) + 1))))))))$, by (11.287),
$178 = c + (b + ((1 \wedge (a + c)) + (-(1 \wedge (1 + (1 + (-a + -(1 \wedge (c + b)))))) + -(c + (b + (-1 + (0 \wedge (1 + (-c + -b))))))))))$, by (Scomm),

$179 = c + (b + ((1 \wedge (a + c)) + (-(1 \wedge (1 + (1 + (-a + -(1 \wedge (c + b)))))) + (-(b + (-1 + (0 \wedge (1 + (-c + -b))))) + -c))))$, by (NegS),
$180 = c + (b + ((1 \wedge (a + c)) + (-(1 \wedge (1 + (1 + (-a + -(1 \wedge (c + b)))))) + ((-(-1 + (0 \wedge (1 + (-c + -b)))) + -b) + -c))))$, by (NegS),
$181 = c + (b + ((1 \wedge (a + c)) + (-(1 \wedge (1 + (1 + (-a + -(1 \wedge (c + b)))))) + (((-(0 \wedge (1 + (-c + -b))) + - - 1) + -b) + -c))))$, by (NegS),
$182 = c + (b + ((1 \wedge (a + c)) + (-(1 \wedge (1 + (1 + (-a + -(1 \wedge (c + b)))))) + (((-(0 \wedge (1 + (-c + -b))) + 1) + -b) + -c))))$, by (DN),
$183 = c + (b + ((1 \wedge (a + c)) + (-(1 \wedge (1 + (1 + (-a + -(1 \wedge (c + b)))))) + (((1 + -(0 \wedge (1 + (-c + -b)))) + -b) + -c))))$, by (Scomm),
$184 = c + (b + ((1 \wedge (a + c)) + (-(1 \wedge (1 + (1 + (-a + -(1 \wedge (c + b)))))) + ((-b + (1 + -(0 \wedge (1 + (-c + -b))))) + -c))))$, by (Scomm),
$185 = c + (b + ((1 \wedge (a + c)) + (-(1 \wedge (1 + (1 + (-a + -(1 \wedge (c + b)))))) + ((1 + (-b + -(0 \wedge (1 + (-c + -b))))) + -c))))$, by (11.267),
$186 = c + (b + ((1 \wedge (a + c)) + (-(1 \wedge (1 + (1 + (-a + -(1 \wedge (c + b)))))) + (-c + (1 + (-b + -(0 \wedge (1 + (-c + -b)))))))))$, by (Scomm),
$187 = c + (b + ((1 \wedge (a + c)) + (-(1 \wedge (1 + (1 + (-a + -(1 \wedge (c + b)))))) + (1 + (-c + (-b + -(0 \wedge (1 + (-c + -b)))))))))$, by (11.267),
$188 = c + (b + ((1 \wedge (a + c)) + (1 + (-(1 \wedge (1 + (1 + (-a + -(1 \wedge (c + b)))))) + (-c + (-b + -(0 \wedge (1 + (-c + -b)))))))))$, by (11.267),
$189 = c + (b + ((1 \wedge (a + c)) + (1 + (-c + (-(1 \wedge (1 + (1 + (-a + -(1 \wedge (c + b)))))) + (-b + -(0 \wedge (1 + (-c + -b)))))))))$, by (11.267),
$190 = c + (b + ((1 \wedge (a + c)) + (1 + (-c + (-b + (-(1 \wedge (1 + (1 + (-a + -(1 \wedge (c + b)))))) + -(0 \wedge (1 + (-c + -b)))))))))$, by (11.267),
$191 = c + (b + (1 + ((1 \wedge (a + c)) + (-c + (-b + (-(1 \wedge (1 + (1 + (-a + -(1 \wedge (c + b)))))) + -(0 \wedge (1 + (-c + -b)))))))))$, by (11.267),
$192 = c + (b + (1 + (-c + ((1 \wedge (a + c)) + (-b + (-(1 \wedge (1 + (1 + (-a + -(1 \wedge (c + b)))))) + -(0 \wedge (1 + (-c + -b)))))))))$, by (11.267),
$193 = c + (b + (1 + (-c + (-b + ((1 \wedge (a + c)) + (-(1 \wedge (1 + (1 + (-a + -(1 \wedge (c + b)))))) + -(0 \wedge (1 + (-c + -b)))))))))$, by (11.267),
$194 = c + (1 + (b + (-c + (-b + ((1 \wedge (a + c)) + (-(1 \wedge (1 + (1 + (-a + -(1 \wedge (c + b)))))) + -(0 \wedge (1 + (-c + -b)))))))))$, by (11.267),
$195 = c + (1 + (-c + ((1 \wedge (a + c)) + (-(1 \wedge (1 + (1 + (-a + -(1 \wedge (c + b)))))) + -(0 \wedge (1 + (-c + -b)))))))$, by (11.270),
$196 = 1 + (c + (-c + ((1 \wedge (a + c)) + (-(1 \wedge (1 + (1 + (-a + -(1 \wedge (c + b)))))) + -(0 \wedge (1 + (-c + -b)))))))$, by (11.267),
$197 = 1 + ((1 \wedge (a + c)) + (-(1 \wedge (1 + (1 + (-a + -(1 \wedge (c + b)))))) + -(0 \wedge (1 + (-c + -b)))))$, by (11.268),
$198 = 1 + ((1 \wedge (a + c)) + (-(1 \wedge (1 + (1 + (-a + -(1 \wedge (c + b)))))) + -((-c + -b) + (1 \wedge -(-c + -b)))))$, by (11.289),
$199 = 1 + ((1 \wedge (a + c)) + (-(1 \wedge (1 + (1 + (-a + -(1 \wedge (c + b)))))) + -((-c + -b) + (1 \wedge (- - b + - - c)))))$, by (NegS),
$200 = 1 + ((1 \wedge (a + c)) + (-(1 \wedge (1 + (1 + (-a + -(1 \wedge (c + b)))))) + -((-c + -b) + (1 \wedge (b + - - c)))))$, by (DN),
$201 = 1 + ((1 \wedge (a + c)) + (-(1 \wedge (1 + (1 + (-a + -(1 \wedge (c + b)))))) + -((-c + -b) + (1 \wedge (b + c)))))$, by (DN),

$202 = 1+((1\wedge(a+c))+(-(1\wedge(1+(1+(-a+-(1\wedge(c+b))))))+-((-c+-b)+(1\wedge(c+b)))))$, by (Scomm),
$203 = 1+((1\wedge(a+c))+(-(1\wedge(1+(1+(-a+-(1\wedge(c+b))))))+-(-c+(-b+(1\wedge(c+b))))))$, by (Sass),
$204 = 1+((1\wedge(a+c))+(-(1\wedge(1+(1+(-a+-(1\wedge(c+b))))))+(-(-b+(1\wedge(c+b)))+--c)))$, by (NegS),
$205 = 1+((1\wedge(a+c))+(-(1\wedge(1+(1+(-a+-(1\wedge(c+b))))))+((-(1\wedge(c+b))+--b)+--c)))$, by (NegS),
$206 = 1+((1\wedge(a+c))+(-(1\wedge(1+(1+(-a+-(1\wedge(c+b))))))+((-(1\wedge(c+b))+b)+--c)))$, by (DN),
$207 = 1+((1\wedge(a+c))+(-(1\wedge(1+(1+(-a+-(1\wedge(c+b))))))+((b+-(1\wedge(c+b)))+--c)))$, by (Scomm),
$208 = 1+((1\wedge(a+c))+(-(1\wedge(1+(1+(-a+-(1\wedge(c+b))))))+((b+-(1\wedge(c+b)))+c)))$, by (DN),
$209 = 1+((1\wedge(a+c))+(-(1\wedge(1+(1+(-a+-(1\wedge(c+b))))))+(c+(b+-(1\wedge(c+b))))))$, by (Scomm),
$210 = 1+((1\wedge(a+c))+((c+(b+-(1\wedge(c+b))))+-(1\wedge(1+(1+(-a+-(1\wedge(c+b))))))))$, by (Scomm),
$211 = 1+((1\wedge(a+c))+(c+((b+-(1\wedge(c+b)))+-(1\wedge(1+(1+(-a+-(1\wedge(c+b)))))))))$, by (Sass),
$212 = 1+((1\wedge(a+c))+(c+(b+(-(1\wedge(c+b))+-(1\wedge(1+(1+(-a+-(1\wedge(c+b))))))))))$, by (Sass),
$213 = 1+(c+((1\wedge(a+c))+(b+(-(1\wedge(c+b))+-(1\wedge(1+(1+(-a+-(1\wedge(c+b))))))))))$, by (11.267),
$214 = 1+(c+(b+((1\wedge(a+c))+(-(1\wedge(c+b))+-(1\wedge(1+(1+(-a+-(1\wedge(c+b))))))))))$, by (11.267),
$215 = 1+(c+(b+((1\wedge(a+c))+(-(1\wedge(c+b))+-((1+(1+(-a+-(1\wedge(c+b)))))+(0\wedge(-(1+(1+(-a+-(1\wedge(c+b)))))+1)))))))$, by (11.288),
$216 = 1+(c+(b+((1\wedge(a+c))+(-(1\wedge(c+b))+-((1+(1+(-a+-(1\wedge(c+b)))))+(0\wedge((-(1+(-a+-(1\wedge(c+b))))+-1)+1)))))))$, by (NegS),
$217 = 1+(c+(b+((1\wedge(a+c))+(-(1\wedge(c+b))+-((1+(1+(-a+-(1\wedge(c+b)))))+(0\wedge(((-(-a+-(1\wedge(c+b)))+-1)+-1)+1)))))))$, by (NegS),
$218 = 1+(c+(b+((1\wedge(a+c))+(-(1\wedge(c+b))+-((1+(1+(-a+-(1\wedge(c+b)))))+(0\wedge((((--(1\wedge(c+b))+--a)+-1)+-1)+1)))))))$, by (NegS),
$219 = 1+(c+(b+((1\wedge(a+c))+(-(1\wedge(c+b))+-((1+(1+(-a+-(1\wedge(c+b)))))+(0\wedge(((((1\wedge(c+b))+--a)+-1)+-1)+1)))))))$, by (DN),
$220 = 1+(c+(b+((1\wedge(a+c))+(-(1\wedge(c+b))+-((1+(1+(-a+-(1\wedge(c+b)))))+(0\wedge(((((1\wedge(c+b))+a)+-1)+-1)+1)))))))$, by (DN),
$221 = 1+(c+(b+((1\wedge(a+c))+(-(1\wedge(c+b))+-((1+(1+(-a+-(1\wedge(c+b)))))+(0\wedge((((a+(1\wedge(c+b)))+-1)+-1)+1)))))))$, by (Scomm),
$222 = 1+(c+(b+((1\wedge(a+c))+(-(1\wedge(c+b))+-((1+(1+(-a+-(1\wedge(c+b)))))+(0\wedge(((-1+(a+(1\wedge(c+b))))+-1)+1)))))))$, by (Scomm),
$223 = 1+(c+(b+((1\wedge(a+c))+(-(1\wedge(c+b))+-((1+(1+(-a+-(1\wedge(c+b)))))+(0\wedge(((a+(-1+(1\wedge(c+b))))+-1)+1)))))))$, by (11.267),
$224 = 1+(c+(b+((1\wedge(a+c))+(-(1\wedge(c+b))+-((1+(1+(-a+-(1\wedge(c+b)))))+(0\wedge((-1+(a+(-1+(1\wedge(c+b)))))+1)))))))$, by (Scomm),

$225 = 1 + (c + (b + ((1 \wedge (a + c)) + (-(1 \wedge (c + b)) + -((1 + (1 + (-a + -(1 \wedge (c + b))))) + (0 \wedge ((a + (-1 + (-1 + (1 \wedge (c + b))))) + 1)))))))$, by (11.267),

$226 = 1 + (c + (b + ((1 \wedge (a + c)) + (-(1 \wedge (c + b)) + -((1 + (1 + (-a + -(1 \wedge (c + b))))) + (0 \wedge (1 + (a + (-1 + (-1 + (1 \wedge (c + b))))))))))))$, by (Scomm),

$227 = 1 + (c + (b + ((1 \wedge (a + c)) + (-(1 \wedge (c + b)) + -((1 + (1 + (-a + -(1 \wedge (c + b))))) + (0 \wedge (a + (-1 + (1 \wedge (c + b))))))))))$, by (11.271),

$228 = 1 + (c + (b + ((1 \wedge (a + c)) + (-(1 \wedge (c + b)) + -(1 + ((1 + (-a + -(1 \wedge (c + b)))) + (0 \wedge (a + (-1 + (1 \wedge (c + b)))))))))))$, by (Sass),

$229 = 1 + (c + (b + ((1 \wedge (a + c)) + (-(1 \wedge (c + b)) + -(1 + (1 + ((-a + -(1 \wedge (c + b))) + (0 \wedge (a + (-1 + (1 \wedge (c + b)))))))))))$, by (Sass),

$230 = 1 + (c + (b + ((1 \wedge (a + c)) + (-(1 \wedge (c + b)) + -(1 + (1 + (-a + (-(1 \wedge (c + b)) + (0 \wedge (a + (-1 + (1 \wedge (c + b))))))))))))$, by (Sass),

$231 = 1 + (c + (b + ((1 \wedge (a + c)) + (-(1 \wedge (c + b)) + (-(1 + (-a + (-(1 \wedge (c + b)) + (0 \wedge (a + (-1 + (1 \wedge (c + b)))))))) + -1)))))$, by (NegS),

$232 = 1 + (c + (b + ((1 \wedge (a + c)) + (-(1 \wedge (c + b)) + ((-(-a + (-(1 \wedge (c + b)) + (0 \wedge (a + (-1 + (1 \wedge (c + b))))))) + -1) + -1)))))$, by (NegS),

$233 = 1 + (c + (b + ((1 \wedge (a + c)) + (-(1 \wedge (c + b)) + (((-(-(1 \wedge (c + b)) + (0 \wedge (a + (-1 + (1 \wedge (c + b)))))) + - - a) + -1) + -1)))))$, by (NegS),

$234 = 1 + (c + (b + ((1 \wedge (a + c)) + (-(1 \wedge (c + b)) + ((((-(0 \wedge (a + (-1 + (1 \wedge (c + b))))) + - - (1 \wedge (c + b))) + - - a) + -1) + -1)))))$, by (NegS),

$235 = 1 + (c + (b + ((1 \wedge (a + c)) + (-(1 \wedge (c + b)) + ((((-(0 \wedge (a + (-1 + (1 \wedge (c + b))))) + (1 \wedge (c + b))) + - - a) + -1) + -1)))))$, by (DN),

$236 = 1 + (c + (b + ((1 \wedge (a + c)) + (-(1 \wedge (c + b)) + (((((1 \wedge (c + b)) + -(0 \wedge (a + (-1 + (1 \wedge (c + b)))))) + - - a) + -1) + -1)))))$, by (Scomm),

$237 = 1 + (c + (b + ((1 \wedge (a + c)) + (-(1 \wedge (c + b)) + (((((1 \wedge (c + b)) + -(0 \wedge (a + (-1 + (1 \wedge (c + b)))))) + a) + -1) + -1)))))$, by (DN),

$238 = 1 + (c + (b + ((1 \wedge (a + c)) + (-(1 \wedge (c + b)) + (((a + ((1 \wedge (c + b)) + -(0 \wedge (a + (-1 + (1 \wedge (c + b))))))) + -1) + -1)))))$, by (Scomm),

$239 = 1 + (c + (b + ((1 \wedge (a + c)) + (-(1 \wedge (c + b)) + ((-1 + (a + ((1 \wedge (c + b)) + -(0 \wedge (a + (-1 + (1 \wedge (c + b)))))))) + -1)))))$, by (Scomm),

$240 = 1 + (c + (b + ((1 \wedge (a + c)) + (-(1 \wedge (c + b)) + ((a + (-1 + ((1 \wedge (c + b)) + -(0 \wedge (a + (-1 + (1 \wedge (c + b)))))))) + -1)))))$, by (11.267),

$241 = 1 + (c + (b + ((1 \wedge (a + c)) + (-(1 \wedge (c + b)) + (-1 + (a + (-1 + ((1 \wedge (c + b)) + -(0 \wedge (a + (-1 + (1 \wedge (c + b)))))))))))))$, by (Scomm),

$242 = 1 + (c + (b + ((1 \wedge (a + c)) + (-(1 \wedge (c + b)) + (a + (-1 + (-1 + ((1 \wedge (c + b)) + -(0 \wedge (a + (-1 + (1 \wedge (c + b)))))))))))))$, by (11.267),

$243 = 1 + (c + (b + ((1 \wedge (a + c)) + (a + (-(1 \wedge (c + b)) + (-1 + (-1 + ((1 \wedge (c + b)) + -(0 \wedge (a + (-1 + (1 \wedge (c + b)))))))))))))$, by (11.267),

$244 = 1 + (c + (b + ((1 \wedge (a + c)) + (a + (-1 + (-(1 \wedge (c + b)) + (-1 + ((1 \wedge (c + b)) + -(0 \wedge (a + (-1 + (1 \wedge (c + b)))))))))))))$, by (11.267),

$245 = 1 + (c + (b + ((1 \wedge (a + c)) + (a + (-1 + (-1 + (-(1 \wedge (c + b)) + ((1 \wedge (c + b)) + -(0 \wedge (a + (-1 + (1 \wedge (c + b)))))))))))))$, by (11.267),

$246 = 1 + (c + (b + ((1 \wedge (a + c)) + (a + (-1 + (-1 + ((1 \wedge (c + b)) + (-(1 \wedge (c + b)) + -(0 \wedge (a + (-1 + (1 \wedge (c + b)))))))))))))$, by (11.267),

$247 = 1 + (c + (b + ((1 \wedge (a + c)) + (a + (-1 + (-1 + -(0 \wedge (a + (-1 + (1 \wedge (c + b))))))))))))$, by (11.268),

$248 = 1+(c+(b+(a+((1\wedge(a+c))+(-1+(-1+-(0\wedge(a+(-1+(1\wedge(c+b))))))))))$, by (11.267),
$249 = 1+(c+(b+(a+(-1+((1\wedge(a+c))+(-1+-(0\wedge(a+(-1+(1\wedge(c+b))))))))))$, by (11.267),
$250 = 1+(c+(b+(a+(-1+(-1+((1\wedge(a+c))+-(0\wedge(a+(-1+(1\wedge(c+b))))))))))$, by (11.267),
$251 = 1+(c+(a+(b+(-1+(-1+((1\wedge(a+c))+-(0\wedge(a+(-1+(1\wedge(c+b))))))))))$, by (11.267),
$252 = 1+(a+(c+(b+(-1+(-1+((1\wedge(a+c))+-(0\wedge(a+(-1+(1\wedge(c+b))))))))))$, by (11.267),
$253 = a+(1+(c+(b+(-1+(-1+((1\wedge(a+c))+-(0\wedge(a+(-1+(1\wedge(c+b))))))))))$, by (11.267),
$254 = a+(c+(b+(-1+((1\wedge(a+c))+-(0\wedge(a+(-1+(1\wedge(c+b)))))))))$, by (11.271),

$263 = a+(c+(b+(-1+((1\wedge(a+c))+-(-1+(a+((1\wedge(c+b))\wedge(-a+--1))))))))$, by (11.290),
$264 = a+(c+(b+(-1+((1\wedge(a+c))+-(-1+(a+((1\wedge(c+b))\wedge(-a+1))))))))$, by (DN),
$265 = a+(c+(b+(-1+((1\wedge(a+c))+-(-1+(a+((1\wedge(c+b))\wedge(1+-a))))))))$, by (Scomm),
$266 = a+(c+(b+(-1+((1\wedge(a+c))+-(-1+(a+((1+-a)\wedge(1\wedge(c+b))))))))$, by (Wcomm),
$267 = a+(c+(b+(-1+((1\wedge(a+c))+-(-1+(a+(1\wedge((1+-a)\wedge(c+b)))))))))$, by (11.275),
$268 = a+(c+(b+(-1+((1\wedge(a+c))+-(-1+(a+(1\wedge((c+b)\wedge(1+-a)))))))))$, by (Wcomm),
$269 = a+(c+(b+(-1+((1\wedge(a+c))+-(-1+(a+(1\wedge(c+(b\wedge(-c+(1+-a))))))))))$, by (11.279),
$270 = a+(c+(b+(-1+((1\wedge(a+c))+-(-1+(a+(1\wedge(c+(b\wedge((1+-a)+-c))))))))))$, by (Scomm),
$271 = a+(c+(b+(-1+((1\wedge(a+c))+-(-1+(a+(1\wedge(c+(b\wedge(1+(-a+-c))))))))))$, by (Sass),
$272 = a+(c+(b+(-1+((1\wedge(a+c))+-(a+(-1+(1\wedge(c+(b\wedge(1+(-a+-c))))))))))$, by (11.267),
$273 = a+(c+(b+(-1+((1\wedge(a+c))+(-(-1+(1\wedge(c+(b\wedge(1+(-a+-c))))))+-a)))))$, by (NegS),
$274 = a+(c+(b+(-1+((1\wedge(a+c))+((-(1\wedge(c+(b\wedge(1+(-a+-c)))))+--1)+-a)))))$, by (NegS),
$275 = a+(c+(b+(-1+((1\wedge(a+c))+((-(1\wedge(c+(b\wedge(1+(-a+-c)))))+1)+-a)))))$, by (DN),
$276 = a+(c+(b+(-1+((1\wedge(a+c))+((1+-(1\wedge(c+(b\wedge(1+(-a+-c))))))+-a)))))$, by (Scomm),
$277 = a+(c+(b+(-1+((1\wedge(a+c))+(-a+(1+-(1\wedge(c+(b\wedge(1+(-a+-c))))))))))$, by (Scomm),
$278 = a+(c+(b+(-1+((1\wedge(a+c))+(1+(-a+-(1\wedge(c+(b\wedge(1+(-a+-c))))))))))$, by (11.267),

$279 = a+(c+(b+(-1+(1+((1\wedge(a+c))+(-a+-(1\wedge(c+(b\wedge(1+(-a+-c)))))))))))$, by (11.267),
$280 = a+(c+(b+(-1+(1+(-a+((1\wedge(a+c))+-(1\wedge(c+(b\wedge(1+(-a+-c)))))))))))$, by (11.267),
$281 = a+(c+(b+(1+(-1+(-a+((1\wedge(a+c))+-(1\wedge(c+(b\wedge(1+(-a+-c)))))))))))$, by (11.267),
$282 = a+(c+(b+(-a+((1\wedge(a+c))+-(1\wedge(c+(b\wedge(1+(-a+-c)))))))))$, by (11.268),
$283 = c+(b+((1\wedge(a+c))+-(1\wedge(c+(b\wedge(1+(-a+-c)))))))$, by (11.271),
$284 = c+(b+((1\wedge(a+c))+-(1\wedge(b+(c\wedge(-b+(c+(1+(-a+-c)))))))))$, by (11.284),
$285 = c+(b+((1\wedge(a+c))+-(1\wedge(b+(c\wedge(-b+(1+(c+(-a+-c)))))))))$, by (11.267),
$286 = c+(b+((1\wedge(a+c))+-(1\wedge(b+(c\wedge(-b+(1+-a)))))))$, by (11.269),
$287 = c+(b+((1\wedge(a+c))+-(1\wedge(b+(c\wedge((1+-a)+-b))))))$, by (Scomm),
$288 = c+(b+((1\wedge(a+c))+-(1\wedge(b+(c\wedge(1+(-a+-b)))))))$, by (Sass),
$289 = c+(b+((1\wedge(a+c))+-(1\wedge(b+((1+(-a+-b))+(0\wedge(c+(-(1+(-a+-b))+(0\wedge(1+(-a+-b)))))))))))$, by (11.292),
$290 = c+(b+((1\wedge(a+c))+-(1\wedge(b+((1+(-a+-b))+(0\wedge(c+((-(-a+-b)+-1)+(0\wedge(1+(-a+-b)))))))))))$, by (NegS),
$291 = c+(b+((1\wedge(a+c))+-(1\wedge(b+((1+(-a+-b))+(0\wedge(c+(((--b+--a)+-1)+(0\wedge(1+(-a+-b)))))))))))$, by (NegS),
$292 = c+(b+((1\wedge(a+c))+-(1\wedge(b+((1+(-a+-b))+(0\wedge(c+(((b+--a)+-1)+(0\wedge(1+(-a+-b)))))))))))$, by (DN),
$293 = c+(b+((1\wedge(a+c))+-(1\wedge(b+((1+(-a+-b))+(0\wedge(c+(((b+a)+-1)+(0\wedge(1+(-a+-b)))))))))))$, by (DN),
$294 = c+(b+((1\wedge(a+c))+-(1\wedge(b+((1+(-a+-b))+(0\wedge(c+(((a+b)+-1)+(0\wedge(1+(-a+-b)))))))))))$, by (Scomm),
$295 = c+(b+((1\wedge(a+c))+-(1\wedge(b+((1+(-a+-b))+(0\wedge(c+((a+(b+-1))+(0\wedge(1+(-a+-b)))))))))))$, by (Sass),
$296 = c+(b+((1\wedge(a+c))+-(1\wedge(b+((1+(-a+-b))+(0\wedge(c+((a+(b+-1))+-b))))))))$, by (11.294),
$297 = c+(b+((1\wedge(a+c))+-(1\wedge(b+((1+(-a+-b))+(0\wedge(c+(a+((b+-1)+-b))))))))$, by (Sass),
$298 = c+(b+((1\wedge(a+c))+-(1\wedge(b+((1+(-a+-b))+(0\wedge(c+(a+(b+(-1+-b)))))))))$, by (Sass),
$299 = c+(b+((1\wedge(a+c))+-(1\wedge(b+((1+(-a+-b))+(0\wedge(c+(a+-1))))))))$, by (11.269),
$300 = c+(b+((1\wedge(a+c))+-(1\wedge(b+((1+(-a+-b))+(0\wedge(a+(c+-1))))))))$, by (11.267),
$301 = c+(b+((1\wedge(a+c))+-(1\wedge(b+((1+(-a+-b))+((1\wedge(a+c))+(-(1\wedge(1+(1+(-a+-c))))+(-1+(1\wedge(-(1\wedge(a+c))+(--1+(1\wedge(1+(1+(-a+-c)))))))))))))))$, by (11.297),
$302 = c+(b+((1\wedge(a+c))+-(1\wedge(b+((1+(-a+-b))+((1\wedge(a+c))+(-(1\wedge(1+(1+(-a+-c))))+(-1+(1\wedge(-(1\wedge(a+c))+(1+(1\wedge(1+(1+(-a+-c)))))))))))))))$, by (DN),

$303 = c+(b+((1\wedge(a+c))+-(1\wedge(b+((1+(-a+-b))+((1\wedge(a+c))+(-(1\wedge(1+(1+(-a+-c))))+(-1+(1\wedge(1+(-(1\wedge(a+c))+(1\wedge(1+(1+(-a+-c))))))))))))))$, by (11.267),
$304 = c+(b+((1\wedge(a+c))+-(1\wedge(b+((1+(-a+-b))+((1\wedge(a+c))+(-(1\wedge(1+(1+(-a+-c))))+(-1+(1\wedge(1+(1+(-a+-c))))))))))))$, by (11.298),
$305 = c+(b+((1\wedge(a+c))+-(1\wedge(b+((1+(-a+-b))+((1\wedge(a+c))+((-1+(1\wedge(1+(1+(-a+-c)))))+-(1\wedge(1+(1+(-a+-c)))))))))))$, by (Scomm),
$306 = c+(b+((1\wedge(a+c))+-(1\wedge(b+((1+(-a+-b))+((1\wedge(a+c))+(-1+((1\wedge(1+(1+(-a+-c))))+-(1\wedge(1+(1+(-a+-c)))))))))))$, by (Sass),
$307 = c+(b+((1\wedge(a+c))+-(1\wedge(b+((1+(-a+-b))+((1\wedge(a+c))+(-1+0)))))))$, by ($m_0$-Re),
$308 = c+(b+((1\wedge(a+c))+-(1\wedge(b+((1+(-a+-b))+((1\wedge(a+c))+(0+-1)))))))$, by (Scomm),
$309 = c+(b+((1\wedge(a+c))+-(1\wedge(b+((1+(-a+-b))+((1\wedge(a+c))+-1))))))$, by (SU),
$310 = c+(b+((1\wedge(a+c))+-(1\wedge(b+((1+(-a+-b))+(-1+(1\wedge(a+c))))))))$, by (Scomm),
$311 = c+(b+((1\wedge(a+c))+-(1\wedge(b+(-1+((1+(-a+-b))+(1\wedge(a+c))))))))$, by (11.267),
$312 = c+(b+((1\wedge(a+c))+-(1\wedge(b+(-1+(1+((-a+-b)+(1\wedge(a+c)))))))))$, by (Sass),
$313 = c+(b+((1\wedge(a+c))+-(1\wedge(b+(-1+(1+(-a+(-b+(1\wedge(a+c))))))))))$, by (Sass),
$314 = c+(b+((1\wedge(a+c))+-(1\wedge(b+(1+(-1+(-a+(-b+(1\wedge(a+c))))))))))$, by (11.267),
$315 = c+(b+((1\wedge(a+c))+-(1\wedge(b+(-a+(-b+(1\wedge(a+c))))))))$, by (11.268),
$316 = c+(b+((1\wedge(a+c))+-(1\wedge(-a+(1\wedge(a+c))))))$, by (11.270),
$317 = c+(b+((1\wedge(a+c))+-(-a+(1\wedge(a+c)))))$, by (11.291),
$318 = c+(b+((1\wedge(a+c))+(-(1\wedge(a+c))+--a)))$, by (NegS),
$319 = c+(b+((1\wedge(a+c))+(-(1\wedge(a+c))+a)))$, by (DN),
$320 = c+(b+((1\wedge(a+c))+(a+-(1\wedge(a+c)))))$, by (Scomm),
$321 = c+(b+(a+((1\wedge(a+c))+-(1\wedge(a+c)))))$, by (11.267),
$322 = c+(b+(a+0))$, by (m_0-Re),
$323 = c+(b+(0+a))$, by (Scomm),
$324 = c+(b+a)$, by (SU),
$325 = c+(a+b)$, by (Scomm),
$326 = (a+b)+c$, by (Scomm). □

Bibliography

[1] Marco Abbadini, *Equivalence 'a la Mundici for commutative lattice-ordered monoids*, Algebra Universalis, 2021, pp. 1–42.

[2] Marlow Anderson, and Todd Feil, Lattice-ordered groups: an introduction, vol. 4, Springer Science and Business Media, 2012.

[3] Garret Birkhoff, Lattice theory, vol. 25, *American Mathematical Soc.*, 1940.

[4] Bruno Bosbach, *Concerning bricks*, Acta Mathematica Academiae Scientiarum Hungaricae 38, 1981, pp. 89–104.

[5] H.W. Buff, *Decidable and undecidable MV-algebras*, Algebra Universalis 21, 1985, pp. 234–249.

[6] Chen Chung Chang, *Algebraic analysis of many-valued logics*, Trans. Amer. Math. Soc. 88, 1958, pp. 467–490.

[7] — *A new proof of the completeness of the Łukasiewicz axioms*, Trans. Amer. Math. Soc. 93, 1959, pp. 74–90.

[8] — *The writing of the MV-algebras*, Studia logica (Special issue on Many-valued logics) 61, (Mundici, D., Ed.), 1998, pp. 3–6.

[9] Roberto L. O. Cignoli, Itala M. L. D'Ottaviano, and Daniele Mundici, Algebras of Łukasiewicz Logics, (in Portuguese). Second Edition. Editions CLE, State University of Campinas, Campinas, S.P., Brazil. (Coleção CLE/UNICAMP, v. 12), 1995.

[10] — Algebraic Foundations of Many-valued Reasoning, Kluwer Academic Publishers & Springer Science, Dordrecht 2000 [*Trends in Logic - Studia Logica Library 7*].

[11] Roberto L.O. Cignoli, and Daniele Mundici, *An elementary proof of Chang's completeness theorem for the infinite-valued calculus of Łukasiewicz*, Studia Logica 58, 1997, pp. 79–97.

[12] Anatolij Dvurečenskij, *Pseudo-MV algebras are intervals in l-groups*, J. Austral. Math. Soc. 72, 2002, pp. 427–445.

[13] Josep Maria Font, Antonio J. Rodríguez, and Antoni Torrens, *Wajsberg algebras*, Stochastica 8 (1), 1984, pp. 5–31.

[14] George Georgescu, and Afrodita Iorgulescu, *Pseudo-MV Algebras: a non-commutative Extension of MV Algebras*, The Proceedings of the Fourth International Symposium on Economic Informatics, INFOREC Printing House, Bucharest, Romania, May 1999, pp. 961–968.

[15] —*Pseudo-MV algebras*, Mult. Val. Logic (A special issue dedicated to the memory of Gr.C. Moisil), Vol. 6, Nr. 1-2, 2001, pp. 95–135.

[16] Afrodita Iorgulescu, *New generalizations of* BCI, BCK *and Hilbert algebras*, Parts I, II (Dedicated to Dragoş Vaida), J. of Multiple-Valued Logic and Soft Computing 27 (4), 2016, pp. 353–406, and 407–456. (A previous version available from December 6, 2013, at `http://arxiv.org/abs/1312.2494`.)

[17] — Algebras of logic as BCK algebras, Academy of Economic Studies Press, Bucharest, 2008.

[18] — *Quasi-algebras versus regular algebras*, Part I (Dedicated to Sergiu Rudeanu), Scientific Annals of Computer Science 25 (1), 2015, pp. 1–43 (doi: 10.7561/SACS.2015.1.ppp).

[19] — Implicative-groups vs. groups and generalizations, First ed.: Matrix Rom, Bucharest, 2018; Second ed.: *Studies in Logic 112*, College Publications, 2025.

[20] — *Algebras of logic vs. Algebras*, Landscapes in Logic, Vol. 1 Contemporary Logic and Computing, Editor Adrian Rezuş, College Publications, 2020, pp. 157–258.

[21] — BCK algebras versus m-BCK algebras. Foundations, *Studies in Logic 96*, College Publications, 2022.

[22] — Non-commutative algebras. Pseudo-BCK algebras versus m-pseudo-BCK algebras, *Studies in Logic 107*, College Publications, 2024.

[23] Kiyoshi Iséki, *An algebra related with a propositional calculus*, Proc. Japan Acad. 42, 1966, pp. 26–29.

[24] Kiyoshi Iséki, and Shôtarô Tanaka, *An introduction to the theory of* BCK-*algebras*, Math. Japonica 23 (1), 1978, pp. 1–26.

[25] Yuichi Komori, *Super Łukasiewicz propositional logics*, Nagoya Math. J. 84, 1981, pp. 119–133.

[26] Francesco Lacava, *Alcune proprietá delle Ł-algebre e delle Ł-algebre esistenzialmente chiuse*, Bollettino Unione Matematica Italiana, A(5), 16, 1979, pp. 360–366.

[27] Saunders Mac Lane, Categories for the Working Mathematician, *Graduate Texts in Mathematics, vol. 5*, 2nd edn. Springer, New York, 1998.

[28] William W. McCune, *Prover9 and Mace4*, available at http://www.cs.unm.edu/ mccune/Prover9.

[29] Piero Mangani, *On certain algebras related to many-valued logics* (Italian), Boll. Un. Mat. Ital. (4) **8**, 1973, pp. 68–78.

[30] Daniele Mundici, *MV-algebras are categorically equivalent to bounded commutative BCK-algebras*, Math. Japonica 31, No. 6, 1986, pp. 889–894.

[31] — *Interpretation of AF C*-algebras in Łukasiewicz sentential calculus*, J. Funct. Anal. 65, 1986, pp. 15–63.

[32] Antonio J. Rodríguez, Un Estudio algebraico de los Cálculos proposicionales de Łukasiewicz, Universidad de Barcelona (Ph.D. Thesis), 1980.

[33] Esko Turunen, Mathematics Behind Fuzzy Logic, Physica-Verlag, 1999.

[34] Mordchaj Wajsberg, *Beiträge zum Metaaussagenkalkül*, Monat. Math. Phys. 42, 1935, p. 240.

Index

www.ingramcontent.com/pod-product-compliance
Lightning Source LLC
Chambersburg PA
CBHW070348200326
41518CB00012B/2172
* 9 7 8 1 8 4 8 9 0 4 9 6 5 *